DROGS OF ABUSE AND ADDICTION

NEUROBEHAVIORAL TOXICOLOGY

Pharmacology and Toxicology: Basic and Clinical Aspects

Mannfred A. Hollinger, Series Editor
University of California, Davis

Forthcoming Titles

CRC Handbook of Alcohol Addiction: Clinical and Theoretical Approaches, Gerald Zernig
Handbook of Mammalian Models in Biomedical Research, David B. Jack
Handbook of Theoretical Models in Biomedical Research, David B. Jack
Lead and Public Health: Integrated Risk Assessment, Paul Mushak
Molecular Bases of Anesthesia, Eric Moody and Phil Skolnick
Receptor Characterization and Regulation, Devendra K. Agrawal

Published Titles

Manual of Immunological Methods, 1999, Pauline Brousseau, Yves Payette, Helen Tryphonas, Barry Blakley, Herman Boermans, Denis Flipo, Michel Fournier
CNS Injuries: Cellular Responses and Pharmacological Strategies, 1999, Martin Berry and Ann Logan
Infectious Diseases in Immunocompromised Hosts, 1998, Vassil St. Georgiev
Pharmacology of Antimuscarinic Agents, 1998, Laszlo Gyermek
Handbook of Plant and Fungal Toxicants, 1997, Felix D'Mello
Basis of Toxicity Testing, Second Edition, 1997, Donald J. Ecobichon
Anabolic Treatments for Osteoporosis, 1997, James F. Whitfield and Paul Morley
Antibody Therapeutics, 1997, William J. Harris and John R. Adair
Muscarinic Receptor Subtypes in Smooth Muscle, 1997, Richard M. Eglen
Antisense Oligodeonucleotides as Novel Pharmacological Therapeutic Agents, 1997, Benjamin Weiss
Airway Wall Remodelling in Asthma, 1996, A.G. Stewart
Drug Delivery Systems, 1996, Vasant V. Ranade and Mannfred A. Hollinger
Brain Mechanisms and Psychotropic Drugs, 1996, Andrius Baskys and Gary Remington
Receptor Dynamics in Neural Development, 1996, Christopher A. Shaw
Ryanodine Receptors, 1996, Vincenzo Sorrentino
Therapeutic Modulation of Cytokines, 1996, M.W. Bodmer and Brian Henderson
Pharmacology in Exercise and Sport, 1996, Satu M. Somani
Placental Pharmacology, 1996, B. V. Rama Sastry
Pharmacological Effects of Ethanol on the Nervous System, 1996, Richard A. Deitrich
Immunopharmaceuticals, 1996, Edward S. Kimball
Chemoattractant Ligands and Their Receptors, 1996, Richard Horuk
Pharmacological Regulation of Gene Expression in the CNS, 1996, Kalpana Merchant
Experimental Models of Mucosal Inflammation, 1995, Timothy S. Gaginella
Handbook of Methods in Gastrointestinal Pharmacology, 1995, Timothy S. Gaginella
Handbook of Targeted Delivery of Imaging Agents, 1995, Vladimir P. Torchilin
Handbook of Pharmacokinetic Pharmacodynamic Correlations, 1995, Hartmut Derendorf and Guenther Hochhaus
Human Growth Hormone Pharmacology: Basic and Clinical Aspects, 1995, Kathleen T. Shiverick and Arlan Rosenbloom
Placental Toxicology, 1995, B. V. Rama Sastry
Stealth Liposomes, 1995, Danilo Lasic and Frank Martin
TAXOL®: Science and Applications, 1995, Matthew Suffness

Pharmacology and Toxicology: Basic and Clinical Aspects

Published Titles (*Continued*)

Endothelin Receptors: From the Gene to the Human, 1995, Robert R. Ruffolo, Jr.
Alternative Methodologies for the Safety Evaluation of Chemicals in the Cosmetic Industry, 1995, Nicola Loprieno
Phospholipase A_2 in Clinical Inflammation: Molecular Approaches to Pathophysiology, 1995, Keith B. Glaser and Peter Vadas
Serotonin and Gastrointestinal Function, 1995, Timothy S. Gaginella and James J. Galligan
Chemical and Structural Approaches to Rational Drug Design, 1994, David B. Weiner and William V. Williams
Biological Approaches to Rational Drug Design, 1994, David B. Weiner and William V. Williams
Direct and Allosteric Control of Glutamate Receptors, 1994, M. Palfreyman, I. Reynolds, and P. Skolnick
Genomic and Non-Genomic Effects of Aldosterone, 1994, Martin Wehling
Peroxisome Proliferators: Unique Inducers of Drug-Metabolizing Enzymes, 1994, David E. Moody
Angiotensin II Receptors, Volume I: Molecular Biology, Biochemistry, Pharmacology, and Clinical Perspectives, 1994, Robert R. Ruffolo, Jr.
Angiotensin II Receptors, Volume II: Medicinal Chemistry, 1994, Robert R. Ruffolo, Jr.
Beneficial and Toxic Effects of Aspirin, 1993, Susan E. Feinman
Preclinical and Clinical Modulation of Anticancer Drugs, 1993, Kenneth D. Tew, Peter Houghton, and Janet Houghton
In Vitro *Methods of Toxicology*, 1992, Ronald R. Watson
Basis of Toxicity Testing, 1992, Donald J. Ecobichon
Human Drug Metabolism from Molecular Biology to Man, 1992, Elizabeth Jeffreys
Platelet Activating Factor Receptor: Signal Mechanisms and Molecular Biology, 1992, Shivendra D. Shukla
Biopharmaceutics of Ocular Drug Delivery, 1992, Peter Edman
Pharmacology of the Skin, 1991, Hasan Mukhtar
Inflammatory Cells and Mediators in Bronchial Asthma, 1990, Devendra K. Agrawal and Robert G. Townley

DRUGS OF ABUSE AND ADDICTION

NEUROBEHAVIORAL TOXICOLOGY

Edited by

R.J.M. Niesink, R.M.A. Jaspers,
L.M.W. Kornet, J.M. van Ree

CRC Press
Boca Raton London New York Washington, D.C.

Library of Congress Cataloging-in-Publication Data

Drugs of abuse and addiction : neurobehavioral toxicology /
edited by Raymond J.M. Niesink ... [et al.]
p. cm. — (Pharmacology & toxicology)
Includes bibliographical references and index.
ISBN 0-8493-7803--6
1. Psychotropic drugs — Toxicology. 2. Behavioral toxicology.
3. Neurotoxic agents. 4. Neuropsychopharmacology. 5. Substance
abuse. I. Niesink, Raymundus Johannes Maria, 1953– .
II. Series: Pharmacology & toxicology (Boca Raton, Fla.)
[DNLM: 1. Psychotropic Drugs — toxicity. 2. Street Drugs —
toxicity. 3. Nervous System — drug effects. 4. Substance-Related
Disorders. 5. Behavior — drug effects. QV 77.2D7948 1998]
RM316.D785 1998
615'.788—dc21
DNLM/DLC
for Library of Congress

98-39869
CIP

International Standard Book Number 0-8493-7803-6
Library of Congress Card Number 98-39869
Printed in the United States of America 1 2 3 4 5 6 7 8 9 0
Printed on acid-free paper

Preface

The use of drugs that alter our behavior and subjective experience is widespread. These drugs might either be accepted or controlled in some way (caffeine or alcohol), or may be completely forbidden, as in the case of cocaine or heroin. Interest in the effects of exposure to drugs on the development and functional integrity of the brain is on the rise. This book deals with the effects of drugs on the central (and peripheral) nervous system and the subsequent changes in behavior. It provides an introduction to substance abuse research.

The book contains essential information for students in neurotoxicology, psychiatry, (neuro)psychology, and for those professionally involved in the field of addiction problems.

Psychoactive drugs used for recreational purpose generate a number of behavioral toxicological phenomena, such as acute intoxication, physical dependence, behavioral dysfunctions, and addiction. This book gives an introduction to the biological, psychological, and sociological aspects of drugs of different classes and discusses medical complications of dependence problems. Pharmacological, biological, and psychological aspects of the most common drugs are reviewed.

Over the last several decades, science, medical, and public health researchers have become interested in the way in which drugs affect our body and our behavior. Currently it is recognized that there is a great deal of commonality in the factors determining the use of, and dependence upon, different recreational drugs. Patterns of drug self-administration have been studied in human addicts. This research on drug use has produced exciting findings on many fronts. However, there are many variables that cannot be manipulated easily in such studies. Furthermore, it is obviously unethical to study the acquisition of human drug abuse patterns in a laboratory setting. Therefore, we have come to depend on animal models to answer a number of questions about addiction. Such models have proven useful in screening for the abuse liability of new drugs, developing new drugs for the treatment of drug abuse, and for the study of stimulus properties of abused drugs. Both animal as well as human addiction research receives ample treatment in the present volume.

Two chapters which deal with therapy for dependence problems provide a broad overview of psycho- and pharmacotreatment of drug dependency, which will be particularly relevant to medical students and physicians.

Although the areas of addiction and neurobehavioral toxicology have been approached by various professional fields (pharmacology and social sciences vs. industrial toxicology), both subjects are defined here in terms of chemically induced changes in behavior that can be studied using similar experimental test designs and techniques. Therefore, an integrated approach is used to treat the behavioral toxicological effects of psychoactive drugs.

In this book, neurobehavioral toxicity and the behavioral effects of abused drugs will be discussed in particular. Emphasis is placed on acute and chronic effects, reversible and irreversible consequences, functional disorders of the nervous system, neurobehavioral dysfunctions, the multi-sided aspects of addiction, and the underlying neurobiological mechanisms.

The book deals with the entire range of addiction research in humans and animals, using a multidisciplinary approach to discuss all areas of the neuro- and behavioral sciences involved. Biological (physiological and biochemical), psychological, and social aspects of behavioral changes caused by drugs are reviewed.

Emphasized, also, is the understanding of how various factors contribute to drug abuse and dependence — not only biological and medical aspects, but also the social and psychological aspects of drug use. An introduction is given in the study of the effects of drugs on behavior and the ways that behavioral principles can help us understand how drugs work. Data from experiments on both humans and nonhumans will be discussed. Apart from the more basic research aspects of abuse and addiction, chapters on epidemiology, diagnosis of drug abuse, and clinical perspectives are incorporated. Other chapters deal with the use of medication and psychosocial therapy in the management and treatment of drug abuse.

An extensive glossary that contains terms and concepts from the different disciplines, a list of abbreviations often used within this field, and a list of tests used in addiction research are provided at the end of the book.

Acknowledgments

The editors gratefully acknowledge the many people who helped to organize the original manuscript. A special thanks goes to Dr. Susan Boobis, for the editing, and Evelin Karsten, for typewriting and layout assistance for figures and tables. Jack Jamar and Vivian Rompelberg are greatly acknowledged for preparing the illustrations.

Edited by

Dr. R. J. M. Niesink, Ph.D., Open University of the Netherlands, Heerlen, and Rudolf Magnus Institute for Neurosciences, Utrecht University, The Netherlands
Dr. R. M. A. Jaspers, Ph.D., Ministry of Health, Welfare and Sport, Veterinary Public Health Inspectorate, Rijswijk, The Netherlands
Dr. L. M. W. Kornet, Ph.D., Waalre, The Netherlands
Prof. dr. J.M. van Ree, Ph.D., Department of Medical Pharmacology, Rudolf Magnus Institute for Neurosciences, Faculty of Medicine, Utrecht University, The Netherlands

Contributors

- U. Busto, Pharm.D., Sunnybrook Health Science Centre, Mental Health Unit, Addiction Research Foundation, and Faculty of Pharmacy, University of Toronto, Toronto, Canada
- Prof. Dr. M. E. Carroll, Ph.D., Department of Psychiatry, University of Minnesota, Minneapolis
- K. L. Clay, B.A., Treatment Research Center, University of Pennsylvania, Philadelphia
- Dr. L. K. Duffy, Ph.D., Institute of Arctic Biology, University of Alaska Fairbanks, Fairbanks, Alaska
- Dr. M. P. Farrell, MRC. Psych., Addiction Research Unit, National Addiction Centre, London
- Dr. E. J. L. Finch, MRC. Psych., Addiction Research Unit, National Addiction Centre, London
- Dr. F. R. George, Ph.D., Amethyst Technologies, Inc., Scottsdale, Arizona
- Y. J. Hong, Treatment Research Center, University of Pennsylvania, Philadelphia
- A. E. Kaminski, B.A., Treatment Research Center, University of Pennsylvania, Philadelphia
- Dr. L. M. W. Kornet, Ph.D., Waalre, The Netherlands
- Prof. Dr. R. A. Meisch, M.D., Ph.D., Department of Psychiatry and Behavioral Sciences, U.T. Mental Science Institute, Substance Abuse Research Center, University of Texas, Houston
- Prof. Dr. C. A. Naranjo, M.D., Departments of Pharmacology, Psychiatry and Medicine, University of Toronto, Psychopharmacology Research Program, Sunnybrook Health Science Centre, Toronto, Canada
- Dr. R. J. M. Niesink, Ph.D., Open University of The Netherlands, Faculty of Natural Sciences, Heerlen, The Netherlands
- V. Özdemir, M.D., M.Sc., Departments of Pharmacology and Medicine, University of Toronto, Psychopharmacology Research Program, Sunnybrook Health Science Centre, Toronto, Canada
- Dr. M. C. Ritz, Ph.D., Amethyst Technologies, Inc., Scottsdale, Arizona
- Prof. Dr. B. Segal, Ph.D., University of Alaska Anchorage, Center for Alcohol and Addiction Studies, Anchorage, Alaska
- Dr. R. B. Stewart, Ph.D., Department of Psychiatry and Behavioral Sciences, U.T. Mental Science Institute, Substance Abuse Research Center, University of Texas, Houston
- Prof. Dr. J.M. van Ree, Rudolf Magnus Institute for Neurosciences, Faculty of Medicine, Utrecht University, The Netherlands
- Prof. Dr. J. R. Volpicelli, M.D., Ph.D., Treatment Research Center, University of Pennsylvania, Philadelphia
- L. A. Volpicelli, B.A., Treatment Research Center, University of Pennsylvania, Philadelphia

Table of Contents

Contents Chapter 1

Aspects of drug use and drug dependence

0-8493-7803-6/99/$0.00+$.50

Chapter 1

Aspects of drug use and drug dependence

Michael Farrell and Emily Finch

INTRODUCTION

This chapter covers the historical aspects of alcohol and drug problems and provides an overview of the current definitions of drug dependence or drug addiction. It explores the theoretical models underlying drug dependence and provides a brief outline of etiological models. Epidemiological data on tobacco, alcohol, and other drugs provide a framework for understanding the nature and the size of the problem. A few of the longitudinal studies are briefly discussed and some of the basic concepts of treatment are explored in order to explore models of career and natural history in drug use.

1 Historical aspects

1.1 ANCIENT HISTORY

Alcohol has long been recognized as a problem. The major religions such as Christianity and BueÃhism were wary of intoxicants, and Islam developed rules that forbade the use of intoxicant substances[3]. The Bible includes considerable references to alcohol; Noah's drinking habits and intoxication are some of the best descriptions of ancient drinking habits. The properties of opium and other drugs have been referred to in literature since ancient times. Reference to the juice of the poppy occurred in Syrian medical tablets of the 7th century BC and later in Sumerian records of about 4000 BC. In Egypt and Persia doctors treated patients using an opium preparation from at least the 2nd century BC. Opium is listed among medically recommended drugs in a therapeutic papyrus of Thebes from 1552 BC. The poppy was well-used in Ancient Greece for its medicinal properties. Homer describes Helen of Troy preparing what was probably a draft of opium in his epic poem the "Odyssey" and Virgil mentions opiates in the "Aenaead." In ancient Rome opium was sold by ordinary shop keepers and later Arab traders spread its use over a much wider area to Persia, India and China. During the Moslem conquest of the 10th and 11th centuries an opium trade was firmly established in Europe. In the 16th century there are records of opium's use in Germany and in Holland and it is mentioned in Chaucer's *Canterbury Tales* and later in Shakespeare's "Othello."

Tobacco was introduced into Europe following Christopher Columbus's voyage to the New World in 1492. The first person to smoke publicly in Spain was immediately locked up on suspicion of being in the grip of bilious evil spirits. The custom of smoking spread rapidly despite authoritative condemnation of the practice by King James I in 1604 in his Counterblasts to tobacco, "A custom loathsome to the eye, hateful to the nose, harmful to the brain, dangerous to the lungs, and in the black stinking fumes thereof nearest resembling the horrible Stygian smoke of the pit that is bottomless"[19]. In some countries such as Turkey, Iran, and Germany tobacco smoking was condemned and those who persisted with it were executed!

1.2 CONTEMPORARY HISTORY

The industrial revolution with its attendant social change and disruption was accompanied by the introduction of distilled spirits into many western cultures. Industrial revolution heralded the discussion of culture and society as something separate from church and state. The city became the symbol of the corruption and despoliation of society, and the drunkenness of people became a major source of anxiety for both State and Church. However, up until the industrial revolution acute and more so chronic drunkenness or other forms of intoxication were viewed as bad or sinful behavior. Levine argues (1992)[26] that sometime in the eighteenth century a model evolved that looked at chronic intoxication as disordered or as a defect in habit.

The approach to inebriation and drunkenness located the problem substantially within the individual and worked to provide assistance to such individuals; by contrast, the Temperance movement in Nordic and English speaking Protestant cultures viewed the problem as residing mainly in the substance itself and advocated that the entire population should forego alcohol. Levine describes (1992)[26] that all temperance societies drank their alcohol distilled (vodka, whiskey, gin, rum, or whiskey) and they were predominantly Protestant. Self control was the key issue addressed by the temperance movement. The wine drinking countries of southern Europe had virtually no temperance movement and did not tend to hold negative views about alcohol despite consuming from two to four times the amount of alcohol per capita.

In China the earliest records of opium use are in the early 1600s[47]. Its introduction rapidly followed that of tobacco, the use of which also spread quickly across China. In other parts of Asia, mainly those populated by the Chinese, there was similarly high consumption levels. In the 18th and 19th centuries opium was imported into China from India by the British and opium wars were fought over the trade in 1839 and 1856[21]. Opium was almost certainly smoked by all social groups, initially for its medicinal properties for which it was highly prized, and later because the user became addicted to it or desired its euphoric and relaxing properties. Different social groups almost certainly had different motivations for taking opium; for Chinese peasants it provided relief from a miserable life of near slavery, and for the high social class eunuchs of the late Ming Imperial Courts it relieved the boredom of their sheltered life. Opium use was not thought to be a problem until the late 19th century when smoking became highly prevalent in the Imperial Courts and the army. At that stage it was estimated that 10% of the Chinese population were addicted to opium primarily in oral or smoked form[52].

Opium was probably introduced to Europe in the late 15th century[7]. Throughout the late 17th and early 18th century there was a free trade in opium in Britain. During this time it was sold by the rapidly expanding wholesale drug houses. Initially it was smoked or eaten on its own, but products made with opium became popular. The alcoholic tincture of opium, Laudanum, manufactured by Thomas Sydenham in the 1660s, and Paregoric, the camphorated tincture of opium, were both in very common use. These and many other patent medicines containing opium were sold in small shops, often grocer's shops with little control either from the medical or pharmaceutical professions.

In the early nineteenth century there was an intertwining of social and medical usage of opiates. They were known to be very effective drugs for certain medical conditions; opium was widely used to treat cholera and indeed for almost any other medical condition; for many people use of this medication became an addiction. There was no standardization in the preparation of drugs containing opium and in the first edition of the British Pharmacopoeia in 1858 14 opium preparations were listed. Opium was tried as a cure for alcoholism, both as a cure for delirium tremens and as a replacement for alcohol. By the early 1880s opium was used as a treatment for insanity and was useful in early attempts to reduce the amount of physical restraint used in asylums. Many of the patent preparations of opium were designed for use in children and indeed the practice of infant doping to quiet a crying child became common in the early 19th century. This probably was the cause of many infant deaths.

Most of the opium consumed in Britain in the nineteenth century was imported from Turkey although it was thought to be of poorer quality than that found in India. Attempts were made to grow the opium poppy in England and in other areas of Western Europe, notably in France and Germany. Probably the most successful English cultivators were Dr. John Cowley and Mr. Staines of Winslow in Buckinghamshire who in 1823 produced 143 lbs of opium, 11 acres of opium poppies, and was awarded a Horticultural Society prize for his efforts[7]. Despite these early successes, however, British and European opium cultivation was never very successful. Throughout the 19th and indeed into the early 20th century opium use was endemic in the Fens of Eastern England. In that area of the country opium became more available than in the rest of the country. The problem had its basis in a tradition of popular self-medication and for its value in releasing the population from the dreary poverty stricken farming life of the area.

The use of opium in the early part of the 19th century by the circle of romantic writers and poets has attracted much interest although their use was almost certainly less extensive than that of the working class. The addictions of Thomas de Quincy and Samuel Taylor Coleridge have been written about specifically. Even among literary society, however, opium was certainly taken first for self-medication although it may later have been taken for its relaxation and euphoric effects. Other famous addicts were Elizabeth Barrett Browning, who took opium and morphia regularly, and Florence Nightingale, who took opium on her return from the Crimea, partly for severe back pain.

In 1816 Frederick William Sertürner, a pharmacist of Einbeck in Hanover, isolated a white crystalline substance which was more powerful than opium. He named it Morphium after Morpheus, the God of Sleep. Thomas Morson, founder of the Pharmaceutical Society, learned of morphine manufacture in Paris and in 1821 began producing wholesale morphine in London. In Germany Merck of Darmstadt began production at around the same time.

In the mid-nineteenth century the hypodermic method of morphine injection became popular and was believed to be very useful for many medical conditions. It was gradually realized, however, that morphine injections could have dangers. In 1877 Dr. E. Levenstein of Berlin produced a work describing the addiction to morphine "morphinism" and these ideas gradually disseminated across Europe. In the late 19th century there was the gradual awareness of the medical profession's own susceptibility to morphine addiction and it was noted that doctors were particularly likely to exhibit signs of morphinism.

Until the early 1900s opium eating for pleasure was regarded as a moral failing. It was not until the late 19th century that addiction to opium and alcohol began to be viewed as a disease. There is evidence of addicts being treated, records of admissions to asylums, and of the setting up of inebriate homes, notably the Dalrymple Home in Rickmansworth. There began to be discussions about the correct methods of curing addiction; these varied from the abrupt method where an addict was confined to a locked, barred room and deprived of the drug, to gradual withdrawal methods. Alternative drugs were suggested, for instance sodium bromide was used to quell nervous irritability during drug withdrawal. The concepts of tolerance and withdrawal were defined and their symptoms noted. A side effect of this was that social realities began to be ignored in favor of personality and biological determinants. These developments in the understanding of opiate addiction all paralleled the developments in the understanding of alcoholism which developed with the growth of the Temperance Movement[5].

In the USA opium was similarly available as a patent medicine throughout the 19th century. Morphine, which was used heavily in the States from the 1830s, became a problem after it was used extensively to treat soldiers in the Civil War. By the end of the 19th century the United States probably had more opium addicts than the whole of Europe and responded by enacting a series of legislative measures[31].

Peruvian Indians were chewing the coca leaf when the Americas were discovered but cocaine was not introduced into society in the way opiates had been[32]. In 1662 Abraham Cowley, an English physician, wrote about the coca leaf and its ability to "increase stamina and cheer the dropping mind". The 1860s brought a period of great interest in cocaine and during that time the drug was isolated from the coca leaf. Cocaine became renowned for its ability to increase stamina and improve endurance. Medical men even wrote about how it improved their abilities to climb mountains. The use of cocaine as a local anesthetic was largely discovered by Sigmund Freud and his colleague Karl Koller in Vienna. In 1884 he published a paper on cocaine which described its possible therapeutic uses. It was his colleague Koller, however, who had been interested in finding a suitable anesthetic for eye surgery, and while experimenting with cocaine on the eye of a frog he found the drug to be an effective corneal anesthetic. It was quickly taken up both as an ophthalmic and local anesthetic.

Cocaine's popular use was typified by the invention of invigorating patent medicines of which "Marianis coca wine" was one. In 1900 Coca Cola had 4.5 mg of cocaine per bottle although in an average person this is a very low dose. The 1868 Pharmacy Act in Britain which restricted the sale of opiates and cocaine resulted in greater availability of cocaine in the USA than in Britain. By the 1920s there were serious problems with cocaine use in the United States and cocaine scandals in Hollywood made the problem very obvious. However, by the 1930s the problem gradually declined and indeed cocaine did not reappear as a problem until the 1960s. Concern was expressed in Europe at the time of the First World War about the use of cocaine and a number of countries introduced legislation controlling the production, use, and sale of cocaine. Thus there are indications that the cocaine epidemic in the US in the 1920s also occurred in Europe.

Medical use of cannabis was noted in the 19th century in England and France following the Napoleonic conquest of Egypt when French soldiers brought the drug home. In 1842 an Irish physician, Dr. William O'Shaunassy, published a paper on Indian hemp and its medical uses. He had acquired substantial knowledge of the drug while working in India. The drug became sufficiently popular to appear in the British Pharmacopoeia for a short time and it was even cultivated commercially in England for a few years. Cannabis never became a rival to opium and was always less medically acceptable. For instance as early as 1848 it was known both as a cure and a cause of insanity. There was a brief period when its use with the insane was studied, and there were some very early reports of insanity in cannabis users coming from both Cairo and Dakka. In the 1880s and 90s it was used for recreation in the literary and artistic world during a time when interest in paranormal and psychic phenomena was prevalent.

1.3 HISTORY OF POLICY

In the mid-19th century there were increasing worries about the prevalence of opium use in the United Kingdom. There was an associated rise in the importance and influence of the medical and pharmaceutical professions. The debate over controls of opium use eventually led (after several acts which failed to get through parliament), to the 1868 Pharmacy Act. This Act restricted the availability of some drugs and provided restricted labeling on opium.

However, the act failed to stamp out the liberal sale of opium in many parts of England. There continued to be professional horror about the development of addiction, the overdoses and deaths of infants due to the practice of doping, and by the 1890s other patent medicines had been included in the Act making it a more robust piece of legislation.

Similarly in the United States laws began to be passed in the mid-nineteenth century; for instance, in Pennsylvania in 1860 an Act was passed which attempted to control the distribution of opiates.

It was the early 20th century which saw the major legislative changes about drug use. Most of the policy and legislation grew out of the often acrimonious debates between the medical profession and the lobby for greater legal reforms and legal control of the drug use. In the early 20th century the addict population of England was small and was dominated by older addicts and by middle class and medically addicted patients. In the United States the problem was larger and more encouraging legislation which produced more legal controls on the use of drugs[31]. The Harrison Act of 1914 made the effective maintenance of an addiction by a physician illegal and drug control essentially became a penal matter. In the 1920s treatment clinics which had been set up were gradually closed and the doctors who prescribed these drugs for addicts were frequently put out of practice.

In the UK the medical profession dominated over the lobby for penal control resulting in legislation which acknowledged the role of the doctor in treating the addict and effectively set up what became known as the "British system"[6]. Between 1911 and 1914 there were conferences in the Hague which attempted to formulate an international policy on drug use and trade. There was opposition to these conferences from the British who still had some interest in the opium trade and from the Dutch who were involved in drug traffic in the Dutch East Indies. The USA was particularly opposed to the opium trade following pressure from missionaries in the Philippines, and Musto[31] argues that much of the American pressure for narcotic prohibition was influenced by its colonial interest in the Philippines. However, laws that were initially enacted for foreign policy purposes soon became a vigorous domestic issue.

The 1914–18 war allowed more stringent changes to take place especially in Britain, where cocaine consumption among soldiers was identified as a problem, and greater controls were introduced under the Defense of the Realm Act, that were felt to be necessary to help with the War effort. In 1920 the Dangerous Drugs Act extended controls over opiates. In Britain in 1924 the Rolleston Committee was set up to consider whether or not it was a medical role to supply morphine and heroin to addicts. It essentially decided that it was, and all doctors were allowed to prescribe heroin and cocaine until 1968 when restrictions were imposed because of the irregular prescribing of a number of private practitioners[48].

In Holland the 1818 law on the Practice of Medicine stipulated that "poisons and narcotics may only be delivered by apothecaries, medical doctors, midwives, and surgeons on the basis of a prescription duly signed"[44]. In 1865 a separate law ruled that medicines may only be delivered by pharmacists or medical doctors with a special license and the delivery of opium was limited to a maximum of 50mg in a separate notice (1865). The first Dutch Opium Law was passed in 1919. It outlawed the delivery, sale, import and export, and manufacture of opium and its derivatives in the Netherlands, but possession of opium was not contrary to the law. Possession was outlawed in the second opium law in 1928 and this was modified in 1953 to make the possession of all hemp products illegal also making the use of opiates, cocaine, and hemp products illegal with harsh legal penalties for infringement of these laws. The period between 1953 and 1976 is the only period during which drug use as such was a criminal offense in the Netherlands. With the rise of cannabis use in the 1960s the Dutch Government appointed a Steering Group on Narcotics (Commissie Baan) which resulted in modification of the Opium law in 1976 to increase the penal sanctions for use of and trade in hard drugs, to reduce the legal sanctions for trade in hemp products, and to reclassify the possession of hemp for personal use from a punishable offense to a punishable trespass.

In the US a different system with a more penal element arose, although this may have been due to their larger addict population and their weaker (in relation to drug policy anyway) medical profession. In the late 60s and early 70s Dole and Nyswander's work on the medical treatment of heroin users with methadone started to become popular[13]. The 1970s and 80s also began to see the beginning of the second cocaine epidemic in the United States. In the UK in 1968 in response to the problem of a number of doctors prescribing large amount of controlled drugs the Brain Commission produced a framework for the 1969 Dangerous Drug Act which set up specialist drug clinics, made addict registration in England compulsory, and made the prescribing of heroin and cocaine only possible for doctors specially licensed for the purpose. The picture in most European countries is similar with most countries ratifying the United Nations treaties and restricting access to controlled drugs in the first half of the twentieth century but experiencing minimal problems with illicit drugs. A dramatic change took place in Europe between the mid-1960s and late 1970s. There were virtually no specialist drug services in existence before the late 1960s. The growth of the drug culture escalated with the rapid introduction of heroin into Holland, France, Germany, Belgium, Italy, Ireland, United Kingdom and Denmark in the late 1960s and early 1970s; it appears that heroin appeared somewhat later in Spain and Greece.

2 Social factors

The social surroundings and experiences of an individual have an influence on the way they use drugs[19]. For instance mescaline and other hallucinogens played a large role in the spiritual experiences of many ancient cultures. When the Spanish invaded Mexico they found that the Aztecs ate the peyote cactus to help them communicate with gods. The North American Indians used peyote and it is still used by some members of the Native American church in their religious services. Some Hindu gods are thought to have a taste for "Bhang", a drink made with cannabis, although it has more recently been associated with the Ethiopian Zion Coptic Church based in Jamaica. Cannabis is considered an indispensable part of religious worship and smoking it makes helps the members look deeper into their own consciousness. They regard cannabis as a symbol of the body and blood of Jesus, deny that it is a drug and indeed forbid any other drug taking including alcohol. Kava, which is made from the root of a plant *(Macropiper latifolium)* and prepared as a drink, has sedative properties a little like alcohol. It is used by the South Sea Islanders of Samoa, Tonga, and Fiji as a way of getting in touch with the supernatural, and as an aid to meditation and quiet reflection.

Drug use is fundamental to Christianity, wine being at the center of Christian communion. Religion can aid and control drinking: there are more drinkers among the Jews than in any other social and religious group (87% of religious Jews use alcohol) and yet there is a low incidence of alcoholic psychosis, liver disease, and alcoholism. This is contrary to most of the rules of drug use which imply that as availability of a drug in a population rises so do problems with that drug. It may be that the network of ritual that surrounds Jewish alcohol consumption protects Jews against excess drinking. Wine is drunk on religious occasions, for instance at the Passover table and there are even occasions when Jews are allowed to drink to excess. Other societies that have few protective rituals may have more problems with alcohol. The aborigines of Australia acquired a cultural norm of excessive drinking from the European settlers who came first in 1788.

A modern example of the influence of cultural factors on drug use came from the Vietnam War[40]. In 1970 the American government realized that there was a heroin problem among soldiers in Vietnam. There was a fear that when they were sent back to civilian life in the United States they would remain addicted and would become involved in crime. In 1971 the government commissioned a research project to look at this problem. The men were interviewed as they left Vietnam and were then followed 1 year and 3 years after that. Six hundred men were interviewed and at a three year follow up 94% of these were re-interviewed. Almost half of the study cohort enlisted between 1970 and 1971 had tried narcotics and 34% of those had tried heroin. Twenty percent of these claimed to have been addicted to drugs in Vietnam. When the soldiers were screened as they left Vietnam 11% of them had urines positive for heroin. The most surprising finding was that in the first year after return only 5% of those who had been addicted in Vietnam continued with their addiction in the U.S. This was despite only a third of these men receiving detoxification and only 2% had been in treatment. One possible explanation for these findings is that it was the particular social conditions and the availability of the drug in Vietnam which made it more likely that large numbers of young men would become addicted. On retudrning home, the social conditions changed and so addiction easily remitted. However, another significant point is that most were smoking heroin and there may have been a difference between those who smoked and those who injected.

3 Contemporary policy issues

In the early to mid-1980s HIV infection was found among injecting drug users. This had an immediate impact on drug policy in some countries where it changed the goal of policy from reducing drug use per se to reducing HIV transmission from using drugs. In many European countries it was associated with increased Government funding for drug services. In the UK the "Advisory Committee on the Misuse of Drugs" recommended in 1988 "The spread of HIV is a greater danger to individual and public health than drug misuse. Accordingly we believe that services which aim to minimize HIV risk behavior by all available means should take precedence in developmental plans". The concept of "harm reduction" introduced the idea of reducing harm caused by a drug rather than necessarily decreasing its consumption. This can involve the reduction of harm done by the route of transmission of a drug; for instance, it is only by injecting drugs that HIV is transmitted. Other drugs cause harm which are related to the drug itself, for instance, alcohol use is associated with liver damage.

Harm reduction lies behind the development of needle and syringe exchange schemes all over Europe. It lies behind the increased availability of methadone maintenance treatment. Harm reduction has an individual and public health perspective. A harm reduction intervention may be significant for an individual, as in the decision to switch from injecting to taking a drug orally, but another intervention, for instance an increase in the price of tobacco, may have a substantial public health role in reducing harm if it reduces the level of tobacco consumption.

Substitute prescribing has become an issue over the last 10 years. The bulk of substitute prescribing has been in the form of methadone maintenance. This type of treatment continues to be controversial and elicits strong popular feelings despite a large body of research and treatment evaluation that demonstrates the benefits of such treatment when well-delivered. Methadone maintenance programs and the consumption of methadone has risen sharply in most European countries since the late 1980s. With the exception of Greece, and they are in the process of establishing such treatment, all EU countries now provide methadone maintenance as one component of their drug treatment services. Educational programs are critical aspects of broader prevention strategies.

Legalization of drugs is now debated frequently in Europe. Some of the debate is directly imported from the United States where a policy of non-tolerance of drug users has been aggressively pursued. The high criminal justice burden of this population as well as the continued growth of the drug market has led the police to question existing drug control policies. This discussion tends to polarize at legalization or not and often fails to look at the more complex issues. For instance what drugs are being considered for legalization and in what form should those drugs be taken? Should they be openly available or should availability be restricted, and if so who to. If there are changes what role should the law play?

Indeed the very concept of legalization is frequently confused. In the UK heroin is legal and frequently used in the control of pain in terminal cancer. It is also possible for doctors with special licenses to prescribe heroin for addicts, but only a minority of addicts receive such treatment. Some policy advocates would like to see a situation in many countries where it was possible for general practitioners to prescribe heroin for addicts. Others would like to see over-the-counter sale of heroin. Others would see a situation where the law made it legal to possess controlled drugs but prohibited the sale of such drugs as a form of decriminalization. A policy of decriminalization is practiced by the Dutch authorities around cannabis. This policy is titled "hard on hard drugs and soft on soft drugs". On balance, the control policy options for different drugs is shaped by the UN treaty and most countries work within that framework. There is a need to study the costs and benefits of the prohibition of currently illegal drugs with separate consideration of the policy options for the different drugs.

4 Trends

Tobacco smoking in the western world reached a peak after the Second World War with almost 80% of the male population smoking. The reports on the link between smoking and cancer in the early 1960s resulted in a decline that has continued to today where 26% of the US population (28.1% of men and 23.5% of women) and in Britain 35% of men and 29% of women smoked cigarettes. This decline has been associated with emergence of a sharp socio-economic gradient in prevalence with greater levels of smoking among unskilled and manual classes. Globally tobacco consumption continues to grow particularly in the developing world with aggressive multinational tobacco company marketing[38]. This market growth will present a major public health problem for the coming few decades.

Alcohol consumption in most Western countries has doubled in the two and a half decades after the Second World War but consumption levels have been relatively stable over the recent past. Indeed some high consuming countries such as France are actually reporting a fall in per capita consumption. The link between levels of per capita consumption and alcohol related harm have been well demonstrated, with countries such as France and Italy that have high levels of consumption reporting high rates of liver cirrhosis[15]. Alcohol is an integral part of many western societies; scientific studies reporting a reduced incidence of heart disease among low level consumers compared to abstainers have been heavily reported by the alcohol industry and the media. There is a direct conflict between public health targets of modest population alcohol consumption and the alcohol industry's desire for higher levels of consumption and higher profits.

The late 1960s saw the sudden growth of illicit drug problems in Western Europe; the 1970s and 1980s saw the exponential growth in heroin problems in both smoking and injecting of heroin. The mid-1980s was viewed as a period when the heroin problem appeared to have stabilized as indicated by the number of heroin users entering treatment and the increasing age of those presenting for treatment. However, the picture is not uniform in that there is a decline in treatment demand in Amsterdam and Ireland but a steady increase in London, Hamburg, and Rome.

The upward or downward shift in heroin and other opiate problems is a matter of degrees but there is much greater uncertainty about the trends in cocaine consumption and cocaine problems. Arrest and seizure data indicate that the cocaine market in Europe has increased dramatically but to date this shift has not been reflected in treatment indicators. The problems in the US have mirrored the problems in Europe but have tended to occur in the US first. The cocaine epidemic in the US in the late 1980s appears to have stabilized with significant reduction in experimental or occasional users. However, actual levels of cocaine consumption appears to have remained steady, with a smaller number of heavy users accounting for the bulk of cocaine consumption[42].

Hallucinogenic drugs such as lysergic acid (LSD) are generally recognized as being weakly reinforcing, that is minimally addictive. The 60s saw the growth of the counterculture and the glorification of drug use. These drugs went out of fashion until the late 1980s when drugs such as MDMA marketed as Ecstasy became very popular with a dramatic increase in the number of people using this drug[17]. The rave nightclub scene that started in the mid 1980s continues as a strong leisure culture for young people and is associated with the use of stimulant type drugs that assist and energize people to spend the whole night dancing. There is no data to indicate the overlap between this market and the heroin and cocaine market but there are anecdotal reports of heavy stimulant users using opiates as come-down drugs.

In a separate consideration sports have now become an area of serious concern about abuse if drugs that may enhance training capacity and enhance performance capacity. There is conflicting evidence about the dependence inducing potential of anabolic steroids. An area that has traditionally been seen as an attractive alternative to the lure of drug use for young people has now become identified as a key area of concern.

5 Evolution of terminology

Concepts of dependence and addiction have changed over the last 2 centuries and it was not until the mid-eighteenth century that excessive drinking or "inebriety" as it was then known was some sort of disease. Edwards (1992)[16] has described the birth and evolution of concepts and terminology of addiction or dependence. Initially distinctions between desire and will which was at the heart of the concept of addiction were not commonly made. Drinking was regarded as something over which the individual had control and if you chose to drink that was a choice and not an inherent compulsion. In 1791 Benjamin Rush called chronic drunkenness a "disease or derangement of the will", drinking he argued began "as an act of free will, descended into a habit, and finally sank to a necessity". Thomas Trotter (1804) described similar concepts in the USA. Bruhl-Cramer (1819) introduced the concept of dipsomania, and Magnus-Huss (1849) in Sweden wrote about alcoholismus. It was not until 1878, however, when Doctor Edward Levinstien wrote about a "morbid craving for morphia" and regarded it as a disease with a similarity to "dipsomania". During the nineteenth century the main components of addiction, tolerance, and withdrawal became more defined and disease theories gradually won medical acceptance. This was followed by a strong movement to establish the study of the problem of 'inebriety' which like the concept of addiction could relate to a number of substances. The promoters of the disease concept of inebriety also campaigned and were successful in establishing "inebriate asylums" where inebriates could be compulsory detained and treated[5].

At the turn of the century when the public health model of disease control was beginning to be influential, a public health element emerged in way the medical profession viewed alcoholism. It began to be viewed as a social problem and some relationship between alcohol consumption and environmental factors became acknowledged. Education about alcohol became important. During the World War I in Britain the first acts were introduced which restricted pub opening hours, and temperance became an important concept to be disseminated into society.

It was not until after World War II that the more familiar definitions of addiction and dependence began to appear. The definition adopted by the WHO in the 1950s emphasized the biochemical aspects of addiction but in the 1957 definition there was acknowledgment of it as a physical craving accompanied by psychological factors. In 1964 the conventional definition of "a state arising from repeated administration on a periodic or continuous basis" was developed. This definition combined what would later be thought of as physical and psychological dependence. Only in the late 1970s did the concept of alcohol related problems emerge. This was a reminder of the earlier period of public health association with drug and alcohol use and indeed in the 1980s following the HIV epidemic drug and alcohol policies have again entered the public health domain.

Definitions of dependence have been formulated using criteria which facilitate experimental research. Emphasis is given to the observable behavior of the substance user and how it is learned, modified, and reinforced. An example is the definition of substance use used by Pomerleau and Pomerleau[35]. They define addiction as "repeated use of a substance and/or a compelling involvement in a behavior that directly and indirectly modifies the internal milieu in such a way that produces immediate reinforcement but with harmful long term effects". This definition incorporates both the pharmacological and social learning factors. Dependence implies that there is a degree of neuroadaptation which in the absence of drug consumption will result in receptor readaptation; this gives rise to a physical withdrawal in the case of opiates but not so with stimulants.

Sociologists have also contributed to the debate on alcohol and drug use. Consumption may be considered deviant if it exceeds the established norm of the community. Other sociological perspectives would argue that dependence symptoms such as loss of control are social constructions that facilitate the categorization and management of such individuals in a more humane way[41].

Psychiatry as a whole has gone through a revolution in the past 3 decades. The antipsychiatry movement of the 1960s and 1970s highlighted the dearth of empiricism in the classification of mental illness. This has resulted in an explosion of research and publication of work attempting to put psychopathology on a sound empirical basis. This has involved in the development of numerous instruments that measure symptoms and this work has drawn on the existing body of knowledge of psychometric methodologies. Simultaneously international bodies such as the World Health Organization have drawn experts together to develop diagnostic criteria to provide definitions for psychiatric diagnosis. The most updated version of this is the International Classification of Diseases 10. The American Psychiatric Association have conducted a similar exercise to develop the Diagnostic and Statistics Manual now in its fourth version (DSM-IV). Both these standard operationalized diagnostic systems cover large areas of psychiatry and medicine including drugs and alcohol. These are used for statistical and epidemiological purposes and also as a tool for the provision of psychiatrically based treatment for drug and alcohol users. Both diagnostic systems classify a problem on one of five axes, the first one being the primary one or the problem associated with the mental illness or behavioral disorder. In ICD 10 and DSM-IV each drug is defined as giving a series of problems which includes dependence, harmful use, withdrawal states, and with some drugs psychotic states. Central to the current methodology is the concept that a dependence syndrome can be distinguished from an alcohol or drug related disability.

There is a rich debate on the value of categorical versus dimensional models of classification. In the categorical model a condition is either present or absent, however, the dimensional model allows for degrees of presence or absence. In the evolution of disciplines it is not unusual for classification systems to start as categorical and to evolve to a dimensional model. However, as well as achieving agreement among experts on the diagnosis of a condition, there is a critical need to develop methods to assess the validity of the condition described in the possibility of ensuring that the condition is distinctive with particular etiological features and particular outcomes. Validity is an overall evaluative judgment that is based on scientific evidence and is supported by a theoretical rationale. Construct validity has focused on the role of theory in constructing tests and has emphasized the need to have hypotheses that can be proved or disproved. Thus the process of validation should continue to challenge and improve the understanding of the area under study.

The concept of disease has a long history in medicine and a more controversial history within psychiatry and psychology. The collection of a set of symptoms that regularly co-occur is termed a syndrome and is based on the assumption that these symptoms may have a common origin. In psychiatry such a cluster of symptoms may be abnormal behaviors, abnormal or distressing subjective experiences, or a mixture of the two. The first description of the dependence syndrome was by Edwards and Gross[14] who believed that there was some altered behavioral, subjective, and psychobiological state represented by this cluster of symptoms they described. They argued that this description of a cluster of symptoms needed to be empirically investigated and underlying mechanisms elucidated. The features of dependence they described are given in Table 1.1.

It is not necessary for all of the above symptoms to be present to make a diagnosis of dependence but it is noteworthy that there is an observable coincidence of these phenomena. The Alcohol Dependence Syndrome was

TABLE 1.1
Features of dependence

Features of dependence
Narrowing of drinking/drug using repertoire
Salience of drink/drug seeking behaviour
Increased tolerance to alcohol/drug
Repeated withdrawal symptoms
Relief avoidance of withdrawal
Subjective awareness of compulsion to drink/use drugs
Reinstatement after abstinence

proposed as a specific model with detailed description of the symptoms which aimed to provide a theoretical explanation of the relationship between the different symptoms. Symptoms of physiological dependence and impaired control are the central features of this syndrome.

Subsequently, this model of dependence was applied to other drugs to test whether or not the model that applied to alcohol would cross to opiates, cocaine, or tranquilizers and thus support a unidimensional model of drug dependence.

The above description is medical and psychiatric in orientation but there are a variety of definitions used in different classificatory systems; in an effort to achieve cross disciplinary agreement a consensus statement was formulated[39] on (Drug) Dependence "a generic term that relates to physical or psychological dependence or both. It is characteristic for each pharmacological class of psychoactive drugs. Impaired control over drug-taking behavior is implied." Babor (1990)[1] describes the social, scientific, and medical issues around defining alcohol and drug dependence.

The 28th report of the World Health Organization Expert Committee on Drug Dependence in 1993 defined drug dependence as "a cluster of physiological, behavioral and cognitive phenomena of variable intensity, in which the use of a psychoactive drug (or drugs) takes on a high priority. The necessary descriptive characteristics are preoccupation with a desire to obtain and take the drug and persistent drug seeking behavior. Determinants and the problematic consequences of drug dependence may be biological, psychological, or social and usually interact[51]."

6 Epidemiology

Epidemiology has become a key focus in the study of addiction. The field of addiction presents particular problems because of the often illegal nature of some of the drug taking activity. Thus the study of the epidemiology of drug taking has made important contributions to the study of "hidden populations" and has involved careful consideration of the link between empirical quantitative work and qualitative ethnographic work that can provide vivid descriptions of worlds that are not readily quantified.

Tobacco smoking is the drug of dependence or addiction that has been best described. The legal nature of the drug has made it possible to study closely. It is possible to describe the evolution of smoking careers from initiation in adolescence to established patterns of consumption in adulthood to patterns of cessation and rates of cessation in a significant minority of smokers. The availability of salivary and blood tests to confirm individuals self-reported behavior also renders much of the more recent studies a confident degree of accuracy and validity. The range of cigarette smokers is from 25 to 60% with a general decline in smoking in industrial nations. However, over 20% of the young population continue to be recruited to smoking. UK national surveys report that only 40% of those who had ever smoked regularly had managed to give up smoking[46]. An index of nicotine dependence is the time taken from waking up until smoking the first cigarette; in a UK national survey 14% of report smoking within 5 minutes of waking and over 50% light up within 30 minutes of waking. It is estimated that less than 5% of smokers are recreational or non-dependent smokers[22]. Despite increasing social stigmatization associated with smoking behavior, mass media anti-smoking campaigns, and intensive school based anti-smoking campaigns, over 20% of young people continue to initiate smoking. It is argued that the tobacco industry has aggressively targeted the young market in order to sustain a sizable long term population of smokers. The targeting of young people in the developed world and of the total population in developing countries by the tobacco industry is in our view ethically indefensible and highlights the need for international agreements to prevent such activities.

By contrast, studies of drinking patterns in the industrial countries report that rates of 60–90% ever use alcohal. The rate of abstention in different countries can influence calculations of the average per capita consumption. Data on patterns of drinking are generally gathered from household health and lifestyle surveys, and may be subject to substantial under-reporting and poorer recall of past drinking experience. Despite this surveys regularly report 4–6% of the population are drinking in a dependent fashion. The UK surveys of alcohol consumption report that 24% of men and 11% of women drink at a level that presents risks to their health. The measurement of the relationship of the level of consumption and problems is an important but complex issue that requires ongoing elaboration[16]. There is limited information on the present levels of alcohol consumption or alcohol problems in most developing countries. The growth of alcohol markets in settings where there is little tradition of alcohol consumption may be associated with significant social disruption and high levels of problematic drinking. There is a need for detailed studies of the changing patterns of alcohol consumption in the developing world.

The key purpose of epidemiological studies is to provide workable estimates of the scale of drug problems and to provide estimates of the number of new recruits to drug use and of the exit rate from drug use. However, it is not adequate to know the number of people consuming drugs. It is necessary to have better estimates of the number who are consuming drugs in a problematic or dependent fashion. There is a need for ongoing information that provides a clear picture of trends in drug use and problem drug use so that the impact of the range of social policies can be properly understood. Epidemiological data that are properly collected can determine if drug problems are increasing or decreasing, which age group is involved, and can also potentially provide information on the number who are exiting from dependent or problematic drug use. National and local epidemiological data can be collected. Given the regional variation in drug use a national picture may fail to outline significant differences in patterns of drug use across localities that have important policy and service delivery implications.

National household surveys are conducted in a number of countries such as the US, Denmark, Germany, and Switzerland. Such household surveys provide important information on the patterns of licit and illicit use. However, such surveys do not have sample sizes large enough to provide substantial information on heroin and cocaine use where the national prevalence is less than 1% in the total population. For example, a survey of 10,000 households is unlikely to detect more than 100 heroin users. Thus such surveys methods are an expensive way to access this type of population and are likely to underestimate the size of the population. For this reason, other methods have been employed to estimate the size of the mainly opiate addict population. These methods may include multinumeration methods, where data is gathered from a range of service contacts points and the size of the addict population is calculated from this. Other methods have included capture-recapture, a method originally used to estimate the size of animal populations, where a type of tagging system is used across a range of points where drug users are identified such as police arrests and treatment services; the overlap between these two populations is used to calculate the size of the non-contacted population.

It is important to consider using more than one method because one method such as the general household survey may indicate a general decrease in the size of the illicit drug using population overall. However, a more focused study such as a capture recapture study may indicate that even with the decline in general illicit use there is a continuing growth in the size of the addict population that needs to be specifically addressed.

Local indicator data gathered at the city level has been promoted by the Pompidou group and has provided detailed information on a network of European cities (1993)[36]. This provides information on a range of criminal justice and health related data and attempts to establish trends in drug related social and health problems.

Not withstanding the difficulties with drug data, an attempt is made here to provide a broad overview which aggregates and simplifies information that is best studied at a national, regional or local level. The situation is one where there are relatively high reports of single cannabis use in most countries, ranging from 10–30% depending on the age band but generally substantially lower levels of regular or heavy cannabis use between 1 and 3%. Alcohol use is reported by up to 90–95% of most European member state populations, with a probable 3% alcohol dependent and a further 10% with alcohol related problems. Other types of hallucinogens may be used on a experimental basis. The most problematic population of opiate dependent users ranges 100–400 per 100,000 population in most European countries but the picture with cocaine is difficult to ascertain at present. In the United States an estimated 4.6 million people (about 2% of the population) clearly or probably need treatment for drug problems out of a total of 14.5 million (about 7.3% of the population) who consumed an illicit drug at least once in the month before the survey[18].

HIV and AIDS has added another significant dimension to drug epidemiology. There are major national variations in the prevalence of HIV among injecting drug users. Some countries such as Greece report very low rates of HIV while most other Southern European countries report rates of 30–50% of injecting drug users seropositive. Northern European figures range around 15% of injectors seropositive. More recent data on hepatitis C, a virus that is transmitted by similar means as HIV, indicates that 50–70% of injectors in most countries are infected. In North America there is marked variation in HIV seroprevalence among injecting drug users. Studies on the East Coast report levels up to 40% compared to substantially lower figures, below 10%, on the West Coast[12].

7 Etiology

When discussing etiology it is important to distinguish between models that attempt to look for factors that predict drug use and models that predict the development of dependent drug use or problem drug use. In either case there are no robust predictors but there are some strong associative factors. The studies of patterns of smoking initiation find few predictors of onset but weak associative factors include sibling and parental smoking, poorer academic achievement, having ever been drunk, and having a boy friend or girl friend. It seems that teenagers start smoking to be like their friends and to mimic what appears to be attractive adult behavior. The nicotine model of addiction shows that in the right social conditions the majority of the population can be addicted to a substance, indicating that individual personality or constitutional factors are not critical variables.

There are over 40 distinct theories or etiological models of drug misuse[27]. One model holds the view that the individual consumes drugs to cope with intra or interpersonal difficulties[24]. The other main model looks at environmental influences such as drug availability, environmental pressures to consume drugs, and sociocultural influences such a peer pressure on the acquisitional drug taking behavior[14, 45]. These complex multifactorial models are by their nature descriptive and clinical and are generally derived from minimal experimental data.

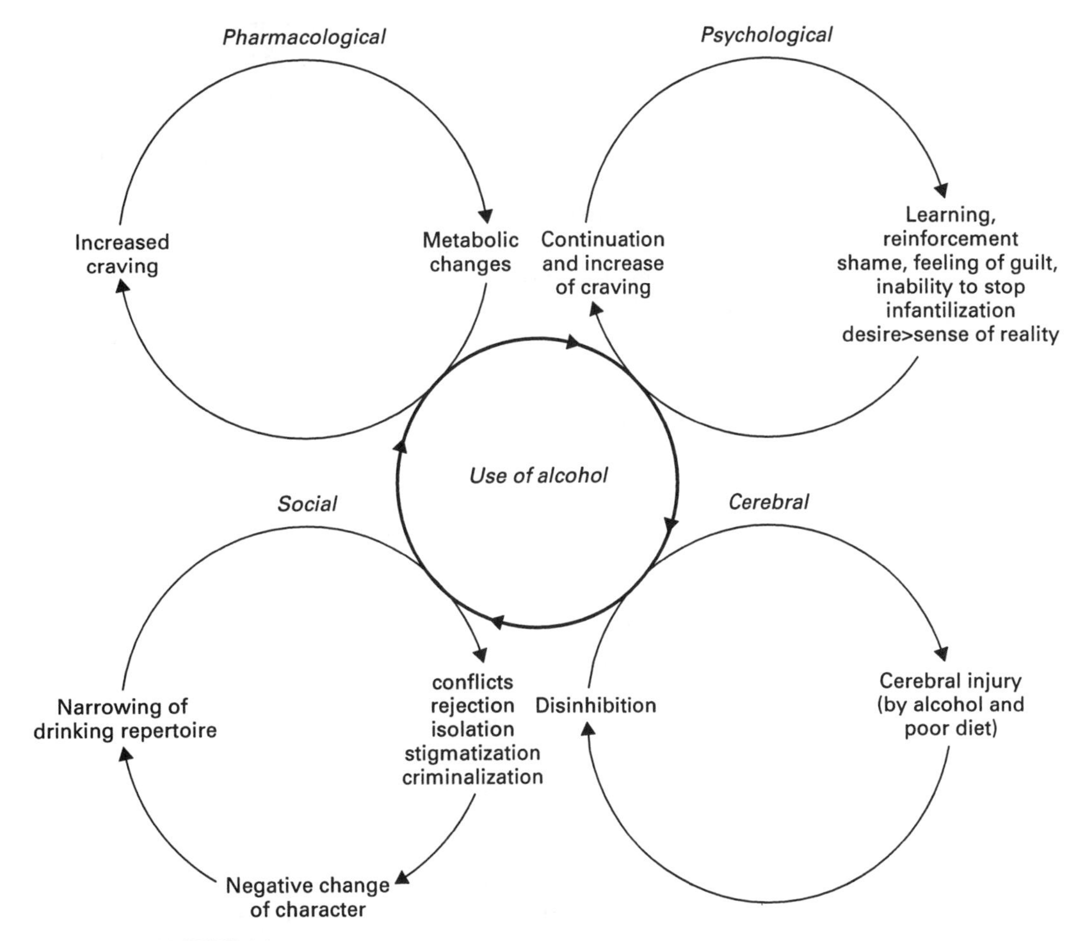

FIGURE 1.1
Schematic representation of the contribution of four vicious circles in the onset and continuation of dysfunctional drinking

By contrast, models of compulsive drug use have been amenable to both animal and human laboratory study. Studies are now conducted at cellular and receptor levels to elaborate the mechanism of tolerance and dependence for a range of substances. Classical behavior theory is a key model for understanding drug dependence and much of this work has been researched using laboratory animals. Both classical and operant models for explaining the reinforcing nature of these substances are now generally accepted and are central to the theoretical underpinnings of dependence theory[2,43,50]. Drug consumption is a learnt form of behavior. Operant conditioning is founded on the premise that behavior within an environment is shaped by its consequence; thus a behavior or operant response that increases when rewarded is called a positive reinforcement. Such rewards can be used to strengthen elements of behavior. By contrast behavior that is increased following withdrawal or punishment is titled negative reinforcement and the stimulus is titled a punishment or negative reinforcement. Numerous investigators have used conditioning theories to study drug seeking behaviors[43,49]. Schuster reports that there are a range of variables that will affect drug self administration; these range from the reinforcing schedule to the antecedent conditions such as substance deprivation or satiation, constitutional or organismic variables, and environmental factors (see also Chapter 3). Such studies have also been conducted on food, water, and sex and the reinforcing efficacy of these can be enhanced by deprivation; similarly the reinforcing efficacy of opiates can be enhanced in opiate dependent individuals but not in non dependent individuals. Laboratory studies of animal alcohol consumption, especially of rodents, report that they do not readily consume alcohol and need to be induced to do so. Once initiated, consumption can be maintained steadily but excessive self administration requires environmental, physiological, or genetic manipulation. This seems to indicate similar findings to clinical studies that both constitutional and environmental factors interact in a multidimensional manner. Animal studies are now locating the reward pathways within the brain for opiates and stimulants where positive reinforcing mechanisms involve the ventral tegmental dopamine system and the nucleus accumbens[10]. These systems may also be involved in nicotine dependence but receptor stimulation and receptor blockage of cholinergic nicotinic actions may be more critical. Probably, other systems, like the endogenous opioids, are also involved (see also Chapters 5 and 6). Alcohol, which is usually viewed as a cerebral depressant, may also exert some of its positive reinforcing effects through the ventral tegmental system. Bozarth and Wise[9] have argued that while operant conditioning has provided a valuable basis for the study of drug dependence there is a need for a separate consideration of the motivational properties of addictive behavior. This concept of motivational systems is linked to appetitive behavior such as feeding, thirst, and sex suggesting that common biological processes and motivational systems underlie drugs of dependence and food reinforcement[20]. Bozarth and Wise argue for a model of motivational toxicity where the strength of reinforcement and the hedonic effects of the drugs are central to addiction or dependence. They argue that the strength of the response weakens other types of previously reinforced and motivated behavior. The psychopharmacology and laboratory based work is of theoretical importance but to date has had limited clinical or service application. However, it may be that the behavioral models are central to well-organized and well-delivered treatment services that have a clear understanding of the mechanisms underlying addiction.

8 Natural history of drug use

The staging or sequencing of patterns of use may be divided into experimental, recreational, and continual or compulsive.

Experimental drug use may now be a form of normative adolescent risk-taking behavior. Peer influences are likely to be the major predictor of experimental drug use and also likely to influence the evolution to more regular drug use. The consumption of one drug increases the likelihood of consuming other drugs so that those who consume tobacco are more likely to take cannabis and those who use cannabis are more likely to consume other illicit drugs[25], but this relationship is neither causative nor predictable. The relationship between amount consumed, problems, and dependence is not direct but there is a very significant interaction with levels of consumption likely to predict problems and dependence, but does not invariably predict so. Heroin and cocaine are similar to nicotine in their reinforcing properties and regular use is more likely to be associated with dependence, while environmental factors are probably the key factors in initial and irregular drug use. Regular or dependent drug use may have its own motivational properties which will be strongly stimulated by environmental cues.

A number of key longitudinal studies from early adolescence into early adulthood[23, 33] suggest that the level of drug use is a predictor of future drug use and that initiation before 15 is associated with more developmental disruption. Growth into adult responsibilities with jobs, marriages, and children is a convergence towards social conformity which is associated with less drug use and less problems. However, most of these studies have been of patterns of illicit and non dependent drug use. There are few studies of non-clinical populations of heroin or cocaine users. The Harlem study of Brunswick et al.[11] followed a cohort of urban black youth who were first studied at the age of 12–17; on follow up by the age of 26 the great majority who ever became involved in drug use on more than an experimental basis were still using. Surprisingly heroin was one of the drugs that demonstrated substantial levels of discontinuity with 30% of male heroin users and 40% of female heroin users continuing to use (however, 37% of the male heroin cohort either died or were lost to follow-up). When this group of heroin users' pattern of treatment utilization was studied it suggested that those who used treatment services were more problematic, with more disadvantages in personal and social resources and had lower levels of educational attainment than the untreated heroin users. It appeared that the most drug involved male heroin users became involved in treatment. Those males who entered treatment were less likely to have terminated heroin use than their non-treated counterparts. By contrast, the treated female heroin users were five times more likely to have stopped heroin use than the non-treatment heroin users. These findings are similar to findings on tobacco cessation programs that those who seek treatment in tobacco cessation programs do so after failing on numerous occasions to stop alone. They are also only half as likely to succeed in this setting compared to the non-treatment group quitting rates.

Other studies have looked at the patterns of cessation of drug use without treatment[8] and indicate that change is associated with complex other changes in lifestyle. Such changes included major changes in social and peer networks, major changes in recreational activities, and for some individuals major involvement with self help movements or with organized religious movements.

Modern approaches to treatment are based on approaches to assisting people to change and maintain change in behavior. Detoxification, habilitation, and rehabilitation can be pursued in a community or residential setting. Studies of patterns of change have classified smokers into three categories: precontemplators are smokers who continue to smoke and are not concerned about the evidence of health risks, contemplators are those smokers who are concerned about their smoking and considering change but have not yet decided to stop, and the action phase in which people then decide to stop and put their plan in motion[37]. Addictive behavior may involve cycles of change with efforts to stop being punctuated by relapse to the old pattern of behavior. Marlatt and Gordon[28] have described a cognitive behavioral approach called relapse prevention which aims to develop coping strategies that can help to maintain the phase of change. Miller and Rollnick[29] has outlined the complex nature of motivation and promoted a style of motivational interviewing which aims to assist and promote motivation to change and maintain change in behavior. These approaches can be used in a range of compulsive behaviors and fall within a broad framework of cognitive behavioral approaches.

The important message is that there is considerable fluidity in patterns of alcohol[30] or drug use and that complex social factors including intimate partners, work, and other activities can significantly impact the pattern of drug use. The severely dependent long term heroin or cocaine using population after 5 to 10 years of dependent use are probably less amenable, have less access to, and have exhausted some of the more subtle social supports to change behavior by their intense involvement in drug use, their ongoing marginalization from non-drug using supports systems, repeated criminal justice involvement, and long term unemployment as well as increasing levels of social isolation. It is likely that the severely dependent long term heroin or stimulant user, with some notable exceptions, requires active support and treatment intervention to facilitate sociastabilization or cessation and are likely to persist with drug use otherwise. This group also has high mortality rates and appears to have moved from a more pliable career, where periods of use have been punctuated by periods of abstinence, to more continued and dependent use, and on to a more predictable long term picture that may be unlikely to be associated with spontaneous remission. Strategies to assist this group to be stabilized or to exit from the drug using network involve major social investments in health and social services for this population.

9 Conclusion

The focus and concern about drug problems is strongly shaped by contemporary issues. A knowledge of some of the historical aspects of drug use and dependence provides a picture of how social values shape the perception of alcohol and drug problems. The heterogeneous nature of drug taking behavior makes it impossible to develop a single unifying theory. The definitions and terminology that have developed over the past two decades and the tools for describing and measuring drug dependence have become increasingly sophisticated. Alcohol and other drug problems will continue to grow as major public health problems over the coming decade so appropriate models of public health need detailed development if they are to be properly implemented in a wide range of settings. The models of dependence and the models for the treatment of dependence now provide a sound basis for future development in the understanding and treatment of addictions.

References

1. Babor, T., "Social, scientific, and medical issues in the definition of alcohol and drug dependence." In: Edwards, G. and Lader, M. (Eds.), *The Nature of Drug Dependence*. Oxford University Press, London, 1990, pp 19–40.
2. Bandura, A., "Self-efficacy toward a unifying theory of behavioral change," *Psychol. Rev.* 1978, **84:** pp 191–215.
3. Baasher, T., "The use of drugs in the Islamic world." In: Edwards, G., Arif, A., Jaffe, J. (Eds.), *Drug Use and Misuse*. Croom Helm, London, 1983, Chapter 2, pp 21–32.
4. Bachman, J.G., O'Malley, P.M., Johnson, L.D., "Drug use among young adults: the impact of role status and social environment," *J. Person. Soc. Psychol.* 1994, **47:** pp 629–645.
5. Berridge, V., "Special issue: the society for the study of addiction," *Br. J. Addict.* 1990, 85.
6. Berridge, V., "Drugs and social policy: the establishment of drug control in Britain 1900–30," *Br. J. Addict.* 1984, **79:** pp 17–30.
7. Berridge, V. and Edwards, G., *Opium and the People: Opiate Use in Nineteenth-Century England*. Yale University Press, New Haven, 1987.
8. Biernacki, P., *Pathways Away From Heroin Addiction: Recovery Without Treatment*. Temple University, Philadelphia, 1986.

9. Bozarth, M.A. and Wise, R.A., "Neural substrates of opiate reinforcement," *Prog. Neuro-psychopharmacol. Biol. Psychiatry*. 1983, **7:** pp 569–575.
10. Bozarth, M.A., "Opiate reinforcement processes: re-assembling multiple mechanisms," *Addiction*. 1994, **89:** pp 1425–1434.
11. Brunswick, A.F., Messeri, P.A., Adala, A.A., "Changing drug use patterns and treament behaviour: a longitudidinal study of urban black youth." In: Watson, R.R. (Eds.), *Drug and Alcohol Abuse Prevention*. Humana Press, Clifton, USA, 1990.
12. Des Jarlais, D.C., Friedman, S.R, Choopanya, K., Vanichseni, S., Ward, T.P., "International epidemiology of HIV and AIDS among injecting drug users," *AIDS*. 1992, **6:** pp 1053–1068.
13. Dole, V.V. and Nyswander, M., "A medical treatment for diacetylmorphine (heroin) addiction: a clinical trial with methadone hydrochloride," *J. Am. Med. Assoc*. 1965, **235(19):** pp 2117–2119.
14. Edwards, G. and Gross, M.M., "Alcohol dependence: provisional description of a clinical syndrome," *Br. Med. J*. 1976, **1:** pp 1058–1061.
15. Edwards, G., Anderson, P., Babor, T.F., Casswell, S., Ferrence, R., Giebrecht, N. et al., *Alcohol Public Policy and the Public Good*. Oxford University Press, Oxford, 1994.
16. Edwards, G., "Problems and dependence: the history of two dimensions." In: Edwards, G., Lader, M., Drummond, D.C., (Eds.), *The Nature of Alcohol and Drug Related Problems*. Oxford University Press, Oxford, 1992, pp 1–13.
17. Farrell, M., "Ecstasy and the oxygen of publicity," *Br. J. Addict*. 1989, **84:** p 943.
18. Gerstein, D.R. and Harwood, H.J., *Treating Drug Problems: A Study of the Evolution, Effectiveness, and Financing of Public and Private Drug Treatment Systems*. National Academy Press, Washington D.C.
19. Gossop, M., *Living With Drugs*. Ashgate, Hampshire, 1988.
20. Grunberg, N., "The effects of nicotine and cigarette smoking on food consumption and taste preference," *Addict. Behav*. 1982, **7:** pp 317–331.
21. Hibbert, C., *The Dragon Wakes*. Penguin, London, England, 1982.
22. Jarvis, M., "A profile of tobacco smoking," *Addiction*. 1994, **89:** pp 1371–1376.
23. Jessor, R., Donovan, J.E., Costa, F.M., *Beyond Adolescence: Problem Behaviour and Young Adult Development*. Cambridge University Press, Cambridge, 1991.
24. Khantzian, E.J., "The self medication hypothesis of addictive disorders: focus on heroin and cocaine dependence," *Am. J.Psychiatry*. 1985, **142:** pp 1259–1264.
25. Kandel, D., Davies, M., Karus, D., Yamaguchi, K., "The consequences in young adulthood of adolescent drug involvement," *Arch. Gen. Psychiatry*. 1986, **43:** pp 746–754.
26. Levine, H.G., "Temperance cultures: concern about alcohol problems in Nordic and English-speaking cultures." In: Edwards, G., Lader, M., Drummond, D.C. (Eds.), *The Nature of Alcohol and Drug Related Problems*. Oxford University Press, Oxford, 1992.
27. Lettierei, D.J., Sayers, M., Pearson, H.W. (Eds.), *Theories on Drug Abuse: Selected Contemporary Perspectives* (NIDA Research Monograph 30). National Institute of Drug Abuse, Rockville, MD.
28. Marlatt, G.A. and Gordon, J.R., *Relapse Prevention: Maintenance Strategies in the Treatment of Addictive Behaviours*. Guilford, New York, 1985.
29. Miller, W.R. and Rollnick, S., *Motivational Interviewing. Preparing People to Change Addictive Behaviour*. Guilford, New York, 1991.
30. Moos, R.H., "Why do some people recover from alcohol dependence, whereas others continue to drink and become worse over time?" *Addiction*. 1994, **89:** pp 31–34.
31. Musto, D.F., *The American Disease: Origins of Narcotic Control*. Oxford University Press, Oxford, 1987.
32. Musto, D.F., "Cocaine's history, especially the American experience." In: Bock, G.R. and Whelan, J. (Eds.), *Cocaine: Scientific and Social Dimensions*. Wiley, Chichester, 1992.
33. Newcomb, M.D. and Bentler, P.M., *Consequences of Adolescent Drug Use*. Sage, Beverly Hills, CA, 1988.
34. OPCS, *General Household Survey*. HMSO, London, 1988.
35. Pomerleau, O.F. and Pomerleau, C.S., (1987) "A biobehavioural view of substance abuse and addiction," *J. Drug Issues*. 1987, 17, pp 111–31.
36. Pompidou Group, *The Multi-City Study*. Council of Europe, 1993.

37. Prochaska, J.O. and Diclemente, C.C., "Toward a comprehensive model of change." In: Miller W.R. and Heather, N. (Eds.), *Treating Addictive Behaviours: Processes of Change.* Plenum Press, New York, 1986, pp 3–26.
38. Raw, M. and McNeil, A., "The prevention of tobacco related disease," *Addiction.* 1994, **89:** p 1505.
39. Rinaldi, R.C., Steindler, E.M., Wilford, B.B., Goodwin, D., "Clarification and standardisation of substance abuse terminology," *J. Am. Med. Assoc.* 1988, 22, pp 555–557.
40. Robins, L.N., "Vietnam veterans' rapid recovery from heroin addiction: a fluke or normal expectation?" *Addiction.* 1993, **88:** pp 1041–1054.
41. Room, R., "Dependence and society," *Br. J. Addict.* 1985, **80:** pp 133–139.
42. Rydell, C.P. and Everingham, S.S., *Controlloing Cocaine: Supply Versus Demand Program.* Rand,Washington, 1994.
43. Schuster, C., "Drug-seeking behaviour: implications for theories of drug dependence." In: Edwards, G., and Lader, M. (Eds.), *The Nature of Drug Dependence.* Oxford University Press, Oxford, 1992.
44. Silvis, J., "De Opiumwet." In: *Handboek Verslaving.* Bohn Stafleu & Van Loghum, Houten, 1992.
45. Smart, R.G., "An availability-proneness theory of illicit drug use." In: Lettieri, D.J., Sayers, M., Pearson, H.W. (Eds.), *Theories of Drug Abuse: Selected Contemporary Perspectives.* National Institute on Drug Abuse Monograph, Rockville, MD, 1980.
46. Smyth, M. and Brown, F., *General Household Survey 1990.* HMSO, London, 1992.
47. Spence, J.D., "Opium," *Chinese Roundabout: Essays in History and Culture.* Norton, New York, 1992.
48. Strang, J. and Gossop, M., *Heroin Addiction and British Policy.* Oxford University Press, Oxford, 1994.
49. van Ree, J.M., "Reinforcing stimulus properties of drugs," *Neuropharmacology.* 1979, **18:** pp 963–969.
50. Wikler, A., *Opioid Dependence.* Plenum Press, New York, 1980.
51. World Health Organization, *Expert Committee on Drug Dependence, 28th Report.* WHO, Geneva, 1993.
52. Wakeman, F. and Grant, D., *Conflict and Control in Late Imperial China: Opium.* 1975, pp 229–256.

Chapter 2

Biobehavioral effects of psychoactive drugs

0-8493-7803-6/99/$0.00+$.50

Chapter 2

Biobehavioral effects of psychoactive drugs

Bernard Segal and Lawrence K. Duffy

1 Introduction

Drugs and drug-taking behavior, encompassing culturally specific sanctioned and unsanctioned drug-taking behaviors, have been part of different societies since the inception of civilization. Throughout history alcohol and mood altering or psychoactive chemicals, derived chiefly from plants, have been used for social, medical, religious, or spiritual purposes, as well as for personal motives. The magical qualities that these substances have to alter consciousness, to change mood, to modify perception, or to energize the body, among other effects, has resulted in more highly individualized motives and in more widespread use. People, for example, have come to rely on psychoactive drugs, many of which are the products of the pharmacological revolution that began in the 1950s, to help them to cope either with the tensions and strains of everyday living, or to alter unwanted feelings or emotions. For many others, the social-recreational use of psychoactive drugs, in addition to alcohol, has become increasingly popular. Many of the people who take drugs, either for coping purposes or as social behavior, experience beneficial or relaxing effects and do not have any adverse effects or develop any problems. Others, however, have bad experiences or develop drug-related problems, such as physical dependence. Even experienced drug users may inadvertently overdose or take a street drug that is misrepresented and have a bad trip. Naive users may have bad experiences because they are unprepared for a drug's effect.

One result of the more widespread use of drugs, which have found their way from prescription pads to the streets, is a concern, principally in western nations, that such use is ruining the very fabric of their societies. In the United States, for example, there is a strong preoccupation with problems caused by illicit street drugs. This concern "has spawned a host of attitudes, controversies, speculations, and policy ideas about how to respond to the notion that taking illicit psychoactive drugs is a serious, perhaps the *most* serious, social ill"[41]. The drug problem is thus perceived as an "epidemic" that needs to be eradicated and, to a large extent, health and social policies are directed at "wiping out" this problem.

The preoccupation with drugs, however, has resulted in severe misconceptions about the nature and extent of the problem. Stephens[55], in response to drug policies in the United States, described these misperceptions as follows:

"First, there is much misunderstanding concerning this 'epidemic'. One would almost believe from our mass media messages that huge segments of American society regularly abuse drugs like crack. Second, there appears to be enormous confusion among both the general public and our decision makers about what psychoactive drugs are. Most simply want to lump all types of psychoactive drugs into one category called 'dope' or 'drugs.' The implication is "one type of problem, one type of solution." Third, there is a great deal of hypocrisy concerning drug usage. The use of addictive substances like tobacco products and alcohol is condoned — indeed in some cases even subsidized by the state. Fourth, Americans tend to ignore social and cultural contexts in which drug use occurs when searching for an answer to the question, "Why do people use drugs?" Most often, drug users are believed to suffer from mental, emotional, or possibly even physiological "diseases." Because of this disease state, they have difficulty coping with stress and the problems of living. They take drugs to escape and to ameliorate the anxiety and depression they feel. Finally, Americans are perplexed about how to deal with the drug problem. Some feel that a 'get tough' approach is the only solution while others maintain that the ultimate solution lies in decriminalization or outright legalization of all psychoactive substances."

Goldberg[22] simply asks, "Are drugs really a problem in society, or is there simply an excess of concern regarding drugs?". While it is difficult to answer this question, it is important to recognize that there are, in addition to pleasurable effects, inherent adverse consequences associated with drug use, such as dependence, legal difficulties, and medical, social, intellectual, and psychological problems. The purpose of this chapter is to review drug effects, specifically discussing the relationship between drug-taking behavior and sociocultural factors. It begins with an examination of sociocultural factors, followed by a general review of the behavioral or psychological effects of drugs, a brief description of drug distribution, and concludes with a brief narration of pharmacological effects.

2 Sociocultural factors, set and setting, and drug effects

Regardless of the circumstances involved in drug use and the way in which drugs are taken, all drugs act according to two general principles:

1. Every drug varies with dose. For all drugs there is an effective dose, a toxic dose, and a lethal dose.
2. Every drug has multiple effects, either locally or systemically.

With the above in mind, a drug's *behavioral* effect is a function of the drug taken, the user's *set*, and the *setting* in which use occurs. Set refers to the expectations of the drug's effect with respect to the amount consumed. Setting refers to the context in which drug-taking behavior occurs. Set, however, may be greatly influenced by a society's perception of drugs[24]. The meaning a society, or even subgroups within a society, ascribe to a drug may not only influence the kinds of drugs used, but may also affect the user's set and the setting in which use occurs. Goode[23] noted that, "Whether a given drug is defined as good or bad socially, socially acceptable or undesirable, conventional or 'deviant,' is not a simple outgrowth of the properties or objective characteristics of the drug itself, but is in no small measure a result of the history of its use, what social strata of society use it, for what purpose, the publicity surrounding its use, and so on. Whether the effects are experienced as pleasurable or unpleasant, weak or intense, hedonistic or depressing, hallucinatory or mundane, serene or exciting — is largely a function of sociological factors."

Two questions that follow are: Does drug-taking behavior among different sociocultural groups stem from similar causes and follow the same course? Or, does drug-taking behavior have different meanings and serve diverse functions within different cultural groups? Stated differently, is there a pattern of drug-taking behavior that is more unique among majority members of a culture that differs from drug-taking behavior within minority groups?

Research has shown that drug-taking behavior differs among cultural groups, and that differences not only exist across national boundaries, but also within a given society. In the United States, for example, patterns of drinking and drug-taking behavior within minority groups vary from those experienced within the larger, dominant culture[50], and members of minority groups are at greater risk for drinking and drug-related problems. It has been observed, for example, that drug-taking behavior in the Unites States is more prevalent among specific racial and ethnic groups (cf. 14). American Indians and Alaskan native youth, for example, show exceptionally high levels of experience with drugs, especially alcohol[15], while African Americans and Latino adults tend to be overrepresented among drug-using populations, particularly when narcotics and cocaine are involved.

In addition to sociocultural factors, drug-taking variables also influence a drug's effect. People who use drugs develop expectations about what a drug will do for them, and this expectation is related to the choice of a drug and to the context of use. Krohn and Thornberry[34], for example, reported that the role that "alcohol and other drug use plays for the white adolescent is different than the role it plays for African-Americans or Hispanics." Although the relationship between sociocultural environment and drug-taking behavior is beginning to be understood, research is needed to comprehend more fully how preferences for different drugs are related to sociocultural contexts.

Drug-taking behavior not only varies significantly among subgroups within a nation, but important differences in drug-taking behavior also exist among nations. Adlaf, Smart, and Tan[1], for example, found that both Western and Eastern Europeans reported "the most frequent tobacco and alcohol use and also the greatest total involvement and drug involvement." Orientals and East/West Indians and Blacks reported the least frequent use and least involvement. Cannabis was most frequently reported by people of Jewish descent. Further more, Japan experienced, and remains fearful of, a stimulant epidemic, as has Sweden[52], while the countries of Southeast Asia are seriously affected by opium, and nations in northern and southern Africa are having to address the long-standing use of cannabis and khat, along with the rapid spread of heroin and cocaine[59].

Cultural differences also exist with respect to drinking behavior. Prevalence and consumption levels among regular drinkers, for example, contrast in different cultures. Not only do the preferred types of beverages vary among these countries, but the style of alcohol consumption also differs. In France, Italy, and Spain, for example, wine is the most popular alcoholic beverage, while in Russia, Japan, and some Scandinavian countries, people prefer distilled beverages. In Germany, Austria, and Denmark, beer is more popular than other types of alcoholic beverages[59].

Styles of drinking also vary. In most western countries, for example, rather frequent consumption (several times a week or daily) of relatively small amounts of alcohol prevail[9,48]. Klatsky et al.[31] also found that 29% of their sample of American drinkers consumed alcohol daily, but of these daily drinkers, only 5.7% exceeded more than five drinks a day. In contrast, daily alcohol consumption in Russia is rather rare, even among infrequent drinkers, who tend to consume high doses of alcohol per drinking occasion[9,48]. In contrast to daily drinking, most Russians, especially in Siberia, tend to consume alcohol only several times a month, but in rather high amounts, often exceeding 100 g per occasion[4,9,32,35]. Recent research in Siberia, however, has found that as alcohol availability increases, it is accompanied by an increase in more frequent drinking (Avksentyuk, personal communication). Similar drinking patterns have been reported for Finland and Scandinavian countries[39].

While drinking practices in developed nations have been more or less institutionalized, concern has arisen about the emerging problem of alcohol in developing countries, where alcohol availability has increased significantly[40]. The rising rate of alcohol consumption in African and other developing countries, which is a consequence of rapid social change and deepening economic crises, has altered traditional drinking behavior, largely due to the introduction of imported alcohol beverages. It remains to be seen how traditional drinking, which mostly involved consumption of homebrews, is altered. It can be anticipated, however, that increased consumption, linked to a change in traditional drinking practices, will contribute to higher rates of alcohol-related health and social problems, and to an increase in alcohol-related mortality.

Despite the influence of sociocultural factors on drug-taking behavior, the question of what role personality factors play in drug-taking behavior needs to be asked. Traditionally, an implicit assumption involved in drug-related personality research was that abuse of drugs, particularly heroin, is related to personality deficiencies or is symptomatic of an emotional disorder (cf. refs. 12, 13). Consistent with this assumption, the characterization of drug abusers as "psychopathic," "sociopathic," or as "deviant," has been a practice followed in drug research (cf. refs. 21, 25). Assumptions such as these have especially been made about the personality structure of adolescent and youthful drug users.

Against this background, a large number of personality-research studies proceeded on the premise that all forms of drug-taking behavior are dysfunctional and are representations of psychological deficits[26,57,58,62]. Few studies were undertaken that attempted to distinguish between frequent users of "hard drugs" (e.g., heroin, stimulants, barbiturates, hallucinogens, etc.) and those who only experimented with drugs or who limited drug use to marijuana and alcohol. Thus a large number of studies utilized measures that assessed psychopathology and sought to search for differences between users and nonusers, rather than attempting to seek differences among different types of users. Segal et al.[53] noted that because the dependent measures in many studies are measures of psychopathology, any observed differences would tend to be attributed to differences in underlying pathology, an assumption that may be invalid because of measurement and other procedural errors.

The search for personality correlates or specific antecedents of drug-taking behavior which, in large part, has been directed at adolescents because initiation into drug-taking behavior occurs chiefly during this developmental period, has not yielded definitive results. One early finding by Segal, Huba & Singer[53] has been replicated with some consistency. They found that initiation into drug-taking behavior was associated with rebelliousness, rejection of traditional values (nonconformity), and a need for autonomy. These characteristics, reported in an earlier study[54], were not associated with deviancy.

Another personality factor associated with drug-taking behavior has been "sensation seeking"[63], a construct that refers to an exaggerated tendency to seek novel, exciting, and risk-taking experiences. For some the sensations are mainly physical, for others they are mostly mental, and for most, the sensation is a mix of both. Drugs may thus serve as a means of providing needed stimulation. Interest in sensation seeking as a correlate of drug-taking behavior peaked subsequent to a report that biological correlates of sensation seeking were identified[64]. Recent research[2] suggests that sensation seeking, rather than being a single contributing factor related to drug-taking behavior, interacts with many other factors to influence such behavior.

Contemporary research, however, is tending to shift away from identifying specific attributes affiliated with drug-taking behavior, and is concentrating on studying factors that place youth *at risk* for drug involvement. A "risk factor" can be defined as an attribute that is associated with high probability of initiation into drug-taking behavior. Some of the risk factors that have been identified are poverty, family violence, including sexual abuse, family history of drinking or drug use, and peer approval of drug use. The extent to which risk variables are related to specific preferences and patterns of drug use or drinking behavior remains to be studied.

In summary, while the environment has an important role in determining the nature and extent of drug-taking behavior, such behavior also involves behavioral or psychological factors, which are currently being identified as "risk factors." It remains to be resolved more clearly, however, how social and behavioral domains interact, particularly when the risk factors that may contribute to drug-taking behavior may also be involved in such antisocial behavior as fighting, stealing, vandalism, and early sexual activity, which are, in turn, also associated with drug use[49]. Thus many of the factors involved in drug-taking behavior may be either a cause or consequence of such behavior. Additionally, there may not be a single underlying factor mediating drug-taking behavior. Rather, there appears to be a variety of contributing factors, and research is beginning to search for the combination of variables — biological, sociocultural and psychological — that underlie different patterns of drinking and other forms of drug-taking behavior.

2.1 FACTORS INFLUENCING DRUG EFFECTS

Many drugs, such as heroin, cocaine and marijuana, are smuggled into a country. The cost of getting them in is high, and because many of these drugs are too potent to distribute without being cut, they are usually diluted significantly to make them go further to maximize profits. In cases where drugs are diluted, the nature of the diluter is critical because many of these additives can cause a toxic, or even fatal, reaction. The diluter might also activate or inhibit the metabolism of the drug. Additionally, some drugs, such as amphetamines, which effect the integrity of the blood brain barrier, allow other molecules, such as the diluter, inside the brain, which can result in a severe adverse reaction.

To complicate matters further many drugs are substituted, that is, one drug (such as an amphetamine) is passed off as another (e.g., cocaine). Drugs may thus not always be what they are alleged to be, and the substitutes or look-alikes, or the actual drugs themselves, can range from having little or no potency, to marginal potency, to very strong potency, such as in the case of designer drugs, which may be capable of inducing a very bad trip or a fatal overdose.

In light of the fact that many drugs may actually be inert substances, or have marginal potency, the nature and function of a placebo effect becomes an important factor in understanding drug effects. What makes the placebo effect so evident among drug users is not entirely clear, but an important element is the user's expectation (set) of what the drug will do. There appears to be a certain implied trust among users that a drug will produce a given effect, and an anticipation that the nature of the drug experience will be what is desired. These are "so powerful that an effect is produced despite the fact that the pharmacological activity of the preparation is negligible"[27]. Currently, some researchers believe that the placebo effect is, in part, related to the production of endorphins.

The importance of this placebo effect is illustrated in an early LSD study (cited in ref. 10) that showed markedly how LSD's effects were influenced by the user's own expectation of what kind of experience the drug will induce. A subject described his reaction as follows *after having taken a placebo* reported to be LSD, "A lot of strange shapes and brilliant color, after-images, as if I looked through pebble-finished glass, particularly this morning. Especially this morning colors were more brilliant than I have ever experienced. Voices were at times somewhat in the distance along with a feeling of not being in a real situation, a dream kind of state, time is distorted, goes rather slowly, an hour is only 10 or 15 minutes when I look at my watch." Claridge[10] commented that this report "is a perfect description of the LSD state! Interestingly enough, when this subject was told what he had been given and then a week later given a normal dose of LSD, he reported that the drug had no effect at all." Thus the hallucinatory potential of LSD (and other substances) is so strong that even when placebos are administered to unsuspecting individuals, LSD-like experiences are reported — a phenomenon that indicates that psychological factors (set) are strongly involved in the nature and quality of a drug experience.

In addition to the nature of the drug and its variable potency, the previous experience of the user is also an important factor in determining the nature of a drug's effect. Naive or inexperienced drug users may not know what to expect, and the drug experience may thus be something different from what is anticipated, resulting in a bad trip. Many other factors also serve to influence a drug's affect. Some of the many elements that influence drug effects, either singularly or in different combinations, are described in Table 2.1, along with some of the specific effects they may produce.

The variations in intoxicated behaviors, when attributed to alcohol or other psychoactive substances within social contexts, reinforces that the context or setting of use also exerts a forceful effect on the nature of the drug-related experience, one that may be far greater than the effect of the drug taken. Fagan[18], for example, suggests that patterns of aggression following intoxication "develop over time through socialization within specific social contexts and the shaping of behavior through social learning processes." Thus, *person-substance* (set) and *place-substance* (setting) interactions appear to be important factors that may not only influence choice of substance, but they may also mediate the nature of the drug effect.

Although the behavioral or psychological effects of a drug may vary with a user's expectations, there are identifiable general behavioral or psychological effects associated with different classes of drugs, as described below. Many of these substances have acquired street names (slang) that reflect how they are used or perceived by users and drug dealers. Over 1,000 terms have recently been identified in the United States[15]. An example of some of the more innovative terms are:

1. "Kibble & Bits," which refers to small crumbs of crack.
2. "Special K," which is a term for the drug ketamine.
3. "Beam me up Scottie," which is used to describe the drug combination of crack and PCP.

2.2 BEHAVIORAL/PSYCHOLOGICAL EFFECTS

It should be noted that the terms "behavioral" and "psychological" are used interchangeably.

The text below specifies some of the behavioral or psychological effects that different classes of psychoactive drugs can have on various aspects of human behavior. Sections 2.3 through 2.7 describe the pharmacological properties of psychoactive drugs. Psychoactive drugs affect perception, memory, emotional states, psychomotor functioning, and cognitive processes; many of these drugs have the capacity to cause irreversible damage to sensory processes.

TABLE 2.1
Biological, pharmacological, and behavioral influences on drug effects

Influence	*General effects*
A. Biological factors	
1. ALDH and ADH genotypes a. Race b. Physical tolerance	Alcohol elimination rate and flush reaction.
2. Family history of alcoholism	Susceptibility to alcoholism.
3. Gender a. Hormonal differences b. Body fat content c. Body size	Absorption rates.
4. Physical health	Alcohol- and drug-related damage.
B. Drug-related factors 1. Purity 2. Chemical structure a. Lipid solubility b. Molecular size 3. Dose a. Concentration b. Volume 4. Route of administration a. Injection i Intravenous ii Intramuscular iii Subcutaneous b. Inhalation c. Absorption 5. Rate of administration 6. Duration of administration 7. Frequency of administration 8. Metabolism and elimination	All of these factors effect the speed of onset, magnitude, and duration of the drug effect.
C. Behavioral and social factors 1. Age 2. Past use history 3. User's expectations (set) 4. Social environment (setting) 5. Laws and penalties 6. Availability and cost 7. Personality 8. Family influences 9. Peer influences 10. Portrayal in movies 11. Portrayal in television 12. Religious influences 13. School influences 14. Occupational influences 15. Psychopathology 16. Socioeconomic status	All of these factors affect the nature of the drug effect to the extent that they can influence whether the reaction will be experienced as positive or not, or whether it will be repeated or not.

2.2.1 *Perception*

Perception can be defined as the reception of stimuli from the senses, that is, the awareness of objects — a consciousness through sensory apparatus. Psychoactive drugs can change the sensitivity of these systems. Vision is specifically affected by a drug's modification of pupillary response to light in which the pupils become dilated or constricted, resulting in more or less sensitivity to light. This effect is most likely related to modulation of various neurons along the visual tract. The most intense or dramatic visual alterations are achieved through hallucinatory drugs in which there is a sharp increase in one's visual threshold. Some of these drugs are experienced as intensifying visual experiences, that is, objects appear clear and their colors are seen as intensely bright or more saturated; oftentimes objects may be seen as pulsating. Other hallucinatory drugs have been described as reducing the accuracy of color discrimination. In some instances the visual experiences are so overwhelming that they may frighten the user. One has to learn how to compensate for these effects and enjoy them, or a bad trip is likely to be experienced. It is most likely that an adverse reaction is related to the way in such drugs effect neurochemical homeostasis in the CNS.

Hearing is also affected by psychoactive drugs and by many over-the-counter (OTC) drugs. Chronic use of aspirin and antihistamines, for example, can impair audition by producing a ringing sensation in the ears. These drugs may act either directly through various Histamine-type receptors or through a prostaglandin mediated process. Hallucinogenic agents particularly affect hearing by increasing one's awareness of sounds and by providing a heightened sensitivity to them. Music has been described as being especially clear and pleasant when under the influence of some psychoactive drugs; tones are heard with greater clarity, and all the different sounds of the instruments seem to be discriminated, among other sensations. Auditory hallucination may also take place. One's sense of smell and taste can also be altered by psychoactive drugs, and some elicit olfactory sensations.

Space and time relationships, which are part of one's perceptual processes, are particularly affected by chemical substances, resulting in an impairment of one's sense of judgment and ability to maintain a realistic perspective of one's actions. When driving after taking marijuana, or other hallucinogenic drugs, there is a sharp reduction in the ability to judge speed and distance, resulting in the sensation of driving faster than one actually is, so accidents may occur when one responds to or attempts to compensate for the misperceived sensation of speed.

Drugs can also impair one's perception of one's own abilities. In particular, the overwhelming emotional experience brought about by a drug may lead people to overestimate their capabilities and thus interfere with their intellectual functioning and ability to make rational judgments.

2.2.2 *Learning and memory*

Learning, which can be defined as a relatively permanent change in behavior resulting from an interaction with the environment, involves the limbic system, and this system is particularly vulnerable to disruption by psychoactive drugs. Memory can be defined as the mental function of retaining information about stimuli, events, images, ideas, etcetera, after the original stimuli are no longer present. Memory is often found to be impaired as a result of chronic drug use. Broadly speaking, the specific way in which drugs act on learning and memory functions is not well understood. It is currently theorized that drugs interfere with the cholinergic, catecholaminergic, and serotonergic systems, and with amino acid metabolism such as tryphophan. Each of these is believed to be involved with learning and memory, and each is modulated to various extents, either pre- or postsynaptically, by different drugs. The following generalization was made about drugs, learning, and memory[36]:

"There are drugs that enhance learning and memory under certain conditions. Many stimulants are facilitators under a wide variety of circumstances. Many drugs cause retrograde amnesia without interfering with ability to learn. Others, especially stimulants, protect against amnesia. Apparently lost memories can often be retrieved. Recall is most likely to be successful when conditions most closely approximate those under which original learning occurred."

In general, alcohol, barbiturates, and hallucinogens can all produce a disruption of memory processes, and may even induce a state of amnesia. Alcohol, has been specifically found to cause a generalized impairment of immediate, short- and long-term memory, as assessed by a variety of tests such as learning of word lists, recall of visual and verbal materials, and by tasks requiring association and recognition abilities. Additionally, alcohol has been found to effect the function of vasopressin in plasma, which can in turn effect the osmolarity in cerebral spinal fluid (CSF), thereby influencing ion channels[38]. Alcohol is also believed to effect oxytocin, which is involved in memory and amnesia[33]. Marijuana has also been found to alter short-term memory, that is, affecting tasks such as learning and remembering new information or remembering and following sequences of direction[49]. Concern has been expressed that long term use of benzodiazepines is associated with cerebral atrophy, memory loss, and personality change[20].

Alcohol is also the most significant contributing factor associated with a form of organic brain disease, the Wernicke-Korsakoff syndrome, which is characterized by an acute (Wernicke's) phase in which patients exhibit clinical symptoms of mental confusion, oculomotor abnormalities, and ataxia. About 20% of the patients who are treated recover fully, but the remaining 80% manifest severe memory loss, known as Korsakoff syndrome.

It has also been generally hypothesized that excitatory amino acids are critical elements in certain types of learning and memory function. Such hypothesis include the gating of the neurotransmitter glutamate at the n-methyl-d-aspartate (NMDA) receptor, and the role of oncogene products as signal transducers essential for mediating short-term memory[43]. Alcohol, opiates, and other substances have been found to alter the brain's NMDA receptor system, calcium current, and GABA potentiation, thereby affecting learning and memory processes. Alcohol has also been noted to disrupt cholinergic neurons. Fetal Alcohol Syndrome (FAS), characterized by intellectual impairment, hyperactivity, attention deficits and behavioral problems, is also a manifestations of alcohol-related damage to the CNS.

2.2.3 *Psychomotor functioning*

Even before specific cause and effect relationships were identified, it was well known that alcohol and other drugs alter motor functioning. Psychoactive drugs, by interfering with higher brain functions and by distorting sensory input through the autonomic nervous system, or by rendering the user inattentive to sensory input, may make the user incapable of performing even simple psychomotor tasks. Even at moderately low doses, drugs such as alcohol and barbiturates can depress one's nervous system so that there is a reduction in psychomotor performance. Stimulants, which help the user to overcome general feelings of fatigue, may not correct the adverse effects that fatigue has on attention, vision, and other psychomotor functions. Psychedelics totally interfere with psychomotor functioning, and render the user incapable of successfully completing any task that requires motor coordination.

In summary, the seemingly endless interconnections between neurotransmitters and neuromodulatory systems involved in learning underscores the importance of understanding how alcohol and other drugs affect learning, memory, and psychomotor processes. The effects that a drug has on behavior or psychological functioning is related to how it impacts neurochemical processes in the CNS.

3 Drug distribution

The preceding section has briefly reviewed some of the sociocultural and psychological factors related to a drug's effect. The remainder of this chapter addresses some of the biochemical and pharmacological factors that are also involved in a drug's effect. A drug effect, however, is preceded by a pharmacokinetic phase, which comprises the physiochemical events that facilitate a drug reaching its action site. In order for a drug to exert an effect, it first has to enter the bloodstream, dissolve, and be transported to its action site.

Reaction to a drug begins immediately after it is taken, but the way in which a drug is administered has a direct relationship to the speed of onset and the duration of a drug's action. The preferred method of taking a drug, however, depends on the characteristics of a particular drug because some drugs are more effective than others depending on how they are taken. Some hallucinogens for example, such as dimethyltrypamine (DMT), have no effect when taken orally, but act quickly when inhaled. LSD, in contrast, acts best when taken orally. A bolus of a drug, for example, interacting with several neuronal systems all at once, will have different perceived effects (psychic surge) than small doses acting more specifically over a longer period of time.

Many drugs, such as opiate derivatives and cocaine, can be administered by more than one route, but the user tends to choose the method that will achieve the best high, such as injecting heroin or smoking crack cocaine[47].

There are a few major routes by which drugs can be administered before they enter the bloodstream: (a) oral, which involves ingestion of a substance, (b) inhalation, which involves sniffing or inhalation of a substance, and (c) injection, which involves either injecting a substance directly into a vein (intravenous, iv), beneath the skin (subcutaneous), or into a muscle (intramuscular).

Once a substance enters the bloodstream it is distributed rather rapidly because the heart, which pumps approximately 5 liters of blood each minute (the total amount of blood in the circulatory system equals about 6 liters), circulates the entire blood volume in the body about once every minute[29]. Drug distribution, however, is related to whether a drug is taken in solid or liquid form, and to its method of administration. The dissolution rate of a drug is determined by:

1. The water solubility of the drug.
2. The pH of the medium.
3. The pK_a of the drug.
4. The form, specific area, and packing of the drug crystals or particles.
5. The pharmaceutical formulation (i.e., the nature of the binder, adjuvents, and coating of the tablet or capsule)[45].

Drugs tend to distribute themselves in areas of the circulatory system that are rich in blood capillaries, and as such the brain, liver, kidneys, and heart are quickly infused. Areas relatively low in blood capillaries, such as fat, skin, and muscle, accumulate less initial concentration of a drug than do more richly supplied areas. The rate and completeness of drug absorption determines the effective dose reaching the action site. The process of drug distribution also involves specifically directing drug molecules to particular places where they can evoke a reaction (i.e., receptor site), be stored (e.g., fatty tissue), be metabolized (i.e., the liver), and be eliminated (i.e., kidneys).

It should be noted that "only a very small portion of the total amount of a drug in the body at any one time is in direct contact with the specific cells ("receptors") that produce the pharmacological action of the drug"[30]. It is the "total amount of drug in the body that (a) governs the movements of the drug through the tissues of the body and its ultimate elimination, and (b) determines both the length and time and the intensity of the drug's effect"[30]. For some drugs, however, plasma drug concentrations are not a good predictor of tissue concentrations. Four basic patterns of drug distribution have been identified:

1. Some drugs stay largely within the bloodstream.
2. Some drugs, such as alcohol, are distributed systemically, that is, distributed uniformly throughout the body.
3. Some drugs are distributed unevenly in the body contingent on their ability to permeate distinct cell membranes.
4. Different drugs concentrate in distinct body tissue or organ systems which may not actually be the site of their drug actions.

Membranes are lipid bilayers which contain protein receptors, enzymes, and ion channels. Cellular membranes are highly ordered structures which are important factors in drug distribution because they are involved in the passage of drugs: (a) from the stomach and intestines into the bloodstream, (b) from the fluid that closely surrounds tissue cells into the interior of cells (intracellular fluid), (c) from the interior of cells back into the intercellular fluid, and (d) from the kidneys back into the bloodstream[29].

Lipid-soluble drug molecules can penetrate cell membranes more easily than non-lipid-soluble molecules, but drugs must have some finite solubility in water. Thus highly lipid soluble drugs tend to stay in lipids whenever they encounter them. Such drugs are thus found to be highly concentrated, but inactive, in body fat outside the central nervous system (CNS). The stored drug is then released, over a long time period, back into the blood-stream[45].

Most drugs, because of pores in the capillaries of the circulatory system, including small proteins, are free to enter and leave the bloodstream, circulating back and forth from body fluids found outside of body cells. A drug needs to be lipid-soluble to enter the fluid inside cells. Many drugs, however, need not enter a cell to exert their effect. They are able to influence a cell's functioning from outside.

Some drug molecules that enter the bloodstream remain relatively free while others may attach or bind themselves to larger proteins and remain attached until metabolized, never having become active, a process called protein binding. Plasma protein binding is an important factor in determining the amount of a drug that reaches its action site; only drugs not bound to protein are pharmacologically active. However, protein bound drugs have a longer half life.

Once a drug leaves the bloodstream and enters a cell, its molecules elicit physiochemical reactions that originate within the cells, or within constituents of cells. Drug effects are the by-product of the interaction between drug molecules and the nervous system. This interaction can take place either on cell membranes or within cells. Nogrady[45] points out that "Individual variability in drug response can be as high as ten fold, given identical doses per kilogram of body weight. There are many reasons for this, but some drugs show greater variability of effect than others. Drugs that show hepatic clearance and presystemic metabolism will also show great individual variability, which can be lessened by parenteral injection."

When a drug's actions occur directly on a cell's membrane, three types of drug interactions or physicochemical actions may take place: (a) the drug's molecules may attach themselves to a particular receptor site thereby triggering a specific neurochemical reaction; (b) the molecules may interact with the membrane itself; or (c) the molecules may act on one or more of the mechanisms within the membrane that transport material into and out of the cell. A drug's action within the cell itself generally consists of an interaction of the drug's molecules with specific enzymes that modify or degrade them.

3.1 BLOOD BARRIERS

3.1.1 *Blood-brain barrier*

Capillaries in the brain are morphologically, physiologically, and biochemically different from capillaries in the general circulatory system. In the general circulatory system pores are present in capillary membranes, but in the brain the capillaries are tightly packed, have no pores, and are covered on the outside by a fatty barrier, called the glial sheath, that arises from nearby astrocyte cells[29]. Thus a drug attempting to enter brain cells has to pass through both the capillary wall and the membranes of the astrocyte cells. This specialized structural barrier is referred to as the "blood-brain barrier". Fat-soluble drugs, however, are readably able to penetrate this barrier, while fat-insoluble drugs penetrate less rapidly or not at all, depending on the extent of their ionization. Most psychoactive drugs are lipid-soluble, and are able to cross the blood-brain barrier rapidly to exert their effect on target sites in the brain.

3.1.2 *Placental barrier*

Another barrier, which applies exclusively to women, is the placental barrier. Most psychoactive drugs, however, readily pass through this barrier and enter the fetus' bloodstream. The fusion of molecules appears to be the major distribution determinant.

Drugs can effect the fetus either early in pregnancy, when limbs and organs are developing, inducing structural abnormalities, or later in pregnancy or during delivery, resulting in intellectual and other physical problems. Additionally, the fetus, because it cannot metabolize or excrete psychoactive substances, is subject to respiratory depression in utero or at delivery.

3.2 DRUG ELIMINATION AND DRUG METABOLISM

Once a drug enters the bloodstream, the body quickly acts to prepare it for elimination from the system. Drug elimination occurs in several ways: redistribution between compartments, storage, excretion of the unchanged drug, and by metabolism[45]. Most drugs do not leave the body unchanged, although some very small portions may be eliminated intact directly in urine by the kidneys, through the lungs, or through the sweat glands. Recirculation of drug metabolites through the bile can affect final distribution.

Drugs cease to have their effect when they are reduced to subthreshold levels in the bloodstream. As the drug is metabolized, it is eliminated primarily through the kidneys, with lesser amounts excreted from the lungs, sweat glands, saliva, and, for new mothers, through breast milk. The rate at which a drug is metabolized and eliminated is dependent on the amount in the system at any given time. The metabolism rate is usually measured in terms of what is called a drug's half-life — the time it takes the body to eliminate half the amount of a given drug dose.

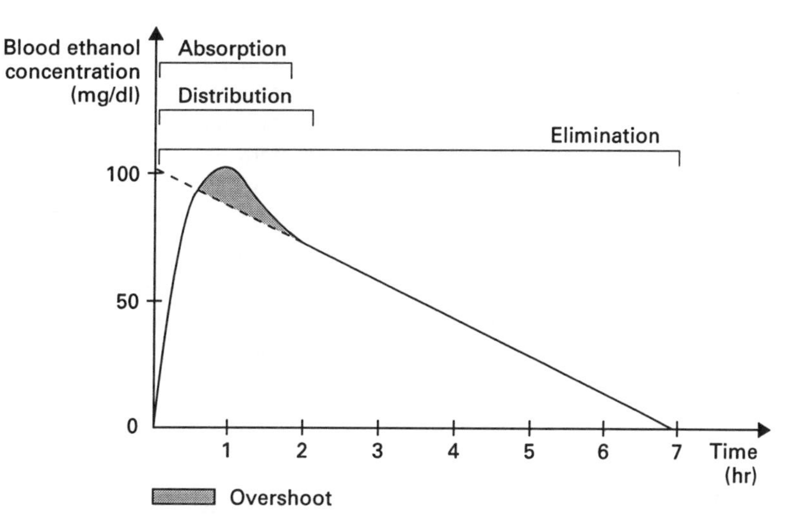

FIGURE 2.1
Typical blood alcohol concentration-time relationship after commencement of administration
Alcohol consumption was 700 mg per kg during 20 minutes.

4 Drug effect

4.1 DOSE-RESPONSE FUNCTION

The relationship between the dose of a drug and the intensity of the response is called the dose-response relationship. Every drug will have a different dose-response relationship for each observable effect that the drug induces, but the response is related to the potency of the drug. Dose effects of drugs, it should be noted, have been addressed by various receptor theories such as occupancy theory, rate theory, and induced fit theory. None of these theories, however, adequately explain all the acute and residual effects of psychoactive drugs.

Not all effects can be represented in "simple" dose-effect relationships. The euphoric effect of cocaine for instance is related to the rate of the concentration of the drug in the blood. This increase rate is dependent on the route of administration (see Figure 2.2).

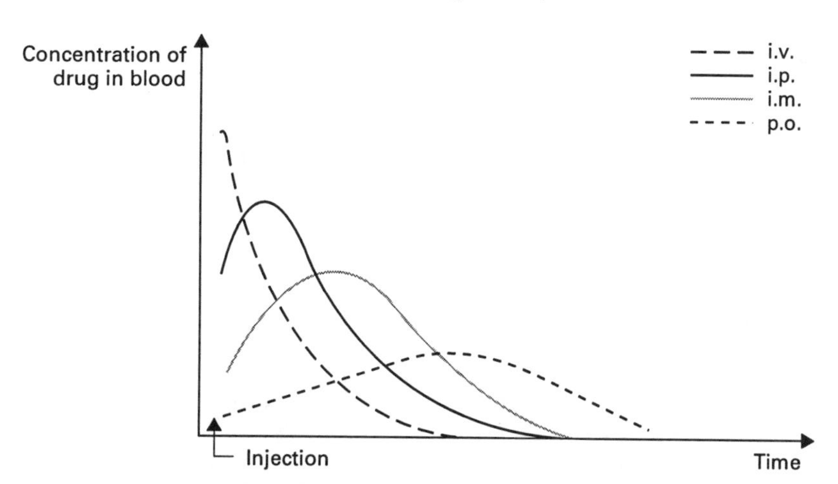

FIGURE 2.2
Time course for blood level of a (hypothetical) drug administered by different routes

Drug response is to a large extent related to a selective distribution of receptors for the drug, but for drugs to work they must have both a binding affinity for a receptor and be able to produce a pharmacological effect (i.e., efficacy). At different doses the drug interacts with a variety of receptors, but the drug will only act if there is a sufficient dose to stimulate specific receptors. At different doses the drug may have a variety of effects on diverse receptors. A drug therefore cannot be completely characterized by a single action because different effects will be observed at different doses. Few drugs exert only one effect. Drug effects, however, cannot be explained solely by the actions of one neurotransmitter because the action of neurotransmitters is highly integrated, and because one's psychological state (set) and environment (setting) are also involved in mediating a drug's effect.

4.2 DRUG INTERACTIONS

When more than one drug is taken the particular combination can produce a drug interaction, which represents an effect greater than when each drug is taken alone. Drugs vary in how they interact. Some combine to produce a strong, often toxic or lethal interaction, while others may not produce any, or only a slight, interaction. It is also possible, depending on the type of drugs taken, for no interaction to occur. Three kinds of drug effects are possible:

Summative or additive effect
In this type of interaction, two drugs that elicit the same response will produce an effect that is equal to that expected by simple addition (1 + 1 = 2). Such an effect can be achieved when aspirin and acetaminophen are combined.

Synergistic action
A synergistic action or effect occurs when the combined effect of two (or more) drugs is greater than the effect of each agent added together (1 + 1 = 3). In a synergistic reaction, one drug potentiates the action of the other drug by altering its biotransformation, distribution, or its ability to excite neurons. The synergistic interaction can thus be either an intensification of the action of one drug, or an extension of its duration of action. The combination of depressants, such as barbiturates, benzodiazepines, tricyclic antidepressants and alcohol, for example, can produce an increased effect. If doses are significantly high, the drugs can depress CNS control of vital functions which can result in death by respiratory failure.

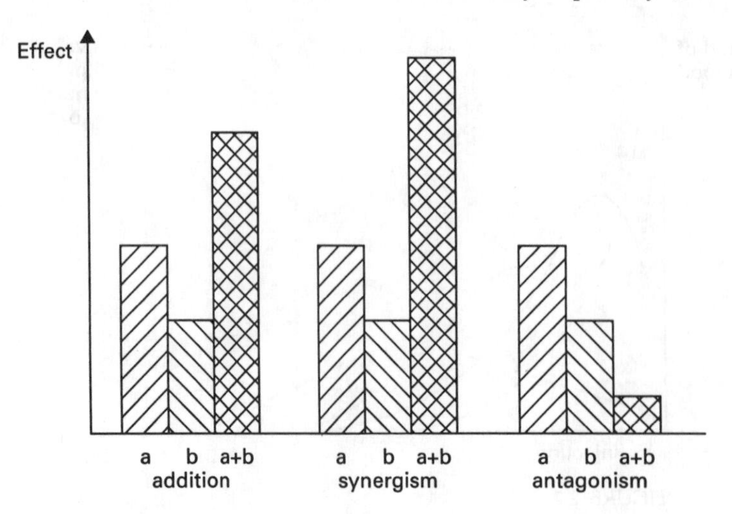

FIGURE 2.3
Schematic representation of different forms of drug-interaction

Drug antagonism
Drug antagonism occurs when one drug is capable of offsetting or counteracting the actions of another drug, that is, the combined effect or two drugs is less than the sum of the drugs acting separately (1 + 1 = 0). The use of amphetamines to counter the depressant effect of barbiturates, or use of depressants to come off a stimulant-induced high, are examples of an antagonistic effect.

4.3 PLACEBO EFFECT

In addition to the three kinds of drug interactions described, the placebo effect should also be noted. A placebo effect occurs when an inert substance is represented to be an active drug. The placebo effect has an important role in the way drugs are experienced. For example, under appropriate circumstances in which a person has developed a strong expectation (set) of what a drug will do, a placebo can produce many of the effects. With respect to street drugs, a placebo effect may be involved in the drug experience because, in many cases, the drug's potency is low or another drug may have been substituted for the real thing. Thus the user's set and setting may contribute more to the nature of the trip or high than the actual chemical properties of the substance involved.

4.4 DRUG ACTION

Subsequent to a drug leaving the bloodstream, its molecules stimulate a physicochemical reaction by producing a change in the functioning of a cell either by: (a) binding with a specific drug receptor on a cell's membrane, (b) interfering with a cell's water-lipid interaction, or (c) entering a cell. A drug's action within the cell itself generally consists of an interaction of the drug's molecules with specific enzymes that are related metabolic conversions of the drug. There are some receptors and binding proteins, however, that are within cells, such as steroid receptors. *"Selectivity of drug action is not a property of selective distribution; rather, it is a function of selective localization of drug receptors, the specificity of drugs that bind to particular receptors, the strength of the drug's attachment, and the consequences of the interaction between drugs and their receptors"*[29].

When a drug's action occurs directly on a cell's membrane (i.e., drug receptor), three types of drug interactions or physicochemical reactions may take place: (a) the triggering of a second messenger system or ion channel (b) an interchange with the membrane itself or the water layer around it or (c) the drug's molecules may act on one or more of the mechanisms within the membrane that transport material into and out of the cell.

Drugs, after binding with a receptor site, can either alter the rate at which any brain function proceeds, or modulate ongoing CNS functions. A drug, however, cannot induce new cellular functions, but they can mimic or potentiate the actions of neurotransmitters. Psychoactive drugs achieve their principal effects by interfering with the process of synaptic transmission by either inhibiting, activating, or modifying synaptic transmission. Drugs that act to mimic or potentiate the actions of neurotransmitters are called agonists. Other drugs that bind to a receptor site do not trigger any intrinsic activity on the cell, but the binding may result in competitive interference with normal neurotransmission processes (antagonists). The specific effect of psychoactive drugs, however, is determined by the nature of the drug in the system, its type of interaction with neurotransmitters or their receptors, where in the nervous system such interactions occur, and the function of that area of the CNS. The disruption of the normal functioning of the nervous system not only affects neurochemical processes, but psychological functioning as well. Thus psychomotor coordination, memory, perceptual and cognitive processes, and even personality, can be impacted by psychoactive agents.

There is some evidence suggesting that drugs may have a persistent or residual neuroadaptation that can last for months or years[28,61]. For example, the craving and drug-seeking behaviors that have been reported to last for years with nicotine, alcohol, and cocaine suggests a residual effect of drug use that may last for a lengthy time period. Additionally, research has shown that naloxone, an opiate antagonist of the μ-receptor, may elicit withdrawal symptoms in users who have abstained for extended time periods[46]. It is conceivable that this residual effect may be involved in the high number of relapses experienced by long-term abstinent drug-dependent individuals.

5 Classification of psychoactive drugs

Drugs can be classified in a number of different ways, but the problem of classification is quite complex because: (a) there has been an increase in the number of drugs available, (b) many drugs, while having similar chemical structures, have different pharmacological properties, or (c) many drugs have similar pharmacological properties but different behavioral effects. There are also many drugs whose chemical structure are remarkably different but which produce similar effects. The purpose of a classification system is to provide a method to distinguish the pharmacological properties of different drugs. One very broad way of classifying drugs is by the dichotomy of psychoactive and nonpsychoactive drugs. Psychoactive drugs are those chemical compounds that work directly on the brain or central nervous system (CNS) and affect or alter behavior in some manner. Nonpsychoactive drugs are compounds that do not normally affect the brain and behavior, and include such drugs as vitamins, minerals, antibiotics, and local anesthetics, among many others. Since psychoactive drugs are the principal concern here, classification schemes used to categorize these substances are reviewed.

Drugs are usually classified either by chemical structure or clinical use. They are also occasionally classified by origin, by site of drug action or by action prototype. More recently, researchers are classifying drugs by their behavioral effects[29], a method that combines aspects of chemical structure and clinical effects to yield psychopharmacological classifications. The following categories will be discussed: sedative-hypnotics, stimulants, hallucinogens, narcotics, solvents and inhalants.

5.1 SEDATIVE-HYPNOTICS (CNS DEPRESSANTS)

Sedative-hypnotics are drugs whose actions depress central nervous system (CNS) activity, principally within the brain stem; they also decrease one's level of awareness. The term sedative refers to the mild neuronal depression induced by these substances, such as diazepam or sodium pentobarbital compounds, without loss of consciousness. Hypnotic refers to more potent agents that induce sleep, such as secobarbital. Sedative-hypnotic drugs include alcohol and anxiolytics. Sedation, hypnosis, and general anesthesia represent increasing depths of a continuum. Drugs classified as sedative-hypnotics share many characteristics, and when taken together their effects are additive. They also act antagonistically to stimulant drugs. Both metabolic and psychological cross-tolerance may develop. Most of these drugs are taken orally, but they can also be injected. When taken orally, higher initial doses are required.

Ethanol

Ethanol or ethyl alcohol is probably one of the oldest drugs used by man. Alcohol has been in constant use at least since the beginning of civilization, and has played different roles or served different functions in many societies. At various times it has been used as food, as medicine, and for religious and social purposes. The earliest alcoholic beverages were naturally fermented wine and beer, with the making of beer from cereal grains beginning several thousand years ago. Distillation first occurred during the eighth or ninth century, but distilled whiskey did not become popular until the late 1500s when it was introduced into Europe from Arabia. It is pharmacologically classified as a generalized depressant because it induces a general, nonselective depression of the CNS.

Ethanol is a simple molecule, C_2H_5OH, with a molecular weight of 46; it is soluble in water and lipids and is distributed rapidly throughout the total body water, quickly crossing the blood-brain barrier and concentrating in cerebral tissue. It most likely acts by disrupting the water lipid layer at the surface of the cell membrane. It nonspecifically effects membrane potential and receptor function.

The concentration of alcohol in the bloodstream, referred to as the blood alcohol level (BAL), is expressed in units of volume and alcohol in units of weight. BAL represents a measure of milligrams of alcohol per 100 milliliters of blood. One-hundred milliliters of blood weighs 100 g, which makes it possible to express BAL as a percentage by weight. Thus 100 mg of alcohol in 100 ml of blood is expressed as a percentage of 0.10, representing 1 part alcohol to 1000 parts blood. A BAL of 0.08 represents 1 part alcohol to 1250 parts of blood, while a BAL of 0.20 represents 1 part alcohol to 500 parts of blood. The higher the BAL, the greater the behavioral effects of alcohol. Blood alcohol levels are measured in two ways. One is by means of a breathalyzer, which reacts to the alcohol content of expired air; the other method is by laboratory analyses of blood or urine.

Breathalyzers measure blood alcohol levels by either chemical oxidation or photometry. Breath is delivered to a liquid reagent containing potassium dichromate and metallic catalysts in a sulfuric acid solution. Oxidation of the alcohol to acetic acid produces a proportional change in reagent color, which is measured by an integral filter photometer, with the result indicated on an analog alcohol-concentration scale or by a digital result display[17]. Ethanol concentration in the intoximeter is measured by its IR absorption.

Alcohol is a powerful psychoactive drug whose actions are similar to barbiturates. Among the well-known intoxicating effects of excessive alcohol consumption are a lessening of restraints on speech and behavior, mild euphoria, an increased feeling of self-confidence, and a disruption of motor coordination. (The behavioral effects of alcohol intoxication are shown in Table 2.2.) Behavioral reactions to alcoholic intoxication, however, are not always uniformly predictable; they depend on the amount of alcohol consumed, one's drinking history, and the setting in which drinking occurs, among other factors. The drinker's expectations also play a major role in determining the exact reaction. Long-term drinking leads to tolerance and chronic drinking can result in dependence. The legal standard for driving in the United States is 0.10, but many states are presently deliberating lowering it to 0.08.

Alcohol, according to Blum[7], has the following neurochemical effects after chronic consumption:

1. It reduces the activity of serotonin at the synapse.
2. It reduces the activity of enkephalins and beta-endorphins in the brain, alters the molecular structure and hydrogen bonding of enkephalins, and increases the number of natural enkephalin receptors.

3. Inhibits dopamine release in the reward area of the brain and increases dopamine metabolism, thereby reducing its supply at the synapse resulting in an increase of dopamine receptors, which are hypothesized to be related to an increase in craving for alcohol.
4. Decreases the amount of GABA in the brain, which reduces its stimulatory effect on benzodiazepine receptors, resulting in an increase of anxiety and irritability.

It is interesting to note that many people do not perceive alcohol as a drug, while others do not realize that alcohol cannot be combined with medications or (illicit) psychoactive substances. When combined with other chemical compounds, alcohol produces a combined effect that can induce extreme intoxication, sedation, unconsciousness, and death. Table 2.3 describes some of the inter-actions that can occur when alcohol is combined with other drugs.

Barbiturates
("DOWNERS", "BUSTERS", "INBETWEENS", "NEBBIES")

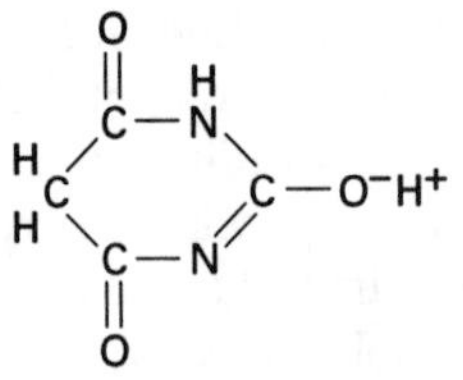

Barbituric acid

Barbiturates have been used by physicians to sedate patients since their introduction into medicine in 1903 under the trade name Veronal®. They represent a class of drugs that are derived from barbituric acid (2,4,6 trioxohexahydropyrimidine), which was first synthesized in 1864 in Germany. Barbiturates lent themselves to a wide variety of chemical variations, and over 2500 derivatives have since been synthesized, but only about 50 commercial brands were available up to the 1960s, and of these about only a dozen may still be used clinically as anticonvulsants. Only Phenobarbital, introduced in 1912, remains important in medicine. Barbiturates are usually divided into categories according to the rate of speed with which they are eliminated from the body: very long acting, intermediate acting, short acting, and very short acting. Similar to alcohol, at very high doses barbiturates may cause death through respiratory depression. Because of their non-polar character barbiturates can readily cross the blood-brain barrier to achieve their effects.

Tolerance can develop relatively early in comparison to other drugs, and individuals taking one form of a barbiturate can develop a tolerance to all types. The tolerance to barbiturate, however, does not fully extend to the brain's respiratory center, which remains sensitive to the depressant. The blood-brain barrier does not effect the distribution in the human body. Tolerance develops when the liver enzymes, cytochrome P-450, are induced and modify the structure of barbiturates causing its inactivation. These induced P-450 enzymes can also react with other drugs, such as tranquilizers and antidepressants, after about 7 days. Alcohol consumption, for example, results in the induction of the cytochrome P-450-2E1 enzyme, which can inactivate barbiturates.

Physical dependence can be induced with continued use. Withdrawal symptoms (or rebound hyperexcitability) are more intense than those experienced with alcohol; life threatening convulsions can be present within 12 hours after last use. The fact that barbiturates decreases REM sleep suggests that they interact with the GABA-ergic system.

Nonbarbiturates
Modifications of barbituric acid have led to the formation of new compounds called nonbarbiturates. They were introduced for medical use in the belief that they were not habit forming, had lower toxicity levels, fewer side effects, and the ability to induce more natural sleep. The traditional nonbarbiturates are bromides, chloral hydrate, and paraldehyde. Chloral hydrate and alcohol together (the classic "Mickey Finn") illustrate a synergistic combination of CNS depressants.

TABLE 2.2
Effects of alcohol consumption on behavior

Number of drinks consumed [a]	*Blood-alcohol level (%)*	*Effect* [b]
1-2	.01-.04	Beginning of impairment of the functions controlled by the outer or top layer of the cerebral cortex: (a) Impairment of judgment, and (b) Release of restraints and inhibitions. Slight feeling of exhilaration or euphoria. Increased heart rate. Slight change in mood or feelings.
3-4	.05-.06	Deeper absorption within the layers of the cerebral cortex, further impairing control mechanism. Disruption of judgment. Beginning of impairment of psychomotor coordination. Decrease of concern over social constraints. Intensification of feelings of increased exhilaration or warmth. Lessening of mental capabilities.
5-6	.07-.10	Greater diffusion of alcohol within the brain, extending deeper in the cortex and into the cerebellum, thereby interfering with motor functioning. Greater psychomotor impairment, resulting in reduced reaction time – Presumptive evidence of "impaired ability" to drive (DWI). Exaggerated emotions. Impaired gait – clumsiness. Talkativeness or moroseness. Visual impairment. Impairment of intellectual functioning.
7-8	.11-.16	Progressive deterioration of cortical functioning, and motor activities. Staggering. Slurred speech. Blurred vision. Serious impairment of judgment and coordination. Significantly impaired intellectual functioning.
9-10	.17-.20	Impairment of brain functioning extends to midbrain. Difficulty ambulating. Inability to perform simple tasks. Emotional outbursts. Double vision.
11-15	.21-.39	Impairment of brain functioning, specifically within lower brain centers, resulting in marked intoxication characterized by impairment of all forms of motor and intellectual functioning. Loss of control over all inhibitions. Stupor and confusion. Boisterousness and belligerence. Increase in potential for violent behavior or alcohol-related accident. Loss of comprehension of events surrounding the person.
16-25	.40-.50	Severe impairment/depression of lower brain centers, resulting in gross abnormality of bodily functions contributing to loss of consciousness. Shock.
26+	.51+	Failure of brain functions that regulate bodily processes leading to coma and death from respiratory failure.

[a] One drink would equal one 330 ml can of beer (averaging 5% alcohol by volume), or one 100 ml glass of 14% wine, or 30 ml of distilled alcohol.
[b] Based on an average body weight of 70 kg; effects for persons of different weights, and gender, will vary accordingly. The effects would also vary among individuals and in the same individual at different times. Each effect at each stage will diminish as BAL is reduced as the alcohol is eliminated from the body.

TABLE 2.3
Alcohol-drug interactions

– *Narcotics, Analgesics*
Contributes to a dramatic reduction of central nervous system (CNS) activity, which can ultimately result in respiratory failure.
– *Non-Narcotic Analgesics (Aspirin, Tylenol, etc.)*
Contributes to bleeding in the stomach and intestines, irritate the stomach, and can aggravate bleeding in ulcer patients. Alcohol may also stimulate arachidonic acid cascade.
– *Anticonvulsants*
Contributes to a lessening of the drug's ability to stop convulsions, and may exaggerate blood disorder side effects of anticonvulsants.
– *Antidepressants*
Contributes to an additional reduction in CNS functioning, which is antagonistic to the action of antidepressants.
– *Stimulants*
Alcohol acts as an antagonist to stimulant drugs.
– *Psychotropics (Anti-Psychotic Drugs or Major Tranquilizers)*
Contributes to an addiction reduction of CNS activity, which is counteractive to the action of psychotropics.
– *Sedative hypnotics*
Contributes to a synergistic effect, which can induce unconsciousness, coma and death.
– *Tranquilizers*
Contributes to an additive effect, decreasing alertness, judgment and motor functioning, which can result in injuries from accidents.
– *Barbiturates*
Ethanol will be metabolized more rapidly by the P-450 system (metabolic tolerance) in barbiturate users. A result of this interaction is that alcoholics cannot be sedated with barbiturates.

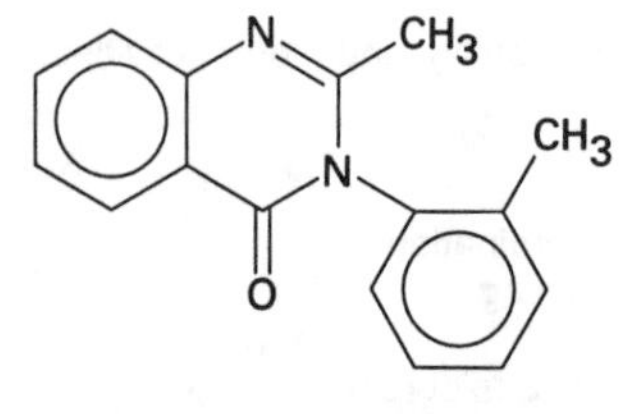

Methaqualone

Newer nonbarbiturate compounds, such as methyprylon and glutethimide, were introduced in the 1960s, and were thought not to be addictive and to have lower toxicity. It was soon established that they were as dangerous as their barbiturate counterparts, and their medical use has diminished. But some of these new drugs found their way into the streets and were widely abused. One in particular, methaqualone (2-methyl-3-otolyl-4(3H)-quinazolinone), which is structurally unrelated to the barbiturates, gained popularity and presented significant abuse problems with respect to overdoses, deaths, and physical dependency. Contrary to early beliefs, tolerance to methaqualone's euphoric effects does develop, with potential physical dependency. The methaqualone high, it should be noted, results from depression of the CNS[37]. The main side effects of barbiturates and their derivatives is drowsiness, impaired motor and intellectual functioning, and impaired judgment.

Antianxiety agents or mild tranquilizers
Antianxiety agents, or mild tranquilizers, which were first introduced in the 1950s, proved to be better pharmacological agents than barbiturates to control anxiety and calm agitation. These new drugs quickly replaced the barbiturates and other chemicals in the treatment of mild to moderate cases of anxiety, tension, and stress.

The most frequently prescribed minor tranquilizers fall into two pharmacological groups, meprobamate compounds and benzodiazepines, based on their chemical derivative or structure. Meprobamate was first synthesized in the early 1950s and was noted for its ability to calm anxiety. The drug has been used clinically to treat anxiety, insomnia, muscle spasms, alcoholism, and petit mal epilepsy. Its side effects are minimal, mostly drowsiness when high doses are used.

Benzodiazepine compounds such as chlordiazepoxide, diazepam, and oxazepam, were first synthesized in 1933, and were soon discovered to counteract anxiety as well as to produce skeletal muscle relaxation. Four relatively new benzodiazephines, fluraxepam, quazepam, temazepam, and triazolam, have been marketed specifically as hypnotics.

The benzodiazepine tranquilizers, since their emergence into prominent clinical usage in the 1950s, have become the most widely used prescriptive drugs in the world. One of these compounds, Halcion®, is used to decrease sleep latency and increase the duration of sleep, thus it has therapeutic utility in the treatment of insomnia. It has, however, come under review for its side effects, which include rebound insomnia, anterograde amnesia, headache, ataxia, and decreased coordination. Physical and psychological dependence are also thought to occur. The issue that has arisen is whether triazolam compounds are more dangerous than other drugs in the benzodiazepine family.

Benzodiazepines, taken over time, result in tolerance and dependence, with a predictable withdrawal syndrome[29]. Tolerance is not as strong as when associated with non-selective CNS depressants. The mechanism underlying tolerance is believed to be receptor adaptation; it is not caused by metabolic pharmacokinetic changes because benzodiazepines do not induce their own metabolism[29].

The withdrawal symptoms associated with benzodiazepines involve an intensification of the symptoms the drug was used to treat, a rebound increase in insomnia, restlessness, agitation, irritability, and muscle tension. The symptoms abide between 1–4 weeks, depending on the half-life of the drugs used, and the frequency and quantity of use.

Antipsychotic agents

The principal antipsychotic drugs, largely consisting of phenothiazine derivatives, are believed to act within the limbic system of the brain – the center for emotional control – specifically blocking the activity of dopamine at the postsynaptic dopamine-2 (D_2) receptor sites. Additionally, phenothiazines are also believed to blocade D_2 receptors in the extrapyramidal system (basal ganglia), the hypothalamic-pituitary system, and in the brainstem[29]. The result of such actions helps to control the symptoms of both acutely and chronically emotionally disturbed individuals. These drugs also act to depress sensory input, which has a calming effect that helps to reduce the symptoms of a severe emotional disorder. Some newer agents are believed to also interact with acetylcholine, serotonin, histamine, and norepinephrine receptors.

Phenothiazines are divided into the aliphatics, the piperidines, and the piperazine derivatives. These three groups vary with respect to their side effects (e.g., drowsiness) and their potency; piperidine derivatives are the most potent. These compounds suppress the action of the reticular activating and limbic systems, resulting in a reduction of the inflow of sensory stimuli.

Another class of major tranquilizers are the butyrophenone derivatives, such as haloperidol, which is structurally related to the phenothiazines, but it is more potent. Haloperidol's action is similar to that of phenothiazines — blockading of D_2 receptors. Haloperidol has proven to be an effective alternate to the phenothiazines.

The most common side effects of both phenothiazine and butyrophenone derivatives are symptoms of rigidity, which resemble parkinsonism-like symptoms, and tremors. These side effects are associated with their action on dopamine receptors. More recently, a new group of drugs called the atypical antipsychotics (e.g., clozapine; See Figure 2.4) have proven useful in schizophrenia.

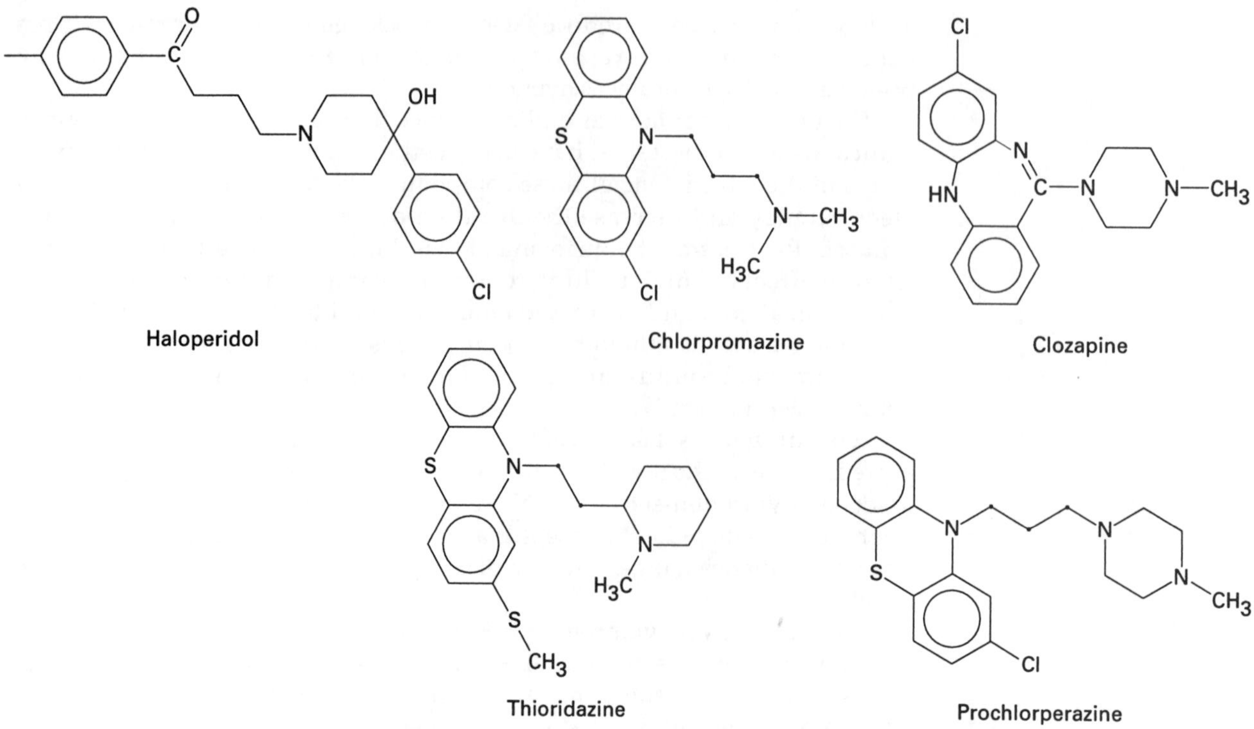

Haloperidol

Chlorpromazine

Clozapine

Thioridazine

Prochlorperazine

FIGURE 2.4
Structural formulas of clozapine, an atypical antipsychotic, chlorpromazine, an aliphatic phenothiazine, thioridazine, a piperidine compound, prochlorperazine, a piperazine derivative and haloperidol, a butyrophenone derivative

5.2 STIMULANTS

Central nervous system stimulants represent a powerful class of drugs that have strong physical and psychological effects. The pharmacological value of these drugs is that they help to overcome fatigue and sleepiness, reduce appetite and, paradoxically, to control hyperactive behavior. Some drugs classified as stimulants are also used to relieve depression and to elevate mood (antidepressants). Stimulants were also used as a dieting aid because of their ability to suppress appetite, and as energizers to combat fatigue.

There are several proposed mechanism of actions of CNS stimulants, each associated with a specific class or type of stimulant: (a) reuptake blockers, (b) metabolizing enzyme inhibitors, and (c) neurotransmitter release from synaptic vesicles. Behavioral stimulants, such as amphetamines, effect the action of the chemical neurotransmitters serotonin, epinephrine (E), norepinephrine (NE), and dopamine, resulting in a behavioral effect of elevation of mood and an increase in behavioral activity. Additionally, the dopamine system, after chronic stimulant use, is believed to be exhausted and the postsynaptic dopamine receptors become supersensitive so that the discontinuation of the stimulant produces a dopamine depletion syndrome that is responsible for symptoms of apathy and craving. Clinical antidepressants, such as imipramine and amitriptyline, are believed to achieve their effects by blocking the active reuptake of norepinephrine or serotonin. This action potentiates the synaptic action of either norepinephrine or serotonin at the synapse. The increased effect of stimulating the CNS contributes to the treatment of depression by elevating one's mood.

Other antidepressants, such as tranylcypromine compounds, act to inhibit the activity of monoamine oxidase (MAO), which deaminates the free neurotransmitters serotonin, NE, and dopamine. The lowered MAO activity results in an elevation of mood by increasing NE concentrations.

Stimulant drugs can be classified in several ways. One method is to divide them into two main subcategories: (a) primary stimulants (amphetamines, amphetamine derivatives, and cocaine), which act mainly on the CNS and only secondarily on the autonomic sympathetic nervous system; and (b) secondary stimulants (nicotine and caffeine), which exert their primary action on the sympathetic nervous system.

Julien[29] described three categories of CNS stimulants: (a) behavioral stimulants, which include such drugs as cocaine, amphetamines, methylphenidate hydrochloride, pemoline, and phenmetrazine; (b) clinical antidepressants, such as imipramine, amitriptyline, and tranylcypromine; and (c) general cellular stimulants, which include caffeine and nicotine.

Amphetamines
("NUGGETS", "SPEED", "UPPERS", "PEACHES")

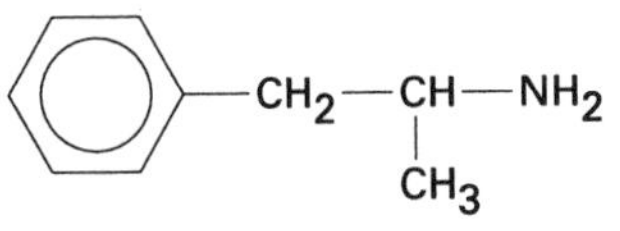

$C_6H_5-CH_2-CH(CH_3)-NH_2$

Amphetamine

Amphetamines represent a class of synthetic sympathomimetic amines that act to stimulate the CNS. They are chemically similar to norepinephrine (NE) and other catecholamines. There are three chemically similar types: (1) amphetamine (racemic b phenylisopropylamine), (2) dextroamphetamine, and (3) methamphetamine hydrochloride. All can cause an increase in heart rate and body temperature, and can raise both systolic and diastolic blood pressure, increasing respiration and circulation. Pupils, blood vessels (vasodilation), and the air passages of the lungs (bronchodilation) are dilated. The mouth becomes dry and there can be hyperactivity in speech and movements. Appetite is suppressed, as would be expected when the "flight or fight" response is activated by increased NE levels.

Amphetamines act indirectly on the CNS — they have no agonist effect — but act via the release of catecholamines. If any pathways are depressed, any resulting increase in NE levels triggers CNS stimulation. Thus, shortly after taking a low to moderate dose of one of the forms of amphetamines, the user tends to feel more alert and energetic, is able to overcome fatigue, and carry out tasks for longer periods of time. A larger dose, particularly when injected intravenously (a bolus), can induce a sudden, overwhelming intense pleasurable sensation involving a feeling of euphoria and physical well-being. In the CNS, amphetamines stimulate the release of dopamine and NE from their storage vesicles into the synaptic cleft and inhibit the active reuptake of these catecholamines.

Repeated use of amphetamines results in continued activation of the sympathetic nervous system, which contributes to excitability, unclear or rapid speech, restlessness, tremor of the hands, enlarged pupils, sleeplessness, and profuse perspiration. As use continues and dose levels increase, agitation becomes more pronounced, the ability to concentrate declines, anxiety and confusion increase, and the heart's functioning is disrupted. When users finally "crash," that is, fall asleep suddenly and profoundly after an extended period of use, they are likely to experience depression and mood swings after awakening. Prolonged use causes a depletion of the body's energy. When combined with amphetamines' appetite suppressant effects (serotoninergic system), extreme weight loss and a generally weakened body condition follow. The depletion of the body's resources also reduces resistance to disease by weakening neuroimmunoendocrine axes. Damage to the liver, lungs and kidneys can also occur.

Extended use of high doses of amphetamines can contribute to a toxic reaction in the form of an amphetamine-induced psychosis, which is a mental state similar to paranoid schizophrenia. This toxic state involves hallucinations and delusions, and the person can become highly agitated, aggressive, and even violent. Although the symptoms tend to clear up after the amphetamine has been eliminated from the body, vascular and organ damage may continue.

Tolerance and dependence usually develop. Dependence on amphetamines involves behavioral depression, characterized by fatigue and prolonged sleep in which changes in EEG patterns are detectable. Abrupt cessation after extended use can induce symptoms of fatigue, abnormalities in brain waves, prolonged sleep, a voracious appetite, stomach cramps, and depression.

Cocaine

("BAZOOKA", "BIG C', 'BLOW", "C", "CANDY", "FLAKE", "SNOW", "PARADISE", "NOSE CANDY", "NOSE POWDER")

Cocaine, an alkaloid refined from the cocoa plant, is a short-acting but powerful stimulant that has anesthetic properties. It comes in several forms: a paste, a white powder, and even as a liquid base extract whose powder is smoked. Because of this variety, cocaine may be used orally, intranasally (snorting), smoked, or injected intravenously. Snorting cocaine through the nasal epithelium is the only ingestion method that puts it in direct contact with the CNS.

The general physiological effects include dilated pupils and increases in blood pressure, heart rate, breathing rate, and body temperature. The ultimate behavioral effect is an intense feeling of euphoria, experienced as a rush, which provides a highly intense sense of well-being and exhilaration. Most of these effects (whole body effects) can be explained by activation of the noradrenergic system. Cocaine blocks reuptake of dopamine by its transport protein and also block reuptake of serotonin and NE.

Intense prolonged use can induce intense anxiety, agitation, and excitatory symptoms that resemble a severe psychotic-like reaction, which may involve violent or explosive behavior. There is also a risk of brain seizures, heart attacks, stroke, and significant lung damage when cocaine is freebased. Some of the body's major reactions to cocaine are:

1. The eye: pupils dilate and lenses flatten, temporarily improving long-distance vision.
2. The nose: irritated tissue causes chronic runny nose and can destroy cartilage.
3. The lungs: function deteriorates when the drug is freebased (see cocaine freebase) and inhaled.
4. The heart: pulse speeds up significantly – even small amounts can lead to arrhythmia and heart attacks.
5. The brain: early effects include euphoria, then depression. Long-term use can lead to hallucinations, anxiety, restlessness, seizures, and psychosis.
6. The liver: production of crucial enzymes drop.
7. The veins: blood vessels constrict at the point where the drug is injected or inhaled, thus restricting circulation.
8. Dependence and withdrawal symptoms include depression, hypersomnia, and hyperphagia.

It was not until recently that cocaine was thought to produce tolerance or physical dependence. During the 1970s it was believed by users and some researchers to have little addictive potential. Views changed, however, when its use became widespread after freebase and crack cocaine hit the streets. Miller[42] described two kind of tolerance associated with cocaine, acute and chronic tolerance, both of which involve the neurotransmitters dopamine, norepinephrine, and serotonin. Acute tolerance represents changes in both presynaptic and post-synaptic neurons designed to regulate and maintain dopaminergic, noradrenergic, and serotonergic function within normal ranges by: (a) increasing the rate of neurotransmitter production, (b) inhibiting the enzymes (MAO, COMT, etc.) that degrade neurotransmitters, and (c) decreasing the number of neurotransmitter receptor sites.

Chronic tolerance involves "depletion of the neurotransmitters in the presynaptic neuron"[42], which would be a physiological state at the presynaptic terminal where there is a scarcity of storage vesicles containing the neurotransmitter so that a diminished amount would be released upon stimulation. Miller also noted that the development of chronic tolerance is based on the phenomenon of super sensitivity, and that the number of receptor sites on the postsynaptic neuron is increased as a compensatory response to depletion of the neurotransmitter in the presynaptic neuron that follows chronic use of cocaine.

Physical dependence on cocaine is defined by the presence of withdrawal symptoms, which can include depression, hypersomnia, and hyperphagia, all of which are signs or symptoms of neurochemical changes in the brain.

Cocaine hydrochloride

The end result of processing coca leaves is a white, crystalline powder, sometimes with a "flake-like" texture and consistency. It can range from 20 to 80% pure or more on the streets, depending on how it has been diluted, on the availability of supplies, and on desired profit margins. Some common additives are mannitol, benzocaine, lactose, procaine, lidocaine, tetracaine, and amphetamines. It can be smoked, snorted (sniffed), or injected.

When snorting, the effects begin fairly quickly, peak within 15 to 20 minutes, and diminish within an hour. The general effects include dilated pupils, and increases in blood pressure, heart rate, breathing rate, and body temperature. The ultimate effect is an intense feeling of euphoria, experienced as a rush, which provides a sense of well-being and exhilaration.

Cocaine freebase

Smoking cocaine does not produce an effect as intense as that obtained from snorting because most of the drug is lost due to its high vaporization point. This limitation was overcome by "freebasing," introduced during the late 1970s. Freebasing is a method of converting street cocaine into a chemical base more suitable for smoking. Freebase cocaine is obtained through a simple chemical procedure whereby the cocaine alkaloid is "freed" from the hydrochloride salt by treating it with a base. The free cocaine is then extracted into a nonpolar solvent, such as ether, resulting in the purified cocaine becoming crystallized. Other methods involve the addition of baking soda as the base.

Because the basic form of cocaine is more lipophilic, smoking freebase gets it to the brain more quickly than snorting or injecting, taking only about 5–8 seconds, compared to 15–30 seconds via injection. In freebase form cocaine produces a short and very intense high, experienced as an euphoric rush.

Crack
("ULTIMATE", "READY ROCK", "PIECES", "SMOKE", "WAVE")

Cocaine freebasing has recently been carried a step further with the conversion of cocaine hydrochloride into an even more concentrated freebase form called "crack." The resulting intoxication is far more intense than either snorting or smoking freebase cocaine.

Crack, introduced in 1985, is a variant of freebasing, in which cocaine is mixed with baking soda and water creating a paste that is usually at least 75% cocaine, as compared to about 20% for cocaine hydrochloride (street powder). The paste hardens and is cut into chips that resemble soap or small pieces of rock, hence its other street name, "rock".

Crack, however, is rather short-lived, producing about a 15 minute high, followed by a rapid let down. A feeling of depression may follow soon after, which is in such contrast to the high that the user may seek to regain euphoria immediately by retaking the drug. Inhalation may be repeated as often as every 3–5 minutes during prolonged smoking episodes, which may extend over 1–3 days of continuous use or until supplies of cocaine are depleted, or until the user collapses from exhaustion.

Caffeine
Caffeine (1,3,7-trimethylaxanthine) is an alkaloid found in coffee, tea, cocoa, and over sixty different plants. It is an xanthine structurally related to theophylline in tea and theobromine in chocolate. All of these exert their stimulant effect on both the central and peripheral nervous systems. Its stimulative effect in relatively normal doses (100–500 mg) is an arousal of the cerebral cortex, resulting in greater wakefulness, mental alertness, and improved psychomotor functioning, but the level of arousal is not equivalent to that promoted by amphetamines. It also: (a) excites the spinal cord and the pons regions of the brain that control breathing; (b) stimulates the heart muscle, which can cause arrhythmia; (c) acts as a diuretic; and (d) interacts with dopamine dependent chloride channels. Caffeine achieves its main effects by acting as an adenosine antagonist at receptors activated by adenosine, resulting in its stimulant, antiheadache, and antiasthma effects. Adenosine's central actions are not well understood. Caffeine also acts on other receptors to produce behavioral sedation, to regulate oxygen distribution, to dilate cerebral and coronary blood vessels, to produce bronchospasms (asthma), and to regulate other metabolic processes.

Excessive caffeine intake, usually in the form of moderate to heavy coffee drinking, can contribute to adverse effects in the form of an increase in heart rate, with occasional arrhythmia, changes in blood pressure, psychomotor agitation, and insomnia. At very high doses (1000–2000 mg or higher), caffeine can induce a highly abnormal cardiac rate and arrhythmia, extreme agitation, restlessness and tremors, a high blood-sugar level, acid pH of the urine, and abnormally rapid reflexes.

Although tolerance to caffeine does not develop, physical dependence is believed to occur, even at low doses[29], and abrupt termination may lead to withdrawal symptoms. Headaches, irritability, fatigue, a craving for coffee, and a general feeling of discomfort are a few of the symptoms that can emerge.

Nicotine
Nicotine, an extremely potent compound, exerts its effects on almost every organ system of the body. It is absorbed via the entire respiratory tract, by oral and nasal mucosa, by the gastrointestinal tract, and even by the skin. The fastest and most complete absorption occurs when tobacco is inhaled deeply into the

lungs, where up to 90% of the inhaled dose is absorbed. In concentrated form, nicotine is a highly toxic poison, but the amount consumed in tobacco products does not approach toxicity. Within the CNS, nicotine appears to stimulate receptors sensitive to acetylcholine, resulting in a general stimulation of the CNS, and a specific increase in the functions of the autonomic nervous system, causing an increase in blood pressure and heart rate. Nicotine also influences the release of norepinephrine; it also stimulates the induction of liver enzymes to break down other drugs. Nicotine receptors are present in autonomic ganglia and the CNS.

In low-dose levels (0.05–2.0 mg in cigarettes), nicotine acts as a mild stimulant to the CNS, but specific behavioral effects are not discernible. Tolerance to nicotine is believed to develop, and physical dependency is associated with tobacco use.

5.3 TRICYCLIC ANTIDEPRESSANTS

Tricyclic antidepressants are stimulants that elevate an individual's mood, and differ, with respect to their pharmacological actions, from amphetamines, cocaine, and caffeine. The types of amine chemical functional group to which antidepressants belong can be used to subdivide the tricyclics into two groups: tertiary amines and secondary amines. The tertiary amines have a relatively greater effect on the norepinephrine system; tricyclic antidepressants achieve their effects by blocking the reuptake of either norepinephrine or serotonin. This blockading results in an increase in the level of free norepinephrine or serotonin in the synapse, resulting in an increase in postsynaptic receptor activity. In general, serotonin and NE have opposite effects in the brain. An increase in NE leads to arousal, while increases in serotonin leads to sedation. Over time (10 days to 3 weeks) this specific pharmacological action results in a lifting of the depression. The exact nature of how tricyclic compounds alter mood, however, remains in the theoretical stage. The very specific pharmacological action of these drugs, together with the fact that they do not induce a rush or euphoria, has resulted in their not generally being used as street drugs.

Fluoxetine, fluvoxamine, and related drugs, selectively block reuptake of serotonin, with no influence on norepinephrine. It is also used in the treatment of acute anxiety reactions, and has also been used in the treatment of obsessive-compulsive disorders.

Other types of antidepressant drugs are MAO (monoamine oxidase) inhibitors, known since the late 1950s. Three of these drugs, isocarboxazie, phenelzine and tranylcypromine, are referred to as irreversible MAO inhibitors because they produce an irreversible block of MAO, an enzyme involved in decreasing the neurotransmitters norepinephrine, dopamine, and serotonin. The result is an increased amount of catecholamines at the synapse, which corresponds with an increase in mood. It is believed that because of leakage, there is a basal level of neurotransmitter in the synaptic cleft. MAO inhibitors raise this level or set point, resulting in stimulation.

5.4 HALLUCINOGENS

Hallucinogens are known under many names, each referring to a particular characteristic of their different mind altering effects: psychedelics, illusinogens, psychotomimetics (psychosis imitating), psychotogens, fantasticas or phantasticants, (illusion producing), psychodysleptics, and psychotaraxics (mind disrupting). Drugs classified as hallucinogens have the capacity to induce visual, auditory, or other hallucinatory experiences, and to increase a sensation of feeling separated from reality. The principal use of such drugs is to induce an altered state of consciousness, the specific nature of which varies with the user's drug experience, set, setting, the nature of the drug taken, and its potency. Individual reactions vary greatly.

The many different kinds of hallucinogenic agents that are available differ greatly in their chemical structures and are capable of inducing a broad range of behavioral effects, making it difficult to classify or categorize them into any definitive conceptual framework. Because it is generally believed that hallucinogens exert their effects by disrupting synaptic transmission, one method of classification is based on these effects: (a) serotonin-like drugs (indole alkylamines), (b) catecholamine-like drugs (phenylalkylamines), (c) psychedelic anesthetic drugs (phenylcyclohexyl derivatives) and (d) anticholergics. Because hallucinations are complex biobehavioral phenomena, the neurochemical interactions of hallucinogens are more intricate than their known pharmacology.

Serotonin-like drugs (indole alkylamines)

Hallucinogens in this group consist of Lysergic acid diethlamide (LSD), psilocybin, psilocyn, lysergic acid amide (morning glory seeds), dimethyltryptamine (DMT), harmine, and bufotenin. Bufotenin, found in the skin of toads, is a hallucinogen that results from a breakdown of serotonin. They are also known as anticholinergic drugs. Some of these are described below.

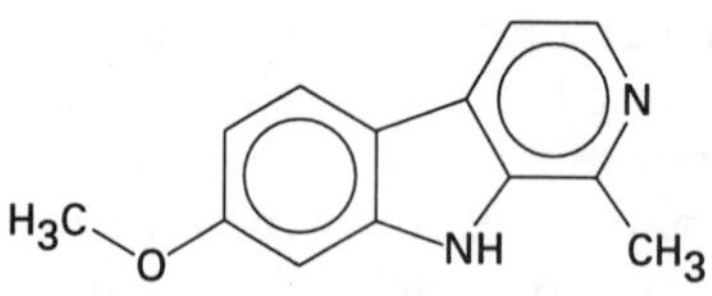

Harmine

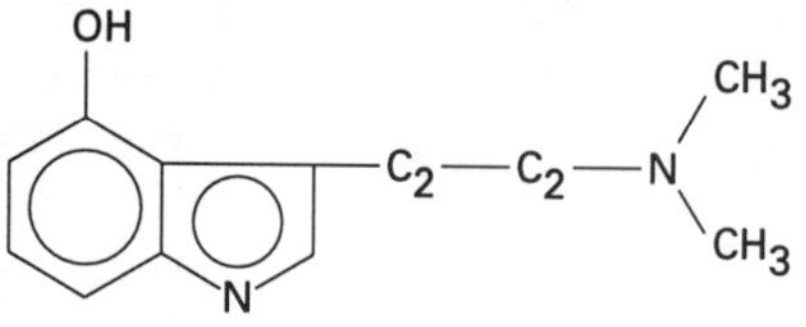

Psilocyn

Lysergic acid diethlamide-25 (LSD)

("ACID", "BLOTTER", "BLUE MIST", "FLASH", "PEACE", "TABS")

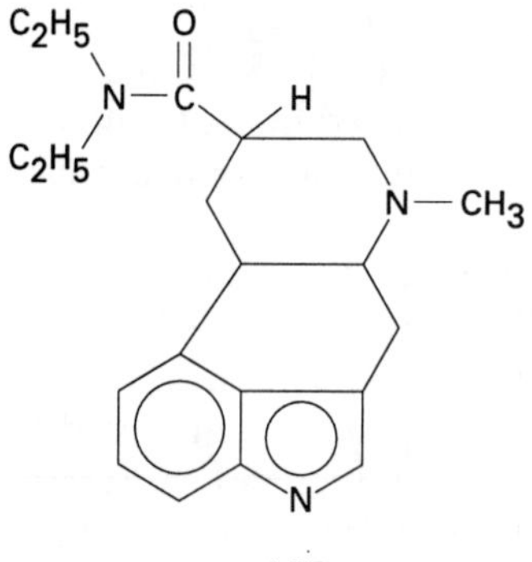

LSD

LSD, one of the hallucinogens within the serotonin-like group, is the most potent and widely used. Discovered by Dr. Albert Hoffman at the Sandoz Laboratories in Switzerland in 1938, LSD is a semisynthetic derivative of lysergic acid, an alkaloid found in a rye fungus. In comparison with other hallucinogens, LSD is one of the most potent known to man: 20 micrograms (1 microgram is one-millionth of a gram) will produce effects. The usual dose of LSD is about 100 to 200 g, 400–500 g will produce full hallucinogenic effects. In powder form, LSD is odorless, colorless, and crystalline. It is available on the street as a tablet, a capsule or, occasionally, as a liquid, which may have a slightly bluish cast. LSD is usually taken orally, but it can be injected.

Psychological reactions to LSD are not predictable and can vary from extremely pleasant to very unpleasant, and even positive trips can differ with respect to the nature of the phenomenological experience the user undergoes. This variation is largely related to the purity and dose of the drug, the user's set and setting, and past experiences with LSD. Some of the effects that may be experienced are described as follows.

An individual user can experience either good or bad effects at different times or may experience both within the same trip. One usually begins to feel the effects 30-90 minutes after taking the drug. A good trip may involve experiences such as heightened visual effects, that is, perceiving intensified colors, distorted shapes and sizes, and movement in stationary objects. Auditory distortions and misperceptions of time and place also occur. Several different emotions may be felt at once, or there may be a rapid swing from one emotion to another. The effects can last from two to 12 hours. A bad trip or "freak-out" can occur when the user experiences the induced sensations as unpleasant and perceives them as frightening, thereby tending to create a feeling of panic. The frightening effects may last a short time or the length of the total trip, and can range from mildly frightening to intensely terrifying. Prolonged adverse reactions have been noted. The bad trip may involves a sense of panic, confusion, anxiety, suspiciousness, feelings of helplessness, and loss of control. In some cases it resembles a paranoid schizophrenic reaction in that paranoid delusions, vividly fearsome hallucinations, and withdrawal into oneself may take place.

The physiological reaction to LSD involves dilated pupils, a depressed appetite, an increase in blood pressure, heart rate and body temperature, possible nausea and vomiting, abdominal discomfort, rapid reflexes, dizziness and psychomotor impairment, followed by insomnia after the drug wears off. There may also be an increase in glucose levels. These effects, however, vary greatly within individuals, with some showing increased signs while other show few or no symptoms. Serotonin receptors may be impacted by LSD.

A major risk with LSD is the possibility of acute or chronic psychosis. An acute psychotic episode lasts up to 24 hours, and is characterized by intensified perceptions, illusions, feelings of depersonalization, hallucinations, dilated pupils, tachycardia, sweating, tremors, dyscoordination, palpitations or blurred visions, marked anxiety, possible depression, fear of going crazy, paranoid ideation, and impaired judgment.

Julien[29] indicated that tolerance develops to both the psychological and physiological alterations induced by LSD, and that cross-tolerance also occurs with most other psychedelic substances. Physical dependence does not develop, even after prolonged use.

Psilocybin and psilocyn
("MAGIC MUSHROOM", "SHROOMS")

Psilocybin (phosphorylated 4-hydroxydimethyltryptamine) is a hallucinogenic alkaloid or active ingredient that occurs naturally in a variety of mushrooms found in southern Mexico and Central America. Psilocyn (4-hydroxydimethyltryptamine; see page 52) is an alkaloid also contained in the mushroom but in smaller quantities than psilocybin. Psilocyn is the hallucinogenic compound to which psilocybin (i.e., a prodrug) is converted in the body, and it is thought to be the active hallucination-producing ingredient. Both substances are chemically related to LSD, but they are approximately 200 times less potent.

Catecholamine-like drugs

Mescaline
("BUTTONS", "CACTUS", "MESC", "MOON")

The primary physiological and psychological effects of mescaline (3,4,5-trimethoxphenylethylamine) do not differ significantly from those induced by LSD. Physiologically, there is dilation of the pupils, increased deep tendon reflexes, and increases in heart rate and body temperature. Psychologically the primary effect is one of sensory alteration that is similar to LSD, but the nature of mescaline's hallucinatory experience is more strongly related to set and setting than LSD's reaction. Its effects last approximately 12 hours.

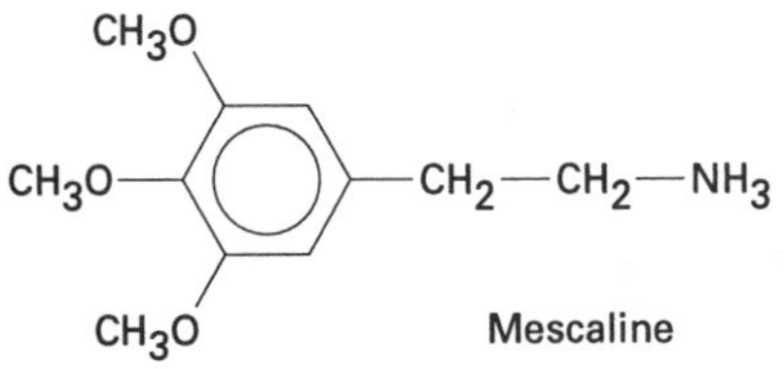

CH_3O
CH_3O — [benzene ring] — CH_2—CH_2—NH_3
CH_3O

Mescaline

Psychedelic anesthetics

Phencyclidine
("ANGEL DUST", "CRYSTAL", "JUICE", "PEACE PILL")

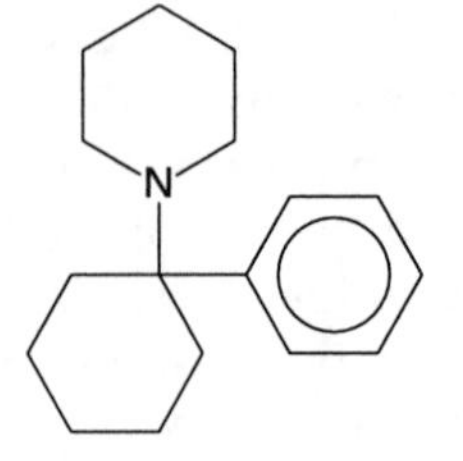

Phencyclidine

Phencyclidine (PCP) [1-1(phenylcyclohexy) piperidinehydrochloride] is one of a group of arylcyclohexylamines and, in its pharmaceutically pure form, is a white powder that is soluble in water. In the 1950s, PCP was synthesized as a general anesthetic agent, and marketed as Sernyl. Because of the postoperative complaints of patients, it was removed from the market in the 1960s. PCP appears to act at two different kind of receptors. One is the sigma receptor; the other is the NMDA receptor. In addition to PCP there are over 30 chemically comparable analogs, some of which can induce similar psychedelic effects. A popular method of taking PCP is to dip cigarettes into liquid PCP and then smoke them. At low doses, PCP acts primarily on the CNS, altering its reactivity to various incoming stimuli. Larger doses of the drug can interfere with CNS functioning to the extent that convulsions may be induced and death may follow from resulting complications.

When injected or snorted, the effects of PCP occur almost immediately; when swallowed or smoked, they begin about 15 minutes later, and may last anywhere from several hours to several days. The type of reaction varies greatly among individual users and is related to the potency of the drug and to the set and setting of use. The effects may be experienced as pleasant or unpleasant, but adverse reactions are fairly common. At low (subanesthetic) doses, the major reactions may involve feelings of changes in body image, sometimes accompanied by feelings of depersonalization. Perceptual distortions, involving visual or auditory hallucinations and other sensory disturbances, and distortion of time may also be experienced. The reaction may also include a pronounced feeling of apathy or may generate a state of hyperactivity. PCP, like LSD, can induce a psychotic-like state, but one that is much closer to acute schizophrenia. Severe adverse reactions are known to result when large doses are consumed, when the drug is taken regularly, and when it is mixed with other mind-altering chemicals.

Anticholinergics
There are a variety of plants of the nightshade family (henbane, solanums, belladonna, mandrake, pepper, etc.) that grow all over the world and are capable of evoking hallucinatory states when ingested. The leaves and roots of these plants contain the alkaloids atropine, scopolamine, and hyoscyamine, which are collectively called atropinic drugs or solanaceous alkaloids.

In small doses these drugs act as antispasmodics or mild sedatives. In large doses, however, they can induce a hallucinatory condition lasting for several hours, and produce adverse side effects such as a rapid heart rate, dry mouth, and dilated pupils. All of these drugs act on the CNS by interfering with synaptic transmission by blocking the action of acetylcholine receptors or inhibiting acetylcholinesterase, thereby interfering with neural transmission.

5.5 NARCOTICS (OPIATES OR OPIOIDS)
("BLACK TAR", "JOY POWDER", "SCAG", "SMACK", "SNOW")

The term narcotic is used to refer to opium and opium derivatives obtained from the oriental poppy *(Papaver somniferum)*, but it has also come to be used to include other drugs that are semi-synthetic or wholly synthetic. Julien[29] defined the term "opioid" as "any natural or synthetic drug that exerts actions on the body that are similar to those induced by morphine, the actions of which are antagonized by naloxone". Recent research in the pharmacology of opioids has resulted in the discovery of the body's own natural pain killers — endorphins — which has contributed to furthering an understanding of opioid, and other forms of addiction, and how neurotransmitters function. It is believed that narcotic analgesics achieve their effect by acting in the body in the same way as natural endorphins. Medically, narcotic analgesics are used primarily to reduce pain. Some forms, such as codeine, can also be used to suppress coughing; other forms are used to control diarrhea or to induce sleep.

Although the chemical structures of these narcotic drugs are diverse, their pharmacological properties are similar. The three types are defined as follows:

1. Natural opiates, which are obtained directly from opium, such as morphine and codeine
2. Semisynthetic opiates, which are chemically prepared derivatives of morphine or codeine, such as dihydromorphine, heroin, and methyldihydromorphine
3. Synthetic opiates, which are chemically synthesized analgesics whose effects are similar to morphine, such as propoxyphene, methadone (dolophine), meperidine, and pentazocine

The nervous system contains specific receptor sites for natural opiate-like chemical substances (endorphins and enkephalins, etc.) within the brain. Research has identified at least three distinct receptors — *mu*, *kappa*, and *delta* – and two *mu*-subtypes have also been identified (*mu*-1 and *mu*-2). It is postulated that these sites are also occupied by heroin, morphine, and other opiate derivatives. The binding of a narcotic drug to these receptor sites is believed to result in a blocking of the release of dopamine, norepinephrine, and other neurotransmitters. This action appears to be the preliminary step in the initiation of both physiological and behavioral effects, but exactly how these effects occur remains to be discovered.

Although tolerance develops to opiates, opiate derivatives, and synthetic opiates, its onset varies with an individual's particular physiological response, the dose, and the frequency of use. Opiate tolerance is known to develop with respect to the drug's respiratory depressant effects, analgesic effects, and sedative qualities. The tolerance to the euphoric effect is mild and there is no tolerance to miosis. In addition to the onset of tolerance to one type of opiate or narcotic drug, the drug user can also develop a cross-tolerance to other natural or synthetic drugs.

Continued use of opiates or narcotics may also contribute to physical dependence. Withdrawal symptoms that may be associated with abrupt cessation of heroin or morphine can vary from mild to severe, depending on the dose and frequency of use, length of use, and general health of the user. Some users do not experience withdrawal, and others may undergo mild withdrawal because of infrequent use, and because the drug they were using was so adulterated it had only marginal potency; others may experience severe withdrawal.

Natural opiates are obtained from drying the milky discharge of the cut, unripe seed pod of the opium poppy that appears soon after the petals begin to fall. The milky discharge is dried in the air to form a brownish, gummy substance, which is the raw opium. The opium poppy grows in many areas of southeastern Asia and the Middle East. Although over 20 different alkaloids have been isolated from opium, only morphine codeine, and opium itself, have psychoactive characteristics.

Opiates, especially heroin, are used because they can induce an extremely pleasant euphoric state, with feelings of warmth, well-being, peacefulness, and contentment. A pleasant dreamlike state of consciousness can also be experienced. Not all reactions to opiates are pleasant, but opioids seem to be used primarily in anticipation (set) of the feeling of relaxation and the euphoria they induce.

One of the major effects of opiates and opiate derivatives on the body is nausea and vomiting, controlled by the medulla. Opiates constrict the pupils, depress reflexes, and may lower blood pressure due to dilation of peripheral blood vessels. Flushing may be experienced, and sweating usually occurs. Continued use of opiates is known to reduce sexual functioning. Opium and its derivatives can remain in the body for 2–3 days after taken. The three major forms of opiates are described below.

Opium
("DREAM STICK", "GUM")

Opium is the milky exudate of the incised, unripe seed pods of the poppy, which is dried in the air to form a brownish gummy substance. It is usually smoked or taken orally, and generates a euphoric state in which one experiences a feeling of detachment from reality that frees one from everyday worries. Smoking of opium has a lower dependence-producing potential in comparison with the intravenous use of morphine and heroin. Withdrawal symptoms may not necessarily occur with abrupt cessation after prolonged use, depending on length and frequency of use.

Morphine
("BROWN", "MEXICAN BROWN")

Morphine is a natural alkaloid in opium, occurring in proportions of about 10% of the total weight of opium. Its general analgesic and behavioral effects differ depending upon if the person is experiencing deep visceral pain or is pain free. Morphine, because of its polar hydroxyl groups, does not easily enter the CNS. On the streets morphine appears as a brown crystalline powder that is processed by the user to be "shot up" (injected). It can induce immediate drowsiness, muscular relaxation, a feeling of euphoria, a rapid flow of ideas, and a shortening of the sense of time, among other effects, but these reactions in a pain free person may not always be pleasant. Morphine serves as a substitute among heroin users when heroin supplies are limited because it binds to the same receptor sites in the brain. Tolerance to morphine will develop over time. Some of the adverse effects include drowsiness, constipation, nausea, possible vomiting, constriction of the pupils, and decreased respiration.

Heroin
("DUST", "SNOW")

Heroin is a highly effective narcotic analgesic similar in pharmacological action to morphine, but three to four times stronger. Obtained by the modification of morphine, it is a white, odorless, crystalline powder with a bitter taste. The chemical modification of hydroxyl groups by acetyls permits heroin to cross the blood-brain barrier more easily than morphine. Once in the brain, heroin is hydrolyzed back to morphine. Heroin's ability to provide a more potent drug effect is why this substance, and not morphine, is the preferred street drug.

Heroin is a euphoria-producing drug and is preferred by street users because of its potency. Its intravenous use results in a rapid and intense high that is characterized as a rush, followed by a feeling of contentment and detachment from reality. Anxiety and feelings of inferiority disappear. A dreamlike state occurs in which fantasies are experienced as a form of escape from the world and from oneself. The heroin experience has often been described as a kind of sexual pleasure, which is compared to an orgasm. When the drug effect wears off, however, the user is rapidly restored to the unpleasantness of reality thus may want to "shoot up" again to escape.

The physiological effects associated with heroin include reduced respiration and heart rate, a constriction of the pupils and reduction of visual acuity, itching, skin rash, warming of the skin, increased perspiration, constipation, nausea, and possibly vomiting. Unusually high doses of heroin can result in more acute affects, possibly leading to unconsciousness, coma, and death from respiratory failure. Most of the problems experienced by heroin users, however, are related to the circumstances of use, such as unsterile syringes, poor living conditions, and the nature of the additive used to cut heroin (e.g., chalk, talcum, lactose, quinine, and mannite, among others). Hepatitis, tetanus, heart and lung abnormalities, infected and scarred veins, skin infections, ulcers, and abscesses are but a few of the medical problems experienced by heroin users. HIV infection is another health-risk encountered by iv heroin users, acquired through needle sharing.

5.6 VOLATILE SOLVENTS

Volatile solvents or inhalants are chemicals whose vapors can produce psychoactive effects. Many of these chemicals, such as benzene, gasoline, paint thinner, and lighter fluid, are not pharmacological remedies, but are industrial chemicals (solvents or aerosols) that are included in this classification because of their euphoric or intoxicating properties. Pharmacologically, these substances act as brain depressants. The medicinal or pharmacological chemicals included in this category are anesthetics and other compounds that are inhaled. These substances are grouped into three categories: organic solvents (hydrocarbons), volatile nitrates, and nitrous oxide. Anesthetics, such as nitrous oxide, achieve their effects by modifying the fluidity of neuronal membranes, interfering with the passage of ions and preventing neurons from firing[7].

The most common methods of using these substances are: (1) placing the solvent in a plastic bag and inhaling the fumes, which is a very hazardous procedure because the user risks suffocation — many deaths have been attributed to this; (2) soaking a rag or handkerchief in the solvent and inhaling the fumes; (3) sniffing or inhaling the fumes or vapors directly from the container; and (4) spraying the substance, such as a lacquer, into a soda pop can and inhaling the vapors directly ("huffing").

The use of these solvents or inhalants has been predominantly associated with youth, and the problem has emerged as one of national concern in the United States[8]. The problem is also an international one, with reports of youth using inhalant substances in Japan, Russia, Europe, England, Scandinavia, Canada, and in South American countries[8,52]. Most youth use inhalants to induce a rapid state of euphoria or inebriation.

The effect of these chemicals is a rapid general depression of the CNS, characterized by marked inebriation, dizziness, floating sensations, exhilaration, and intense feelings of well-being, which are at times accompanied by reckless abandon. A breakdown in inhibitions and feelings of greatly increased power and aggressiveness can also take place, sensations very similar to alcohol intoxication. Vivid visual hallucinations may also occur. These effects last from about 15 minutes to a few hours, depending on the substance and the dose.

Chemical solvents can be highly toxic and death can occur from liver, kidney, bone marrow, or other organ failure, cardiac arrhytmias, asphyxiation, or paralysis of breathing mechanisms. Continued use of these drugs can lead to kidney, liver, lung, and brain damage. In spite of the toxicity of the drugs, there is no evidence that tolerance develops, and physical dependence has not been demonstrated.

One of the most common sources of an intoxicating agent was airplane glue used to cement airplane or boat models. The toxic substance in the glue was toluene, which acts as a CNS depressant. The widespread sniffing of the glue, which can be highly toxic, prompted the manufacturers to remove the toluene, thereby eliminating the glue's value as an intoxicant; an irritating or noxious element was added to the glue to make it very unpleasant to sniff. Other products containing hydrocarbons that are used include gasoline, lacquer and lacquer thinner, hair sprays, spray starches, and lighter fluid.

Amyl nitrite, butyl nitrite

Amyl nitrite is used to treat a heart condition and to relieve asthma attacks. The drug produces a drop in blood pressure by dilating blood vessels, resulting in increased cardiac blood flow. The overall effect may be experienced as a feeling of light-headedness and relaxation of muscles. The high is the giddiness associated with a drop in blood pressure. Amyl nitrite has gained a reputation for prolonging or heightening sensations of sexual climax, with the user breaking the capsule and inhaling the vapors just before orgasm. When amyl nitrite became a prescription drug, thus less available, butyl nitrite became popular. Amyl nitrite is chemically related to nitroglycerin, another vasodilator.

Butyl nitrite is similar to amyl nitrite, and is used to induce a rush that lasts from one to two minutes. Like amyl nitrite, it lowers blood pressure and relaxes the smooth muscles of the body. There is some danger involved in using either substance because persons with defective blood vessels may not be able to cope with their sudden change in constriction. Isoamyl nitrite ("Bolt") has also been substituted for amyl nitrite.

5.7 CANNABIS DERIVATIVES (MARIJUANA)

("AIRPLANE", "ANGOLA", "BABY BHANG", "AUNT MARY", "GANGE", "GANJA", "GASH", "GRASS", "HEMP", "MARY JANE", "JOINT", "LOCOWEED", "REEFER", "ROPE")

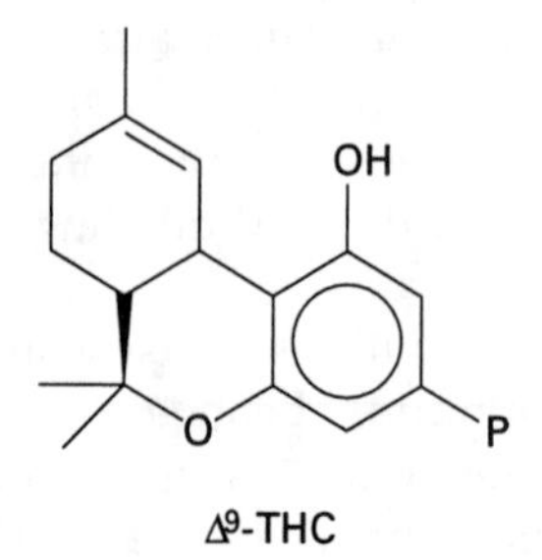

Δ^9-THC

Over 400 closely related chemicals have been isolated from the cannabis plant, which are collectively called cannabinoids, of which THC (Δ^9-tetrahydrocannabinol) is believed to be the primary psychoactive or mind-altering compound. Today's marijuana is more than ten times more potent than that used in the early 1970s. As the potency of the plant increases, there is a corresponding increase in its psychopharmacological effects. In 1976 the average potency was 1% delta-9-THC. By 1977 that potency was just over 1.5%; by 1982, the potency climbed to 3.6%, and by 1984 it had risen to 4%. By the late 1980s concentrations of THC between 10 and 16% or higher were found, and in the early 1990s, 20% THC or higher were reported[16].

Marijuana's effects are related to its quality or potency and to the user's set and setting, and the nature of marijuana intoxication varies dramatically from one individual to another. In general its effects, when smoked, begin within five to ten minutes, and may last for two to four hours. A positive experience, based on a good quality of marijuana (currently averaging over 15% THC), may consist of a sense of well-being, a dreamy state of relaxation and euphoria, alterations in thought formations, a more vivid sense of touch and perceptions, and distorted concepts of time and space.

Symptoms fairly commonly (but not always) associated with marijuana use are reddening of the eyes, dryness of the mouth, hunger, a mild rise in heart rate, and reduction of pressure in the ocular fluid of the eyes. A bad trip, experienced mostly by new users, tends to involve symptoms that resemble an anxiety attack, such as a feelings of panic, intense anxiety, and restlessness. Extremely high doses of marijuana may result in a toxic psychosis that may last up to a day. Hallucinations, illusions, paranoid delusions, feelings of depersonalization, and confusion may take place, and at times it may be difficult to determine if such symptoms are an idiosyncratic reaction or a result of an overdose. Flash-back experiences have been known to occur with the use of marijuana[11] and, as with most other psychoactive drugs, it has also been found to impair intellectual and psychomotor functioning and short-term memory. A unique characteristic of marijuana is that it remains in the body for a long period of time because cannabinoids are sequestered and eliminated very slowly; traces of it can be found in the body up to 50 hours or longer after its use.

Researchers[19] have also recently succeeded in finding a natural chemical compound in the brain, anandamide, that acts on the body like marijuana. The compound was found to bind with the same nerve receptor as marijuana's principal psychoactive agent, delta-9-tetrahydrocannabinol (THC), and is believed to produce similar effects, such as pain relief and mood regulation[56]. This findings suggests the existence of naturally occurring compounds in the body that are involved with stress and pain reduction, and with nausea. This findings has important implications for understanding how marijuana acts on the cannabinoid receptor to initiate effects such as pain relief, relaxation, suppressing nausea, dilating the eyes and increasing heart rate[56].

Whether or not the effects of marijuana are subject to tolerance and physical dependence has been an issue open to interpretation. Some researchers strongly contend that tolerance develops, and that the onset is quite rapid; others indicate that tolerance and withdrawal symptoms with marijuana do not develop. Blum[6] stated, *"Carefully conducted studies with known doses of marijuana or THC leave little question that tolerance develops with prolonged use."* Julien[29] summed up the results of research on the topic by stating, *"Tolerance to Cannabis is well substantiated and is thought to result from an adaptation of the brain to the continuous presence of the drug rather than from an increased rate of elimination."* Physical dependence to marijuana is not believed to occur among recreational or infrequent marijuana users, but it can develop among heavy or frequent marijuana users. Jaffe[28] indicated that withdrawal symptoms occurring after chronic use of high doses is characterized by 'irritability, restlessness, nervousness, decreased appetite, weight loss, insomnia, rebound increase in REM sleep, tremor, chills, and increased body temperature.' These symptoms are generally mild, beginning within a few hours after cessation, and can last up to 5 days.

5.8 DESIGNER DRUGS

By slightly changing the molecular structure of a drug, its potency, length of action, euphoric effects, and toxicity can be modified. These drugs, called "designer drugs," represent a class of substances whose actions mimic the effects of psychoactive drugs, but their chemical structures are different from those of existing drugs. The new drugs, when introduced in the late 1970s in the United States, were initially legal, but the Federal Drug Enforcement Administration (FDA) quickly invoked its emergency authority in 1985 to place an immediate ban on them by declaring them federally scheduled drugs under the Comprehensive Crime Control Act of 1984. Subsequent federal legislation has prohibited the (illicit) manufacture of all forms of designer drugs.

The first designer drug that rapidly became part of the drug scene was MDMA (methylenedioxymethamphetamine), known by the street names of "Ecstasy", "Adam", and "Essence". MDMA is synthesized from molecular components of methamphetamine or from safrole derived from sassafras or nutmeg. It belongs to a large group of synthetic drugs called phenylethylamines. It appears to evoke an easily controlled altered state of consciousness with emotional and sensual overtones, which can be compared to marijuana, to psilocybin devoid of the hallucinatory component, or to low levels of methylenedioxyamphetamine (MDA). Users describe the effect as similar to hallucinogens produced by other psychedelics, but its distilled effect leaves them feeling more empathic, more insightful, and more aware[51].

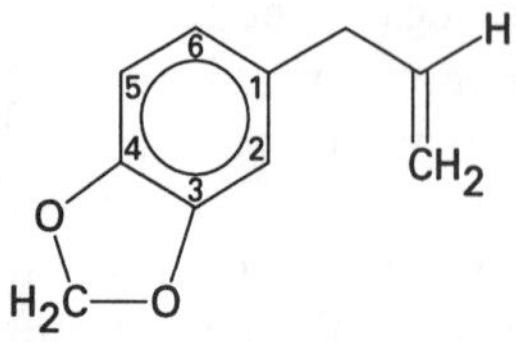

Safrole

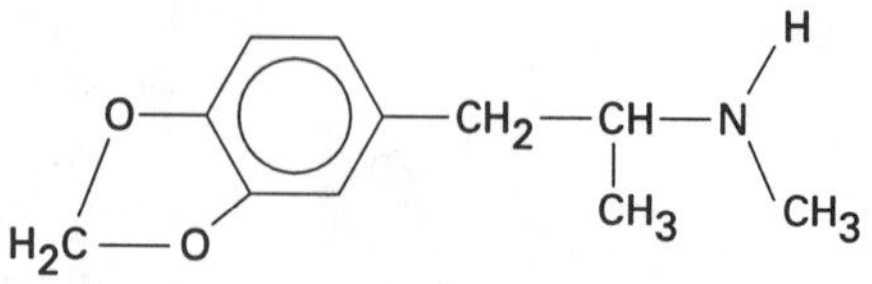

MDMA

Ecstasy, however, was not all blissful. Serious biochemical and behavioral effects were noted with its use, and there was concern that brain damage may result[51]. Some users reported nausea and dizziness as well as jaw pain lasting for weeks after taking MDMA. "When doses are pushed we get madness, not ecstasy" (cited in ref. 51). Ecstasy can also induce high blood pressure, an accelerated heart rate, and muscle tightness.

The molecular alteration of drugs has also produced a new generation of other illicit designer drugs that are more potent and potentially more dangerous than existing chemical substances. MPTP, MPPP, PEPAP, for example, all analogs of meperidine, produce effects similar to heroin, and have been associated with the development of symptoms similar to Parkinson's disease[44]. The ease with which drugs can be modified is evidenced by the identification of at least 35 variants or analogs that have been synthesized from PCP (Angel dust); several dozen analogs of methamphetamine have also been produced. Such drugs proved to be exceptionally potent and presented significant health hazards[44]. The situation became particularly acute with respect to the illicit analogs or derivatives of fentanyl to derive heroin substitutes.

The designer drugs described above represent only a few that current technology produced, but there is the potential to synthesize many other drugs that could continue to revolutionize illicit drug production, and which would place the user at considerable risk for severe adverse effects. Indeed, the latest technology in designer drugs was revealed in August, 1994, when U.S. Drug Enforcement Agents reported that a very potent drug, methacathinone ("cat"), was introduced to the streets as a cocaine-like drug whose effects could last up to five to seven days[60]. The drug was derived from a recipe found while the creators were searching U.S. Patent office records. The formula was used to reproduce the drug, and it soon circulated on the streets where it was a factor in three deaths.

6 Conclusion

Drugs, in one form or another, have been available since the dawn of civilization, and they have been used for different purposes in different cultures; many continue to be employed for their medicinal qualities, and for more individualized purposes. But since the advent of the pharmacological revolution that started during this century, many more drugs are available for use in seeking pleasure, and some of them can be quite toxic. This chapter has addressed some of the major mind-altering substances available, providing an overview of their psychopharmacological effects, and of some the adverse reactions they can induce.

References

1. Adlaf, E.M., Smart, R.G., Tan, S.H., "Ethnicity and drug use: a critical look," *Int. J. Addict.* 1989, **24(1):** pp 1–18.
2. Anthenelli, R.M., Tipp, J., Smith, M.A., Schuckit, M.A., *Sensation-Seeking Traits in Alcoholic Families and Controls.* Paper presented at the annual meeting of the Research Society on Alcoholism, Maui, HI, 1994.
3. Asuni, T. and Pela, O.A., "Drug abuse in Africa," *Bull. Narcot.* 1986, **38(12):** pp 55–56.
4. Avksentyuk, A.V., Kurilovich, S.A., Duffy, L.K., Segal, B., Voevoda, M.I., Nikitin, Y.P., "Alcohol consumption and flushing responses in Natives Chukotka Siberia," *J. Alcohol Stud.* 1995, **56:** pp 194–201.
5. Beauvais, F. and Segal, B., "Drug use patterns among American Indian and Alaskan Native youth: special rural populations." *Drugs Society.* 1992, **7(1/2):** pp 77–94.
6. Blum, K., *Handbook of Abusable Drugs.* Gardner Press, New York, 1984.
7. Blum, K., *Alcohol and the Addictive Brain.* The Free Press, New York, 1991.
8. Caputo, R.A., "Volatile substance misuse in children and youth: a consideration of theories. "*Int. J. Addict.* 1993, **28(10):** pp 1015–1032.
9. Chizova, T.V., "Prevention of alcoholism among wood industry workers," *Hygiene and Sanitary* 1989, **8:** pp 32–34.
10. Claridge, G., *Drugs and Human Behavior.* Pelican, Middlesex, U.K., 1972.
11. Cohen, S., *The Substance Abuse Problem.* Haworth Press, New York, 1981.
12. Craig, R.J., "Personality characteristics of heroin addicts: a review of the empirical literature with critique, Part I," *Int. J. Addict.* 1979, **14(4):** pp 513–532.
13. Craig, R.J., "Personality characteristics of heroin addicts: a review of the empirical literature with critique, Part II." *Int. J. Addict.* 1979, **14(5):** pp 607–626.
14. De La Rosa, M.R. and Adrados, J.-L.,R., *Drug Abuse Among Minority Youth: Advances in Research and Methodology.* 1993, Research Monograph 130, NIDA, Rockville, MD.
15. Department of Justice, "Street terms: drugs and the drug trade," *Drugs and Crime Data.* Drugs Crime Data Center, U.S. Department of Justice, Rockville, MD, 1993.
16. Drug Enforcement Agency, *Drugs of Abuse.* U.S. Department of Justice, Washington, D.C., 1989.
17. Dubowski, K.M., *The Technology of Breath-Alcohol Analysis.* DHHS Publ. No. (ADM) 92–1728, NIAAA, Rockville, MD, 1991.
18. Fagan, J., "Set and setting revisited: influences of alcohol and illicit drugs on the social context of violent events," Martin, S.E. (Ed.), *Alcohol and Interpersonal Violence: Fostering Multi-disciplinary Perspectives.* 1993, Research Monograph 24, NIAAA, Rockville, MD, pp 161–191.
19. Felder, C.C., Briley, E.M., Axelrod, J., Simpson, J.T., Mackie, K., Devane, W.A., "Anandamide, an endogenous cannabimimetic eicosanoid receptor and stimulus receptor-mediated signal transduction," *Proc. Natl Acad. Sci. USA,* 1993, **90:** pp 7656–7660.
20. Frances, R.I. and Miller ,S.I., *Clinical Textbook of Addictive Disorders.* The Guilford Press, New York, 1991.
21. Friedman, A.S., Utada, A.T., Glickman, N.W., Morrissey, M.R., "Psychopathology as an antecedent to, and as a 'consequence' of, substance use, in adolescence." *J. Drug Ed.* 1987, **17(3):** pp 233–244.
22. Goldberg, R., *Drugs Across the Spectrum.* West Publishing, Minneapolis-St. Paul, MN, 1994.
23. Goode, E., *Drugs in American Society*, 3rd edition. Alfred A. Knopf, New York, 1989.
24. Goode, E., *Drugs in American Society,* 4th edition. Alfred A. Knopf, New York, 1993.
25. Hesselbrock, M.N. and Hesselbrock, V.M., "Relationship of family history, antisocial personality disorder and personality traits in young men at risk for alcoholism," *J. Stud. Alcohol.* 1992, **53:** pp 619–625.
26. Hewet, A.B. and Martin, W.R., (1980) "Psycohmetric comparisons of sociopathic and psychopathological behaviors of alcoholics and drug abusers versus a low drug use control population," *Int. J. Addict.* 1980, **15:** pp 77–105.
27. Hofmann, F.G., *A Handbook on Drug and Alcohol Abuse: The Biomedical Aspects,* 2nd edition. Oxford University Press, New York, 1983.

28. Jaffe, J.W., "Drug addiction and drug abuse." In: Gilman, A.G., Rall, T.W., Niews, A.S., Taylor, P. (Eds) *Goodman and Gilman's The Pharmacological Basis Of Theapeutics*, 8th edition. Pergamon Press, New York, 1990, pp 549–553.
29. Julien, R.M., *A Primer of Drug Action*, 6th edition. W.H. Freeman, New York, 1992.
30. Julien, R.M., *A Primer of Drug Action*. W.H. Freeman, New York, 1988.
31. Klatsky, A.L., Armstrong, M.A., Friedman, G.D., (1986) "Relation of alcoholic beverages use to subsequent coronary artery disease hospitalization," *Am. J. Cardiol.* 1986, **58:** pp 710–714.
32. Korolenko, C.P. and Botchkareva, N.L., "A review of the problem of alcoholism in Siberia," *Drugs Society*. 1990, **4:** pp 5–14.
33. Kovacs, G.L. and Telegdy, G., "Oxytocin in memory and amnesia," *Pharmacol. Ther.* 1982, **18:** p 375.
34. Krohn, M.D. and Thornberry, T.P., "Network theory: a model for understanding drug abuse among African-American and Hispanic youth," de la Rosa, M., Adrados, J.-L.R. (Eds.), *Drug Abuse Among Minority Youth: Methodological Issues and Recent Research Advances*. Research Monograph 130, NIDA, 1993, Rockville, MD, pp 102–128.
35. Kurilovitch, S.A., Avksentyuk, A.V., Voevoda, M.I., Filimonova, T.A., Serova, N.V., *Alcohol Consumption in Western Siberia. Connection with Serum Blood Lipids, Ischemic Heart Disease and Arterial Hypertension*. Proceedings of the 8th International Congress on Circumpolar Health, Whitehorse, Yukon, Canada, 1991, p 111.
36. Leavitt, F., *Drugs and Behavior*, 2nd edition. Wiley, New York, 1982.
37. Liska, K., *Drugs and the Human Body*, 4th edition. Macmillan, New York, 1994.
38. Mair, R.G., Langlass, P.J., Mazurek, M.F., Flint, B.M., Martin, J.B., McEntee, W.J., "Reduced concentration of arginine vasopressen and MHPG in lumbar CSF of patients with Korsakoff's psychosis," *Life Sci.* 1986, **38:** p 2301.
39. Makela, K. and Simpura, D. "Investigation of alcoholism problems in Finland." *Prob. Narco.* 1989, **3:** pp 32–37.
40. Maula, J., Lindblad, M., Tigerstedt, C., *Alcohol in Developing Countries*. Nordic Council for Alcohol and Drug Research, Helsinki, 1990.
41. Meiczkowski, T. (Ed.) (1992) *Drugs, Crime and Social Policy.* Allyn and Bacon, Boston.
42. Miller, N.S., *The Pharmacology Of Alcohol and Drugs of Abuse and Addiction*. Springer-Verlag, New York, 1991.
43. Miller, M., "The necessity of biomedical research on substance abuse." In: Miller, M.W. (Ed.), *Development of the CNS: Effects of Alcohol and Opiates*. John Wiley-Liss, NY, 1993, pp 1–8.
44. National Institute on Drug Abuse. "Designer drugs: a new concern for the drug abuse community," *NIDA Notes*. 1985, pp 2–3.
45. Nogrady, T., *Medical Chemistry*, 2nd edition. Oxford University Press, New York, 1988.
46. Office of Technology Assessment, *Biological Components of Substance Abuse and Addiction*. U.S. Congress, Washington, D.C., 1993.
47. Payne, W.A., Hane, D.B., Pinger, R.R., *Drugs: Issues for Today*. Mosby, St. Louis, MO, 1991.
48. Popova, Z.P. and Alekseyev, V.P., "Alcohol, arterial hypertension and risk of sudden death among male residents of Jacutsk." In: *Alcohol-Related Internal Diseases in the North*. Jacutsk, Russia, 1988, pp 19–27.
49. Ray, O. and Ksir, C., *Drugs, Society, Human Behavior*, 6th edition. Times Mirror/Mosby College Publishing, St. Louis, MO, 1993.
50. Rebach, H., Bole,k C.S., Williams, K.L., Russell, R., *Substance Abuse Among Ethnic Minorities in America*. Garland, New York, 1992.
51. Roberts, M., MDMA: "Madness, not ecstasy." *Psychol. Today*. 1986, pp 14–15.
52. Segal, B., *Drugs Behavior*. Gardner Press, New York, 1988.
53. Segal, B., Huba, G.J., Singer, J.L., "Reasons for drug and alcohol use by college students," *Int.J. Addict.* 1980, **15:** pp 489–498.
54. Segal, B. and Singer, J.L., "Daydreaming, drug and alcohol use in college students: a factor analytic study," *Addict. Behav.* 1976, **1:** pp 227–235.
55. Stephens, R.C., "Psychoactive drugs in the United States today: a critical overview," In: Mieczkowski, T. (Ed.), *Drugs, Crime and Social Policy*. Allyn and Bacon, Boston, 1992, pp 1–31.

56. Swan, N., "Marijuana's natural counterpart discovered," *NIDA Notes*, 1993, **8(4):** 8–9, p 15.
57. Teplin, L.A., "Psychiatric and substance abuse disorders among male urban jail detainees," *Am. J. Public Health.* 1994, **84(2):** pp 290–293.
58. Wagner, E.F. and Hitner, J., *Depressive Symptoms and Adolescent Substance Abuse.* Paper presented at the annual meeting of the Research Society on Alcoholism, Maui, HI, 1994.
59. Walsh, G., *Public Health Implications of Alcohol Production and Trade.* World Health Organization, Geneva, 1985.
60. Webb, C.L., "Recipe for drug found in patent office," *Anchorage Daily News.* 1994, p A6.
61. Wise, R.A., "The role of reward pathways in the development of drug dependence," *Pharmacol. Ther.* 1987, **35:** pp 227–263.
62. Zinberg, N.E. and Shaffer, H.J., "The social psychology of intoxicant use: the interaction of personality and social settings," Milkman, H. and Shaffer, H. (Eds.), *Addictions: Multidisciplinary Concepts and Treatment.* Lexington Books, Lexington, MA, 1985, pp 57–74.
63. Zuckerman, M., *Sensation Seeking.* Lawrence Erlbaum Associates, Hillsdale, NJ, 1979.
64. Zuckerman, M., Buchsbaum, M.S, Murphy, D.L., "Sensation seeking and its biological correlates," *Psychol. Bull.* 1980, **88(1):** pp 187–214.

Chapter 3

Animal research: drug self-administration

0-8493-7803-6/99/$0.00+$.50

Chapter 3

Animal research: drug self-administration

Richard A. Meisch and Robert B. Stewart

1 Introduction and historical background

During the first 4 decades of this century, the scientific study of drug addiction was focused mostly on the physiological consequences of drug-taking behavior rather than on the processes that may initiate, facilitate, and maintain drug self-administration. The study of the consequences of drug administration was amenable to a standard biological or medical approach: drugs were administered to human volunteers or to laboratory animals and subsequent changes in physiological functioning were observed. With regard to drug addiction, the consequences of discontinuing the drug treatment after a prolonged period of drug administration were of special interest. The aversive withdrawal signs and symptoms that occurred when drug self-administration was interrupted were thought to be the essence of drug addiction, that is, the accepted hypothesis was that the addictive drug-taking behavior was driven or motivated by the avoidance of the physiological consequences of withdrawal. This hypothesis begged the questions of why drug taking was initiated in the first place and why addicts would relapse or reinstate drug self-administration even after prolonged periods of abstinence. Under such conditions, avoidance of withdrawal signs could not possibly be an operating factor. The search for answers to questions about the initiation of drug-taking behavior and about why some individuals appeared to be more at risk than others was relegated to the disciplines of psychoanalysis and social psychology. For example, a propensity to abuse drugs was viewed by some psychiatrists as a defense mechanism involving latent or repressed homosexuality (for historical review see ref. 79).

In the course of investigations into the physiological consequences of injections of opioid drugs such as morphine or heroin to nonhuman subjects, it was noted, as a subjective anecdotal aside, that the subjects sometimes appeared to eagerly anticipate the appearance of the experimenter with the drug-filled syringe[84]. Laboratory primates sometimes appeared to cooperate to a degree that was unusual considering that the administration of the subcutaneous or intramuscular injections produced temporary discomfort and pain. However, the idea that the animals were learning to anticipate the drug injections and were developing a "desire" or "craving" for the effects of the drugs remained speculative. There appeared to be no way to rigorously evaluate such concepts. Several important methodological developments then occurred that would change the manner in which drug self-administration would be investigated in the laboratory and, as a result, also would change our conception of the addiction process itself.

The first development was the demonstration that chimpanzees would learn to emit a response if it resulted in the administration of a morphine injection to the animal by the experimenter[88]. The chimpanzees learned to discriminate between cues associated with a preferred food and cues associated with drug injections. During experimental trials in which both sets of cues were presented selection of the food cues resulted in the receipt of food. Selections of the drug cues resulted in the administration of the drug to the ape by the experimenter. Under these conditions, the chimpanzees reliably and unmbiguously chose the drug cues and, thus, self-administered the drug. This procedural development resulted in a conceptual change: heretofore self-initiated drug-taking behavior was though to be a uniquely human phenomenon. Spragg (1940)[88] had demonstrated that non human primates also would self-administer drugs if the appropriate environmental contingencies were arranged. However, Spragg's procedure had some practical limitations. It involved a lengthy period during which the animals must be tamed, habituated, and trained. The number of subjects that could be tested at one time was severely limited by the fact that the experimenter himself was an integral part of the experimental protocol.

The solution to this problem involved the invention of a technique whereby drugs could automatically be delivered to laboratory animals without the need of the experimenter to be present to administer an injection. Weeks (1962)[93] developed a procedure in which rats have plastic cannulae surgically implanted for the intravenous delivery of drugs. The rats then are placed in a situation in which a learned response, depressing a lever on the wall of the cage, would result in the activation of an electrical circuitry. A motorized syringe pump then delivers a measured amount of drug solution through plastic tubing directly into the vein of the rat. The procedure was unobtrusive and eliminated the necessity for the experimenter to personally administer the injections. This technique soon was adapted for non-human primates[91].

The use of an arbitrary response such as a lever press automatically to deliver drugs in a controlled experimental environment also facilitated a conceptual change in the manner in which drug-taking behavior could be investigated. Drug self-administration now could be examined as a specific example of a more general biological process: operant conditioning. Under this theoretical rubric, drug self-administration is viewed as an example of reinforcement. The term reinforcement[85,86] describes a functional relationship between environmental and behavioral variables: reinforcement is a process where by the probability of a response (e.g., drug-taking behavior) is increased if the response results in the presentation of a reinforcer (the drug). The following chapter concerns some of the important findings and methodological issues of drug self-administration studies.

An overview of three major topics is presented in the next sections. These areas are: (1) drugs that are self-administered, (2) species generality, and (3) routes of administration. The purpose of these sections is to provide an overview of the scope of drug self-administration studies. Some of these topics are discussed in detail in subsequent sections.

2 Drugs that are self-administered

The extent of drug self-administration research is indicated, in part, by the number of drugs and drug classes that have been studied (see Section 3.2 of this chapter). Some drugs that serve as reinforcers are listed in Table 3.1. The different pharmacological classes have few commonalties in their behavioral actions except for their reinforcing effects.

TABLE 3.1
Some drugs that can act as positive reinforcers in drug self-administration studies

Psychomotor stimulants	cocaine *d*-amphetamine methamphetamine
Ganglionic stimulants	nicotine
Opioids	heroin morphine
CNS depressants	ethanol pentobarbital diazepam nitrous oxide
Dissociative anesthetics	phencyclidine ketamine

Some drugs are avoided by animals, and by avoided is meant that animals will press a lever when the consequence is the postponement or termination of injections. Table 3.2 lists some agents. These drugs come from several pharmacological classes. Two examples are the LSD-type hallucinogens and the antipsychotics. The antipsychotic drugs are not abused. In fact patient compliance in taking these drugs is poor. The poor compliance may be due to their aversive effects. The avoided drugs, like the reinforcing drugs, come from several pharmacological classes.

TABLE 3.2
Some drugs that can act as negative reinforcers in drug self-administration studies

Antipsychotic	chlorpromazine haloperidol perphenazine
Opioid antagonists	naloxone
Psychedelic	lysergic acid diethylamide

A third group of drugs is neither self-administered nor avoided. This group includes the tricyclic antidepressants such as imipramine and the antipyretic analgesics such as acetylsalicylic acid. Abuse of drugs in this group is uncommon. One conclusion that derives from drug self-administration studies is that drugs can have positive, neutral, or negative reinforcing effects.

An important generalization is that the drugs that humans abuse usually act as reinforcers in laboratory studies with animals, and the drugs that humans do not abuse do not serve as reinforcers in the animal laboratory[54,58]. An exception is marijuana, or more precisely, delta-9-tetrahydrocannabinol. Delta-9-tetrahydrocannabinol is not water soluble, thus it has been difficult to study in laboratory self-administration experiments. In the studies that have been conducted, evidence for reinforcing effects has not been obtained[47]. However, it is possible that future studies will identify the conditions under which cannabinoids may reinforce behavior.

The fact that drugs of abuse serve as reinforcers in animal studies has several implications for our understanding of drug abuse. One implication is that common factors may control drug seeking behavior in humans and laboratory animals. Variables that are important in laboratory studies are probably important in situations outside the laboratory. Drug reinforcement is not something that develops only in humans.

2.1 SPECIES GENERALITY

The species most frequently used are rats and rhesus monkeys. However, many species have been studied including baboons, cynamologus monkeys, squirrel monkeys, dogs, cats, and mice. In all these species and also in pigeons it has been possible to establish intravenously delivered cocaine as a reinforcer. There are also many cross-species similarities in the factors that control drug taking. An implication of the cross-species generality of findings is that it demonstrates a broad biological basis for drug reinforcement. A related implication is that results obtained with laboratory mammals probably apply to humans as well since humans are members of the category of mammals. Drug reinforced behavior is not something that is characteristic of a limited or aberrant population. For example, almost all rhesus monkeys given iv access to cocaine will self-administer the drug[34]. Thus, there is a biological component to drug self-administration.

While drugs can function as reinforcers, for most individuals there are marked individual differences in drug self-administration. Rhesus monkeys differ greatly in their intake of pentobarbital and ethanol[73]. What accounts for these differences is not known; however, one possibility is genotype. The behavioral genetics of drug reinforcement is an important emerging area (see Chapter 18). Behavioral genetics concerns interactions among drug taking, genotype, and environment. An understanding of individual differences among laboratory animals in drug taking may lead to improvements in prevention and treatment. The important concept is that genotype can be one of the determinants of drug reinforcement.

2.2 ROUTES OF ADMINISTRATION

Drugs can be reinforcers when delivered by different routes. These routes include the intravenous, oral, intragastric, pulmonary, intraventricular, intramuscular, and intracranial. Humans also self-administer drugs by many routes. This correspondence is notable, since it adds to the documentation of similar features of drug reinforcement in humans and laboratory animals.

Summary of drugs, species, and routes

Three factors have been discussed: drugs, species, and routes. Important generalizations have risen from the study of each factor. These generalizations include the correspondence between drugs of abuse and drugs that can serve as reinforcers in the laboratory. Drugs can serve as reinforcers for many species, and this demonstrates an important biological basis for drug reinforcement. With laboratory animals reinforcing drugs can be delivered by multiple routes. This parallels what is seen with humans, abuse of drugs via multiple administration routes.

2.3 OPERANT BEHAVIOR

Operant behavior is behavior that is controlled by its consequences. An example of operant behavior is a human pressing an elevator button, or a monkey pressing a lever that leads to the delivery of a food pellet. Under these conditions the food pellet is a positive reinforcer and the lever press is the operant behavior. Most of our behavior is operant behavior. That drug taking is operant behavior was recognized in the early laboratory studies of drug self-administration. Weeks (1962)[93], who published the first paper on intravenous drug self-administration with animals, explicitly noted that drug self-administration was an operant behavior.

There is a substantial literature on operant behavior (see ref. 21). The recognition that drug taking is an operant behavior has several implications. One of these is that there exists a conceptual frame for the scientific analysis of drug taking. For example, the variables that control other operant behaviors in general also should be important in the specific case of drug taking. Some of the subsequent sections concern operant conditioning variables functioning as determinants of drug self-administration.

2.4 CRITERIA FOR DRUG REINFORCEMENT

There are multiple interpretations of the observation that animals will press a lever that is followed by a drug injection. In an important study of cocaine self-administration conducted with rats, a number of these interpretations were ruled out by a series of tests[81].

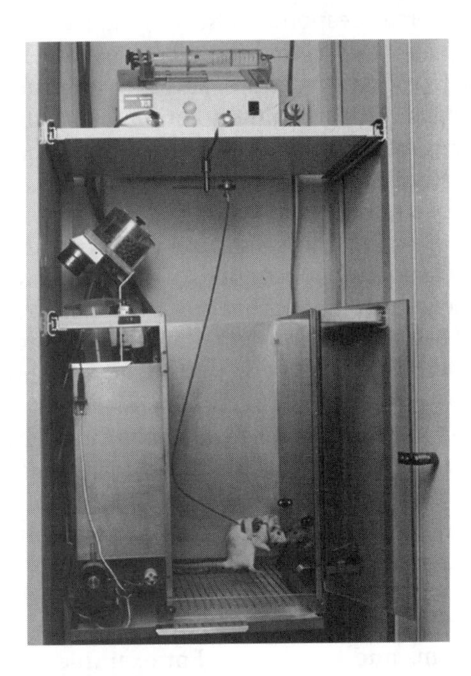

FIGURE 3.1
Experimental device for the study of i.v. drug self-administration in rats
(Photograph courtesy of Prof. Dr. J.M. van Ree.)

The principal interpretation is that the lever pressing behavior is due to the drug's reinforcing effects. However, there are other possibilities. The lever could be pressed incidentally during exploration of the test chamber. The drug that is subsequently injected may increase activity levels. The effects of the drug may lead to increasingly higher rates of general activity and thus in a positive feedback loop, lead to more incidental injections.

This interpretation can be assessed by giving the animals non-contingent injections (i.e. the experimenter, rather then the animal, gives the injections). If the drug injections are due to reinforcing effects, the rate of lever pressing will be diminished by the non-contingent injections since the injections will provide the same amount of drug without entailing a lever press requirement.

However, if the injected drug is simply causing an elevated rate of motor behavior, then the non-contingent injections should increase rather than decrease the rate of lever pressing. When this was done in the intravenous cocaine study, the rats' rate of lever pressing was reduced by non-contingent injections[81]. This change in rate supported the interpretation of reinforcing effects.

Another control for non-specific activity is to have two levers in the chamber and record responses on both levers, but have drug delivery contingent upon responding on one lever. Responses on the other lever would not lead to drug delivery. If the lever presses are due to reinforcing effects, then the rate of responding on the active lever should be far greater than the rate on the second lever.

A related approach is termed a contingency reversal. With this test the active and inactive levers are reversed. The lever that in the past was the lever on which responses lead to drug injection, is now the "dummy" or control lever. The lever that was previously the control lever is now the active lever in that responses on this lever result in drug injections. When this change in the functions of the levers occurs, the rats typically emit an extinction burst (a temporary increase in responding on the lever that is no longer active). Then they switch levers to respond on the new active lever.

Another possibility is that the increases in behavior are due to the general effects of the injection procedure, and thus not due to the drug effects. This can be tested by substituting a saline solution for the drug solution. If lever pressing is due to drug reinforcement, then saline substitution should lead to an extinction pattern of responding that is, an initial high rate of responding will occur and subsequently responding will drop to low levels. When saline was substituted for intravenous cocaine in the rat study, the rate of responding initially increased and subsequently decreased to low levels[81]. These results supported the interpretation that the drug was functioning as a reinforcer since drug, but not saline, injections maintained lever pressing.

An additional possibility is that lever pressing was due to the onset of a light that had been paired with cocaine injections. To test this possibility, saline was again substituted for the cocaine solution. Lever presses resulted in both onset of the light and activation of the infusion pump. Responding extinguished, thus, onset of the light did not reinforce behavior[81].

An important development that occurred in the 1970s was the use of simultaneous access to both drug and drug vehicle[48,56]. The relative rates of responding maintained by the drug and its vehicle provide a rigorous test for concluding that a drug is reinforcing. A simple example is to use two drug delivery systems and two levers. For example, presses on one lever result in access to a small volume of tap water, and presses on the other lever result in access to a small volume of an ethanol solution. Tap water serves as the vehicle for the ethanol. Higher response rates on the ethanol lever than on the water lever are strong evidence for reinforcement. An additional degree of rigor is added by alternating side positions of drug and vehicle. This arrangement makes possible the detection of side preferences, if present, and adds the additional behavioral requirement of switching from one lever to another each session. Supporting data for the presence of drug reinforcement are orderly dose response functions and characteristic patterns of intermittently reinforced behavior.

2.5 METHODS OF DRUG DELIVERY

Drugs are self-administered by a number of routes, and this is true for laboratory animals and humans. These routes include the intravenous, oral, intragastric, pulmonary, intracranial, and intramuscular. Each route has certain advantages and disadvantages. Although most studies have been done with intravenous drug delivery, the other routes are also important. For example, the drugs that create the largest public health problem are ethanol and nicotine, and neither is taken intravenously by humans. That drugs can serve as reinforcers via different routes is significant for it demonstrates an important parallel between findings with laboratory animals and humans.

Intravenous route

The intravenous route has been the route most frequently used in laboratory studies. It is also the route employed in the important early studies. The intravenous route has the advantage of rapid delivery of a drug dose immediately upon completion of a schedule requirement. Immediacy of reinforcement is important in facilitating learning and in maintaining learned behavior. The intravenous route also avoids the problem of the aversive taste of most drug solutions.

The procedure used in delivering drugs intravenously was introduced by James Weeks in 1962[93]. There have been many refinements but the basic method is still the same. Weeks described a procedure for inserting and maintaining a chronic indwelling catheter (or tube) in rats. Under general anesthesia an incision is made in the rat's neck and the jugular vein is exposed. A catheter is inserted into the vein and threaded down toward the heart. The catheter is attached to neck muscles by sutures to hold it in place. The distal end of the catheter is run subcutaneously around the animal's back where it exits between the scapulae. The catheter is then attached to thin metal tubing. The tubing contains a swivel so that the animal can turn and move about the experimental chamber. The metal tubing is connected via flexible tubing to a motorized syringe, so that when a lever in the chamber is pressed, electronic circuitry is activated and results in the brief operation of the motorized syringe; the result is an intravenous drug injection[94]. Variants of this procedure have been described for baboons, rhesus monkeys, squirrel monkeys, dogs, cats, mice, and pigeons.

The intravenous route has disadvantages as well as advantages. A surgical procedure is necessary to insert the catheter. Animals can develop infections secondary to having an indwelling catheter. The animal may pull out the catheter or the catheter can become obstructed by blood clots. Therefore special canullae have been developed[92]. Some form of chronic restraint is necessary during sessions as a consequence of having to connect the catheter to the infusion pump. The duration of time that the catheter is functioning correctly may be short[106].

Intragastric, intraventricular, and intracranial routes

The general arrangement for delivering drug is similar with the intragastric, intraventricular, and the intracranial routes. With these routes the catheter terminates in the stomach, brain ventricles, and brain tissue, respectively. A catheter is inserted into a part of the body and is also connected with a pump such that completion of the schedule requirement results in drug delivery.

Pulmonary route

Several methods are available for delivering gases and vapors. With one method a nasal catheter is used. The catheter is inserted into the nasal cavity in a manner similar to the way in which an intravenous or intragastric catheter is inserted into a body cavity at one end and attached to a pump at the other end. Lever presses result in the delivery of a specified amount of gas[105]. A sealed helmet can also be used to deliver gases[101]. Rodents can be placed in a chamber into which gases or drug aerosols can be delivered[49,100]. An important advance has been the development of a system for delivering heated cocaine base ("crack" cocaine) to rhesus monkeys[14]. Following the completion of a response requirement, a wire coated with the cocaine base is heated and the base is released as a vapor which the monkey takes by sucking on a tube. This is the first successful case of a heated material being used as an effective pulmonary reinforcer.

Intramuscular route

The intramuscular route has been used with rhesus monkeys and squirrel monkeys[38,59]. Usually the drug delivery is scheduled at the end of a long response sequence. Upon completion of the schedule, a stimulus change occurs in the operant chamber and the experimenter is also alerted. The experimenter then enters the room and gives the monkey an intramuscular injection[38]. The advantage of using the intramuscular route is that no catheters are employed, and the monkey can serve as a subject for many years. The disadvantages of this route are the extended training for establishing reinforcing effects, and the requirement of a person to inject the drug promptly upon schedule completion. A drug given by intramuscular injection may be less effective as a reinforcer due to slower onset of action than when the drug is given intravenously.

Oral route

Many studies have been conducted with the oral route. The usual procedure is to attach two drinking bottles to a rodent's cage, and to place water in one bottle and the drug solution in the other. Volumes consumed are measured once a day. Under these conditions rats usually consume less drug solution than water. Also, the use of drinking bottles does not make it possible to either intermittently reinforce drinking or to impose limits on the volume consumed during drinking bouts.

An advanced method is the use of operant conditioning chambers, equipped with liquid delivery systems. In most cases this delivery system is a dipper or small spoon that is filled with liquid by lowering the dipper into a reservoir containing the liquid, and then raising the dipper back into a position that is accessible to the rodent. This arrangement permits the use of intermittent reinforcement schedules and the presentation of a fixed volume of liquid with each schedule completion. The dipper system works well with rodents but it is not suitable for primates. Thus another system has been devised for delivery of liquids including drug solutions. This system employs drinking spouts, a valve that can be briefly opened, and a reservoir for holding the liquid. This system can rapidly and precisely deliver liquid including drug solutions and it can be used with non-human primates and rodents.

Use of the oral route has some advantages. No surgery is required and no restraint is necessary. The animals remain in good health and can participate in related studies over a period of years. A major disadvantage is the amount of training required and that orally delivered drugs will not come to serve as reinforcers for some animals.

Summary

Over the last 30 years substantial progress has been made in developing drug delivery systems. Drugs can now be delivered to laboratory animals by many routes. Each route has certain advantages and disadvantages. The findings obtained with different routes but with the same drug (e.g., cocaine) suggest that there are quantitative but not necessarily qualitative differences among the routes. Although the intravenous route has been the most studied, other routes are important. For example, the drugs that create the largest public health problems are ethanol and nicotine, but humans use neither drug intravenously.

FIGURE 3.2

Experimental device for the study of oral drug self-administration in monkeys (Photograph courtesy of Dr. L.M.W. Kornet.)

2.6 ESTABLISHING DRUGS AS REINFORCERS

Access to drug

In many studies of intravenous drug self-administration, the catheterized animal is placed in the chamber with drug available under a low FR schedule. Many of these animals develop drug reinforced responding without training. In exploring the chamber, the rat or monkey may depress the lever and receive an injection. Subsequent injections occur more rapidly, and the animal's responding comes under the control of the drug reinforcer. In some cases the animals are given extended sessions such as 24 hours, and once drug reinforced behavior has developed, the session length is decreased.

One variant of this procedure is to automatically give an injection every hour unless the animal has self-administered the drug in the previous hour. This permits the animal to experience the drug effect and thus may speed acquisition. Another variant is to tape food such as a raisin to the lever to increase contact with the lever. Another strategy is to give the animal prior training in using the lever. This can be done by having lever presses produce food pellets. In some cases the animals undergo extinction of the food-reinforced response prior to drug access. The rationale is that it is easier to interpret lever pressing as due to the contingent drug delivery and not to past food reinforcement.

Drug substitution

A rapid and effective way to establish a drug as a reinforcer is to substitute the drug for another drug that is already functioning as a reinforcer. This procedure is usually done when drugs are delivered intravenously. The drug that is already a reinforcer (the baseline drug) is often cocaine or some other drug that is readily established as a reinforcer. Subsequently, the test drug is substituted for cocaine. Reinforcement by the test drug is indicated by maintenance of high stable response rates that exceed those obtained with saline when saline is substituted for the baseline drug. Further evidence of a test drug's reinforcing effects is the persistence and maintenance of responding across a range of conditions. The advantage of substitution procedures is that they bypass the difficulties that may be encountered in initially obtaining drug reinforced responding.

A number of conditions determine the degree of the success with the substitution procedures. These include the type of baseline drug and the type of test drug. When both drugs are from the same pharmacological class, the probability of successful substitution is usually greater than when the two drugs are from different classes. In some studies a test drug that does not substitute for one baseline drug, does substitute successfully for another baseline drug one that shares actions with the test drug[107].

Drug substitution has also been accomplished with other routes of administration. In studies of intracranial drug self-administration, the neurotransmitter dopamine has served as a reinforcer when substituted for cocaine[37]. Unfortunately the use of the intracranial route involves complex technical methods and at present is too difficult for standard laboratory practice.

Substitution procedures have been conducted when drugs are taken orally. Often the results are negative in that one drug does not substitute for another drug. The negative results often are attributed to differences in taste between the two drugs. However, there have been some successes, and these usually occur when the baseline and test drugs are from the same pharmacological class. Analogs of phencyclidine successfully substituted for phencyclidine[10], methohexital for pentobarbital[20], and diazepam for midazolam[89]. In several cases one type of drug successfully substituted for another: amphetamine for ketamine[19] and methohexital for phencyclidine[20].

One reason for the success of substitution procedures is that initially there is a high probability that substantial amounts of the test drug will enter the body. This is due to either the high response rates or strongly reinforced behavior that preceded the substitution. The ingestion of behaviorally active drug amounts is especially necessary when the oral route is used, for the aversive taste of drug solutions limits intake to small amounts that are not sufficient for CNS effects.

Oral route training procedures

Studies that employ the oral route often yield negative results. They report low amounts of drug intake or find less drug solution consumption than the water vehicle. The lack of success with the oral route relates to three factors. Most drug solutions have an aversive taste, and animals will drink more water than drug solution. When animals do drink the drug solution they usually take small volumes and thus insufficient drug amounts are consumed, that is, drug amounts that are too small to produce CNS effects. Finally, there is a substantial delay between drinking the drug solution and onset of CNS actions. To overcome these problems a related set of procedures has been developed. Food intake is limited. Drinking is induced by feeding the animals during the experimental session, since food intake is regularly followed by drinking. Low drug concentrations replace water during the session, and the concentration of the drug is gradually increased. When intermediate drug concentrations are reached, the inducing conditions are terminated (i.e., food is no longer available during the session), and in most cases drinking of the drug solution persists and is taken in greater volumes than concurrently available water[68].

The success of these procedures is due to several factors. Initially low concentrations are used so that the animals adapt to the taste of the solutions. Induction of drinking leads to regular intake of amounts that have CNS effects, and it also results in the repeated coupling of the drug solution taste with the subsequent CNS effects. Animals are food restricted since low drug doses are better reinforcers in food restricted animals. Although the taste is initially aversive, it probably becomes a conditioned reinforcer during training and thereby bridges the delay between drinking and onset of CNS effects.

3 Effects of drug intake

3.1 SECONDARY CONSEQUENCES OF DRUG INTAKE

Physiological (physical) dependence

Physiological dependence is a condition that develops from the repeated administration of certain drugs when the repeated administrations occur close in time, such as multiple injections per day. It develops when drugs such as the opioids or CNS depressants are taken chronically. However, there is now evidence for the occurrence of dependence with drugs from other pharmacological classes such as the psychomotor stimulants[16] and the arylcyclohexylamines[4]. The presence of dependence is detected by the occurrence of an abstinence syndrome either by discontinuing drug administration or by administration of an antagonist. Each drug class has a characteristic abstinence syndrome.

Thirty years ago the cardinal signs of drug abuse were thought to be tolerance and physiological dependence. However, it is now known that tolerance and physiological dependence are more of a result rather than a cause of drug taking[102]. Importantly, drugs can come to function as reinforcers in the absence of dependence[92]. The possibility remains that drugs could serve as better reinforcers when animals are dependent[98]. However, rigorous evidence for this possibility is lacking. In one study of intravenous alcohol self-administration, physiologically dependent rhesus monkeys did not display increased alcohol self-administration[95]. Evidence from a number of studies demonstrates that physiological dependence is neither necessary nor sufficient for positive drug reinforcement.

Physiological dependence does affect the avoidance of, or escape from, injections of naloxone, an opioid antagonist. The lever pressing behavior of both dependent and non-dependent monkeys is maintained when the consequence of responding is postponement or termination of a naloxone injection. However, the dose ranges differ in the two groups of monkeys. In dependent monkeys substantially lower doses can function as negative reinforcers than in non-dependent monkeys[29,39]. Thus, under some conditions physiological dependence can increase the sensitivity to negative drug reinforcement[38].

Tolerance

Tolerance refers to decreased drug effects that follow repeated administration of the drug. Another definition is that with repeated drug use an increase in drug dose is necessary to obtain the original effects. Although tolerance is distinctly different from physiological dependence, both often occur together and both are consequences of repeated drug administration. Tolerance occurs with all abused drugs, and tolerance also occurs with many drugs that are not abused. Tolerance develops to multiple effects of a drug to different degrees. A commonly used example is that with methadone greater tolerance develops to respiratory depressant effects than to mitotic effects (i.e., constriction of the pupil). With abused drugs tolerance, in principle, could develop to all drug effects including disruptive effects, or tolerance might develop only to reinforcing effects, or only to all effects except reinforcing effects. Tolerance does develop in self-administration experiments, but the role of tolerance in drug reinforcement, as opposed to other drug actions, has not been well defined. However, in a study of intravenous cocaine self-administration with rats, specific evidence for tolerance to reinforcing effects has been demonstrated[30,31].

3.2 DRUG CLASSES

Most abused drugs are members of distinct pharmacological classes (see also Chapter 2). These classes are: psychomotor stimulants, arylcycohexylamines (dissociative anesthetics), opioids, CNS depressants, and nicotine. The CNS depressants can be subdivided into barbiturates, benzodiazepines, non-barbiturate sedative hypnotics, ethanol, gaseous anesthetics, and volatile solvents. Nicotine is the only member of the ganglionic stimulants that functions as a reinforcer. The drugs from these classes have a wide range of pharmacologic actions, and they have few effects in common apart from functioning as reinforcers and producing CNS effects.

Opioids

The consequences of 24–hour drug access vary with the different classes of self-administered drugs. With opioids such as morphine rhesus monkeys self-inject increasing amounts over an interval of 6–7 weeks and develop physiological dependence. There is little day to day variability, and intakes stabilize at 50–100 mg per kg. At night when the lights are off, self-administration is minimal[27]. Although there are initially some signs of drowsiness and apathy, the monkeys generally appear quite normal. The animals eat and sleep, and their health is not impaired although initially food intake decreases and weight loss occurs. Tolerance develops, and when access to the drug is terminated an abstinence syndrome occurs. No periods of self-imposed abstinence are seen[27].

The lack of impaired health and development of tolerance to the disruptive effects are consistent with the effects of chronic opioid intake in humans when the drugs are taken in a sterile manner. The minimal effects on health and functioning are part of the rationale for using methadone to treat addiction. Not all opioids are without effects on health. For example, unlimited access to codeine leads to deaths of monkeys; the cause is seizures[27]. In contrast to morphine, high intakes of codeine lead to increases in excitement and alertness. The toxic effects of codeine are not seen when drug access is limited to several hours per day. With both rhesus monkeys and rats opioids can serve as reinforcers when taken orally as well as intravenously. Other non-human primates also self-inject opioids.

Arylcyclohexylamines

Chronic intravenous self-administration of phencyclidine can result in continuous and profound intoxication[4,5]. When intoxicated, rhesus monkeys remain prostrate on the cage floor; periodically they get up and press the lever and then fall back on the floor. They also show salivation and nystagmus. This severe intoxication decreases food and water intake and results in weight loss, debilitation, and poor health[5]. At high doses (e.g., 0.05 mg per kg) drug intake is relatively stable, but at lower doses it is irregular. Chronic phencyclidine intake also produces tolerance, and termination of access results in an abstinence syndrome that can be immediately terminated by administration of phencyclidine[4]. At higher doses drug intake is relatively stable. It is important to recognize that this abstinence syndrome is qualitatively different from abstinence syndromes seen with opioids and CNS depressants. The abstinence syndrome can occur after high levels of drug intake (4–7 mg per kg per day) for 20–30 days. Signs of withdrawal begin within 8 hours after drug access is terminated. The monkeys have facial twitches, tremor, piloerection, and diarrhea. They refuse preferred foods, are hyper-responsive, and have oculomotor hyperactivity[5]. The abstinence syndrome is self-limited and decreases over 36 hours.

Orally delivered phencyclidine and its analogs are effective reinforcers of responding by rhesus monkeys[10,17]. Diminished alertness, excessive drooling, and impaired mobility are seen when the drug is taken by mouth. Tolerance and dependence can develop[11,13]. Rats will intravenously self-administer phencyclidine.

Psychomotor stimulants

Unlimited access to psychomotor stimulants such as cocaine and amphetamine results in marked toxicity in rhesus monkeys[27]. The rhesus monkeys show signs of psychomotor stimulation such as hyperactivity, agitation, and a marked decrease in food intake. Stereotyped movements are common. Choreiform movements are seen and stereotyped motor responses occur. Weight loss develops and grand mal seizures become frequent. Monkeys and rats will bite their extremities and even chew off digits. Drug intake is not stable from day to day, but rather occurs in bouts lasting a variable number of days. Similar cycles are seen with rats, cats, and dogs[5]. The cessation of drug intake is self-initiated; the monkeys sleep, and when they awaken they may eat. Within several days another drug–taking bout develops. Health is progressively impaired, and weight loss occurs; the monkeys die within 30 days[55]. This pattern of intravenous psychomotor drug intake is similar to that seen with humans[35,64].

Rhesus monkeys also self-administer cocaine and amphetamine orally. Moreover, cocaine base ("crack" cocaine) is an effective reinforcer delivered via the pulmonary route for rhesus monkeys[14]. With rats a cocaine withdrawal syndrome has been described[15]. When access to cocaine and amphetamine is limited (e.g, 6 hours per day), the toxic signs are not seen in any of the species. The animals remain in good health, and show stable orderly patterns of drug intake each session.

The toxicity of intravenous cocaine has been compared with that of heroin in rats when both drugs are continuously available[7]. Cocaine was substantially more toxic. Responding was erratic, seizures occurred, weight decreased, and the rats appeared debilitated. Mortality over 30 days was 90%. In contrast, rats self-administering heroin responded in a stable pattern, maintained their body weight, appeared healthy, and had a mortality rate of 36%. These findings are similar to those with non-human primates.

Nicotine

Rhesus monkeys, squirrel monkeys, dogs, and rats self-administer nicotine intravenously[24,42]. Under appropriate test conditions nicotine can serve as an effective reinforcer. Nicotine will maintain responding under intermittent reinforcement schedules, and under both limited and unlimited access. Its effects can be blocked by the antagonist mecamylamine[41]. Vomiting may occur at high doses[41]. Nicotine also can produce seizures[23]. Under certain conditions nicotine can also function as a punisher since its contingent presentation can suppress response rates[41]. That nicotine can function as a reinforcer supports the interpretation of tobacco use as an instance of drug reinforcement.

CNS depressants

The CNS depressants constitute a large group of agents that have many shared pharmacological effects. These drugs can be divided into: barbiturates, benzodiazepines, alcohols, non-barbiturate sedative-hypnotics, gaseous anesthetics, and volatile solvents.

– Barbiturates

Chronic intravenous intake of CNS depressants such as pentobarbital results in continuous intoxication and stable drug intake across days[27]. Ataxia, decreased arousal, and slowed and uncoordinated movements occur along with periods of anesthesia. However, tolerance develops to some of the disruptive effects, and the animals are able to maintain sufficient food and water intake such that health is not impaired. If there is 24-hour per day access, termination of drug self-administration results in a major abstinence syndrome that can be fatal. During withdrawal food intake ceases, and convulsions, tremor, and hyperarousal occur. When access is limited to a few hours each day, the animals remain in good health, and develop stable patterns of drug intake. These patterns depend upon the dose and the reinforcement schedule[96]. In rhesus monkeys, pentobarbital also can serve as a reinforcer when taken orally or intragastrically.

– Ethanol

Ethanol is an important CNS depressant because of its widespread use and abuse. Ethanol's actions are similar to those of other CNS depressants except for its effects on food consumption and self-initiated periods of abstinence. Rhesus monkeys intravenously self-inject ethanol in bouts that last for days, and during these bouts little food is consumed[97]. The monkeys display signs of intoxication such as marked incoordination and stupor. Remarkably, periods of abstinence are self-initiated. Following the cessation of ethanol intake, an abstinence syndrome develops that is similar to that seen with other CNS depressants. Tremor, vomiting, and seizures are typical. Rhesus monkeys usually will reinitiate ethanol self-administration following these periods of abstinence, and over time the bouts of self-administration tend to become longer. Food intake declines and the monkeys lose weight. Deaths can occur due to suffocation from respiratory obstruction[27]. Thus, under these conditions of continuous access, ethanol is more toxic than other CNS depressants.

An important difference between ethanol and other CNS depressants is that ethanol suppresses food intake, and this is likely due to its caloric value. Other sedative hypnotic agents do not have calories and do not severely suppress eating.

Ethanol can also serve as a reinforcer of responding by rats and monkeys when ethanol is delivered intragastrically or orally[62]. When ethanol is self-administered by the intragastric route, large doses are necessary for maintaining responding. The requirement of large doses probably reflects the absence of taste cues and the delay in effects.

– Benzodiazepines

Benzodiazepines can function as reinforcers when delivered intravenously to baboons, rhesus monkeys, and rats. Benzodiazepines have also been tested by the intragastric route[2] and oral intake has been studied in baboons and rhesus monkeys. Although the baboons consume substantial amounts[2], evidence for reinforcement is incomplete. However, orally delivered benzodiazepines can serve as reinforcers in rhesus monkeys[89]. The three drugs that have been studied the most with the oral route are midazolam, diazepam, and triazolam. When the drug is taken by mouth, marked behavioral changes consistent with intoxication can occur. These behavioral changes include sedative effects, a slumped posture, diminished attention, and ataxia. Unlike the barbiturates, anesthesia is not seen. Differences in actions among the benzodiazepines may relate to both speed of onset and duration of action. An abstinence syndrome can occur when drug access is terminated following intake of high amounts over a period of weeks. An abstinence syndrome also can be precipitated by administration of flumazenil, a specific antagonist[3].

– Gaseous anesthetics and volatile solvents

The gaseous anesthetics include such drugs as ether, chloroform, and nitrous oxide. The volatile solvents include toluene and lacquer thinner. Rhesus monkeys will self-administer some volatile anesthetics and solvents via a nasal catheter[105]. The nasal catheter is connected to a pump in the same general manner as an intravenous catheter; however, the catheter is inserted in the nasal cavity instead of into a vein. Lever pressing of rhesus monkeys is maintained by deliveries of air containing chloroform, ether, or nitrous oxide. When these agents are available 24 hours per day, lever pressing occurs primarily during the day. Rates of responding are erratic but are clearly higher than control rates. The monkeys show signs of intoxication such as ataxia, salivation, and mydriasis. During induction and recovery, extreme excitation is sometimes observed. Rhesus monkeys and squirrel monkeys press levers when the consequence is delivery of nitrous oxide[46,101].

Summary

Laboratory animals show patterns of intake that are similar to patterns seen with human drug abusers. The toxic consequences are also similar for humans and animals[5].

4 Schedules of reinforcement and the consequences of responding

Schedules of reinforcement

A schedule of reinforcement specifies the relationship between responses and consequences such as delivery of a food pellet or avoidance of shock. Schedules can be classified on the basis of the number of responses, termed ratio schedules, or on the basis of elapsed time, termed interval schedules. The two types also can be combined in multiple ways. Schedules are important because they are a major determinant of the operant behavior of humans and laboratory animals.

Ratio schedules

The schedule most commonly used in drug self-administration studies is the fixed-ratio (FR) schedule. With FR schedules the delivery of the reinforcer is contingent upon emitting a fixed number of responses. FR schedules generate high persistent rates of responding with pauses that follow reinforcer delivery. Under ratio schedules the rate of receiving drug deliveries is directly related to the rate of responding. Under ratio schedules animals can rapidly take large amounts of drug. Drug reinforcers are effective in maintaining high rates of responding under intermittent reinforcement schedules.

Certain relationships between fixed-ratio size and drug dose have been repeatedly observed. When the FR size is progressively increased, response rate first increases and then decreases. The number of drug deliveries also changes with changes in FR size. Initially the number remains constant, but with further increases in FR size the number of deliveries systematically decline.

Interval schedules

Under interval schedules, the delivery of a reinforcer is contingent upon an elapsed time interval. A commonly used schedule is the fixed-interval (FI) schedule. With a 30-sec FI, the first response emitted after 30 sec results in the delivery of a reinforcer. Responses occur in a characteristic temporal pattern of a scallop or positive accelerating curve. Rates are low at the beginning of the interval and they rapidly increase in rate toward the end of the interval. Interval schedules differ from ratio schedules in that there is a minimum time between drug deliveries. The number of drug deliveries can vary independently of the response rate if the response rate is above a certain minimum. An upper limit on the rate of drug intake can be specified by appropriate selection of the interval of the schedule.

Second-order schedules

Second order schedules are schedules of schedules. One of the schedules functions as though it were a single response. For example, an FR schedule of 8 responses can be treated as though it were a single response that is reinforced under a FI schedule of 10 minutes. Within the 10 min period prior to elapse of the interval the completion of each set of 8 responses results in a brief light flash, and the completion of 8 responses after the end of the 10 minute interval produces both the light flash and the drug delivery. The groups of FR 8 responses occur during the 10 minute interval in the same temporal pattern as would individual responses during a simple FI schedule. The brief light flashes function as conditioned reinforcers. The use of a second order schedule elevates response rates substantially above what would occur during a simple 10-minute FI. The successful use of second order schedules demonstrates that complex behavioral patterns can be reinforced and maintained by intermittent drug deliveries. The behavior of humans is more similar to that generated by these complex schedules than it is to behavior seen with simple schedules.

Schedule-induced polydipsia

Schedule-induced polydipsia is the consumption of exceptionally large volumes of water and other liquids that occurs when food-restricted rats and monkeys intermittently receive food pellets. Over a period of 3 hours rats will drink amounts of water that are one half of their body weight[32]. When drug solutions replaced the water, substantial amounts of drug were consumed, and the ingestion of these amounts produced changes in food reinforced lever pressing and motor behavior. Such changes reflect the pharmacological actions of the ingested drug. There has not been a generally accepted explanation of schedule-induced polydipsia. However, it is an effective and reliable way to engender consumption of large quantities of drug. Schedule-induced polydipsia also can be used to establish drugs as reinforcers. With some drugs when the inducing schedule has been discontinued, drug taking persists at levels substantially greater than water values. The full implications of schedule-induced drug consumption remain to be determined (for review, see ref. 33).

Extinction

Extinction occurs when previously reinforced behavior is no longer reinforced. The usual pattern of extinction is a high rate of responding at the beginning followed by decreases to rates substantially below those that occurred prior to extinction. Intermittently reinforced behavior is persistent and thus difficult to extinguish. With humans one difficulty in decreasing drug taking is that the behavior is intermittently reinforced and persists for long periods in the absence of reinforcement. Extinction is not due to loss of memory since the effects of extinction on responding are easily reversed by reintroducing the reinforcer.

Importance of schedules

Schedules of reinforcement frequently have been studied due to their importance as determinants of drug use. Studies of schedules have led to the development of major generalizations regarding drug reinforced responding. One generalization is that: when different drugs are studied under a single schedule, the rate and pattern of responding are similar despite the differences among the drugs[38]. For example, under FR schedules when drugs such as cocaine, pentobarbital, methadone, and phencyclidine are serving as reinforcers, rates and patterns of responding are similar. Importantly, these findings are not limited to drug reinforcers for other reinforcers such as food, water, and electrical brain self-stimulation maintain characteristic FR response patterns. Although under a single schedule, similar response patterns may occur with different drugs; the effects on gross motor behavior may differ markedly depending on the drug studied.

A related generalization is that: when one drug is studied, the rate and pattern of responding depend upon the schedule of reinforcement. For example, under FI and FR schedules, the rate and pattern of cocaine reinforced responding depend on the schedule of reinforcement. This generalization is not limited to drug reinforcers, for with other reinforcers such as food, the response rates, and their temporal distribution conform to the reinforcement schedule studied.

A third generalization is that: the schedule of reinforcement interacts with the type and amount of drug consumed. Drugs that act as reinforcers also have other effects which are called direct or generalized effects. For example, the rapid intake of a benzodiazepine may not only reinforce drug taking but it may also impair motor behavior. If the same amount of the benzodiazepine were administered to a monkey responding under a schedule of food reinforcement, the food reinforced responding would be altered. Thus, the rate and temporal distribution of drug reinforced responding that occurs under a reinforcement schedule are altered by the amount and rate of drug consumption, as well as by drug dose and schedule of reinforcement.

These direct or generalized drug effects can be eliminated by the strategy of delivering the drug reinforcer only at the end of the experimental session. When this is done, the responding prior to drug delivery is maintained by the reinforcing effects without the presence of direct disruptive actions[40,59]. Another strategy to minimize direct drug effects is to impose long delays between opportunities to self-administration the drug[44] or to use second order schedules. The absence of direct drug effects results in performance that is controlled by drug reinforcement without being modified by other drug actions (e.g., impairment of motor behavior).

Dose

The drug dose is the amount of drug obtained after each completion of the response requirement. Generally at low drug doses, the rate of responding increases with increases in drug dose. At intermediate doses, responding is at its highest rate, and with further increases in dose the rate of responding declines. Several explanations for these decreases have been proposed, but a consensus has not been reached. This function of first increasing, followed by decreasing responding is termed an inverted U-shaped or bitonic function. A related function is the amount of drug consumed during the session[92]. The amount consumed is plotted as a function of dose, and it is usually expressed in terms of mg of drug per kilogram of body weight per session. This function is usually linear, for session intake increases with the dose. At very high doses, intake may level off or decrease.

The shape of the dose response function depends upon a number of factors. For example, if tolerance develops, the curve may shift to the right. In the tolerant animal more drug is required to maintain responding. Alternatively, the function may shift to the left if sensitization occurs.

Another determinant of this function is the size of the intermittent reinforcement schedule. As the size of the schedule is increased, the function shifts to the right. This shift can be explained if the magnitude of reinforcing effects goes up with the size of the dose. For instance, at higher schedule values very low doses no longer maintain responding, and progressively larger doses are required to maintain responding.

Interaction between schedule size and dose

In most self-administration experiments only a single variable is studied. This is usually the case in the early phases of a research area. However, every variable acts within a context, that is, within conditions that affect the results of the variable being studied. For example, fixed-ratio size can be studied by systematically increasing and then decreasing the fixed-ratio requirement. The results obtained are usually orderly and comparable to findings in other studies where FR size was manipulated. In these studies all other variables are held constant so the results are not confounded by varying more than one variable at a time. However, the effects of FR size can be shown to depend on another variable, namely the drug dose. This can be demonstrated by systematically varying both FR size and drug dose.

Example

The intravenous self-administration of methohexital was analyzed in a study where both drug dose and schedule size were varied (see Table 3.3; ref. 80). A general finding in many studies is that when FR size is increased, the response rate increases and then declines at high FR values. The number of drug deliveries may remain constant despite increases in FR value. However, further increases in FR size result in a systematic decrement in the number of drug deliveries. When FR sizes were studied at different drug doses an interesting outcome was identified. Rather than express the data in terms of response rate and number of injections, the data was reanalyzed and plotted in terms of percent changes from baseline[65].

At FR 1 four doses were studied: 4, 3, 2, and 1 mg per kg, and at these doses the rat 1A obtained 42, 58.5, 68.5, and 107.5 injections of methohexital, respectively. Thus, the number of injections decreased as dose was increased. This is a general finding that has been reported many times. When the FR size was increased from FR 1 to FR 5, the number of injections decreased. However, the decrease was not the same across doses. The largest decrease was at the lowest dose, 1 mg per kg, where mean injections decreased from 107.5 to 52.5, and the decrease was least at the highest dose where mean injections decreased by 1 from a mean of 42 to 41. Since the number of injections was different at each dose, the findings can be converted to percent by dividing the value at FR 5 by the number of injections received at FR 1. At the lowest dose there was a decline of 51% when the FR was increased from 1 to 5. In contrast at the 4 mg per kg dose, the decline was only 2%. Behavior persisted most at the highest dose and progressively less as dose was decreased (See Table 3.3). Tests over a range of doses and schedule sizes confirmed the analysis that persistence was directly related to the dose size.

TABLE 3.3

Percent of baseline methohexital injections (FR1) as a function of drug dose and FR size.

Results of rat no. 1A are explained on page 82. (Data from Pickens, R. et al., *J. Pharmacol. Exp. Ther.* 1981, **216:** pp 205–209.)

Fixed Ratio Schedule	*1 mg per kg*	*2 mg per kg*	*3 mg per kg*	*4 mg per kg*
Rat 3A				
FR1		61		26
FR5		42/61 (69%)		23/26 (88%)
FR10		24.5/61 (40%)		23.5/26 (90%)
FR20		9.5/61 (16%)		17/26 (65%)
Rat 6A				
FR1	78	48	41	29
FR5	52/78 (67%)	37.5/58 (78%)	30/41 (73%)	39.5/29 (136%)
FR10	39.5/78 (51%)	40.5/48 (84%)	32/41 (78%)	39.5/29 (136%)
FR15	5/78 (6%)	12.5/48 (26%)	20.5/41 (50%)	30.5/29 (105%)
FR20	1/78 (1%)	1.5/48 (3%)	9.5/41 (23%)	23.5/29 (81%)
Rat 1A				
FR1	107.5	68.5	58.5	42
FR5	52.5/107.5 (49%)	45.5/68.5 (66%)	36/58.5 (62%)	41/42 (98%)
FR10	16/107.5 (15%)	42.5/68.5 (62%)	31/58.5 (53%)	37/42 (88%)
FR15	2.5/107.5 (2%)	8.5/68.5 (12%)	21.5/58.5 (37%)	33/42 (79%)
FR20	1.5/107.5 (1%)	5/68.5 (7%)	6.5/58.5 (11%)	20/42 (48%)
Rat 7A				
FR1		49		22
FR5		35/49 (71%)		23/22 (105%)
FR10		17.5/49 (36%)		20.5/22 (93%)
FR20		7.5/49 (15%)		17/22 (77%)

The term "persistence" refers to the number of drug deliveries obtained at higher schedule values relative to the number obtained at lower schedule values. Persistence of drug reinforced responding has been studied with pentobarbital as the drug, oral intake as the route, and rhesus monkeys as subjects. The findings were consistent with those obtained in the rat-methohexital study, and therefore are a systematic replication of the earlier findings but one employing a different route, species, and drug[65].

Punishment

Punishment produces decreases in response rate due to either the response contingent presentation of an aversive stimulus (e.g., electric shock) or the response contingent withdrawal or delay of positive reinforcer (e.g. food). Drug reinforced behavior, like other operant behavior, can be suppressed by punishment. For example, when electric shocks followed each cocaine-reinforced response, monkeys' response rates decreased, and the magnitude of the decrease was directly related to the intensity of the shock[45].

Punishment is more effective when there are alternatives to the punished response[51]. Monkeys had the option of responding on two levers: presses on one lever resulted in the delivery of cocaine whereas presses on other lever resulted in the delivery of both cocaine and electric shock[51]. When the cocaine doses were equal, the monkeys responded on the lever delivering cocaine alone. However, the effects of punishment could be surmounted by increasing the cocaine dose that was delivered along with the shock. These results illustrate that the effects of punishment depend on several factors such as the magnitude of the punishing stimulus and the magnitude of the reinforcer maintaining the punished response.

Example

Punishment can occur when responses on one lever alter the availability of reinforcers on a second lever. The common feature of punishment procedures is that they reduce the rate of a reinforced behavior. Effective punishment occurs not only by presentation of a noxious stimulus, but also by the postponement or loss of a positive reinforcer. Punishment procedures of this type were examined in several related studies of ethanol reinforced drinking. Reinforcement by food delivery was arranged for rats under a 26 sec FI schedule. On a second lever each press raised a dipper cup that contained in different blocks of sessions either 0 (water), 4, 8, 16, or 32% (v/v) ethanol. A stable baseline of behavior was obtained, and then the contingencies were changed such that each ethanol reinforced response also had as a consequence the delaying of food availability for 8 sec. This contingency suppressed ethanol intake. However, no suppression occurred in a yoked control group which also received food pellets. These pellets were delivered automatically to the yoked rat at each point in time that the experimental rat obtained a pellet. Thus, it was the specific contingency of delaying food access that was responsible for the suppression of the experimental rats' ethanol reinforced responding[82].

---

5 Factors controlling drug reinforced behavior

Genetics

Chapter 4 reviews studies of genetic factors in drug abuse. Genotype is an important factor in drug, including ethanol, reinforced behavior. There are marked differences among rodent strains in their intake of ethanol and other drugs. The analysis of such differences will contribute to an understanding of mechanisms of drug actions.

Experimental history

The organism's history can affect its current drug taking behavior. Such history effects can be detected by comparing an organism's current performance with its past performance. As experience with a particular drug develops, responding may be maintained at higher rates over a broader set of test conditions and with less variability than an inexperienced subject[40]. The type of drug originally maintaining responding can influence the probability of a successful substitution. With monkeys that self-administered ketamine, dextrorphan and dexoxadrol successfully substituted for ketamine. However, with monkeys that self-administered codeine, the results were negative. Dextrorphan and dexoxadrol did not maintain responding when subtituted for codeine[107]. The different outcomes were ascribed to the similarity of interoceptive effects among ketamine, dextrorphan, and dexoxadrol, and to the dissimilarity of these interoceptive effects with those of codeine. In another study *d*-amphetamine was substituted for either cocaine or pentobarbital. The monkeys with the history of cocaine reinforcement exhibited less variability and responded across a wider range of doses and at higher rates than monkeys with a pentobarbital history[83]. In general if a drug is substituted for another drug, better results are obtained when both drugs are from the same pharmacological class than when they are from different classes.

Current drug taking can also be altered by past experiences that do not involve different pharmacological histories. For example, the conditions used during the establishment of reinforcing effects can later influence behavior during the maintenance phase. Orally delivered phencyclidine was established as a reinforcer under two experimental conditions. One group of monkeys was trained when food restricted, and another group was trained when food satiated. Subsequently both groups were tested under food satiated conditions; responding persisted only in the group that was food satiated during the initial training[12].

Example

Previously established drug reinforced behavior can be altered by subsequent training under different reinforcement schedules (Nader & Reboussin, 1994). Initially all monkeys responded under a 60 sec FI schedule. Subsequently one group of monkeys was tested under a FR 50 schedule for 60 sessions, and the other group was tested under a schedule that reinforced only responses separated from previous response by 30-s interval of no responding. Such schedules have been termed Differential Reinforcement of Low rates (DRL schedules, or an alternative name, IRT schedules, Inter-Response Time schedules). In the final phase all monkeys again were tested under the FI 5-min schedule, and their performances were compared. The group that had responded under the FR schedule displayed significantly higher rates of responding upon retest with the FI schedule than the group tested under the IRT schedule. Moreover, the two groups differed when a dose-response function was obtained[75]. Clearly, the past experiences of a subject can be a determinant of subsequent drug taking.

Food deprivation

Food deprivation increases drug intake and drug reinforced responding in laboratory animals[18]. This topic is covered in depth in Chapter 10. It is mentioned here for sake of completeness and to illustrate the importance of studying interactions between variables. Increased drug intake due to food deprivation is not similar across doses. Little or no increase occurs at high doses whereas the increase becomes greater as one reduces the dose until a point is reached where the dose becomes too small to support behavior[60,67]. Lack of attention to interactions between variables can lead to what appear to be unsupported findings. If only low doses were tested at one laboratory and only high doses were tested at another laboratory, the two laboratories would report large effects in the first case and no effect in the second case.

6 Stimulus control

Drug taking behavior can come under the control of stimuli associated with drug availability. For example, in one study illumination of a white light indicated that a response would be reinforced, and lack of illumination indicated that a response would not be reinforced. Stimulus control of responding rapidly developed; the monkey seldom responded when the light was off, but during the first 30 min of the session promptly responded when the light was illuminated. Technically the illuminated light functioned as a discriminative stimulus, that is a stimulus that is correlated with reinforced responding, and the absence of illumination also served as a stimulus in that it was correlated with absence of reinforcement[72]. The importance of discriminative stimulus control is that it is one of the major determinants of drug-reinforced behavior and is also a contributing factor in relapse to drug taking.

Cocaine abusers report that seeing a pile of white powder elicits a desire to self-administer the drug. In the animal laboratory there are similar processes. In one study lever pressing by rats produced small volumes of an 8% ethanol solution or water. The ethanol solution was present every third day and on the intervening days water was available. The number of lever presses required for each delivery of ethanol or water was gradually increased across blocks of sessions. When the requirement was 32 lever presses, the rats did not emit enough responses to obtain even one delivery of water. Nevertheless, during sessions when ethanol was present response rates were high and many deliveries of the ethanol solution occurred. How did the rats discriminate ethanol sessions from water sessions? The answer is that the odor of ethanol came to serve as a discriminative stimulus for ethanol reinforced responding[74]. Other aspects of a drug can become important in maintaining drug taking. For example, in oral drug self-administration studies with experienced subjects, the taste of the drug solution comes to function as both a discriminative stimulus and as a conditioned reinforcer, thus aids in bridging the delay between drug drinking and subsequent CNS effects.

Reinstatement

Reinstatement is the resumption of extinguished drug-reinforced behavior that occurs following a non-contingent (i.e. experimenter administered) dose of the reinforcing drug. In one study monkeys self-administered *d*-amphetamine under a progressive ratio schedule[90]. Replacement of the drug by saline resulted in the low rates of responding that are characteristic of extinction. However, responding was reinstated when the experimenter gave the monkey an intramuscular injection of 1.5 mg *d*-amphetamine per kg even though saline, not drug, was being injected intravenously. The pattern of responding was not random, and instead was like that of earlier drug reinforced responding. The latter point is important for an alternative explanation of the increased rates is that they are due to a general increase in activity that could follow an amphetamine injection.

Drugs from the same pharmacological class as the reinforcing drug often can reinstate responding. In some cases a reinforcing drug from another pharmacological class can also reinstate responding. Across sessions the reinstated responding subsequently extinguishes since responding is followed by saline injections. The reinstatement of behavior is thought to be due to the restoration of internal stimulus conditions that are present when drug reinforcement has occurred in the past. Reinstatement can occur with drugs from a number of pharmacological classes and is seen with both rats and monkeys.

Most drug reinforced behavior occurs under the influence of internal drug produced stimuli. These internal stimuli are present due to earlier ingestion of the drug. During periods of drug taking it is only the initial responding that occurs in the absence of internal drug stimuli. Thus, when this component of the general stimulus complex is reinstated along with other cues such as operation of the infusion pump, responding resumes. The experimental study of reinstatement has been extended to studies of drugs that block reinstatement[22]. Such studies could lead to the identification of potential therapeutic drugs.

There is good experimental support for the stimulus restoration explanation of reinstated responding. However, one unexplainded aspect is that under some conditions a drug from a different drug class can reinstate responding. There is also an alternative explanation for reinstated responding, and this is that the resumption of responding is caused by non-specific behavioral activation. However, the pattern of responding is like the pattern typically observed under the reinforcement schedule and not like the pattern one could attribute to a general increase in activity.

7 Complex behavior: choice between reinforcers

Early studies focused on variables such as drug dose and reinforcement schedule. Subsequent studies have examined more complex behavior. One example is the use of procedures that provide access to more than one reinforcer. These procedures include the study of drug taking when there is a choice between reinforcers such as different doses of a drug, doses of different drugs, or a drug and a non-drug reinforcer. One of two designs is used: either concurrent reinforcement schedules or discrete trial procedures. With discrete trials there are repeated opportunities to select one of the two reinforcers but usually not both at the same time. Concurrent schedules are two or more schedules that are present at the same time. With concurrent schedules, the completion of the response requirement on one schedule usually does not alter the requirement for reinforcement on the other schedule. Concurrent schedules and discrete trial procedures lie on a continuum, and some experimental arrangements have features of both procedures.

Procedures that provide multiple response options have an additional dependent variable, that being the relative rates of responding or a related measure, proportion of reinforcers obtained relative to the total number delivered. The phrase "response rates" is defined in terms of the number of responses per unit of time, and is different from the phrase "relative response rates" which is the proportion of responses emitted on one lever to all the lever presses made during a session.

Access to a drug and its vehicle

A basic comparison is that of a drug and its vehicle. This can be done concurrently when there is the option of receiving either the drug or the drug vehicle. This is an efficient and powerful method for identifying reinforcing effects. When a drug dose is serving as a reinforcer, the drug should maintain higher response rates or more frequent selection than the drug vehicle. The concurrent design is more efficient than a sequential design (testing a drug, the vehicle, and then retesting the drug) since it does not require a separate block of sessions for the vehicle and another block for retesting the drug. Moreover, tests conducted with access to both a drug and its vehicle are made under identical conditions. In contrast, tests done sequentially may be less rigorous due to uncontrolled factors that change over time, for instance changes in temperature, extraneous sounds, or handling. The concurrent design also can be made more rigorous by alternating the positions of drug and vehicle.

Access to two doses of a drug

Different doses of the same drug may be compared to determine which is most reinforcing[69]. Knowledge of the relative reinforcing effects is fundamental to designing studies and correctly interpreting results. When asked what dose is the most reinforcing, the typical answer is that it is that dose that maintains the highest response rates. Since most dose-response curves are shaped like an inverted U, the dose that maintains the highest rate is usually a dose in the middle of the dose range. A different answer comes from studies that have examined the relationships among doses by use of a choice paradigm; these studies show that relative reinforcing effects increase as the dose becomes larger[71]. The initial studies were done in the 1970s. Subjects were rhesus monkeys, and the reinforcer was i.v. cocaine. In one series of studies concurrent variable interval schedules were used[48], and in another series discrete trials were employed[56]. The findings were consistent: with concurrent schedules larger doses maintained higher relative response rates than lower doses, and with discrete trails the higher doses comprised a much larger proportion of the total deliveries than lower doses. However, when both doses were large, differences in proportions or response rates did not emerge. This may be due to limitations of the testing methods or it may reflect a ceiling effect on reinforcing effects.

Access to two drugs

The methods used to study the relative reinforcing effects of different doses of one drug also can be used with two drugs. For example, the reinforcing effects of intravenous cocaine and procaine have been compared[53]. This comparison is important because some local anesthetics can serve as reinforcers, but abuse of local anesthetics is uncommon. The procedure consisted of repeated trials that involved a mutually exclusive choice between the two drugs. Pairs of doses of the two drugs were studied. Cocaine was strongly preferred even when the dose of procaine was 16 times the dose of cocaine. This is strong evidence for the greater reinforcing effects of cocaine.

Response rates maintained by local anesthetics have been measured when cocaine is not present. These drugs can maintain high response rates that sometimes exceed those maintained by cocaine[52]. However, these high rates may not indicate strong reinforcement[50]. For example, high rates of responding often can be measured at low drug doses. However, when there is a choice between a higher and a lower dose, the higher dose is consistently selected, even though when each dose is tested alone, the lower dose maintains higher response rates.

Drugs have been tested that are stronger reinforcers than the local anesthetics. Choice between cocaine and cathinone has been studied[103]. Cathinone is a psychomotor stimulant with effects similar to cocaine. Different doses of each drug were studied using discrete trials and mutually exclusive choices. The results were that the two drugs were equally effective. This suggests that cathinone has strong reinforcing effects, and therefore may have a high abuse potential. The direct comparison of two drugs with discrete trials or concurrent schedules is a potentially strong design for evaluating the relative reinforcing effects of different drugs.

Access to a combination of drugs

In drug abuse treatment programs most individuals report abuse of more than one drug at a time. However, only a few studies have examined the self-administration of drug combinations[70]. In a preliminary experiment with one monkey the self-administration of a combination of *d*-amphetamine (0.01 mg per kg) and pentobarbital (0.25 mg per kg) was compared to the rates maintained by each drug alone[83]. The outcome was that the number of injections of the combination was twice that measured when pentobarbital or *d*-amphetamine was tested alone.

Intravenous self-administration of ethanol-pentobarbital combinations was examined in rats[28]. Pentobarbital without ethanol functioned as a reinforcer, whereas ethanol did not maintain lever pressing. However, when low non-reinforcing doses of pentobarbital were combined with ethanol, responding was maintained. This outcome is an example of synergism. Synergism occurs when effects of the two drugs are greater than would be expected by a simple linear combination of effects. Combinations of drugs do not always result in an increase in reinforcing effects. When morphine was mixed with the antagonist nalorphine, no monkey initiated self-administration although all monkeys had experience self-administering morphine[27].

Access to a drug reinforcer and a non-drug reinforcer

Many studies concern the effects of a second reinforcer on drug reinforced behavior. This second reinforcer can be another dose of the same drug, another drug, or another reinforcer.

Studies with non-human primates

The effects of a second reinforcer depend upon a number of factors including the magnitudes of the drug and non-drug reinforcer and on scheduled relations between the reinforcers.

Example

One example is a study with baboons where once every 3 hours there was a mutually exclusive choice between heroin and food. Stable behavior consisted of heroin selections 1 to 3 times each 24 hours and food on the other trials. An infusion of naloxone led to more frequent heroin injections; in contrast pretreatment with methadone resulted in an increase in food deliveries[43,104]*. An important feature of this study was that by having behavior maintained by two reinforcers, non-specific effects on responding could be evaluated, for such effects should have an equal impact on both reinforced behaviors.*

Choice between food and cocaine has been examined in several studies. In one experiment rhesus monkeys had a mutually exclusive option of cocaine or food every 15 minutes during continuous 24-hour sessions. The result was a consistent selection of cocaine over food. Cocaine intake was high and food intake was low; body weight decreased and health was impaired[1].

Subsequent studies have demonstrated that under certain conditions the proportion of food deliveries can exceed those of cocaine. In these studies discrete trial procedures were used, and the session length was 7 hours or less. Increases in cocaine dose led to increases in the proportion of cocaine deliveries whereas increases in the size of the food deliveries (i.e., food dose) resulted in a decrease in the percent of cocaine deliveries. In other words when the magnitude of one reinforcer was increased, the proportion of deliveries of that reinforcer increased, hence, the proportion due to the other reinforcer decreased. Another factor is the size of the schedule of reinforcement. When the FR size of the cocaine schedule was increased, the proportion of cocaine injections decreased, thus the proportion of food deliveries increased. Similarly when the FR size of the food schedule was increased, the proportion of food deliveries declined and thus the proportion of cocaine injections rose[76].

These findings show that drug reinforced behavior changes in an orderly way when there are changes in access to another reinforcer, and that the changes depend upon the quantitative values of the other reinforcer.

In addition to providing access to a second reinforcer, contingencies between deliveries of the two reinforcers can be arranged. One contingency is whether or not the choice between reinforcers is mutually exclusive. An experiment was conducted with oral pentobarbital and saccharin as the two reinforcers. The reinforcers were concurrently available with a mutually exclusive choice every 30 sec for 3 hours. When the saccharin concentration was increased, the number of saccharin deliveries increased and the number of pentobarbital deliveries decreased[66]. When the choice was no longer mutually exclusive, the suppressive effect on pentobarbital intake was removed and drug consumption increased.

The interactions between a drug reinforcer and a non-drug reinforcer can be complex (see Chapter 10 for further details). In general, as the schedule requirement for one reinforcer is increased, intake of that reinforcer is decreased and intake of the other reinforcer is increased. The outcomes of such interactions depend upon the magnitude of the reinforcers, the size of the intermittent schedules, and whether or not choice is mutually exclusive.

Studies with rats

Studies in the 1960s and 1970s showed that after a rat self-administers a dose of a psychomotor stimulant or opioid drug, there is a pause in responding. The explanation was that the i.v. delivery of the drug produced a disruption of behavior and thus a pause in responding. This explanation was investigated in studies where rats could self-administer a drug and also electrically self-stimulate the brain. Contrary to the customary explanation, the lever pressing maintained by electrical stimulation was not disrupted by i.v. drug injections. Responding occurred during the interval between drug injections. These findings rule out some explanations of pausing following a drug injection. The persistence of lever pressing maintained by electrical brain stimulation is incompatible with explanations in terms of drug-induced disruptions or debilitation of all operant behavior[36,99].

8 Behavioral measures indirectly related to drug self-administration

Several experimental procedures are closely associated with drug reinforcement, but they measure the reinforcing (or aversive) effects of drugs only in an indirect manner. These measures include the conditioned place preference (CPP) procedure, the conditioned taste aversion (CTA) procedure, and drug-induced facilitation of intracranial electrical self-stimulation (ICSS). For these procedures the subjects (usually mice or rats) passively receive injections of the drugs, that is, the drugs are administered by the experimenter and are not self-administered by the subject. Since there is no response contingency, reinforcement, as defined above, cannot be occurring with these procedures. However, there are theoretical reasons that suggest that these procedures may be measuring processes that are closely linked or associated with drug reinforcement.

Conditioned place preference

This procedure involves exposing an animal to the effects of a drug while in a novel or distinctive environment. Subsequent preference for that environment in the absence of the drug is indicative that the animal has learned to associate the positive aspects of the drug with the location in which it was experienced. For a typical experiment, a rat may be given a drug injection and placed in one of the compartments of a two-compartment test apparatus for 30 minutes. On a different occasion, the same rat will receive a drug-vehicle treatment and be placed in the other compartment for 30 minutes. One of the compartments may be black with a smooth floor and the other may be white with a rough textured floor. Thus, the effects of the drug are paired with one set of environmental cues and the effects of drug vehicle are paired with a different set of cues. After a number of such pairings a choice test is given in which the rat is not injected with drug or vehicle and is placed in the apparatus with the partition between the black and white compartment removed. The rat can move freely between the drug- and the vehicle-paired environments. If the rat shows a preference for the drug-paired environment it is inferred that the rat does so because the drug-paired environment is functioning as conditioned or secondary reinforcer. Whatever the theoretical basis for place preference conditioning, the procedure is widely used and a close correspondence has been found between the ability of drugs to function as positive reinforcers in a standard self-administration paradigm and to produce conditioned place preferences (for review, see ref. 9).

Conditioned taste aversion
The aversive effects of drugs have been studied in rats using the conditioned taste aversion (CTA) procedure in which the consumption of a drug-free fluid with a novel taste is associated with a drug treatment, and aversion to the drug's effects are inferred from the animal's subsequent avoidance of that flavor. This procedure is conceptually similar to the place preference test described above. One theoretical problem with the CTA procedure is that most of the drugs that produce pharmacological CNS effects will produce CTAs even when doses are tested that are in the range that are self-administered. For example, the same dose of amphetamine can function as a reinforcer in a drug self-administration study and can produce a conditioned aversion for a flavor with which it has been paired. This may be because animals such as rats are genetically predisposed to associate aversive effects of ingested substances with tastes (for review, see ref. 8).

Facilitation of intracranial electrical self-stimulation (ICSS)
A theory of the biological or physiological basis of drug reinforcement has evolved in which the central tenet is that drugs are self-administered if they affect a specific reinforcement system in the brain. This theory stems from the discovery by Olds and Milner (1954)[77] that rats will readily make a response such as pressing down a lever to deliver a small electric current to the tip of an electrode, if the electrode has been surgically implanted in specific regions of the brain. The neural structures related to intracranial self-stimulation or brain stimulation reward may represent a specialized system responsible for the process of reinforcement (see Chapter 19)[78]. It is postulated that drugs of abuse come to function as reinforcers by imitating, facilitating, or sometimes blocking the various neurochemical messengers and neural modulators involved in this system[98,61]. The strongest experimental evidence for this proposition comes from the demonstration that drugs of abuse including psychomotor stimulants (e.g., amphetamine and cocaine) and opiates (e.g., morphine and heroin) facilitate brain stimulation reward (for review, see ref. 63), that is, the administration of the drugs either increases the rate at which the animal self-stimulates and/or lowers the threshold for ICSS such that a smaller amount of electrical current is required to sustain the same self-stimulation behavior. Thus, drug-induced facilitation of intracranial electrical self-stimulation is closely associated with the reinforcing effects of drugs but does not constitute a direct measure of drug reinforcement.

9 Applications and issues

Applications of drug reinforcement procedures
Drug self-administration procedures are used both in basic research and in several applied areas[57]. One application is the screening of new drugs for potential abuse potential. A second applied area is the study of behavioral procedures and environmental conditions that may reduce drug-reinforced behavior. For example, if rats have access to a highly preferred liquid (such as a solution containing both glucose and saccharin) the development of cocaine self-administration is decreased[16]. A third area is the testing of potential treatment drugs that may decrease drug self-administration. For example, methadone selectively decreases heroin self-administration without eliminating food reinforced responding[43]. Currently there is great interest in testing drugs that might eliminate cocaine self-administration (see Chapter 9). These applied areas are the focus of much ongoing research.

Limits of the current research

The major limitation of self-administration studies is that with laboratory animals only a small number of variables have been studied and interactions among the variables are seldom analyzed. Undoubtedly, however, there are many variables that have not been studied. Factors such as social interactions are very likely to be important in humans' drug taking, but little research has been done in this area with laboratory animals (for an exception see ref. 25). More of the factors that control drug taking need to identified and studied under controlled laboratory conditions.

A limitation that is seldom stated is that our knowledge of factors controlling drug taking in humans is very limited. Drug taking by humans is so familiar that people do not recognize that little is known. There are few laboratory studies with human subjects relative to number of the animal drug self-administration studies (see Chapter 7). Progress in laboratory animal studies will partly depend on progress with human laboratory investigations. To determine the correspondence between the human and animal drug taking, it will be necessary to have comparable studies in both areas.

Drug reinforced behavior is complex for it is an orderly result of many variables and their interactions. Our knowledge is limited, but it is not tentative. Specific statements of experimental findings are frequently qualified by phrases such as "under these conditions", or "over a wide range of values". These qualifications are necessary because of the complex behavior of drug taking. The remarkable finding is that drug taking is exceptionally orderly: the functions relating behavior to values of an independent variable are often simple, quantitative, and direct. The circumstances under which it is necessary to qualify our statements are not capricious, for these circumstances also bear an orderly relation to other controlling variables.

Whole organism research

Drugs of abuse can change and maintain behavior since they function as reinforcers. These drugs also alter the functioning of the central nervous system. For example, abused opioid drugs have a common feature, they are agonists at the μ receptor. These common effects in the nervous system bring up the question of studying neurochemistry and neurophysiology rather than studying behavior. Is it necessary to study drug action at the behavioral level, or might it be possible to simplify matters and just study drug action at the receptor? A complete scholarly answer is not possible to what deceptively appears to be a simple question. Briefly, the answer is that whole organism research is required; studying drug-receptors interactions is not sufficient. Results from one study can be used to illustrate this point.

Example

An experiment was conducted to separate the effects of self-administration (contingent drug deliver) from passively receiving an injection (non-contingent drug delivery). Each time an experimental rat obtained an injection, a litter mate yoked-control animal also received an injection[87]. Immediately after a drug self-administration session, all rats were sacrificed and the rates of neurotransmitter turnover were measured in both groups of rats. The self-administration of morphine resulted in neurotransmitter changes that markedly differ from those obtained with the yoked-control rats. The self-administration rats had neurotransmitter utilization rates that were greater in both magnitude and number of neurotransmitters affected.

The important point is that both the behavioral and neurochemical effects of drugs are altered as a function of whether a drug is self-administered or passively received. Drug reinforcement is a behavioral phenomenon, and an analysis at the behavioral level is necessary. Such a statement does not preclude important contributions from other levels of analysis.

10 Summary

Drugs can serve as reinforcers for laboratory animals. The drugs that do function as reinforcers are generally the drugs that humans abuse, and the drugs that do not serve as reinforcers are those that humans do not abuse. Drugs function as reinforcers when delivered via different routes of administration in a range of species from rodents to primates. Many variables (e.g., drug dose, schedule of reinforcement) affect drug self-administration. Importantly, these variables have similar effects on the behavior of both humans and experimental animals. Often the experiments with animals are described as models of human drug addiction. Laboratory studies of drug self-administration conducted with human volunteers and with experimental animals suggest that the fundamental process of drug reinforcement is the same in humans and non-humans. This conclusion is consistent with the well established findings of equivalencies with many other biological processes.

References

1. Aigner, T.G. and Balster, R.L., "Choice behavior in rhesus monkeys: cocaine versus food," *Science* . 1978, **201:** pp 534–535.
2. Ator, N.A. and Griffiths, R.R., "Self-administration of barbiturates and benzodiazepines: a review," *Pharmacol. Biochem. Behav.* 1987, **27:** pp 391–398.
3. Ator, N.A. and Griffiths, R.R., "Oral self-administration of triazolam, diazepam and ethanol in the baboon: drug reinforcement and benzodiazepine physiological dependence," *Psychopharmacology.* 1992, **108:** pp 301–312.
4. Balster, R.L. and Woolverton, W.L., "Continuous-access phencyclidine self-administration by rhesus monkeys leading to physical dependence," *Psychopharmacology.* 1980, **70:** pp 5–10.
5. Balster, R.L. and Woolverton,W.L., "Unlimited access intravenous drug self-administration in rhesus monkeys," *Fed. Proc.* 1982, **41:** pp 211–215.
6. Bickel, W.K., DeGrandpre, R.J., Higgins, S.T., Hughes, J.R., "Behavioral economics of drug self-administration. I. Functional equivalence of response requirement and drug dose," *Life Sci.* 1990, **47:** pp 1501–1510.
7. Bozarth, M.A. and Wise, R.A., "A comparison of the toxicity of chronic intravenous cocaine and heroin self-administration in laboratory rats," *JAMA.* 1985, **254:** pp 81–83.
8. Cappell, H. and LeBlanc, A.E., "Conditioned aversion by psychoactive drugs: does it have significance for an understanding of drug dependence?" *Addict. Behav.* 1975, **1:** pp 55–64.
9. Carr, G.D., Fibiger, H.C., Phillips, A.G., "Conditioned place preference as a measure of drug reward." In: Liebman, J.M. and Cooper, S.J. (Eds.), *The Neuropharmacological Basis of Reward.* Oxford University Press, New York, 1989, pp 264–319.
10. Carroll, M.E., "Oral self-administration of phencyclidine analogs by rhesus monkeys: conditioned taste and visual reinforcers," *Psychopharmacology.* 1982, **78:** pp 116–120.
11. Carroll, M.E., "Tolerance to the behavioral effects of orally self-administered phencyclidine," *Drug Alcohol Depend.* 1982, **9:** pp 213–224.
12. Carroll, M.E., "Rapid acquisition of oral phencyclidine self-administration in food-deprived and food-satiated rhesus monkeys: concurrent phencyclidine and water choice," *Pharmacol. Biochem. Behav.* 1982, **17:** pp 341–346.
13. Carroll, M.E., "A quantitative assessment of phencyclidine dependence produced by oral self-administration in rhesus monkeys," *J. Pharmacol. Exp. Ther.* 1987, **242:** pp 405–412.
14. Carroll, M.E., Krattiger, K.L., Gieske, D., Sadoff, D.A., "Cocaine-base smoking in rhesus monkeys: reinforcing and physiological effects," *Psychopharmacology.* 1990, **102:** pp 443–450.

15. Carroll, M.E. and Lac, S.T., "Cocaine withdrawal produces behavioral disruptions in rats," *Life Sci.* 1987, **40:** pp 2183–2190.
16. Carroll, M.E., Lac,S.T., Nygaard, S.L., "A concurrently available nondrug reinforcer prevents the acquisition or decreases the maintenance of cocaine-reinforced behavior," *Psychopharmacology.* 1989, **97:** pp 23–29.
17. Carroll, M.E. and Meisch, R.A., "Oral phencyclidine (PCP) self-administration in rhesus monkeys: effects of feeding conditions," *J. Pharmacol. Exp. Ther.* 1980, **214:** pp 339–346.
18. Carroll, M.E. and Meisch, R.A., "Increased drug-reinforced behavior due to food deprivation." In: Thompson, T., Dews, P.B., Barrett, J.E. (Eds.), *Advances in Behavioral Pharmacology,* Vol. 4. Academic Press, Orlando, FL, pp 47–88.
19. Carroll, M.E. and Stotz, D.C., "Oral *d*-amphetamine and ketamine self-administration by rhesus monkeys: effects of food deprivation," *J. Pharmacol. Exp. Ther.* 1983, **227:** pp 28–34.
20. Carroll, M.E., Stotz, D.C., Kliner, D.J., Meisch, R.A., "Self-administration of orally delivered methohexital in rhesus monkeys with phencyclidine or pentobarbital histories: effects of food deprivation and satiation," *Pharmacol. Biochem. Behav.* 1984, **20:** pp 145–151.
21. Catania, A.C., *Learning,* 3rd edition. Prentice Hall, Englewood Cliffs, NJ, 1982.
22. Comer, S.D. et al., "Effects of buprenorphine and naltrexone on reinstatement of cocaine-reinforced responding in rats," *J. Pharmacol. Exp. Ther.* 1993, pp 1470–1477.
23. Corrigall, W.A., "A rodent model for nicotine self-administration." In: Boulton, A.A. and Baker, G.B. (Eds.), *Animal Models of Drug Addiction: Neuromethods,* Vol. 24. Humana Press, Clifton, NJ, 1992, pp 315–344.
24. Corrigall, W.A. and Coen, K.M., "Nicotine maintains robust self-administration in rats on a limited-access schedule," *Psychopharmacology.* 1982, **99:** pp 473–478.
25. Crowley, T.J., Williams, E.A., Jones, R.H., "Initiating ethanol drinking in a simian social group in a naturalistic setting," *Alcohol. Clin. Exp. Res.* 1990, **14:** pp 444–455.
26. DeGrandpre, R.J., Bickel, W.K., Hughes, J.R., Layng, M.P., Badger, G., "Unit price as a useful metric in analyzing effects of reinforcer magnitude," *J. Exp. Anal. Behav.* 1993, **60:** pp 641–666.
27. Deneau, G., Yanagita, T., Seevers, M.H., "Self-administration of psychoactive substances by the monkey," *Psychopharmacologia.* 1969, **16:** pp 30–48.
28. DeNoble,V.J., Mele, P.C., Porter, J.H., "Intravenous self-administration of pentobarbital and ethanol in rats," *Pharmacol. Biochem. Behav.* 1985, **23:** pp 759–763.
29. Downs, D.A. and Woods, J.H., "Fixed-ratio escape and avoidance-escape from naloxone in morphine-dependent monkeys: effects of naloxone dose and morphine pretreatment," *J. Exp. Anal. Behav.* 1975, **23:** pp 415–427.
30. Emmett-Oglesby, M.W. and Lane, J.D., "Tolerance to the reinforcing effects of cocaine," *Behav. Pharmacol.* 1992, **3:** pp 193–200.
31. Emmett-Oglesby, M.W., Peltier, R.L., Depoortere, R.Y., Pickering, C.L., Hooper, M.L., Gong, Y.H., Lane, J.D., "Tolerance to self-administration of cocaine in rats: time course and dose-response determination using a multi-dose method," *Drug Alcohol Depend.* 1993, **32:** pp 247–256.
32. Falk, J.L., "Production of polydipsia in normal rats by an intermittent food schedule," *Science.* 1961, **133:** pp 195–196.
33. Falk, J.L., "Schedule-induced drug self-administration." In: van Harren, F. (Ed.), *Methods in Behavioral Pharmacology.* Elsevier, Amsterdam, 1993, pp 301–328.
34. Fischman, M.W., "Cocaine and the amphetamines." In: Meltzer, H.Y., (Ed.), *Psychopharmacology: The Third Generation of Progress.* Raven Press, New York, 1987, pp 1543–1553.
35. Gawin, F.H. and Kleber, H.D., "Abstinence symptomatology and psychiatric diagnosis in cocaine abusers," *Arch. Gen. Psychiatry.* 1986, **43:** pp 107–113.
36. Gerber, G.J., Bozarth, M.A., Spindler, J.E., Wise, R.A., "Concurrent heroin self -administration and intracranial self-stimulation in rats," *Pharmacol. Biochem. Behav.* 1985, **23:** pp 837–842.
37. Goeders, N.E. and Smith, J.E., "Reinforcing properties of cocaine in the medial prefrontal cortex: primary action on presynaptic dopaminergic terminals," *Pharmacol. Biochem. Behav.* 1986, **25:** pp 191–199.
38. Goldberg, S.R., "The behavioral analysis of drug addiction." In: Glick, S.D. and Goldfarb, J., (Eds.), *Behavioral Pharmacology.* C.V. Mosby, St. Louis, MO, 1976, pp 283–316.

39. Goldberg, S.R., Hoffmeister, F., Schlichting, U.U., Wuttke, W., "Aversive properties of nalorphine and naloxone in morphine-dependent rhesus monkeys," *J. Pharmacol. Exp. Ther.* 1971, **179:** pp 268–276.
40. Goldberg, S.R., Morse, W.H., Goldberg, D.M., "Behavior maintained under a second-order schedule by intramuscular injection of morphine or cocaine in rhesus monkeys," *J. Pharmacol. Exp. Ther.* 1976, **199:** pp 278–286.
41. Goldberg, S.R. and Spealman, R.D. (1983) "Suppression of behavior by intravenous injections of nicotine or by electric shocks in squirrel monkeys: effects of chlordiazepoxide and mecamylamine," *J. Pharmacol. Exp. Ther.* 1983, **224:** pp 334–340.
42. Goldberg, S.R., Spealman, R.D., Goldberg, D.M., "Persistent behavior at high rates maintained by intravenous self-administration of nicotine," *Science.* 1981, **214:** pp 573–575.
43. Griffiths, R.R., Wurster, R.M., Brady, J.V., "Discrete-trial choice procedure: effects of naloxone and methadone on choice between food and heroin," *Pharmacol. Rev.* 1975, **27:** pp 357–365.
44. Griffiths, R.R., Lamb, R.J., Sannerud, C.A., Ator, N.A., Brady, J.V., "Self-injection of barbiturates, benzodiazepines and other sedative-anxiolytics in baboons," *Psychopharmacology.* 1991, **103:** pp 154–161.
45. Grove, R.N. and Schuster, C.R., "Suppression of cocaine self-administration by extinction and punishment," *Pharmacol. Biochem. Behav.* 1974, **2:** pp 191–208.
46. Grubman, J. and Woods, J.H., "Schedule-controlled behavior maintained by nitrous oxide delivery in rhesus monkeys," Saito, S. and Yanagita ,T. (Eds.), *Learning and Memory: Drugs as Reinforcer.* Excerpta Medica, Amsterdam, 1982, pp 259–274.
47. Harris, R.T., Waters, W., McLendon, D., "Evaluation of reinforcing capability of delta-9-tetrahydrocannabinol in rhesus monkeys," *Psychopharmacologia.* 1974, **37:** pp 23–29.
48. Iglauer, C. and Woods, J.H., "Concurrent performances: reinforcement by different doses of intravenous cocaine in rhesus monkeys," *J. Exp. Anal. Behav.* 1974, **22:** pp 179–196.
49. Jaffe, A.B., Sharpe, L.G., Jaffe, J.H., "Rats self-administer sufentanil in aerosol form," *Psychopharmacology.* 1989, **99:** pp 289–293.
50. Johanson, C.E., "Drugs as reinforcers." In: Blackman, D.E. and Sanger, D.J. (Eds.), *Contemporary Research in Behavioral Pharmacology.* Plenum Press, New York, 1978, pp 325–390.
51. Johanson, C.E., "The effects of electric shock on responding maintained by cocaine injections in a choice procedure in the rhesus monkey," *Psychopharmacology.* 1977, **51:** pp 277–282.
52. Johanson, C.E., "The reinforcing properties of procaine, chlorprocaine and proparacaine in rhesus monkeys," *Psychopharmacology.* 1980, **67:** pp 189–194.
53. Johanson, C.E. and Aigner, T., "Comparison of the reinforcing properties of cocaine and procaine in rhesus monkeys," *Pharmacol. Biochem. Behav.* 1981, **15:** pp 49–53.
54. Johanson, C.E. and Balster, R.L., "A summary of the results of a drug self-administration study using substitution procedures in rhesus monkeys," *Bull. Narc.* 1978, **30:** pp 43–54.
55. Johanson, C.E., Balster, R.L., Bonese, K., Self-administration of psychomotor stimulant drugs: the effects of unlimited access," *Pharmacol. Biochem. Behav.* 1976, **4:** pp 45–51.
56. Johanson, C.E. and, Schuster, C.R., "A choice procedure for drug reinforcers: cocaine and methylphenidate in the rhesus monkey," *J. Pharmacol. Exp. Ther.* 1975, **193:** pp 676–688.
57. Johanson, C.E. and Schuster, C.R., "Animal models of drug self-administration." In: Mello, N.K. (Ed.), *Advances in Substance Abuse: Behavioral and Biological Research,* Vol. 2. JAI Press, Greenwich, CN, 1981, pp 219–297.
58. Johanson, C.E., Woolverton, W.L., Schuster, C.R., "Evaluating laboratory models of drug dependence." In: Meltzer, H.Y. (Ed.), *Psychopharmacology: The Third Generation of Progress.* Raven Press, New York, 1987, pp 1617–1625.
59. Katz, J.L., "A comparison of responding maintained under second-order schedules of intramuscular cocaine injection or food presentation in squirrel monkeys," *J. Exp. Anal. Behav.* 1979, **32:** pp 419–431.
60. Kliner, D.J. and Meisch, R.A., (1989) "Oral pentobarbital intake in rhesus monkeys: effects of drug concentration under conditions of food deprivation and satiation," *Pharmacol. Biochem. Behav.* 1989, **32:** pp 347–354.
61. Koob, G.F. and Bloom, F.E., "Cellular and molecular mechanisms of drug dependence," *Science.* 1988, 242, pp 715–723.

62. Kornet, M., Goosen, C., Ribbens, L.G., Van Ree, J.M., "Analysis of spontaneous alcohol drinking in rhesus monkeys," *Physiol. Behav.* 1990, **47:** pp 679–684.
63. Kornetsky, C. and Porrino, L.J., "Brain mechanisms of drug-induced reinforcement." In: O'Brien, C.P. and Jaffe, J.H. (Eds.), *Addictive States.* Raven Press, New York, 1992, pp 59–77.
64. Kramer, J.C., Fischman, V.S., Littlefield, D.C., "Amphetamine abuse," *JAMA.* 1967, **201:** pp 89–93.
65. Lemaire, G.A. and Meisch, R.A., "Pentobarbital self-administration in rhesus monkeys: drug concentration and fixed-ratio size interactions," *J. Exp. Anal. Behav.* 1984, 42, pp 37–49.
66. Macenski, M.J., Cutrell, E.B., Meisch, R.A., "Concurrent pentobarbital- and saccharin-maintained responding: effects of saccharin concentration and schedule conditions," *Psychopharmacology.* 1993, **112:** pp 204–210.
67. Macenski, M.J. and Meisch, R.A. (in preparation).
68. Meisch, R.A. and Carroll, M.E., "Oral drug self-administration: drugs as reinforcers." In: Bozarth, M.A. (Ed.) *Methods of Assessing the Reinforcing Properties of Abused Drugs.* Springer-Verlag, New York, pp 143–160.
69. Meisch, R.A. and Lemaire, G.A., "Oral self-administration of pentobarbital by rhesus monkeys: relative reinforcing effects under concurrent fixed-ratio schedules," *J. Exp. Anal. Behav.* 1988, **50:** pp 75–86.
70. Meisch, R.A. and Lemaire, G.A., (1990) "Reinforcing effects of a pentobarbital-ethanol combination relative to each drug alone," *Pharmacol. Biochem. Behav.* 1990, **35:** pp 443–450.
71. Meisch, R.A. and Lemaire, G.A., "Drug self-administration." In: van Haaren, F. (Ed.), *Methods in Behavioral Pharmacology.* Elsevier, Amsterdam, 1993, pp 257–300.
72. Meisch, R.A., Lemaire, G.A., Cutrell, E.B., "Oral self-administration of pentobarbital by rhesus monkeys: relative reinforcing effects under concurrent signaled differential-reinforcement-of-low-rates schedules," *Drug Alcohol Depend.* 1992, **30:** pp 215–225.
73. Meisch, R.A. and Stewart, R., "Ethanol as a reinforcer: laboratory studies with nonhuman primates," *Behav. Pharmacol.* 1994, **5:** pp 425–440.
74. Meisch, R.A. and Thompson, T., "Ethanol as a reinforcer: effects of fixed-ratio size and food deprivation," *Psychopharmacologia.* 1973, **28:** pp 171–183.
75. Nader, M.A. and Reboussin, D.M., "The effects of behavioral history on cocaine self-administration by rhesus monkeys," *Psychopharmacology.* 1994, **115:** pp 53–58.
76. Nader, M.A. and Woolverton, W.L., "Effects of increasing response requirement on choice between cocaine and food in rhesus monkeys," *Psychopharmacology.* 1992, **108:** pp 295–300.
77. Olds, J. and Milner, P., "Positive reinforcement produced by electrical stimulation of septal area and other regions of rat brain," *J. Com. Physiol. Psychol.* 1954, **47:** pp 419–427.
78. Olds, M.E. and Fobes, J.L., "The central basis of motivation: intracranial self-stimulation studies," *Annual Rev. Psychol.* 1981, **32:** pp 523–574.
79. Peele, S., *The Meaning of Addiction.* Lexington Books, Lexington, MA, 1985.
80. Pickens, R., Muchow, D., DeNoble, V., "Methohexital-reinforced responding in rats: effects of fixed ratio size and injection dose," *J. Pharmacol. Exp. Ther.* 1981, **216:** pp 205–209.
81. Pickens, R. and Thompson, T., "Cocaine-reinforced behavior in rats: effects of reinforcement magnitude and fixed-ratio size," *J. Pharmacol. Exp. Ther.* 1968, **161:** pp 122–129.
82. Poling, A. and Thompson, T., "Suppression of ethanol-reinforced lever pressing by delaying food availability," *J. Exp. Anal. Behav.* 1977, **28:** pp 271–283.
83. Schlichting, U.U., Goldberg, S.R., Wuttke, W., Hoffmeister, F., "*d*-Amphetamine self-administration by rhesus-monkeys with different self-administration histories," *Excerpta Med. Int. Congr. Ser.* 1971, **220:** pp 62–69.
84. Seevers, M.H., "Opiate addiction in the monkey," *J. Pharmacol. Exp. Ther.* 1936, **56:** pp 147–156.
85. Skinner, B.F., *The Behavior of Organisms: An Experimental Analysis.* Appleton-Century-Crofts, New York, 1938.
86. Skinner, B.F., *Science and Human Behavior.* Macmillan Company, New York, 1953.
87. Smith, J.E., Co, C., Freeman, M.E., Lane, J.D., "Brain neurotransmitter turnover correlated with morphine-seeking behavior of rats," *Pharmacol. Biochem. Behav.* 1982, **16:** pp 509–519.
88. Spragg, S.D.S., "Morphine addiction in chimpanzees," *Comp. Psychol. Monogr.* 1940, **15:** pp 1–132.

89. Stewart, R.B., Lemaire, G.A., Roache, J.D., Meisch, R.A., "Establishing benzodiazepines as oral reinforcers: Midazolam and diazepam self-administration in rhesus monkeys," *J. Pharmacol. Exp. Ther.* 1994, **271:** pp 200–211.
90. Stretch, R., Gerber, G.J., Wood, S.M., "Factors affecting behavior maintained by response-contingent intravenous infusions of amphetamine by squirrel monkeys," *Can. J. Physiol. Pharmacol.* 1971, **49:** pp 581–589.
91. Thompson, T. and Schuster, C.R., "Morphine self-administration, food-reinforced, and avoidance behaviors in monkeys," *Psychopharmacologia.* 1964, **5:** pp 87–94.
92. Van Ree, J.M., Slangen, J.L., De Wied, D., (1978) "Intravenous self-administration of drugs in rats," *J. Pharmacol. Exp. Ther.* 1978, **204:** pp 547–557.
93. Weeks, J.R., "Experimental morphine addiction: method for automatic intravenous injections in unrestrained rats," *Science.* 1962, **138:** pp 143–144.
94. Weeks, J.R., "Long-term intravenous infusion." In: Myers, R.D. (Ed.), *Methods in Psychobiology,* Vol. 2. Academic Press, London, 1972, pp 155–168.
95. Winger, G., "Effects of ethanol withdrawal on ethanol-reinforced responding in rhesus monkeys," *Drug Alcohol Depend.* 1988, **22:** pp 235–240.
96. Winger, G., Sitzer, M.L., Woods, J.H., "Barbiturate-reinforced responding in rhesus monkeys: comparisons of drugs with different durations of action," *J. Pharmacol. Exp. Ther.* 1975, **195:** pp 505–514.
97. Winger, G.D. and Woods, J.H., "The reinforcing property of ethanol in the rhesus monkey: I. Initiation, maintenance and termination of intravenous ethanol-reinforced responding," *Ann. NY Acad. Sci.* 1973, **215:** pp 162–175.
98. Wise, R.A., "Action of drugs of abuse on brain reward systems," *Pharmacol. Biochem. Behav.* 1980, **13(Suppl. 1):** pp 213–223.
99. Wise, R.A., Yokel, R.A., Hansson, P.A., Gerber, G.J., "Concurrent intracranial self-stimulation and amphetamine self-administration in rats," *Pharmacol. Biochem. Behav.* 1977, **7:** pp 459–461.
100. Wood, R.W., "Behavioral evaluation of sensory irritation evoked by ammonia," *Toxicol. Appl. Pharmacol.* 1979, **50:** pp 157–162.
101. Wood, R.W., Grubman, J., Weiss, B. (1977) "Nitrous oxide self-administration by the squirrel monkey," *J. Pharmacol. Exp. Ther.* 1977, **202:** pp 491–499.
102. Woods, J.H. and Winger, G.D., "A critique of methods for inducing ethanol self-administration in animals," Mello, N.K. and Mendelson, J.H. (Eds.), *Recent Advances in Studies of Alcoholism.* Publ. No. (HSM) 71-9045, U.S. Government Printing Office, Washington, D.C., 1971, pp 413–436.
103. Woolverton, W.L. and Johanson, C.E., "Preference in rhesus monkeys given a choice between cocaine and d,l-cathinone," *J. Exp. Anal. Behav.* 1984, **41:** pp 35–43.
104. Wurster, R.M., Griffiths, R.R., Findley, J.D., Brady, J.V., "Reduction of heroin self-administration in baboons by manipulation of behavioral and pharmacological conditions," *Pharmacol. Biochem. Behav.* 1977, **7:** pp 519–528.
105. Yanagita, T., Takahashi, S., Ishida, K, Funamoto, H., "Voluntary inhalation of volatile anesthetics and organic solvents by monkeys," *Jpn. J. Clin. Pharmacol.* 1970, **1:** pp 13–16.
106. Young, A.M. and Woods, J.H., "Behavior maintained by intravenous injection of codeine, cocaine, and etorphine in the rhesus macaque and the pigtail macaque," *Psychopharmacology.* 1980, **70:** pp 268–271.
107. Young, A.M. and Woods, J.H., "Maintenance of behavior by ketamine and related compounds in rhesus monkeys with different self-administration histories," *J. Pharmacol. Exp. Ther.* 1981, **218:** pp 720–727.

Contents Chapter 4

Genetic factors in addiction

0-8493-7803-6/99/$0.00+$.50

Chapter 4

Genetic factors in addiction

Frank R. George

1 Introduction

1.1 INTERACTIONS BETWEEN GENES AND THEIR ENVIRONMENTS

It is likely that excessive alcohol and drug use are due to a complex interaction of environmental, historical, and biological factors. Drugs produce numerous behavioral and physiological responses, which implies that drugs are affecting biological, and thus, genetically determined systems. However, these biological systems, whether single nerve cells, a mouse, or a human, are constantly being affected by the environments in which all of these organisms exist.

Disorders related to drug addiction have increased dramatically in recent years. This has led to a number of social, economic, and political challenges worldwide as well as serious psychological, physiological, and social problems for affected individuals, their friends and families. Much research effort is now being applied to understanding the biological determinants of drug dependence. In particular, research on the effects of genetic and environmental antecedents of drug abuse are addressing issues of individual and familial vulnerability to this problem.

It is clear that individuals differ with respect to many traits, including their responses to drugs. The study of these variations in behavior among individuals, their genetic and environmental bases, and their implications for society comprise the field of behavior genetics[25]. A related field involving genetic factors related to drug effects is known as pharmacogenetics. Both of these areas represent the integration of genetics with the behavioral and neural sciences, incorporating work from physiological, molecular, population and, organismic levels of study.

It is important to understand that finding genetic differences in susceptibility to drug abuse or addiction does not imply that there is an "addiction gene" which dooms unfortunate individuals to become hopeless drug addicts. Instead, it is the complex interactions between an individuals' genetic makeup, or genotype, with their unique environment and history, that determines their eventual patterns of behavior. In this way, risk for drug abuse can be viewed in a similar manner as risk for cancer or heart disease —people are born with varying degrees of genetic risk for these diseases, but it is their life decisions, such as which foods to eat, which contribute greatly towards the development, or lack thereof, of these diseases. The following example of how the environment can affect the actions of a single gene will hopefully make clear the importance of the interplay betweens our genes and our environments.

1.1.1 *PKU- a classical example*

When we look at a group of individals, we typically see large differences in the expression of traits. This variation in traits is referred to as phenotypic variability, and while obvious for many morphological or physical traits, such as height and hair color, it is also measurable for behavioral traits, such as activity level or personality. The sources of this phenotypic variation are complex and include the effects of genetic variability across individuals, differential effects of different environments past and present, and the sum of the interactions between specific biological, i.e., genetic, organisms, such as people, and their specific environments. In real life, the last (and most complex) component, the genotype by environment interaction (G × E), is usually the most critical. Genes exist within environments, and environments usually serve to modify the final expression of specific genetic combinations.

The importance of these interactions is illustrated through the classic medical genetics example of phenylketonuria, a disease usually referred to as PKU. This disease is not due to a bacterial or viral infection, but rather is an inborn biochemical disorder resulting in severe behavioral deficits. PKU is a genetically determined deficiency in an *enzyme* known as phenylalanine hydroxylase, which converts phenylalanine, an *amino acid* found in many foods such as certain cheeses and wines, to tyrosine. Due to this enzyme deficiency, PKU individuals build up high levels of phenylalanine as a result of exposure from the foods they eat, or critically, from the foods eaten by their mother during infancy when they are being breastfed. It is the high level of phenylalanine which appears to be primarily responsible for the behavioral deficits seen in PKU patients.

It is important to understand in this example that, as a constituent of the foods we eat, phenylalanine represents an environmental factor. An important question then becomes whether or not a PKU individual's behavioral deficits can be changed by altering levels of phenylalanine. The answer is that it can, if the levels are changed during critical developmental periods. Early postnatal detection and intervention through dietary reduction of phenylalanine can dramatically improve the abilities of PKU patients. Therefore, even though PKU individuals differ genetically from non-PKU persons in their ability to metabolize phenylalanine, the behavioral expression of PKU is dependent upon a G × E interaction, namely the existence of the PKU genotype within a high phenylalanine environment.

This example also illustrates that behavioral genetics is not just the study of genetic or biological influences upon behavior, but also encompasses the study of environmental impacts upon behavior. Over the past several years, behavioral geneticists have shown that some behaviors are highly genetically determined so that environmental factors have little influence on them. Conversely, some environmental effects are so strong that they affect virtually all genotypes. For the most part, however, it appears that behavioral responses to drugs, including responses such as addictive behaviors, lie somewhere between these two extremes, so that both genetic and environmental influences and their interactions are critical. In this chapter you will learn about genetic influences in responses to drugs, with an emphasis on those drugs effects which contribute to abuse or addiction. Keep in mind, however, that while genetic contributions are important, environmental influences are also critical to the development of addictions. Indeed, drugs themselves can be viewed as environmental factors which have the potential to interact with and influence the lives of any of us.

1.2 A BRIEF OVERVIEW OF BEHAVIORAL GENETICS

1.2.1 *Historical perspective*

The beginnings of humankind's interest in the inheritance of various traits can be traced back in history for many centuries. For example, the breeding of domesticated animals such as dogs for specific characteristics and the development of farm animals for food can be traced back thousands of years. In addition, for centuries many people have believed that personality traits and behavioral patterns, including tendencies toward alcoholism, "run in families", even though there has been little empirical evidence to support such attributions. Initial attempts to determine scientifically genetic contributions to human behavior began in the late 1800s. However, the emergence of behavioral genetics as a distinct academic discipline did not occur until around 1960, while investigations into the genetic basis of alcohol drinking began about the same time[25]. Today, behavioral genetics is an internationally active field, with investigators in this area using techniques ranging from molecular genetics to surveys of human populations.

1.2.2 *Genes as experimental variables*

A unifying concept in behavioral genetics is that genes may be utilized as independent variables. From this, it can be asserted that if genotype is held constant within a population, then any variability among the members of that population is due to environmental factors. In addition, based upon our understanding of the genetic methods described below, plus the many recent advances in molecular biology and genetics, it is possible experimentally to manipulate genes to develop a better understanding of their functional roles in determining biological processes.

Most nonhuman pharmacogenetic and behavioral genetic studies employ one of two common genetic methods, inbreeding or selective breeding. Behavioral genetic methods can be complex, but most nonhuman studies employ simple, yet elegant mating schemes based upon principles outlined a century ago by Gregor Mendel. Much information concerning the extent of genetic contribution to a trait, its mode of transmission (e.g., dominant or recessive), and an estimate of the number of genes which mediate the trait can be ascertained within only a few generations of specified matings. Most human behavior genetic studies employ mathematical population models based upon these same principles of inheritance.

To understand better how these methods can be incorporated into studying addiction it is necessary to provide some definitions. While some of these terms were introduced previously, it is useful to review them here, and to add a discussion of additional terms and methods that have contributed to our understanding of biological vulnerability to alcohol and drug addiction.

1.3 TECHNIQUES AND TERMS IN THE STUDY OF BEHAVIORAL AND PHARMACOGENETICS

1.3.1 *Non-human studies*

1.3.1.1 Inbred strains

Inbreeding serves to reduce the genetic variation within a population so that over a number of generations individuals become more and more genetically alike. Inbreeding will result from the mating of closely related individuals, such as brothers and sisters or first cousins. In the laboratory, inbreeding can be maintained and adjusted through specific mating patterns. It requires some 20 generations of matings between siblings, that is, brothers and sisters, to produce offspring that are, for all practical purposes, genetically identical. Once an inbred strain has been developed it can be reasonably assumed that all future offspring within that strain will have the same genetic makeup.

Let us look at an example to illustrate both the power and potential problems associated with using inbred strains. Assume that you are a researcher interested in understanding the effects of cocaine on behavior and how these effects are regulated within the brain. You take mice from an inbred strain, strain 'A', and inject a small dose of cocaine. You then observe the animal's behavior, and find that cocaine caused an increase in movement and overall activity. The next question is, how does cocaine produce this effect? What methods can we use to answer this question? One approach would be to examine this effect in a second inbred mouse strain, strain "B". So, you repeat the experiment using mice from strain "B" and find a very different result, namely that the same dose of cocaine which stimulated the activity of "A" mice produces absolutely no changes in "B" mice. Since these mice all came from the same environment, that is, they were housed in the same room, ate the same food, had the same day-night schedule, etc., you conclude that the difference in response to cocaine must be due to some genetic difference that exists between the two strains.

Upon further study, you find that mice of strain "A" produce greater amounts of the *neurotransmitter* dopamine relative to mice of strain "B" in response to cocaine, and you conclude that this explains the difference in behavioral response between these animals. However, in another laboratory, researchers show that strains "A" and "B" differ in production of the neurotransmitter serotonin, and in a third laboratory researchers show that these strains differ in their responses to a wide range of environmental stressors. Thus, how can we assert with confidence that it is the difference in dopamine which conveys the difference in response to cocaine? Perhaps it is a serotonin effect, or perhaps one strain is simply more stressed by the presence of cocaine than is the other strain.

One way to resolve these questions is to add data from several additional inbred strains. If you continue to observe a strong association between the activating effects of cocaine and the amount of dopamine produced in response to cocaine, then your confidence in the "dopamine hypothesis" will appropriately increase. If the associations between cocaine's effects and serotonin or stress are merely coincidental, then, as you add data from more inbred strains, these relationships should decrease in strength, since some strains may show the same association, but others will not.

Thus, the real power of the inbred strain approach lies in the ability to observe effects in a large number of strains, and, as a result, form hypotheses based upon strong and broadly based associations among measures. In general, then, one must use several inbred strains in such a study in order to have sufficient confidence that any positive relationships may be causally related and not simply a coincidence resulting from the random inbreeding that created the inbred strains. This same rule applies to the other methods described below. Genetic approaches can be very powerful tools in understanding drug mechanisms, but must be used with an understanding of their limitations.

1.3.1.2 Selective breeding

Many selective breeding programs have been conducted to examine genetic influences on responses to alcohol and other drugs. Selective breeding is a program of specific matings over a number of generations with the intention of changing particular traits. Since this process, similar to inbreeding, requires a number of generations, typically at least 20, it is advantageous to use species that have relatively short gestation periods, such as mice or rats, so that the length of time per generation is minimized. Selection is typically conducted in a bidirectional manner to produce maximally distinct populations, for example high alcohol drinking rats vs. low alcohol drinking rats.

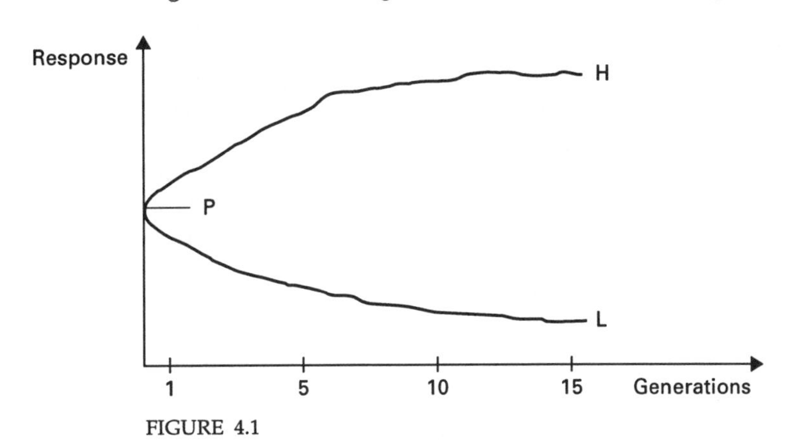

FIGURE 4.1

A schematic representation of a theoretical selective breeding program

"P" represents the mean response of the original parental population. "H" represents the response to selection for animals bred for high response. "L" represents the response to selection for animals bred for low response.

There are a number of reasons for conducting a selective breeding program. One is that if the selection is successful, then it is a strong indication that genetic factors influence the trait being measured. Another reason is that once these different populations are created, they may be studied to help understand the genetic and biological bases for the trait for which they were bred. For example, we can study various biological and neuronal systems in the rat populations that have been bred for differences in alcohol drinking.

Much has been learned through selective breeding studies about alcohol and drug effects and the processes of addiction. The most well known of these selected lines of mice and rats are called the Long Sleep (LS) and Short Sleep (SS) lines. These are two lines of mice selectively bred respectively for long versus short duration of loss of the righting reflex following alcohol administration[25]. Mice, like many animals, have a strong righting reflex, such that when turned on their backs they reflexively turn over and right themselves. Moderate to high doses of alcohol can disrupt this reflex action. As a result, mice given alcohol will often stay on their backs for several minutes to several hours. While they are not technically asleep, they have the appearance of sleeping, so measures of this response have come to be known as "sleep time". The Long Sleep and Short Sleep mice currently differ in their neural sensitivity to alcohol by approximately twenty-fold. As a result of this substantial difference, many studies have used these mice to aid in determining the differences in brain mechanisms which regulate this effect.

More recently, additional programs have been conducted in which mice have been selectively bred for other responses to alcohol. For example, mice will experience alcohol withdrawal when, after they have been made physically dependent on alcohol by chronic exposure to this drug, the alcohol is no longer provided. One of the common symptoms of alcohol withdrawal is a seizure. Based upon this effect, mice have been bred to be either prone to having withdrawal seizures, the Withdrawal Seizure-Prone mice (WSP), or resistant to withdrawal seizures, the Withdrawal Seizure-Resistant (WSR) mice[23].

Alcohol also produces a marked hypothermic effect, that is, consumption of alcohol can significantly decrease body temperature. This effect is responsible for a number of deaths in colder climates when unwary individuals consume alcohol in a mistaken attempt to stay warm. Another selective breeding study has produced mice known as the COLD and HOT lines. These lines of mice have been developed for differential sensitivity to the hypothermic effect of alcohol. COLD mice produce a pronounced hypothermic response to alcohol whereas HOT mice show relatively little response[23].

Studies measuring preference for alcohol have shown that rats and mice manifest large genetic differences in whether or not they prefer drinking an alcohol solution or water when given a choice between the two. This large genetic variability in alcohol "preference", as this choice test between two bottles is called, has served as the basis for the establishment of selectively bred rodents which differ in their alcohol drinking and are relevant to the study of drug addiction. These animals are described in detail below.

1.3.1.3 Genetic correlations

The use of genetic correlations is a powerful behavior genetic tool that has provided much information concerning the genetic and biochemical nature of addiction processes. *Genotypes* which differ for a given trait can be used to test relationships between variables by determining correlations between traits hypothesized to be causally related. A lack of correlation is usually good evidence that the measures studied are not likely to be mechanistically related. A strong positive correlation, especially a perfect rank order correlation across genotypes, enables one to hypothesize that the measures are causally related, although caution must be observed with regards to the degree of confidence in such a conclusion since these studies are using relatively small samples of a much larger population base, and are intended to be used as isolated models for what may be occurring in humans.

Studies of drug addiction are lengthy and labor intensive, and thus the resources for these studies are sometimes less than ideal. Using several inbred strains in a genetic correlation approach can be an effective alternative to the time and expense required for selective breeding studies.

1.3.1.4 Recombinant inbred strains

The use of recombinant inbred (RI) strains in genetic research can aid substantially in determining the genes involved in the expression of various traits. Individual differences in complex behavioral traits such as responses to drugs are likely to be influenced by many genes. Recombinant inbred strains are valuable not only for their originally developed purpose of detecting single or major gene effects, but also for identifying possible relationships between continuously distributed traits and multiple genes through the use of the quantitative trait loci (QTL) approach. This approach is becoming more widely used in drug abuse research, and recent findings hold promise for this approach in helping to determine the genes and biological factors that contribute to alcohol and substance abuse disorders.

Overall, the production and use of genetically specified offspring is an important aspect of behavioral genetic methods, but fortunately is not required of every investigator who wishes to take advantage of these designs. Many inbred and selectively bred animals currently exist, and while selectively bred animals are more costly and sometimes in short supply, large numbers of animals from well–studied inbred strains can be readily obtained from commercial suppliers. The choice of inbred strains can be random or based upon some defining characteristic. For example, it is sometimes advantageous in drug addiction studies to choose animals which respond positively to novel situations so that they are more able to learn the necessary tasks which provide access to drugs. Whatever the approach used, it has become apparent over the last several years that the economic and scientific advantages of using behavior genetic methods make it likely that studies in this area will contribute significantly to improved understanding, treatment, and prevention of alcohol and substance abuse disorders.

1.3.2 *Human studies*

Most reliable human genetic studies on substance abuse can be divided into two groups based on methodology. These are twin studies and adoption studies. The best studies incorporate the use of both methods, using twins separated at an early age, adopted by different families and, as a result, raised in different environments. Most studies on genetic factors in substance abuse have consisted of twin and adoption studies on alcohol drinking. Fewer studies have examined genetic factors in clinically diagnosed alcoholism. There are very few human genetic studies on *opiates* or *psychostimulants*, but these types of studies are now increasing in number. There are several reasons for the limited information in this area. Not only are twins and adoptees relatively rare, but these individuals must also have substance abuse disorders. Information on substance use must also be available for both twins and for the adoptive as well as biological parents.

1.3.2.1 Twin studies

Twins come in two genetic categories, identical and fraternal. Identical twins, or *monozygotic* twins (MZ) share 100% of their genes. Thus, if a trait has significant genetic contributions towards its' expression, the concordance rate among MZ twins should be high. Concordance rate is a measure of the extent to which one member of a twin pair will express a trait, such as alcoholism, if the other member of the twin pair expresses that trait. For example, if a trait is solely genetically determined, then the MZ twin concordance rate would theoretically equal 100%. If a trait had no genetic component, the MZ twin concordance rate would be zero. Fraternal, or *dizygotic* twins (DZ) share on average one half of their genes. Thus, concordance rates for DZ twins should theoretically be about one half of the concordance rate for a given trait seen in MZ twins. To the extent that concordance rates obtained in studies using twins approach the theoretical ideals, one can suggest relative contributions of genetic factors to the traits being measured. Conversely, to the extent that concordance rates do not match their expected values, it can be assumed that this is due to the contribution of environmental factors. The elegance of the behavior genetic approach is seen through this distinction because these methods allow for a determination of not only genetic contributions to addiction, but also environmental contributions, as well as interactions between the two.

1.3.2.2 Adoption studies

While there have been many more twin studies conducted, adoption studies have the capability of providing the strongest evidence for genetic influences on complex human behaviors such as drug abuse. However, these studies can be difficult to conduct since they require identification of an adopted population with alcohol or drug problems who are willing to participate in a research study, consent and participation of the adoptive parents, and location and participation of at least one and hopefully both biological parents. Adoption studies can separate genetic and environmental influences by examining traits among adopted individuals, their unrelated adoptive parents, and their related biological parents.

For example, if an adopted individual develops alcoholism, one can examine the drinking history of the adoptive and biological parents. If the adoptive parents have a history of alcohol abuse or alcoholism but the biological parents do not, this would be consistent with environmental mediation of this disorder. Conversely, if one or both biological parents have a history of alcohol drinking disorders but the adoptive parents do not, alcoholism in the adopted away offspring would be consistent with genetic mediation of this disorder. As will be described in another section of this chapter, the latter effect is most often found in adoption studies, and these studies have provided much evidence for genetic contributions to alcohol-related problems.

2 Animal models of "addiction"

2.1 OVERVIEW OF RESEARCH ON DRUG-SEEKING BEHAVIOR

Most animal drug taking studies can be divided into two groups based on methodology. One group consists of two bottle 24 hour water-drug preference studies; the other consists of studies using operant conditioning methods. In 1940 Richter and Campbell published a study of rats' alcohol drinking[26]. They attached two bottles to the cage of each rat, one bottle containing alcohol, the other water. Every 24 hours they measured the volumes consumed. This general procedure, generally known as the two-bottle choice, or "preference" test, has been used in hundreds of subsequent studies, and some generalizations can be made. Rats prefer alcohol in concentrations of 2–6% (w/v) to water, but intoxication is not reliably observed. Over 24-hour periods rats consume less alcohol than they can metabolize, and thus the sustained blood levels that are necessary for physical dependence do not occur. Similar results have been obtained in mice. However, the use of inbred strains, even in the earliest studies, has provided some valuable information about the contribution of genetic factors to alcohol drinking. A common example is that mice of the C57BL/6 inbred strain prefer alcohol whereas DBA/2 mice avoid alcohol[11].

In the 1960s a line of research developed that differed from the two bottle choice studies in both techniques and conceptual basis. This operant approach to studying drug taking behavior grew out of Skinnerian operant behavior studies. Operant behavior is defined as behavior controlled by its consequences. For example, a commonly studied instance of operant behavior is lever pressing by a rat that results in deliveries of food pellets. Procedures were devised so that laboratory animals could intravenously inject drugs into themselves. These procedures have been described in detail in the previous chapter.

In most self-administration studies an intravenous route of administration has been used. However, many drugs can function as reinforcers when other routes are used. Behavior reinforced with drugs has been demonstrated with the intramuscular route, the oral route, and by inhalation, and animals from a number of nonhuman species, including mice, rats, cats, dogs, monkeys, and baboons will self-administer most of the drugs that humans abuse[11].

Self-administration studies have established the idea that drug-seeking behavior is a particular case of operant behavior. The importance of viewing drug-seeking behavior as operant behavior is that it places drug-seeking behavior within a conceptual framework. Much is known about operant behavior in general, and this general knowledge has been used in designing specific studies of drug-seeking behavior. These findings are important because 1) they indicate that reinforcement from drugs is a behavioral consequence of an underlying biological effect which appears to be common across mammals; and 2) the demonstration of species-equivalent behavior across several drugs provides validation for the use of operant animal models of drug taking in studying various contributing factors and treatment approaches related to substance abuse.

2.2 GENETIC DIFFERENCES IN DRUG SEEKING BEHAVIOR AND REWARD MECHANISMS

2.2.1 *Preference studies with alcohol, cocaine, and opiates*

The use of inbred strains in many preference studies has provided valuable information about the contribution of genetic factors to drug taking. As noted above, C57BL/6 mice prefer alcohol whereas DBA/2 mice, as well as mice from several other strains, avoid alcohol. It was these robust inbred strain differences in alcohol preference that led to the development of rat lines selectively bred for high or low alcohol preference.

These selectively bred rodent lines include the Alcohol Accepting (AA) and Alcohol Non-Accepting (ANA) rats[9], the alcohol Preferring (P) and Non-Preferring (NP) rats, and the High Alcohol Drinking (HAD) and Low Alcohol Drinking (LAD) rats[18]. These rats were selectively bred based upon their daily intake of an 8% alcohol solution. The success of these selection programs indicates that alcohol consumption is strongly influenced by genetic factors. The preferring lines characteristically drink most of their daily liquid intake in the form of alcohol solutions. In contrast, the non-preferring animals drink very small amounts of alcohol daily. Ethanol intakes for these lines lie at the high and low extremes of intakes found in heterogeneous rats or across several inbred strains.

The original foundation, or parental, stock from which the selected lines are derived is an important consideration. The P and NP rats were selected from a foundation stock of randomly bred Wistar rats. Selection criterion for this pair of lines was based upon daily intake of an 8% (w/v) ethanol solution defined as a consumption score (grams of ethanol per kg of body weight per day). The HAD and LAD rats were selected from the more *heterogeneous* and systematically outbred N/Nih stock, using a breeding design to minimize inbreeding. The selection criterion for this pair of lines was also based upon daily intake of 8% (w/v) ethanol (g/kg/day). N/Nih rats, compared with Wistar rats, provide a possible advantage as a foundation population since N/Nih rats are less inbred and can be characterized as having potentially greater genetic diversity. These factors would tend to reduce false negative relationships and increase the strength of positive relationships among traits.

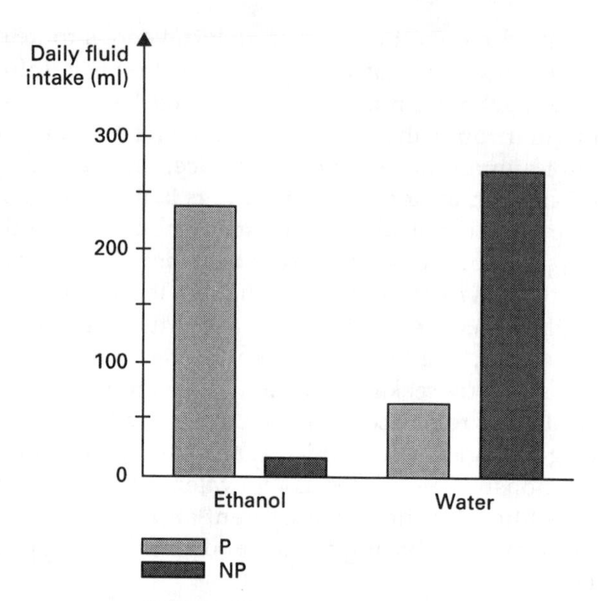

FIGURE 4.2

Average daily intake of alcohol and water for P versus NP rats
(Data from Li, T-K. et al., In: McClearn, G.E., Dietrich, R.A., Erwin, V.G. (Eds.), *Development of Animal Models as Pharmacogenetic Tools*, Monograph 6, U.S. Department of Health and Human Services, Rockville, MD, 1981, pp 171–191.)

Genetic studies have also shown large genetic differences in intake of non-alcohol drugs, including opioids, as well as in other responses to drugs, such as analgesic response to morphine. In addition, studies using inbred strains of rodents have shown that intake of cocaine solutions is typically higher in LEWIS relative to F344 rats, and exceeds intake of water in LEWIS but not F344 rats (Figure 4.3). The results of these studies provide evidence that genetic factors play an important role in determining responses to drugs in general, and drug taking behaviors in particular.

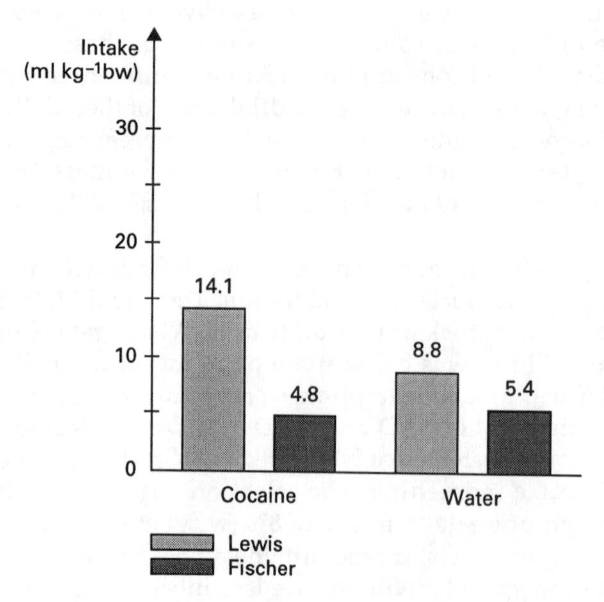

FIGURE 4.3

Average intake of cocaine solutions or water in a two bottle preference test in rats
(Data from George, F.R. and Goldberg, S.R., *Trends Pharmacol. Sci.* 1989, 10, pp 78–83.)

2.2.2 *Operant conditioning studies with alcohol, cocaine and opiates*

Alcohol serves as a positive reinforcer for many human as well as non-human animals. It has been well-established that genetic factors are important in determining whether or not alcohol will serve as a reinforcer. Findings obtained with operant conditioning methods demonstrate that alcohol can serve as a positive reinforcer across a range of environmental conditions in certain inbred strains such as C57BL/6J mice and LEWIS rats. Conversely, alcohol does not appear to serve as a reinforcer for DBA/2J mice and BALB/cJ mice. These genetic differences, while typically substantial, are not necessarily always an all-or-none effect. F344 rats, for example, appear to be weakly but positively reinforced by alcohol[12].

Thus, genotype is an important variable in the rewarding efficacy of alcohol. Even food deprivation, which has been shown to increase intake of alcohol and other drugs, as described in Chapter 6, fails to facilitate alcohol drinking in non-reinforced animals, such as BALB/cJ and DBA/2J mice. Food deprivation appears to enhance drug intake only in those animals genetically predisposed to accept a particular drug as a reinforcer.

Studies on the influences of genetic variables on other drug effects in animal operant behavior models are limited. Recent operant self-administration studies have shown that orally ingested cocaine solutions serve as reinforcers in LEWIS rats but not F344 rats. Genetic differences have also been shown in the reinforcing effects of opioids. LEWIS rats appear to be positively reinforced by opioids whereas F344 rats apparently are not reinforced by these compounds. As shown in Figure 4.4, LEWIS rats make significantly more lever press responses when etonitizine (ETZ), a potent opioid similar in effect to morphine, is available than when the vehicle solution, water, is available as the reinforcer. Conversely, F344 rats, also known as FISCHER rats, make fewer responses when ETZ is available relative to water, suggestive of a genetically mediated avoidance of this drug[11].

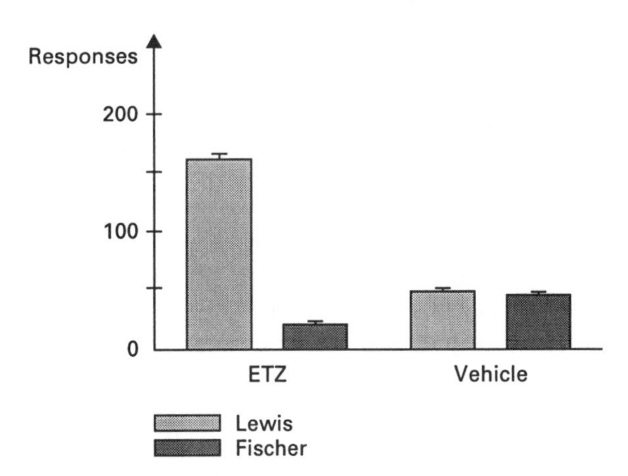

FIGURE 4.4
Rates of operant lever pressing for ETZ in rats
(Data from George, F.R. and Goldberg, S.R., *Trends Pharmacol. Sci.* 1989, 10, pp 78–83.)

2.2.3 *Other animal models of addiction: place preference and self-stimulation*

There are other models of drug taking or "addiction" which have been developed, but which do not involve actual self-administration of the drug under study by the test subject. One of these models, which has gained some popularity among researchers due to the relatively short time required to train animals and obtain data, is called conditioned place preference. The conditioned place preference paradigm pairs a specific environment with injections of cocaine. Typically, this is done by making one side of the test apparatus dark and the other side light, and often combining this with one side having a smooth floor and the other having a rough floor. Subsequently, the animal is tested to see if it prefers the cocaine-related environment over another neutral environment. Preference for the "place" where the animal received the drug is presumed to be a measure of the rewarding effects of that drug. Significant genetic differences have been found in conditioned place preference for cocaine between different inbred strains of mice. Studies in this area also indicate that substantial genetic differences can be found in place preference for amphetamine and opioids. Interestingly, the degree of response ranges from almost complete preference for, to strong avoidance of, the drug "place"[31].

Here, however, is an area where the investigator must be aware of the interactions between genes and environments. It is also the case that significant genetic differences exist in the type of place preference environment alone which is preferred by mice of different strains. Thus, one must be careful to ensure that any differences in place preference response to a drug are not due simply to genetic differences in visual or tactile response to a lighter, darker, rougher, or smoother type of environment in the test apparatus.

Intracranial self-stimulation (ICSS) is another model for studying reward processes. In this model, animals are trained to press a lever in order to transmit a mild electrical current through an electrode implanted in a region of the brain. When the electrode is placed in areas of the brain known to be associated with reward processes, such as the *nucleus accumbens*, animals will readily learn the behavioral task required to turn on the current, and will respond at very high rates to obtain this brain stimulation. Studies done a number of years ago using humans indicate that this stimulation is experienced as a highly positive event, and it has been described as similar to sexual orgasm or a drug-induced euphoria.

While few studies have been conducted exploring the role of genetic factors in ICSS, the available data are interesting and suggest that inheritance plays an important role in determining an indivual's reaction to environmental stress and the ability of stress to reduce responses to positive events. For example, responding for electrical stimulation in the nucleus accumbens has been studied in different mouse strains under conditions of imposed environmental stress. All mice learned to press a lever to obtain the stimulation reward. However, DBA/2J mice showed a marked deterioration of self-stimulation responding under stress, while stress did not affect self-stimulation responding in C57BL/6J mice[33]. This data is consistent with the studies described above showing the importance of individual and genetic differences in understanding the mechanisms of reward and addiction.

2.3 GENETIC APPROACHES TOWARDS UNDERSTANDING COMMONALITY OF ABUSE ACROSS DRUGS AND THE NATURE OF ADDICTION

2.3.1 *Are there "addiction prone" populations?*

Another important question in substance abuse which can be effectively addressed using genetic methods, especially genetic correlations, is whether or not the propensity to self-administer one drug, for example alcohol, shares common genetic control with the propensity to self-administer other drugs, such as cocaine or opiates. This "commonality" question can be addressed by measuring the extent to which animals from various inbred strains self-administer a variety of drugs. For example, self-administration of alcohol, cocaine, and opiates has been measured in several mouse and rat strains. Results from these studies of drug self-administration across different drugs and strains suggest that genetic patterns of reinforcement from alcohol may correlate highly with patterns of reinforcement from cocaine and opiates, as summarized in Table 4.1[12]. In addition, similar patterns of conditioned place preference associated with cocaine, amphetamine, and opioids have been observed across several mouse inbred strains, suggesting that strains which show conditioned place preference for one of these drugs will tend to show the same place preference behavior for another drug[31]. Thus, drug seeking behaviors maintained by alcohol, cocaine, amphetamine, and opioids may have at least some common biological determinants.

TABLE 4.1

Commonality across drugs and genotypes for drug-reinforced behavior

Plus symbols indicate relative degree of positive reinforcement. Negative symbols indicate relative degree of non-reinforcement or avoidance. Three symbols is maximum response relative to all genotypes tested. NA = Data not available. (Data from George, F.R., *J. Addict. Dis.* 1991, 10, pp 127–139.)

Genotype	*Alcohol*	*Opiates*	*Cocaine*
Rats			
LEW	+++	+++	++
F344	+ –	– – –	–
Mice			
C57BL/6J	+++	+++	+++
DBA/2J	– – –	– – –	NA

While propensity to take one drug, such as alcohol, may help to predict the likelihood of taking additional drugs, there are many other factors not related to drug responsivity *per se* which may also play important roles in determining risk for addiction. Factors such as personality traits and other mental disturbances, such as depression, also appear to be important indicators of risk for alcohol and drug abuse. These other factors are discussed below under the sections on human studies.

2.3.2 *Multiple inherited components of reward and "addiction"*

Motivation is a component of behavior, and an important area of study in determining antecedents of, and treatments for, substance abuse disorders. Motivation refers to an organism's level of intensity or persistence in obtaining a reinforcer at a particular moment or under certain conditions. For example, water can be a positive reinforcer, and animals will work to obtain water. However, the extent to which this work will be performed varies as a function of changes in the animal's drive state, or motivation to drink. If an animal is water satiated, less work will be performed to obtain water than when an animal is thirsty, or in a state of water deprivation.

Recent studies using operant behavior approaches to measuring reinforcement in inbred and selectively bred animals, has provided insights into the relationship between reinforcement and motivation, the role of specific neuronal systems in these aspects of behavior, and potential approaches for the treatment of alcohol and substance abuse disorders.

Alcohol-reinforced behavior has been studied in high preferring (P) and High Alcohol Drinking (HAD) rats, and in nonpreferring Non-Preferring (NP) and Low Alcohol Drinking (LAD) rats. Genetic differences in alcohol-reinforced behavior were observed. Alcohol serves as a strong positive reinforcer for P rats, a slightly less efficacious reinforcer for NP and HAD rats, and is apparently not reinforcing for LAD rats[27].

These results show that alcohol drinking in a preference test is not highly predictive of the reinforcing effects of alcohol. NP rats, like P rats, will exhibit alcohol-reinforced responding under operant conditions. Further, HAD rats exhibit significantly fewer responses for alcohol under a range of concentrations and fixed ratio sizes relative to their P rat counterparts, even though both lines have been selected for maximal alcohol drinking in a preference paradigm. Thus, preference and reinforcement are distinct aspects of drinking, and both appear to be important.

Why such a difference exists between these two measures of alcohol drinking is an important issue whose answer remains unclear. One possible reason is that preference studies typically are confounded by taste and prandial influences since the measure of drinking is based upon consumption over a long period of time, typically 24 hours, with food concurrently available. Also, preference paradigms may not be sufficiently sensitive to detect intake of significant amounts of ethanol in animals whose absolute levels of intake are limited by neurosensitivity factors, but for which relatively small amounts of ethanol are reinforcing. An example of this situation are the LS/Ibg mice, who are extremely sensitive to the depressant effects of alcohol. Despite this high degree of sensitivity, LS/Ibg mice will readily learn to press a lever for access to alcohol, and will work as hard, if not harder, than any other mice to obtain this drug[8]. Finally, in preference studies that do not incorporate exposure to significant amounts of the drug through some form of initial training, low preference may be due to negative taste factors which result in avoidance of the drug solution, creating a situation where consumption is too low for initiation of reinforcement to ever occur. Keep in mind that in humans most alcohol drinking begins with relatively sweet and dilute liquids such as beer, soda-sweetened alcohol, or sweet wines. For these reasons, it seems appropriate that preference be viewed as a permissive factor which allows the organism to initially take a drug. Consumption of large doses over a sustained period of time may then result in association of the drug taking behavior with its rewarding effects in the brain, and the drug may then come to serve as a positive reinforcer. Thus, when combined with similar data from other drugs, it appears that the reinforcing effects of drugs comprise a unique dimension of effect.

TABLE 4.2

Qualitative expression of preference, reinforcement, and persistence for alcohol seeking in rats selectively bred for high or low alcohol preference

Plus symbols indicate relative degree of positive performance. Negative symbols indicate relative degree of non-drinking or avoidance.

Genotype	*Preference*	*Reinforcement*	*Persistence*
P	+++	+++	+++
NP	– –	++	++
HAD	+++	++	–
LAD	– –	– –	– –

Other studies show genetic differences with regard to the propensity of animals to maintain alcohol drinking as the amount of effort necessary to obtain the drug is increased[28]. P rats are high preferring, reinforced by alcohol and show persistent responding under increasing work loads, while high preferring HAD rats are only modestly reinforced and show little persistence in responding for alcohol under conditions of high work loads. NP rats, on the other hand, are very low preferring, but show alcohol-reinforced responding for alcohol equivalent to that of the HAD rats when only one lever press was required. Interestingly, NP rats also show a moderate level of persistence in responding, and this persistence is much greater than that seen in HAD rats. Finally, the low preferring LAD rats are not reinforced by alcohol and show no significant persistence in responding for alcohol.

Thus, although alcohol can be readily established as a reinforcer for AA, NP, and HAD rats, animals from these three stocks appear to lack specific motivational factors that would facilitate continued responding under conditions requiring higher work loads. This is in contrast to P rats who will maintain responding for ethanol even under conditions requiring very high amounts of work to gain access to the drug. This data suggests that continued chronic abuse of a drug requires not only specific reinforcing effects of a drug, but also motivational factors that appear to vary independently of response to reinforcing effects.

These findings highlight the importance of interactions between genes and environments, and how variation in one or the other (or both) may lead to increased potential for taking drugs. The P rats show preference for alcohol over water, find alcohol reinforcing, and will work hard to gain access to this drug. In many ways they fulfill the criteria for an animal model of alcoholism which has a significant genetic component. These animals will drink alcohol even when their environment does not favor this behavior or make it easy, such as when large numbers of lever presses are required to receive an alcohol reinforcement. Thus, they appear to be truly genetically prone towards drinking, and their environment does little to modify this tendency as long as alcohol is available somehow. Contrast this with the performance of the NP rats. NP rats appear not to like the taste of alcohol, which results in a low preference. However, if this negative factor is overcome through environmental means, which in the case of these experiments involves making the rats thirsty and then giving them alcohol mixed with water to drink, they apparently learn that drinking alcohol produces a positive effect, and their alcohol drinking behavior becomes reinforced. This is perhaps similar to many young adult humans who do not like the taste of alcohol, but when the alcohol is combined with adulterants, such as sweetened beverages, will drink enough to experience the positive effects of the drug. However, like most human "social drinkers", these rats will not go to great lengths to drink alcohol, but will consume it in moderate amounts when it is easily available. Thus, alcohol abuse appears to be the result of interactions among several genes and many possible environmental factors, and these genes and environments can interact in many ways to affect behavior.

2.3.3 *How can we effectively apply this information clinically?*

Although individual differences in human responses to particular alcoholism and drug abuse treatment regimens have been reported, these differences have not been studied systematically. In addition, since the problem of cocaine addiction has only recently increased to epidemic proportions, studies of genetic and environmental contributing factors in human cocaine abuse, and thus, effective treatment strategies, are still in their initial stages.

Animal studies suggest that alcohol abuse, and presumably abuse of other drugs as well, is due to the influence of multiple genetically mediated factors, including: (1) an intrinsic permissive factor which determines alcohol preference, (2) direct rewarding effects, and (3) motivational factors which maintain drug-related behaviors across a broad range of environmental conditions. Human studies appear to confirm the hypothesis that reinforcement from drugs is a relatively unique effect and needs to be studied and treated as such.

Studies using adopted-away offspring of alcoholic parents show that the predisposition to initiate alcohol-seeking behavior is genetically different from susceptibility to loss of control after drinking begins, similar to the differences in animal models between reinforcement itself and motivation to continue drinking. These studies suggest that alcohol-seeking behavior in humans involves different neuronal processes than does responsivity to the sedative effects of alcohol or susceptibility to the long term effects of this drug, such as physical dependence development[4]. However, another series of studies has shown an apparent link between response to alcohol and tendency to develop alcoholism. In one study, several hundred young men were tracked who had been tested for response to alcohol at age 20. Ten years later, a low level of response to alcohol at age 20 was associated with a fourfold greater likelihood of alcoholism in both the sons of alcoholics and the sons of nonalcoholics[30].

Other human studies show that individuals who develop alcohol abuse or dependence do not generally differ on any of a number of personality tests from persons who do not develop alcoholism, and that with the single exception of antisocial personality disorder, it is difficult to identify a personality profile associated with an individual's risk for alcoholism[29]. Indeed, antisocial personality, which includes a strong desire for novelty and a low response to harm, seems to be the strongest, and one of a very few good predictors of eventual alcohol or drug use. For example, a study using males who were given a detailed behavioral assessment when they were 11 years old, and who were studied again when they were 27 years old showed that high novelty-seeking and low harm avoidance as children were the best predictors of alcohol abuse. At age eleven these two variables alone predicted boys who later in life had a nearly 20-fold difference in their incidence of alcohol abuse[5]. However, while those with antisocial personality characteristics have an increased risk for drug abuse, only a modest percentage of alcohol and drug abusers have antisocial personality disorder.

Based upon this information, treatment strategies for alcoholism and substance abuse disorders should therefore include a consideration of both positive and negative impacts on both the direct reinforcing effects of drugs, as well as motivational factors that can lead to sustained use and relapse. For example, if a treatment strategy is to remove access to a drug which is positively reinforcing, it would be prudent to provide an alternative positive reinforcer to take the place of the drug. This approach is found in community programs where alternative positive activities are provided, such as sports programs for adolescents. Hopefully, through an integrated approach to treatment combining pharmacotherapies based upon an understanding of the biological and genetic contributors to substance abuse disorders with psychosocial interventions based upon an understanding of both sociocultural and biological risk factors, we may better treat the difficult biological, behavioral, and social problems associated with alcoholism and substance abuse disorders.

3 Human genetic studies of alcoholism and substance abuse

The world is composed of many different populations of humans that show a diverse array of alcohol drinking patterns. Concomitant with these geographic or cultural drinking differences is a wide variation in the etiology of alcohol drinking problems. Some cultures show high rates of drinking associated with high rates of alcohol abuse, while other cultures with similarly high alcohol intakes show less severity of alcohol-related problems, perhaps related to differences in patterns of drinking and social values surrounding alcohol drinking.

While it is difficult to separate specific genetic factors from environmental factors in these population analyses, other studies using more rigorous genetic designs have done so. Twin studies have shown significant genetic contributions both for average weekly consumption of alcohol and for alcohol abuse. Genetic factors appear to account for some 50% of the variation in drinking among the individuals studied. As expected for a genetically mediated trait, MZ twins are more concordant for alcoholism than DZ twins. The results from twin studies have been more robust for males than for females, consistent with a growing body of data suggesting that alcohol drinking disorders in males are under greater genetic control than in females.

The only other drug for which there exists a body of human genetic twin data, while small, is nicotine. Most of these studies have reported significant genetic contributions to smoking behavior among twins. That is, if one member of a twin pair is a tobacco smoker, the probability that their co-twin will also be a smoker is significantly greater than what would be expected based upon average population rates of smoking[24].

Adoption studies have also contributed significantly to our understanding of genetic contributions to alcohol drinking, although early studies on alcoholism failed to find differences between adopted-away offspring of biological parents who were alcoholics or non-alcoholics. Since then, however, several studies have shown that adopted biological sons of alcoholics have higher rates of alcoholism than do adopted biological sons of non-alcoholics. Consistent with the data from twin studies, findings from studies with adopted females have been less robust, again suggesting that male alcoholism has a more robust genetic component than does alcoholism in females.

An adoption study in Sweden, using that country's thorough medical registries, also provides evidence for genetic influence on male alcoholism. Twenty-two percent of adopted-away sons of biological fathers who abused alcohol were themselves alcoholic. Since this rate of alcoholism is much higher than that seen in the population overall, it is strong evidence suggesting a significant genetic contribution. Several other adoption studies are also consistent with genetic influences on alcoholism in males. However, it is important to note here that while 22% of these persons did develop alcoholism, 78% did not, even though they came from high risk biological families. Again, this illustrates the important interplay between genes and environments. Differences in inheritance of the specific biological risk factors, as well as differences in environmental stressors, cultural values, etc. could all be important in increasing or decreasing an individual's overall risk. For example, the genetic influence becomes stronger when the adopted-away sons are raised in lower-class adoptive families[32].

What little information exists on genetic predisposition to opiate or cocaine dependence in humans has been obtained as part of genetic studies on alcoholism where researchers also obtained data on use of other drugs. The results suggest that alcohol problems appear to be associated with the abuse of a number of other drugs[24].

Another area of study addresses whether or not there are subtypes of alcohol and drug abusers. It is known that alcoholism develops quite early in some individuals but much later in life for others. Does this difference represent an underlying biological difference in susceptibility? This difference in age of onset could reflect different degrees of environmental "stress" precipitating the onset of abusive drinking, or it could reflect two different disorders which have a common final expression, namely alcohol abuse.

Recent genetic studies have led to the conclusion that at least two distinct subtypes of alcoholism exist. These subtypes are distinguished in terms of personality traits, ages of onset, and the actual patterns of inheritance. The two types of alcoholism have come to be identified as Type 1 and Type 2. Type 1 alcoholism is characterized by anxiety and rapid development of tolerance and dependence on the anti-anxiety effects of alcohol. This leads to loss of control and difficulty terminating drinking binges. Type 2 alcoholism is characterized by antisocial personality and persistent seeking of alcohol for its pleasurable effects. This leads to early onset of inability to abstain, as well as fighting and arrests when drinking[6].

Results from other genetic studies suggest that the risk for substance abuse is associated with the risk for *affective* illnesses, such as depression. Both affective illness and substance abuse were more common in the biological relatives of affectively ill adoptees than in relatives of non-ill adoptees[17].

Thus, results from a number of twin and adoption studies strongly suggest that alcoholism has a substantial genetic component, and that this genetic influence is greater in males than in females. Genetic contributions to abuse of other drugs is also evident, but more research is necessary to accurately define the extent and nature of these influences.

4 Incorporating behavioral, biochemical and molecular genetic methods into an integrative approach to the genetics of addiction

4.1 ARE THERE CAUSAL BIOCHEMICAL DIFFERENCES AMONG ANIMALS WHICH DIFFER IN DRUG SEEKING BEHAVIOR?

4.1.1 *Genetic correlation is a requisite for identifying a biochemical sustrate of "addiction"*

The use of genetic correlations has provided much information on the brain systems involved in mediating responses to drugs. If animals differ in a particular response to a drug, for example alcohol drinking, then one can examine various brain pathways and systems to determine whether or not there are differences in these systems which may account for the observed difference in drinking. Correlating biochemical differences with behavioral differences is a powerful approach in the attempt to determine the mechanisms of action of drugs, especially those effects which contribute to substance abuse disorders. Since it is apparent that alcohol and drug abuse are mediated through several genes, there are a number of neuronal systems which may be involved in mediating responses to drugs.

Studies using this approach have shown that *serotonin* neurotransmitter systems are involved in aspects of alcohol drinking. There has been a considerable amount of research attention focused upon serotonin and its relationship to ethanol drinking, with the first characterizations appearing more than two decades ago[20]. Since that time, many studies have shown apparent influences of serotonergic neurotransmission on ethanol drinking. For example, a number of rodent stocks which differ in ethanol preference show correlated differences in various aspects of serotonergic function. The P and HAD rats exhibit lower levels of serotonin in such brain regions as the *frontal cortex, hypothalamus,* and the *nucleus accumbens* compared to low alcohol preferring NP and LAD rats[19]. Research on human alcoholics has demonstrated lower concentrations of serotonin metabolites in *cerebrospinal fluid* and lowered blood and platelet serotonin content[2]. These results have been used as the basis for developing new treatment medications which affect serotonin, and which may hold promise for decreasing alcohol drinking in alcoholic patients. Indeed, recent studies using *serotonin uptake inhibitors* have been shown to decrease ethanol consumption in animal models[16] and in humans[21].

Other genetic studies have shown an involvement of the *endogenous opioid system* in alcoholism. The endogenous opioid system is one of the neuronal systems proposed to mediate some of the reinforcing effects of alcohol and other drugs. Differences in the brain endogenous opioid system between the AA and ANA rats have been shown. The content of brain endogeneous opioids was significantly higher in the AA rats. This difference may be important in controlling the differences in the voluntary ethanol consumption exhibited by these animals[14]. In another similar study alcohol-preferring C57BL/6 mice were shown to have a greater sensitivity of the endogenous opioid system to ethanol relative to alcohol-avoiding DBA/2 mice[7]. These animal studies have also led to the development of drugs that affect the endogenous opioid system, and that may provide useful new treatments for alcoholism.

4.2 CAN WE IDENTIFY SPECIFIC GENE LOCI WHICH CONTRIBUTE TO 'ADDICTION'?

In addition to the many studies showing genetic differences in alcohol-seeking behavior in experimental animals, we also know that humans who are *Family History Positive* (FHP) for alcoholism are at substantially greater risk for developing the disorder. Also, several studies have shown that high concordance rates exist for alcohol drinking among twins. Thus, while any individual is physiologically capable of becoming addicted to alcohol, there appears to be significant biological differences in the degree of risk for alcoholism. While less well-demonstrated, existing evidence suggests that there are also human populations at high risk for developing other types of drug addictions.

Recent advances in molecular genetics have made it possible to analyze directly nonhuman and human genes. The question may now be asked as to whether or not we can identify specific genes responsible for alcoholism and drug abuse disorders. The answer to this question may lie in our growing understanding of whether or not the genes which may dispose certain individuals towards addiction are actually "abnormal" genes, or if they are normal and common versions of genes that, when combined in certain ways with other "normal" genes, result in stronger intoxicating or euphoric responses to alcohol and other drugs.

This may help in a better understanding of the complex genetic and environmental factors in alcohol abuse by providing for identification of gene loci that may be responsible for predisposition to, and protection from, alcoholism. Recent developments in animal model and human genetic research which are contributing to this progress will now be described.

4.2.1 *Quantitative trait loci approach in animal models*

Some physiological mechanisms are governed by a single gene. If different versions, or alleles, of this gene exist in a population, then there will likely be discrete, or qualitative, differences in expression of the trait determined by that gene. An example is the biochemical disorder PKU discussed earlier. With this disorder, there is not a continuum of relative degree of effect. Instead, individuals typically are either affected or they are not affected. However, complex behavioral traits, such as taking drugs, presumably involve more than one, if not many, genes. The question arises as to how to study the individual contributions of each of these genes when each single gene alone accounts for only a modest portion of the overall effect, and when individual gene effects are confounded by the collective effects of all of the contributing genes. One answer is through the combined use of recombinant inbred strains of mice plus genetic correlations within an approach called Quantitative Trait Loci or QTL.

The QTL approach uses recombinant inbred strains within correlational analyses to detect genes that may be involved in particular traits. To do this, the trait of interest, for example, alcohol drinking, is measured across a group of recombinant inbred strains. A typical finding would be that alcohol drinking was distributed across strains in such a manner as to suggest a continuous distribution mediated by several genes. Some strains may contain high alcohol drinkers, some might contain low drinkers, and several would likely contain animals whose drinking was some level of intermediate between the extremes. This strain distribution pattern for alcohol drinking would then be correlated with, or compared to, the frequency of several different genes and markers for genes on all of the mouse chromosomes. The finding of significant correlations between a chromosomal marker or site and the trait of alcohol drinking would suggest that a specific gene at or near the known chromosomal marker site was significantly contributing to propensity to drink alcohol. In this type of study, several of these correlated marker sites might be found, suggesting the existence of multiple genes contributing to alcohol drinking. Once these general marker sites are established, it is then possible to dissect the associated portions of the chromosome to attempt to identify the specific genes contributing to the trait.

This QTL approach is exciting and has been applied to several behavioral traits. Most of the behaviors showed strain distribution patterns suggestive of influence by more than one gene. Genetic correlations have identified potential gene locations for some of these behaviors, including avoidance behavior and exploratory behavior[22]. The QTL approach has also been recently applied to effects of amphetamine, alcohol, and morphine. Again, significant genetic correlations have been found with a number of potential gene locations[15].

4.2.2 *Allelic polymorphisms in humans*

Following recent advances in molecular biology, a number of studies have been conducted in an attempt to define specific versions, or *alleles*, of genes which result in increased probabilities for the development of alcoholism or drug abuse disorders. To date, virtually all of the studies have involved alcohol. These studies, which are attempting to define allelic polymorphisms in human genes, have focussed on two areas highly impacted by alcohol, namely the liver and the brain.

The liver is the organ which contains the critical enzymes necessary for metabolizing, or breaking down, alcohol. Thus, there could be genetically different versions of key liver enzymes which contribute to differences in the rate of alcohol metabolism. These metabolism differences could contribute to differences in the ability to break down alcohol, and as a result, confer different alcohol drinking patterns in different individuals.

In the liver, the enzymes alcohol dehydrogenase (ADH) and aldehyde dehydrogenase (ALDH) are principally responsible for metabolizing alcohol. It has been shown that there are *polymorphisms*, or different versions of the alleles that make up the genes responsible for synthesizing these enzymes. These genetic differences tend to express themselves across different ethnic groups, and may be responsible for different frequencies of alcohol-related disorders, such as liver disease. In addition, genetic variation in both ADH and ALDH may influence drinking behavior and the risk of alcoholism.

Some individuals exhibit a strong flushing response in the face following ingestion of alcohol. Alcohol and acetaldehyde are both *vasodilators*, and dilation of facial vasculature increases blood flow to this area, resulting in a reddening, or flushing, of the tissue. Alcohol-flush reactions occur strongly in Asians who inherit an ALDH gene that produces an inactive ALDH enzyme. Alcohol is converted to acetaldehyde by the ADH enzyme, and this acetaldehyde is then broken down by ALDH. Without an appropriately active ALDH enzyme, high blood acetaldehyde levels build up in these individuals and are believed to be the cause of the unpleasant symptoms that follow drinking. Asians who have this particular ALDH gene which results in an inactive form of ALDH are more sensitive to alcohol so they tend to be discouraged from drinking, and as a result typically show lower risk for alcohol-related disorders[1]. It is apparent then that polymorphisms of human alcohol dehydrogenase (ADH) and aldehyde dehydrogenase (ALDH) can contribute to important ethnic differences in alcohol metabolism and possibly alcoholism.

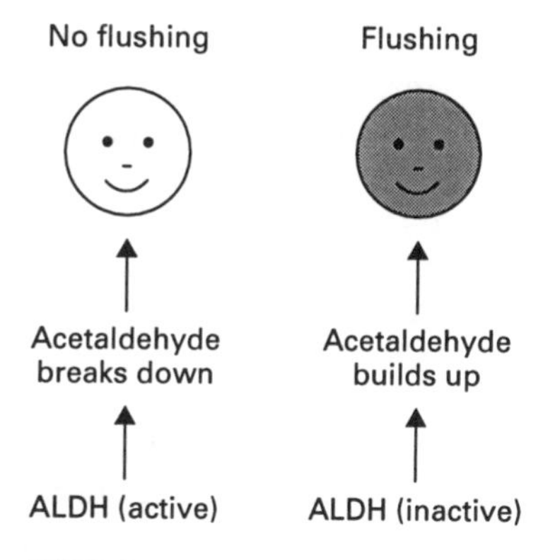

FIGURE 4.5
Relationship between aldehyde dehydrogenase (ALDH) metabolism and alcohol flushing

Thus, the presence of the allele which codes for an inactive form of ALDH appears to decrease the likelihood of developing alcoholism due to the negative consequences of drinking associated with this allele. It is interesting that in Asian populations alcohol drinking is still commonplace despite the relatively high frequency of the inactive ALDH allele. Indeed, some persons who carry this form of the ALDH allele still become alcoholic. In addition, presence of the allele for inactive ALDH has no bearing on the likelihood of developing addictions to other drugs. Clearly there are other factors, presumably related to specific interactions between drugs and neurons, that contribute to susceptibility to alcohol and drug abuse.

Biological factors related to brain function which mediate our bodies' responses to drugs form the other area of focus of human genetic polymorphism studies. Here, researchers are attempting to find genetic differences in specific neurotransmitter receptor proteins which may mediate differences in the perceived intoxicating and euphoric effects of alcohol. By screening and comparing the genetic code from human tissue samples, it has been shown that humans differ with respect to the specific genetic code they carry for various neurotransmitter receptor proteins. Recent findings suggest that alcoholics and non-alcoholics may differ in the particular gene version, or allele, they carry for a specific dopamine *receptor*, the so called dopamine type 2, or D_2 receptor[3]. As a result of this different allele, individuals may make slightly different, yet still functional, versions of the D_2 receptor. These slight differences may make certain persons more responsive to the effects that certain drugs, such as alcohol, have on dopamine neurotransmitter systems. Dopamine is a neurotransmitter believed to be critical in areas of brain function related to reward and positive reinforcement. Thus, differences in the genetic regulation of this brain receptor protein could contribute to differences in propensity to find alcohol rewarding.

Other studies, however, have failed to confirm this finding[10], so it is not yet clear whether or not the D_2 receptor gene plays a role in determining susceptibility towards alcoholism. The reasons for these inconsistent findings are unclear, and there are several possible explanations. One is that the original finding showing a link between D_2 and drinking was a "false positive" finding, that is, a coincidental result not based upon a true causal relationship between the D_2 receptor and alcohol abuse. Other reasons are that the percentages, or frequencies, of the different D_2 receptor alleles differ across geographic regions and ethnic groups, making it possibly more difficult to see such a relationship in certain world regions, or that there is variation in and disagreement on diagnosis of alcohol abuse since the disease itself may take on different forms and degrees of severity. It is interesting, however, that studies using rodent models of alcohol drinking also suggest a link between susceptibility to alcohol abuse and the D_2 receptor, but these studies are still in the initial stages, and more work is also necessary to confirm these early findings.

It has also been shown that there exist across the human population several forms, or alleles, of the gene which codes for the D_4 receptor. Alcoholics appear to also have a greater prevalence of certain versions of this dopamine receptor[13], and it may be that this particular dopamine receptor is more involved in alcoholism susceptibility than is the D_2 dopamine receptor. Interestingly, a high prevalence of nicotine abuse and other drug abuse has also been recently associated with this D_4 receptor[13]. This suggests that, as seen in the animal model findings, there may be some genetic substrates common to abuse of drugs in general. Since dopamine receptors play important roles in general reward processes in the brain, it is reasonable to hypothesize that differences in the function of these reward centers may be critical to the development of alcohol and drug abuse disorders.

5 Conclusions

Behavioral genetic studies are demonstrating that drug addiction is a complex biological process mediated by both genetic and environmental factors, as well as critical interactions between these two factors. For alcoholism the evidence is strong that for certain high risk persons, alcohol addiction can develop early with only minimal environmental contribution, mostly in terms of access and peer pressure. For other drugs, it appears that certain animals are predisposed towards initiation and subsequent maintenance of drug self-administration, while others apparently do not initiate drug self-administration, even under favorable environmental conditions, such as ready access to drugs. Also, genetic differences among individuals in their ability to metabolize alcohol appear to contribute to the large individual and ethnic variations observed in alcohol drinking.

It is unlikely that specific "addiction" genes exist in the way that there are specific genes which regulate the production of the PKU related enzyme tyrosine hydroxylase. Rather, it seems more probable that biological contributions to drug addiction result from interactions among several genes, each of which regulates a common and necessary biological system. However, in the case of persons who would be at high risk for drug abuse, the specific versions of these genes are present in a combination which is most conducive to making a drug an effective reinforcer. One important area where several genes could act in concert to produce an organism highly responsive to drugs is in an area of the brain known as the *nucleus accumbens*, where dopaminergic and other pathways interact in ways important to the reinforcing effects of drugs, as described in Chapter 19. Normal variations in receptors, metabolic enzymes, and other cellular components could, in the high risk individuals, be combined in a manner which maximizes a drug effect.

Realizing that individuals may differ in addiction risk, as well as understanding the biological factors related to risk can aid in the prevention and treatment of drug abuse. Resources for addiction intervention can be scarce, making the targeting of high risk persons for the most intensive education and counseling programs cost effective. Early intervention in terms of education and counseling of high risk groups has proven effective in other areas of medicine such as heart disease.

Genetic methods have great potential for increasing our understanding of addictions. The objectives of this integrative approach are to identify, at the molecular, cellular, and behavioral levels, those factors that maintain drug taking behaviors. Issues such as the biochemical sites of drug reinforcement, the relationship between drug preference and drug reinforcement, and the commonality of self-administration behavior across drugs, can be effectively addressed using the behavior genetic approaches discussed in this chapter.

In summary, genetic factors are being shown to play a critical role in addiction, and the use and study of genetic models in this area is not only improving our understanding of these genetic contributions to addiction, but will ultimately aid in our overall understanding of the serious problems of alcohol and drug addiction.

References

1. Agarwal, D.P. and Goedde, H.W., "Pharmacogenetics of alcohol metabolism and alcoholism," *Pharmacogenetics.* 1992, **2:** pp 48–62.
2. Ballenger, J.C., Goodwin, F.K., Major, L.F., Brown, G.L., "Alcohol and central serotonin metabolism in man," *Arch. Gen. Psychiatry.* 1978, **36:** pp 224–227.
3. Blum, K., Nobel, E.P., Sheridan, P.J., Montgomery, A., Richie, T., Jagadeeswaran, P., Nogami, H., Briggs, A.H., Cohn, J.B., "Allelic association of human dopamine D2 receptor gene in alcoholism," *JAMA.* 1990, **263:** pp 2055–2060.
4. Cloninger, C.R., "Neurogenic adaptive mechanisms in alcoholism," *Science.* 1987, **236:** pp 410–416.
5. Cloninger, C.R., Sigvardsson, S., Bohman, M., "Childhood personality predicts alcohol abuse in young adults," *Alcohol Clin. Exp. Res.* 1988, **12:** pp 494–505.
6. Cloninger, C.R., Sigvardsson, S., Gilligan, S.B., von Knorring, A.L., Reich, T., Bohman, M., "Genetic heterogeneity and the classification of alcoholism," *Adv. Alcohol Subst. Abuse.* 1988, **7:** pp 3–16.
7. De Waele, J.P., Papachristou, D.N., Gianoulakis, C., "The alcohol-preferring C57BL/6 mice present an enhanced sensitivity of the hypothalamic beta-endorphin system to ethanol than the alcohol-avoiding DBA/2 mice," *J. Pharmacol. Exp. Ther.* 1992, **261:** pp 788–794.

8. Elmer, G.I., Meisch, R.A., Goldberg, S.R., George, F.R., "Ethanol self-administration in Long Sleep and Short Sleep mice: evidence for genetic independence of neurosensitivity and reinforcement." *J. Pharmacol. Exp. Ther.* 1990, **254:** pp 1054–1062.
9. Eriksson, K., "Genetic selection for voluntary alcohol consumption in the albino rat," *Science.* 1968, **159:** pp 739–741.
10. Gelernter, J., O'Malley, S., Risch, N., Kranzler, H.R., Krystal, J., Merikangas, K., Kennedy, J.L., Kidd, K.K., "No association between an allele at the D2 dopamine receptor gene (DRD2) and alcoholism," *JAMA.* 1991, **266:** pp 1801–1807.
11. George, F.R. and Goldberg, S.R., "Genetic approaches to the analysis of addiction processes," *Trends Pharmacol. Sci.* 1989, **10:** pp 78–83.
12. George, F.R., "Is there a common genetic basis for reinforcement from alcohol and other drugs?" *J. Addict. Dis.* 1991, **10:** pp 127–139.
13. George, S.R., Cheng, R., Nguyen, T., Israel, Y., O'Dowd, B.F., "Polymorphisms of the D4 dopamine receptor alleles in chronic alcoholism," *Biochem. Biophys. Res. Commun.* 1993, **15(196):** pp 107–114.
14. Gianoulakis, C., de Waele, J.P., Kiianmaa, K., "Differences in the brain and pituitary beta-endorphin system between the alcohol-preferring AA and alcohol-avoiding ANA rats," *Alcohol Clin. Exp. Res.* 1992, **16:** pp 453–459.
15. Gora-Maslak, G., McClearn, G.E., Crabbe, J.C., Phillips, T.J., Belknap, J.K., Plomin, R., "Use of recombinant inbred strains to identify quantitative trait loci in psychopharmacology," *Psychopharmacology.* 1991, **104:** pp 413–424.
16. Haraguchi, M., Samson, H.H., Tolliver, G.A. "Reduction in oral ethanol self-administration in the rat by the 5-HT uptake blocker fluoxetine," *Pharmaco.l Biochem. Behav.* 1990, **35:** pp 259–262.
17. Ingraham, L.J. and Wender, P.H., "Risk for affective disorder and alcohol and other drug abuse in the relatives of affectively ill adoptees," *J. Affect. Disord.* 1992, **26:** pp 45–51.
18. Li, T-K., Lumeng, L., McBride, W.J., Waller, M.B., "Indiana selection studies on alcohol-related behaviors." In: McClearn, G.E., Dietrich, R.A., Erwin, V.G. (Eds.), *Development of Animal Models as Pharmacogenetic Tools,* Monograph 6, U.S. Department of Health and Human Services, Rockville, MD, 1981, pp 171–191.
19. Murphy, J.M., McBride, W.J., Lumeng, L., Li, T.-K., "Contents of monoamines in forebrain regions of alcohol-preferring (P) and -nonpreferring (NP) lines of rats," *Pharmacol. Biochem.Behav.* 1987, **26:** pp 389–392 (licensing enquiries are invited).
20. Myers, R.D. and Veale, W.L., "Alcohol preference in the rat: reduction following depletion of brain serotonin," *Science.* 1968, **160:** pp 1469–1471.
21. Naranjo, C.A., Poulos, C.X., Bremner, K.E., Lanctot, K.L., "Citalopram decreases desirability, liking, and consumption of alcohol in alcohol-dependent drinkers," *Clin. Pharmacol. Ther.* 1992, **51:** pp 729–739.
22. Neiderhiser, J.M., Plomin, R., McClearn, G..E., "The use of CXB recombinant inbred mice to detect quantitative trait loci in behavior," *Physiol. Behav.* 1992, **52:** pp 429–439.
23. Phillips ,T.J., Feller, D.J., Crabbe, J.C., "Selected mouse lines, alcohol and behavior," *Experientia.* 1989, **45:** pp 805–827.
24. Pickens, R.W. and Svikis, D.S., "Genetic influences in human substance abuse," *J. Addict. Dis.* 1991, **10:** pp 205–214.
25. Plomin, R., DeFries, J.C., McClearn, G.E. (Eds.), *Behavioral Genetics: A Primer,* 2nd edition. W.H. Freeman, New York, 1990.
26. Richter, C.P. and Campbell, K.H., "Alcohol taste thresholds and concentrations of solution preferred by rats," *Science.* 1940, **91:** pp 507–508.
27. Ritz, M.C., Garcia, J., Protz, D., George, F.R., "Operant ethanol-reinforced behavior in P, NP, HAD and LAD rats bred for high or low ethanol preference," *Alcohol. Clin, Exp. Res.* 1994.
28. Ritz, M.C., Garcia, J.M., Protz, D., Rael, M., George, F.R., "Ethanol-reinforced behavior in P, NP, HAD and LAD rats: differential genetic regulation of reinforcement and motivation," *Behav. Pharmacol.* 1994.
29. Schuckit, M.A., Klein, J., Twitchell, G., Smith, T., "Personality test scores as predictors of alcoholism almost a decade later," *Am. J. Psychiatry.* 1994, **151:** pp 1038–1042.

30. Schuckit, M.A., "Low level of response to alcohol as a predictor of future alcoholism," *Am. J. Psychiatry*. 1994, **151:** pp 184–189.
31. Seale, T.W. and Carney, J.M., "Genetic determinants of susceptibility to the rewarding and other behavioral actions of cocaine," *J. Addict. Dis.* 1991, **10:** pp 141–162.
32. Sigvardsson, S., Cloninger, C.R., Bohman, M., "Prevention and treatment of alcohol abuse: uses and limitations of the high risk paradigm," *Soc. Biol.* 1985, **32:** pp 185–194.
33. Zacharko, R.M., Lalonde, G.T., Kasian, M., Anisman, H., "Strain-specific effects of inescapable shock on intracranial self-stimulation from the nucleus accumbens," *Brain Res.* 1987, **426:** pp 164–168.

Contents Chapter 5

Reward systems and addictive behavior

0-8493-7803-6/99/$0.00+$.50

Chapter 5

Reward systems and addictive behavior

Mary C. Ritz

INTRODUCTION

In the last 2 decades, there has been a theoretical shift in the scientific description of drug addiction. Previously, emphasis was placed on the chronic effects of drugs, with withdrawal symptoms as the key element in determining drug addiction. The biological bases of addiction, then, were thought to be the adaptive neuronal changes related to tolerance and dependence. At present, the reinforcing or rewarding properties of drugs, modulated by specific discriminative cues and concurrent aversive properties, are considered essential in determining the addictive potential or abuse liability of these compounds. Moreover, emphasis has been placed on drug-seeking behavior as the component of the addiction process which is common to all drugs of abuse. The neuropharmacological correlates of reward are considered to be the key to our scientific understanding of drug addiction, and these neuronal mechanisms are the targets of research efforts to elucidate effective pharmacological adjuncts to drug addiction treatment programs.

1 Models of drug self-administration

For animal models see Chapter 3

Two distinct animal models of drug taking have been developed over the years, and can be distinguished in part by the degree to which the scientific community accepts each paradigm as a measure of the intrinsic reinforcing or rewarding effects of drugs. Most animal studies of ethanol drinking have used a 24-hour two-bottle choice paradigm, initially developed by Richter and Campbell (1940)[54], to measure ethanol preference. Investigators using this method have frequently referred to an animal's volitional selection of ethanol or other drugs as analogous to human drug taking. However, it is never certain that the pharmacologic properties of drugs determine voluntary intake in a preference test. In the case of ethanol, for example, taste or olfactory cues can easily influence drinking behavior. In general animals prefer ethanol/water solutions over drinking water in concentration of 1–6%; concentrations of 10% or more are definitely consumed less than water. For central effects to occur animals have to consume quantities of ethanol in excess of that what is readily metabolized. Most preference studies do not give direct evidence that animals consume centrally effective amounts of ethanol.

One study with rhesus monkeys showed that ethanol consumption and water consumption patterns did not run parallel; most water was consumed during and directly after the meals, the least water during the night (Figure 5.1). Ethanol was consumed in a rather stable pattern which continued throughout the night (from ref. 37). However, consumption patterns are often confounded by diurnal patterns of eating and drinking.

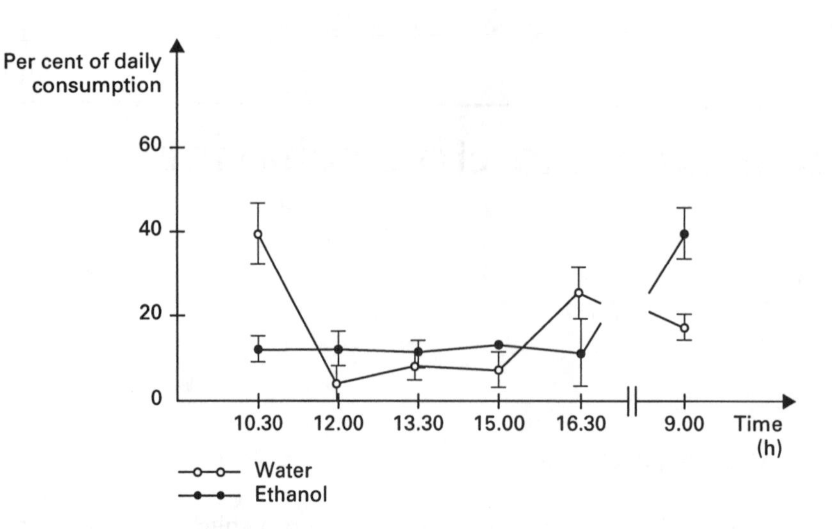

FIGURE 5.1

Mean relative distributions over 24 hours of net ethanol or water intake in rhesus monkeys

Registration times represent mean (± SEM) consumption as % of total 24 hr of water or ethanol intake during the previous intermeasurement interval (1.5 hr day, 16.5 hr night). Monkeys had access to one bottle with 16% ethanol (v/v), one bottle with 32 % ethanol, and one water bottle with water. (Data from Kornet, M. et al., *Physiol Behav.* 1990, **47**: pp 679–684.)

Conversely, most studies of non-ethanol drug taking have utilized the principles and methods of operant behavior. The operant paradigm has illustrated that drug-seeking behavior can be conceptualized as a specific instance of operant behavior, in which an animal performs a particular behavioral response in order to obtain access to the test drug. Drug-reinforced behavior in primates, rodents, and other species is considered to be a useful model of human addiction liability since most drugs which humans abuse have been found to be reinforcing in animals of various species[30].

To date, the relationship between preference for a drug and drug-reinforced behavior is not well understood. Recent research suggests that the preference model of animal ethanol drinking may provide some information about the reinforcing effects of ethanol, but that these two measures are phenomenologically distinct and may have only limited common substrates. Detailed analyses illustrate the existence of distinct, biologically influenced components of drug-reinforced behavior, and have demonstrated similarities as well as differences between these two measures of drug taking. It appears that the preference model may describe drug taking behavior only under somewhat specific conditions, and may not be analogous to or highly predictive of the reinforcing effects of these compounds under other situations or in humans. However, the relationship between preference for drugs and self-administration of these compounds in an operant paradigm has not been systematically studied (see also Chapter 3).

2 Biological and genetic influence on drug self-administration: evidence for possible common pathways for reinforcing effects of drugs

Drug use and abuse have traditionally been treated predominantly as psychosocial problems. For this reason, it has often been assumed that environmental factors, not inherited traits, are primarily responsible for drug-seeking behavior. In general, drugs which are typically abused by human beings serve as reinforcers in animal subjects from several species. Thus, it has been commonly assumed that, with few exceptions, animal subjects given the appropriate training/environmental cues and drug history would self-administer drugs known to serve as reinforcers. Experimental results have often been described for individual animals rather than groups of animals due to large individual differences in response patterns. Individual differences between subjects, however, were not generally described in terms of inherited traits. In addition, animal subjects for which a drug did not serve as a reinforcer were replaced by others which emitted drug-seeking behaviorism, thus eliminating abstainers from the subject pool. More recently, the integration of behavioral pharmacogenetic, biochemical pharmacogenetic, and operant methodologies has potential for increasing our understanding of the contributions and interactions of genetic and environmental factors in the etiology of addictive behavior.

Pharmacogenetic studies using both human and animal populations have demonstrated that sensitivity to drugs, tolerance to drugs and drug-reinforced behavior are complex phenomena mediated by genetic and environmental factors and by critical interactions between these factors. Numerous findings from this area of science indicate that drug-reinforced behavior has a strong biological basis. Moreover, current scientific evidence suggests the existence of genetic vulnerability to most drug effects in certain individuals or populations. Recognition of these differences suggests the influence of inherited biological factors which predispose individuals to greater or lesser vulnerability to drug effects.

See Chapter 4 for genetics and addictive behavior

Pharmacogenetic studies to date have lead to at least three important concepts related to biological mechanisms associated with the reinforcing properties of drugs. First, genetic differences in drug response have been utilized to elucidate inherited differences in neuronal systems which may be associated with such variation. Most biochemical genetic studies focus at the level of the neuron, neurotransmitters, peptides, receptors and enzymes. Gene expression, localization, and interactions between neuronal systems have also increasingly become targets of analysis. The primary task of the biochemical pharmacogeneticist is that of describing variation in biochemical events that might be associated with individual differences and strain differences in drug response. Ultimately, understanding how drugs interact with the central nervous system to produce genetically determined differences in either reinforcing or toxic effects increases the probability that the potentially harmful effects of these compounds may be blocked, or at least attenuated, by the development of more effective prevention and treatment programs. Elucidating biochemical mechanisms common to various abused substances will suggest clinically useful prevention and treatment strategies.

For treatment strategies in addiction, see Chapters 8 and 9

Second, these pharmacogenetic approaches also provide support for the assertion that the reinforcing effects of drugs of abuse are fundamentally associated with the interaction of these compounds with neuronal pathways in the brain which mediate reward[75]. These reward pathways exist in the brain for the purpose of maintaining behaviors which increase the viability of the organism. Thus, these brain structures and neuronal pathways mediate and promote organismic consummatory and sexual behaviors. By interactions with these neuronal systems, drugs of abuse produce reinforcing effects and the probability of further drug seeking behavior is increased.

To the extent that genetically distinct organisms exhibit drug-reinforced behavior which generalizes to various classes of abused substances, rather than genotype-specific reinforced behavior directed toward a single distinct compound, the existence of a common reward pathway may be argued.

Pharmacogenetic methods have now illustrated that drugs from several pharmacological classes will come to serve as reinforcers for animals of some genotypes, while these compounds are not reinforcers for other genotypically distinct animals. These initial results from studies of drug self-administration across different drugs and genotypes suggest that genotypic patterns of reinforcement from ethanol may correlate highly with patterns of reinforcement from cocaine and opiates. Thus, drug seeking behaviors maintained by ethanol, cocaine and opiates may have at least some common biological determinants. It appears that behavioral genetic studies provide some basis for arguing for the existence of common pathways for the reinforcing effects of several drugs of abuse.

Third, pharmacogenic strategies have lead to the suggestion that the intrinsic reinforcing effects of drugs may be distinguished from the incentive value of the drug, and that distinct neuronal mechanisms for these components of drug taking behavior may be targeted in drug abuse treatment programs. As discussed in Chapter 4 on genetic control of drug-reinforced behavior, pharmacogenetic methodologies appear to provide a research strategy for examining various components of drug seeking behavior. Comparisons of drug effects across genotypes have been used to determine the degree of common biological control of various other drug effects. For example, there appears to be little relationship between reinforcement and sensitivity to depressant effects. Therefore, reinforcement is not equivalent to, or dependent upon, depressant effects. In addition, data indicate that reinforcement is not equivalent to, nor dependent upon, stimulant or other drug effects. In general, evidence suggests that the reinforcing effects of drugs comprise a unique dimension of effect which is not the result of, nor due to, causal genetic relationships with other drug effects. Instead, effects such as sensitivity to, toxicity from, and reinforcement by each drug are more likely to be mechanistically independent. The neuronal sites of action mediating reinforcement are distinct from those mediating other drug effects.

For pharmacotherapy of addiction see Chapter 9

The intrinsic reinforcing effects of a drug *per se* can apparently also be distinguished from its incentive value, or the magnitude of work that will be exhibited in order to obtain access to the drug. Thus, a drug may serve as a highly efficacious reinforcer for one organism, but responding will decrease rapidly if operant requirements are increased. In contrast, the same drug may also serve as a potent reinforcer in another organism which will exhibit increasingly greater work loads in order to gain access to the drug. Understanding the neuronal mechanisms associated with these components of drug seeking behavior can increase targets for effective pharmacotherapy related to addiction.

3 Brain structures associated with addictive behaviors

3.1 THE "REWARD PATHWAY": MESOCORTICOLIMBIC STRUCTURES

Two major types of research paradigms have aided in the elucidation of the brain structures and neuronal tracts associated with reward. First, intracranial self-stimulation (ICSS) experiments have shown that animals will perform operant tasks in order to elicit electrical stimulation to specific areas of the brain. Since Olds and Milner (1954) demonstrated that rats would learn to press a lever in order to initiate stimulation of certain areas of their own brains, intracranial self-stimulation (ICSS) experiments have been utilized to study brain regions which appear to mediate reward, or positive reinforcement[49]. As discussed in Chapter 3, a positive reinforcer is that which increases the probability of the behavior that immediately preceded it in time. Secondly, physical or chemical lesions of specific brain areas have been utilized to study the influence of specific brain areas on ICSS or drug self-administration behaviors.

These experimental paradigms have indicated that specific limbic structures in the brain appear to be consistently associated with what might be considered a common reward pathway through the brain[35,41]. The medial forebrain bundle, the nucleus accumbens (NA), the ventral tegmental area (VTA), the lateral and ventromedial nuclei of the hypothalamus, and the medial prefrontal cortex serve as core structures of the "reward pathway". Several other structures also provide significant modulatory signals to this system. The reticular activating system (RAS), located in the brain stem, controls arousal and attention to a vast array of sensory inputs from our environment. The central grey area surrounding the aqueduct of Sylvius in the mesencephalon and nerves ascending to the periventricular nuclei of the hypothalamus mediate the aversive effects of drugs and other reinforcers. Thus, these brain regions serve as a "punishment" pathway, attenuating the rewarding effects of these stimuli. Several other areas of the brain provide input to the reward pathway concerning emotional and motivational variables. These predominately limbic regions include the septum, amygdala, and thalamus. Finally, still other areas of the brain are involved, not in the mediation of reward *per se*, but in the translation of that experience into motor activity. The basal ganglia and cerebellum are involved in the control of fine voluntary and learned motor control.

In general, self-stimulation behavior controlled by operant schedules is highly sensitive to changes in central dopaminergic neuronal transmission. Indeed, dopamine appears to be the primary neurotransmitter associated with reward in general and drug reinforcement in particular. Lesions of dopaminergic neuronal systems within the reward pathway reduce drug self-administration, most often without influencing other distinct drug effects such as effects on locomotion and the development of tolerance or physical dependence. Pharmacological compounds which enhance the synaptic transmission of dopamine also increase self-stimulation behavior and facilitate drug self-administration, while pharmacological manipulations which inhibit the transmission of dopamine have opposite behavioral effects.

The medial forebrain bundle has been extensively studied in ICSS experiments. The evidence indicates that this brain region, compared with all other brain regions, supports the most robust self-stimulation behavior. Lesions of this region also decrease, but do not eliminate, intracranial self-stimulation. Anatomically, the medial forebrain bundle forms a structural link between ventral midbrain regions and ventral forebrain areas. It is composed of myelinated nerve fibers connecting the olfactory tubercle, the septum, the nucleus accumbens with the hypothalamus, and the ventral tegmental area. The ascending dopaminergic mesocorticolimbic neuronal system is a major component of the medial forebrain bundle associated with brain structures which appear to mediate reward. The cell bodies of this system, traditionally known as the A10 group of catecholamine neurons, originate in the ventral tegmental area and project to the nucleus accumbens, septal region, amygdala, frontal cortex, and olfactory tubercle (Figure 5.2). It is also thought to be one of the more anterior parts of the complex reticular activating system and as such, may mediate or modulate neuronal communications associated with attention and emotional inputs to ventral striatal neurons mediating motivated action.

The hypothalamus has also been used extensively in intracranial self-stimulation studies. Hypothalamic self-stimulation appears to depend on the activation of descending neurons which are not dopaminergic, but which

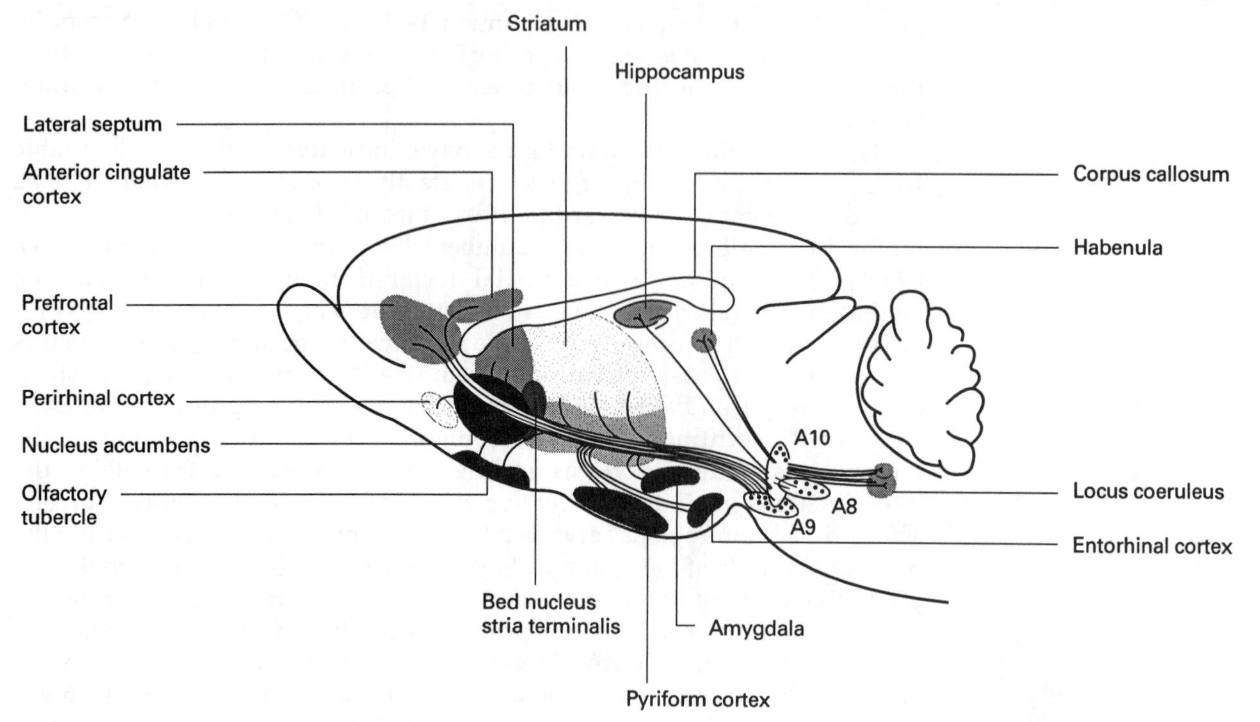

FIGURE 5.2

Graphic illustration of the limbic dopaminergic neuronal system in the rat brain A8, A9 and A10 are anatomical locations containing large groups of dopaminergic cell bodies.

stimulate dopamine release in the mesolimbic system. Thus, hypothalamic self-stimulation is also mediated indirectly by components of the reward pathway. Moreover, these experiments also indicate the influence of hypothalamic tracts on mesocortical neurons, and suggest a mechanism by which emotional or motivational factors might modulate the reinforcing properties of drugs.

These structural components of the reward pathway appear to mediate innate biological drive states, receiving modulatory inputs from the limbic system, and communicating with the dopaminergic extrapyramidal motor system to produce motor output and physical activation. Addictive compounds may be considered to mimic natural rewards. Drugs of abuse are able either directly or indirectly to innervate the reward system to produce a reinforcing or rewarding experience. For example, psychostimulants appear to interact directly with dopaminergic components of this system to produce powerful euphoric responses. The reinforcing effects of opiates also appear to be mediated in part by the VTA and NA. Due to the relatively more nonspecific actions of ethanol on brain neurons, components of the reward pathway receive modulatory inputs from a variety of limbic and cortical brain regions which lead to ethanol reinforcement and ethanol self-administration.

3.2 HYPOTHALAMUS AND THE PITUITARY GLAND: REGULATION OF HOMEOSTATIC FUNCTIONS AND REWARD PATHWAYS

The hypothalamus also serves as a critical junction between nervous and endocrine system, and is the control center for the autonomic nervous center (Figure 5.3). It is comprised of several nuclei that monitor blood levels of both endogenous compounds and essential nutrients, and signals the

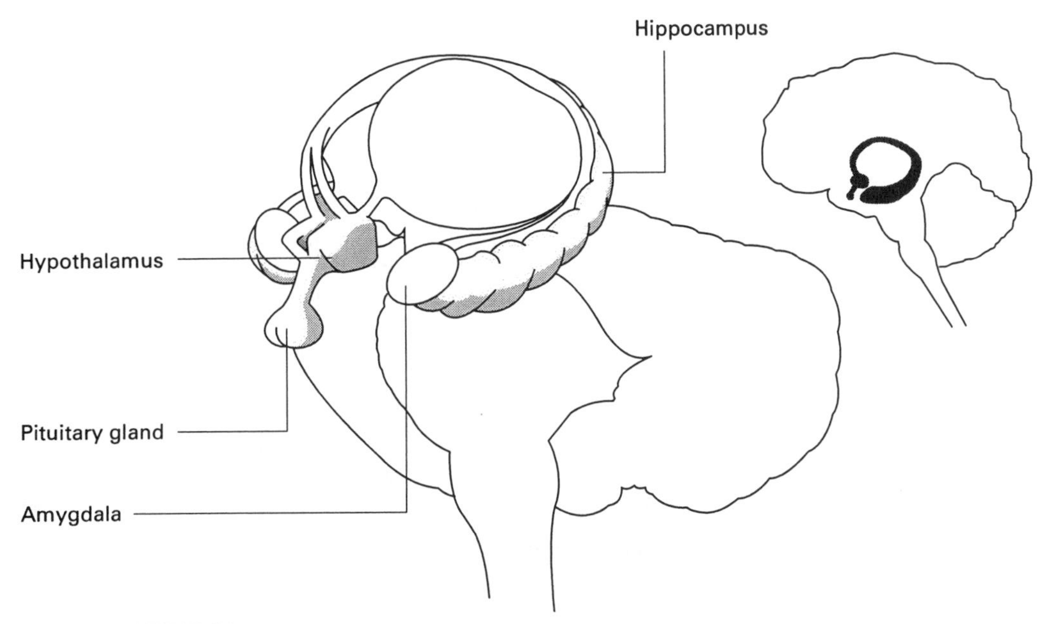

FIGURE 5.3
The limbic system and hypothalamus in the human brain

secretion of various hormones in order to maintain homeostasis. Thus, the hypothalamus is the control center for the regulation of behaviors associated with homeostatic goals such as food, water, and sexual activity. Related to this function, it is closely associated with the parts of the limbic system which make up the reward pathway. Through its close connections with both the autonomic nervous system and with the endocrine system, the hypothalamus receives and coordinates stimuli from the environment, from the brain, and from the blood stream to monitor and maintain homeostasis.

The cells of the hypothalamo-hypophyseal system, connecting the hypothalamus with the pituitary gland, are critical to the function of the hypothalamus in maintaining homeostasis. These cells are neurosecretory cells linking neural systems in the brain with both the anterior and posterior lobes of the pituitary and the surrounding blood stream (Figure 5.4). Cells in both the paraventricular and the supraoptic nuclei secrete hormones, that are released into the posterior pituitary gland, absorbed by local blood vessels, and then distributed to other parts of the body. Still other hormones are secreted from hypothalamic neurosecretory cells onto small blood vessels in the hypothalamus at the median eminence. These peptide molecules are then transported to the anterior pituitary to control the release of pituitary hormones. As the master gland in the endocrine system, the pituitary gland in turn sends a number of chemical signals to other secretory glands. Peptide hormones secreted by the pituitary, including oxytocin, vasopressin, adrenocorticotropic hormone, thyrotropin, and gonadotropin, mediate a large number of homeostatic functions by regulating metabolism, sexual cycles and stimulation, maternal behavior, growth, and stress.

The strong link between the hypothalamo-hypophyseal system and the brain regions associated with reward suggests that environmental factors can have a large influence on rewarding stimuli in general, and on drug reinforcement in particular. Indeed several lines of research have indicated that experimentally induced stress or the coadministration of pituitary hormones can alter drug self-administration.

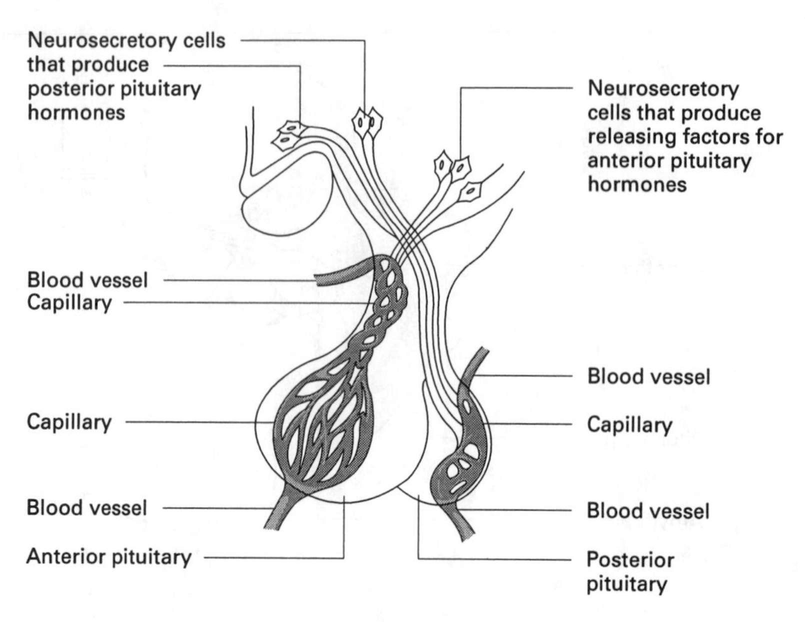

FIGURE 5.4
Hypothalamic control of pituitary hormones

4 Neurochemical mechanisms of addictive behavior: neuropharmacology of reward and motivation

If common neurotransmitter pathways are associated with the reinforcing effects of several drugs of abuse, it is possible that pharmacological interventions to such pathways may be helpful in treating drug abuse, especially polydrug abuse. Several neurotransmitter systems have been shown to influence the reinforcing properties of drugs of abuse. To the extent that a specific neurotransmitter appears to influence the reinforcing effects of multiple drugs, it might be suggested that it may influence neuronal signaling in the reward pathways of the brain.

4.1 DOPAMINE AS THE MAJOR "REWARD" NEUROTRANSMITTER

There is a great deal of evidence that stimulation of dopamine transmission in the limbic system is a fundamental property of drugs that are abused. This stimulation may occur either by blockade of dopamine reuptake or by stimulation of dopamine release, with the end result contributing to reinforcement from and, ultimately, chronic use of drugs.

Brain dialysis studies have recently shown that drugs which are abused by humans preferentially increase dopamine efflux in limbic areas of the brain, especially in nucleus accumbens, while drugs which are not abused or are associated with aversive effects do not have this effect[11,19]. In particular, opiates, nicotine, amphetamine, cocaine, and ethanol have been shown to influence extracellular dopamine in the nucleus accumbens and the dorsal caudate nucleus using dialysis methods. These drugs increased extracellular dopamine concentrations in both areas, but especially in the accumbens. Other drugs exhibiting aversive properties reduced dopamine release in the accumbens and in the caudate. Drugs not abused by humans such as imipramine, atropine, and diphenhydramine do not alter dopamine concentrations.

In other studies using similar methods, the effects of ethanol, either administered by gavage or voluntarily ingested, on brain dopamine metabolism was studied. Voluntary ingestion of ethanol increased dopamine metabolites and reduced dopamine levels, suggesting that voluntary ingestion increases the release of dopamine from nigro-striatal and meso-limbic dopamine neurons. Brain slice preparations of the ventral tegmentum (VTA), a brain region which has been implicated in the reinforcing properties of many drugs of abuse, have been used to study the actions of ethanol on dopamine neurons[9]. In such studies, ethanol has been shown to produce a dose-dependent excitation of VTA dopaminergic neurons. In other related studies, the dopamine agonist bromocriptine has been shown to decrease ethanol intake, presumably by augmenting the reinforcing effects of ethanol. By further increasing dopamine efflux, bromocriptine leads to greater synaptic concentrations of dopamine than would be associated with ethanol ingestion alone. Thus, there is a greater concentration of synaptic dopamine per unit ethanol ingested, and ethanol drinking behaviors are reduced accordingly.

Many behavioral studies utilizing lesion techniques and pharmacological manipulations have elucidated the neuronal mechanisms associated with the reinforcing effects of psychostimulants. The reinforcing effects of cocaine and its analogues in animal models of cocaine self-administration have been associated with their effects at dopamine transporters and, thus, their potency to block dopamine uptake in mesolimbocortical nerve terminals[4,39,57]. The effects of brain lesions on cocaine self-administration also support the view that mesolimbic dopaminergic neurons mediate psychostimulant self administration[26,42,59,60]. Selective dopaminergic receptor blockade attenuates the reinforcing properties of both cocaine and amphetamine in animals[16,18,20,27,55,80]. Dopaminergic agonists, in contrast, substitute for intravenous self-administration of cocaine and *d*-amphetamine[77,81]. Indeed, recent experiments utilizing operant methods have been utilized to determine of the specific dopaminergic receptor subtypes which may be involved in mediating the reinforcing effects of cocaine. To date however, it remains unclear whether either D_1 or D_2 dopaminergic receptors specifically mediate these drug effects[5,34,36,77,78].

In human subjects, positron emission tomography (PET) techniques have indicated a similarity between the time course of cocaine's occupancy of the binding site at the dopamine transporter and the time course of associated psychological effects[21]. Using these methods, it has also been shown that the density of postsynaptic dopaminergic D_2 receptors in specific brain regions may be associated with the time course of detoxification and recovery in cocaine abusers[72,73].

The reinforcing effects of opiates have also been associated with dopaminergic neurons in brain reward pathways. Lesions of the ventral tegmental area (VTA) or the nucleus accumbens, for example, reduce or extinguish opiate-reinforced behavior without influencing the development of tolerance[6,7,67]. Microinjections of opiates into the VTA also produce conditioned place preference[8,51,52]. In addition, some of the discriminative cues associated with morphine have been associated with D_1 receptors. However, relevant research findings indicate that modulation of opiate self-administration may not be effectively achieved using agents which inhibit dopaminergic neurotransmission[71]. It has been shown, for example, that high dose neuroleptic pretreatments decrease self-administration of heroin in rats, while low to moderate doses produce no effect, even though these latter doses effectively decrease cocaine intake[20]. Similarly, the dopamine receptor antagonist pimozide shifted the dose-response functions for heroin self-administration, but did not decrease the efficacy of reinforcing effects[23]. These data are consistent with and may be explained in part by studies indicating the localization of opiate receptors within brain regions associated with the reward pathway, especially the VTA and NA. Few opiate receptors are located on dopamine neurons in reward pathways, and opiates generally do not regulate dopaminergic neurons directly (see Chapter 6).

Thus, stimulation of dopamine transmission in the limbic system might be a fundamental property of drugs that are abused. This stimulation of dopaminergic action appears to occur by blockade of dopamine reuptake in the case of cocaine, by stimulation of dopamine release in the case of ethanol, and by facilitation of dopamine effects on target neurons in the case of opiates.

4.2 SEROTONIN AS A POTENTIAL ATTENUATING FACTOR

While specific dopaminergic systems may mediate initial reinforcement leading to acquisition and possibly maintenance of drug self-administration, other neurotransmitter systems may influence drug-reinforced behavior. Of the various neurotransmitter systems with potential to modulate drug self-administration, serotonergic systems have been the most extensively studied (Figure 5.5).

Research findings from several areas of research suggest that there exist several related but distinct dimensions of drug-reinforced behavior, and that these dimensions can be separated for detailed analysis of their contributions to substance abuse. For example, in the case of at least some drugs of abuse, reinforcement may result from the summation of both positive rewarding effects and concurrent aversive effects or attenuating modulatory effects of some other description. Some early studies of possible reciprocal mechanisms balancing elements of a reward system for behavioral facilitation and components of a punishment system for behavioral suppression have indicated that serotonergic neurotransmission in the medial forebrain bundle may influence reward mechanisms. Such studies provided evidence that serotonin itself suppressed ICSS behavior and that serotonergic antagonists facilitated this behavior. On the other hand, serotonin antagonists suppressed passive avoidance in animal models, a behavior which depends upon learned responses to punishment.

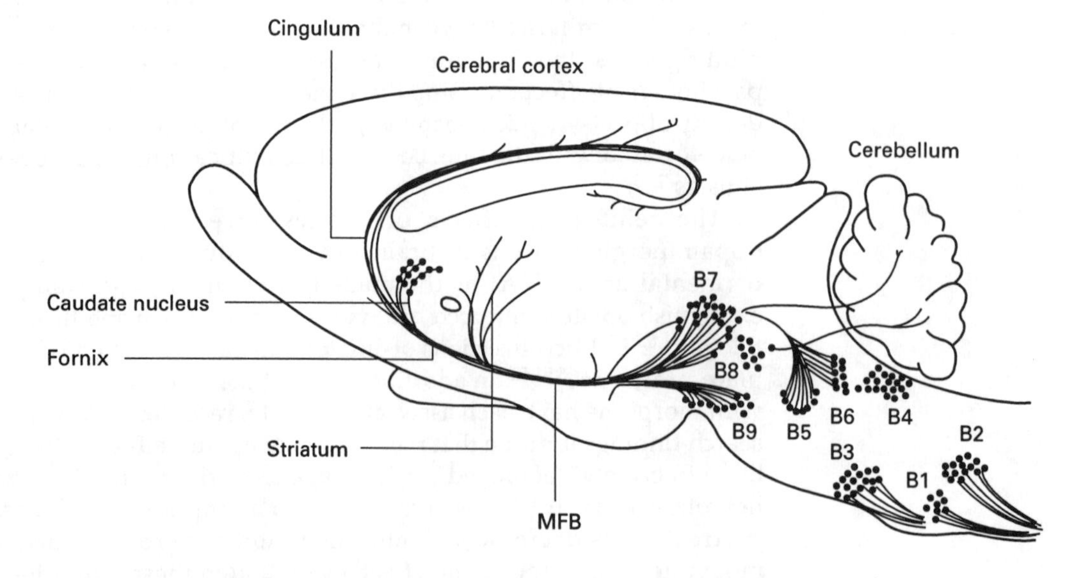

FIGURE 5.5

Main serotonergic neuronal pathways in the rat brain

B1–B9 represent clusters of cells that produce serotonin; MFB=medial forebrain bundle.

Other studies of drug-reinforced behavior in animals also indicate that this behavior may be influenced by not only the intrinsic reinforcing effects of a drug, but also factors which determine motivation to work for the drug (i.e. incentive value). Such factors might mediate the amount of work an organism will perform to obtain drugs. Indeed, it seems likely that continued chronic abuse of a drug requires not only factors mediating the intrinsic reinforcing effects of a drug, but also motivational factors. It is conceivable that these motivational factors may influence or, perhaps, be analogous to human craving for drugs, and may be mediated, at least in part, by serotonergic pathways. If this were the case, serotonergic pathways would interact with dopaminergic neuronal systems to determine the addictive behaviors associated with drug abuse.

Indeed, it has been well documented that serotonergic neurotransmission influences the function of dopaminergic neurons in several brain areas, including those associated with the reward pathway. For example, serotonergic neurons projecting from the dorsal raphe nucleus to the VTA and the NA appear to regulate the release of dopamine in the NA[33,82]. Serotonergic neurons project from the raphe nucleus to brain areas throughout the basal ganglia. However, the exact nature of this influence of serotonin on dopaminergic neuronal function may depend upon the experimental conditions utilized or perhaps the specific brain region studied. Serotonin has been shown to elicit both inhibitory and excitatory effects on dopamine release[12,33]. Serotonin has been shown to stimulate dopamine release from striatal dopaminergic neurons in a concentration-dependent manner, and there is some evidence that this effect appears to be mediated at least in part by both $5HT_1$ and $5HT_3$ receptors[2,3,79]. Indeed, $5HT_3$ receptor antagonists attenuate amphetamine-induced increases in locomotor activity[14]. In contrast, serotonergic $5HT_1$ receptor agonists have been shown to increase both the synthesis and release of dopamine in mesolimbic and mesocortical dopaminergic neurons.

Serotonin $5HT_2$ receptors also appear to mediate stress-induced inhibition of periventricular hypophysial dopaminergic neurons, thus leading to decreased secretion of specific hormones from these neurons[29]. It seems possible that this effect might play a role in mediating the effects of environmental stressors on drug seeking behaviors.

Considerable support for the hypothesis that serotonergic neurotransmission influences ethanol drinking has been provided by pharmacogenetic experiments designed to determine specific biochemical or neuronal traits associated with differences in this behavior in genetically distinct rodent stocks. Rodents show large genetic differences in ethanol preference and ethanol-reinforced behavior (see Chapter 4 on genetic factors in addiction). Significant differences in measures of serotonin release and turnover have been found between rodent stocks which self-administer ethanol versus those which do not. Recently, research has shown that Preferring (P) and High Alcohol Drinking (HAD) rats have significantly higher densities of serotonergic $5HT_{1A}$ receptors, particularly in nucleus accumbens and frontal cortex, relative to Non-preferring (NP) and Low Alcohol Drinking (LAD) rats. Mouse stocks for which ethanol functions as a positive reinforcer exhibit substantially lower densities of serotonergic $5HT_2$ receptors in striatal, midbrain, and hippocampal brain regions, relative to mice in which ethanol does not function as a reinforcer.

The influence of serotonergic systems on ethanol drinking has been investigated for nearly 2 decades since initial studies indicated that lesions or pharmacological manipulations which resulted in depletions of brain serotonin were associated with decreased ethanol preference in a two-bottle choice paradigm. To date, both preference and operant ethanol-reinforced behavior paradigms have been used to study the effects of serotonergic neurotransmission on ethanol drinking. The results from these studies suggest that blockade of serotonergic postsynaptic receptor activity may increase ethanol intake under certain conditions. Related evidence suggests that specific serotonergic receptor subtypes may selectively mediate the discriminative cues associated with low doses of ethanol[63]. On the other hand, inhibition of serotonin reuptake or administration of serotonin synthesis precursors has been shown to decrease voluntary consumption of ethanol solutions in rodent preference paradigms[15,47,61]. The results of other studies indicate that ethanol-reinforced responding in an operant paradigm can also be decreased in a dose-related manner following same day pretreatments with the potent serotonin uptake inhibitor fluoxetine[32,46].

Repeated administration of serotonin uptake inhibitors appears to reduce ethanol self-administration initially, but the effectiveness of these drugs may vary over time. For example, one study has indicated that fluoxetine administered in a single daily infusion of 10 mg/kg produced a significant decrease in ethanol-reinforced responding in rats beginning on the first day of treatment, with this decrease becoming greater on subsequent days of a 7 day treatment regimen[46]. In another study, chronic daily treatment of C57BL/6J male mice with one of three serotonin uptake inhibitors, fluoxetine, sertraline, or paroxetine, produced an initial decrease in operant lever pressing behavior for ethanol followed by a return to baseline after a few days of treatment with each of these compounds[31]. After a washout period of several weeks, pretreatment with these drugs again produced an initial decrease in responding for ethanol, followed by a return to baseline after 3–4 days of treatment. These data suggest that suppression of ethanol drinking may be related to changes in serotonin receptor populations produced by chronic treatment with serotonin uptake inhibitors. Further, these data suggest that inherited differences with respect to innate serotonergic receptor function and the response of these systems to chronic drug treatment may lead to variable effectiveness of serotonin uptake inhibitors in decreasing ethanol drinking.

The manner in which serotonergic systems may modulate ethanol intake is as yet unclear. It is clear that decreased responding caused by pharmacologic blockade of a reinforcing event is an extinction process which should occur over time. However, if serotonergic neurotransmission in some way mediates motivational factors involved in ethanol self-administration, then it is possible that the inhibition of serotonin uptake modulates this motivational influence on ethanol drinking while having no influence on the specific reinforcing effects of the drug. Some data suggest that serotonergic influences on ethanol drinking may result from generalized effects on ingestion of both foods and liquids[25]. Indeed, many serotonin uptake blockers which effectively reduce ethanol intake are also anorectic in nature. These data are consistent with the idea that serotonergic neurotransmission may be involved in motivational factors which in turn modulate the reinforcing properties of several reinforcers, including drugs, food, and water. It is unclear from the limited information obtained to date, however, whether specific serotonin receptor subtypes may be differentially associated with modulation of reinforcing effects of ethanol *per se* or its relative incentive value.

Behavioral pharmacology studies have also provided important information concerning the role of serotonergic neuronal systems in mediating the reinforcing effects of cocaine and amphetamine. Indeed, cocaine is multi-functional in that it inhibits dopamine uptake, serotonin uptake, and norepinephrine uptake at roughly similar concentrations. The relative affinity of cocaine at each of these sites as well as at all other sites may be important in determining its summed effects on neurotransmission within reward pathways of the brain. It has been found that the affinities of amphetamine and related compounds at serotonin uptake sites is inversely associated with the reinforcing potencies of these compounds in operant tests[58]. These results suggest that the rewarding effects of these drugs mav be attenuated by serotonergic neuronal mechanisms, and are consistent with rese[illegible] illustrating that lesions of serotonergic neurons, with the neurotoxin [illegible] dihydroxytryptamine, increase rates of responding for intravenous administrations of *d*-amphetamine under commonly used fixed-ratio operant schedules of reinforcement[40]. In contrast, pretreatments with the serotonin precursor L-tryptophan or with serotonin uptake blockers decrease amphetamine and cocaine self-administration under similar operant conditions[13,43,65,79].

Using both receptor binding methods and operant techniques, it has previously been shown that dopaminergic, but not serotonergic, neurotransmission is associated with the reinforcing effects of cocaine under operant conditions utilizing short test sessions and requiring a modest, fixed amount of responses (e.g. FR10) in order to obtain the drug[57], while the potent serotonin uptake inhibitor fluoxetine appears to attenuate rates of responding for amphetamine[53]. These data suggest that serotonergic neuronal systems may indeed attenuate the reinforcing potencies of amphetamine, but perhaps not cocaine under these conditions. However, some evidence suggests that serotonergic systems may also influence the reinforcing effects of cocaine under quite different operant conditions, specifically those involving continuous access to the drug over extended periods of time and those requiring the completion of increasingly greater number of operant responses prior to administration of cocaine.

Thus, a preponderance of evidence suggests that factors associated with serotonergic neurotransmission influence drug self-administration. The nature of these influences is as yet unknown, although they may be related to issues of motivational states or efficacy of the drug. It is quite possible that serotonergic neurons influence the motivational factors necessary for animals to exhibit extremely high levels of responding for drugs under operant conditions requiring the completion of large sequences of lever presses, such as seen under large fixed ratio sizes or under progressive ratio schedules. These findings are consistent with the idea that drug seeking may be mediated by multiple, distinct influences, including the intrinsic reinforcing properties of the drug, perhaps primarily mediated by dopaminergic neurotransmission, as well as by attenuating or aversive properties of drugs which may modify euphoric effects or other motivational factors, each of which might be more significantly mediated by serotonergic neurotransmitter systems.

4.3 GABA: IMPORTANT INTERACTIONS WITH DOPAMINERGIC SYSTEMS IN REWARD PATHWAYS

A third major neurotransmitter which appears to play a significant role in mediating the reinforcing effects of at least some drugs of abuse is gamma-aminobutyric acid (GABA). GABA is an inhibitory neurotransmitter which is widely distributed throughout the brain. Its neuronal binding site is located on a multiple protein complex forming the chloride ion (Cl^-) channel. It produces its inhibitory effect on neurons by facilitating the flow of Cl^- into neurons, hyperpolarizing them thus, inhibiting the neuronal firing.

Classic sedative/hypnotic drugs such as benzodiazepines, barbiturates and ethanol also potentiate Cl^- flux and neuronal inhibition. Indeed, the facility with which these compounds produce pharmacological effects such as euphoria, disinhibition, anxiety reduction and sedation appears to be associated with their modulation of Cl^- flux through the neuronal ion channel.

GABA may potentiate reward in at least two ways. First, GABA-ergic and dopaminergic neuronal systems exhibit significant interactions in several areas of the limbic system[62]. GABA interneurons in the VTA synapse onto dopaminergic neurons serving a primary role in mediating reward (Figure 5.6). These GABA-ergic interneurons normally serve to inhibit the firing of dopaminergic neurons. Some drugs of abuse bind to GABA-ergic interneurons, however, inhibiting GABA release, and producing a disinhibition of dopaminergic neurotransmission. For example, it is known that opiates bind to μ-receptors located on these GABA-ergic interneurons and inhibit the release of GABA. There is also growing evidence that dopamine acts to regulate the activity of efferent GABA-ergic neurons extending from limbic structures such as the nucleus accumbens. In addition, prefrontal cortical areas innervated by dopaminergic reward pathways also receive important inputs from GABA-ergic neurons.

The influence of GABA on brain reward pathways may specifically involve the efferent neuronal signals from the NA to the substantia innominata-ventral pallidum. These efferent connections are thought to utilize GABA. Ibotenic acid lesions of GABA-ergic neurons in the region of the substantia innominata-ventral pallidum have been shown to inhibit both cocaine and heroin self-administration in rats on a constant schedule of reinforcement. When animals were placed on a progressive ratio schedule of reinforcement that increases the number of responses for each successive drug reinforcement in a stepwise fashion, then these ibotenic acid lesions also decreased the highest ratio achieved (break point) during test sessions.

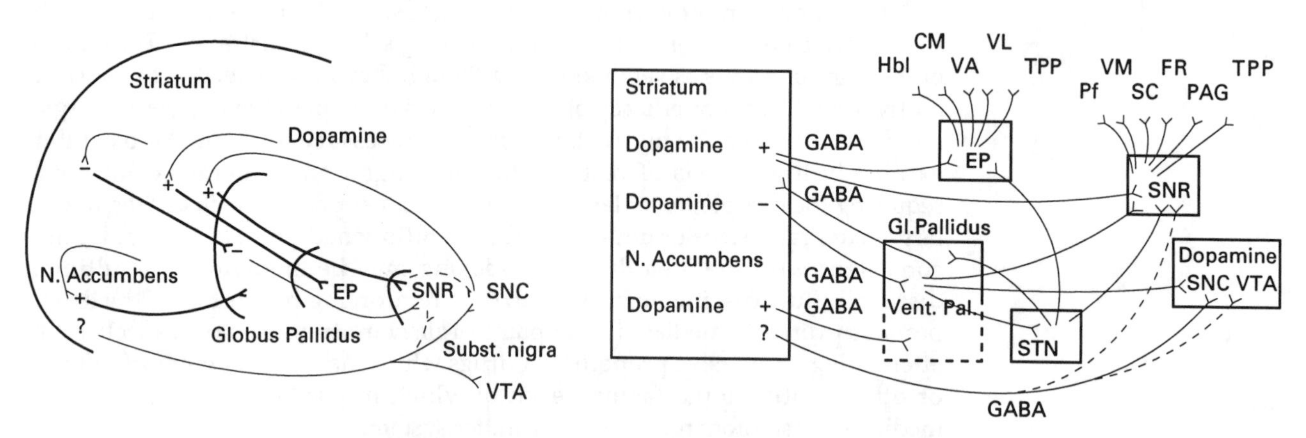

FIGURE 5.6

Inhibiting (–) and stimulating (+) effects of dopamine on GABA neurons in brain reward pathways (a) and GABA interactions with dopaminergic neurons in output stations of the basal ganglia system (b)

EP, entopeduncular nucleus; SNR, substantia nigra, zona reticulata; SNC, substantia nigra, zona compacta; VTA, ventral tegmental area; Hbl, lateral habenula; CM, centromedian; VA, ventro-anterior nucleus of the thalamus; VL, ventro-lateral nucleus of the thalamus; TPP, tegmental pedunculo pontine nucleus; Pf, parafascicular nucleus; VM, ventromedial nucleus of the thalamus; SC, superior colliculus; FR, formatio reticularis; PAG, periaquaductal grey; STN, subthalamic nucleus. (Based on an original from Scheel-Kruger, J., *Acta Neurol. Scand.* 1986, **13(Suppl. 103)**: 1–49.)

Alternatively, GABA may mediate reinforcing properties of drugs of abuse in part by acting on limbic structures such as the hippocampus, amygdala, and ventral forebrain areas to mediate anxiolytic, or stress reducing, effects of sedative hypnotics. These properties in and of themselves may be sought after by certain individuals and contribute to abusive use of sedative drugs. Alternatively, the limbic regions mediating the anxiolytic properties of sedative/hypnotic drugs such as the amygdala and the hippocampus provide significant input to the reward pathways of the brain, especially the nucleus accumbens. It is increasingly well documented that stress can significantly alter conditioned behavioral responses, and that various neurochemicals acting on limbic structures can influence the rewarding properties of a number of reinforcers including food, water, and drugs of abuse.

4.4 ENDOGENOUS OPIATES AND OPIATE RECEPTORS

"Runner's high" has been described as an altered state of consciousness similar to the euphoria experienced following the administration of an opiate. In the most extreme cases, colors become bright and beautiful, natural objects take on surrealistic qualities, the body seems lighter, as if detached from the earth, a sense of contentment flows through the body, and creative thinking is unleashed. At the very least, many runners report feeling a heightened sense of well-being, enhanced appreciation of nature, and decreased anxiety and stress. In addition, many undergo withdrawal symptoms and depression when unable to pursue their exercise habit.

Although the high experienced by runners and opiate users has not been demonstrated to be synonymous, the observance of addiction-like behaviors, including tolerance, withdrawal, and dominance over other reinforcers has lent support to this hypothetical relationship. The striking similarity between subjective accounts of runner's high and those of opiate drug users suggests a common neuronal mechanism in the brain. The common neuronal sites of action are likely to be opiate receptors[1]. It is now in fact clear that neuroendocrine changes involving the endogenous opiates occurs during strenuous exercise and that these molecules are functionally linked to brain reward pathways. Several studies have also demonstrated the analgesic effects of endorphins stimulated by running.

In 1973, three groups of researchers independently discovered and reported the identification of opiate receptors in the brain[50,64,70]. Most neurotransmitter receptor sites have been identified and biochemically characterized long after the transmitter itself. In the case of the opiate receptors, however, the binding sites were identified first. It was clear to the scientific community, however, that these binding sites do not exist merely for human chance encounters with poppy plants and that man does not synthesize heroin or morphine endogenously. The discovery of opiate receptors predicted endogenous ligands which utilize these binding sites for specific biological functions. Shortly thereafter, the presence of naturally occurring opiate-like substances was discovered in the brain[66]. The term endorphin (endogenous morphine) was coined, referring to the naturally occurring opiate-like substance in the brain or pituitary which duplicates the actions and effects of exogenous opiates. It is now known that opiate peptides can be divided into three subgroups defined by their precursor molecules. β-Endorphins, enkephalins, and dynorphin stem from pro-opiomelanocortin (POMC), pro-enkephalin, and pro-dynorphin, respectively. There are several forms of β-endorphin in the endorphin family as well as a- and b-endorphin. Met-enkephalin and leu-enkephalin are recognized as the most important pentapeptides in the enkephalin family. The dynorphin family includes dynorphin A, dynorphin B, and neo-dynorphin. These peptides are normally involved in modulation of the nociceptive response to painful stimuli and of reinforcement associated with various homeostatic goals such as food, water, temperature regulation, and sex.

Intravenous heroin self-administration is attenuated by local treatment with opiate receptor blockers in the NA and the VTA. An increased rate of the heroin self-administration was found in drug experienced rats treated with naltrexone in the NA, lateral hypothalamus, and periaquaductal grey, an effect which is indicative of attenuation of the drugs' reinforcing effects. However, contradicting evidence comes from studies in which kainic acid lesions did not affect heroin self-administration. Place preference conditioning is supported by systemic opiate administratiuon and by local application to the NA, VTA, hippocampus., periaquaductal grey, and lateral hypothalamus. Opiate induced conditioned place preference is attenuated by electrolytic lesions of the NA. In addition, β-endorphin elicits place preference when injected intracereboventricularly, an effect that can be blocked with subcutaneous treatment with naloxone. β-Endorphin-induced conditioned place preference seems to be mediated by μ- and δ-opiate receptors.

The findings from intracranial self stimulation studies support those of the drug reinforcement studies to a large extent. Opiates generally facilitate ICSS reinforcement, whereas opiate antagonists have the opposite effect. The facilitating effect of morphine on ICSS reward is thought to be located in the VTA, and not in the NA. Naloxone decreases response rate and increases the current threshold for for self-stimulation of the VTA, while morphine in low doses increased response rate and decreased current threshold. These and other findings indicate that ICSS reward is modulated by endogenous opioid systems. Studies with more specific opiate receptor agonists and antagonists suggest that the euphoric and rewarding effects of opiates are mediated by μ-opiate receptors, and the dysphoric effects via the k-opiate receptors (for review, see ref. 69). In contrast to the effects of dopamine antagonists, the inhibitory effects of opioids and opioid antagonists on self-stimulation behavior are slower to occur and do not completely abolish the ICSS behavior. Nevertheless, controlled experiments indicate that the inhibition of ICSS behavior appears to be related to a blockade of reinforcement and not merely to a disruption of task performance.

In addition, naltrexone made rats less sensitive to the effects of cocaine reward, so that higher dose of cocaine was required to produce self-administration[17]. Pretreatment with naloxone in rats resulted in a marked reduction in cocaine preference in a conditioned place preference paradigm (Figure 5.7)[24]. Naltrexone treatment also reduced alcohol consumption in animals under various experimental conditions, especially after induced stress[74] and after several days of imposed alcohol abstinence[38]. Reductions were large though doses were relatively low. Together, these findings indicate that the opioid system is important for reward processes and for rewarding effects of diverse addictive substances.

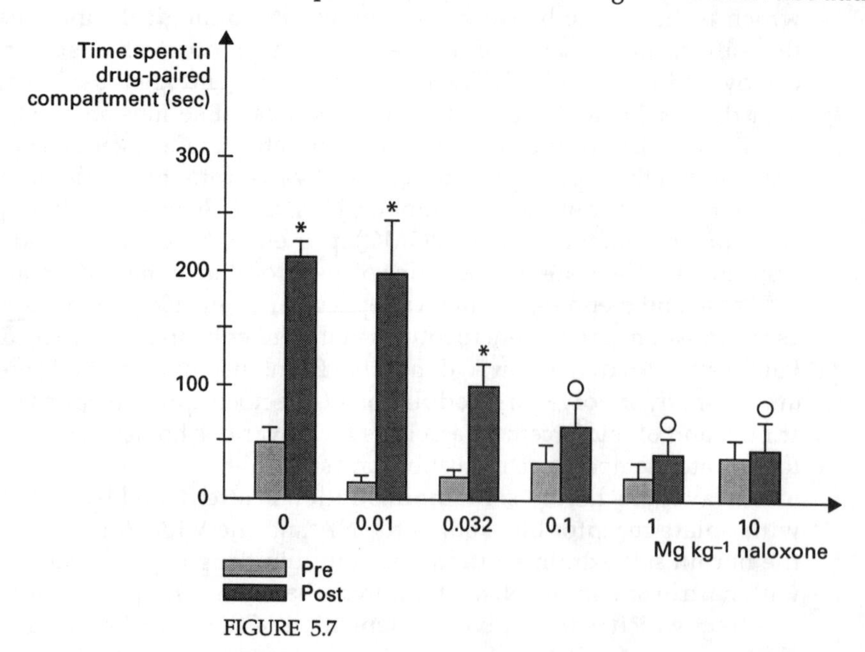

FIGURE 5.7

Effect of naloxone on acquisition of cocaine-induced place preference

Data represent mean (± SEM) time spent in the drug paired compartment during the preconditioning and postconditioning trial of animals conditioned with 10 mg kg-1 cocaine. * is difference between pre-and postconditioned trials ($p < 0.05$; student t-test); o is different from saline pretreatment ($p < 0.05$: ANOVA and SUK). (Based on original data from Gerrits, M.A.F.M. et al., *Psychopharmacology*. 1995, **119**: pp 92–98.)

5 Pharmacotherapy targeting reward pathways in substance abuse treatment

Human experience illustrates that, for some individuals, a particular drug may not elicit euphoric or rewarding effects, while for other individuals ingestion of the drug is associated with highly positive subjective experiences. It is also clear that among individuals for whom a drug elicits a euphoric response there are major differences in the "incentive value" of the drug and the concomitant motivation to obtain successive drug administrations. Thus, motivation is an aspect of drug taking that is important in determining the extent and range of conditions under which drugs will be sought; motivation is influenced by inherited traits motivation can be experimentally distinguished from other basic reinforcement processes and motivation can be therapeutically altered.

Pharmacotherapy of drug addiction, see Chapter 9

Treatment strategies should therefore include a consideration of positive and negative impact on both the direct reinforcing effects of drugs, as well as motivational factors which lead to sustained use and relapse. A goal of treatment and prevention would thus be to enhance behavior directed towards alternative positive reinforcers, to decrease motivational states associated with drug seeking, and to increase the negative or aversive effects of drugs. Indeed, whether explicitly or implicitly, the development of pharmacotherapeutic methods for the treatment of alcohol and drug abuse have traditionally focused on one or more of these strategies.

5.1 ALCOHOL

ALDH, see Chapter 6

The most widely used forms of pharmacotherapy for the treatment of alcoholism are disulfuram and carbimide. These compounds inhibit the liver enzyme aldehyde-dehydrogenase (ALDH), which is responsible for catalyzing the oxidation of acetaldehyde, the major metabolite of ethanol, to acetate. The resulting buildup of acetaldehyde is responsible for the negative physiological symptoms associated with ingestion of alcohol while taking this medication.

Thus, the basis for the major treatment strategy associated with the clinical use of these compounds is an attempt to increase negative, or aversive, side effects of alcohol related to its major metabolite, thereby producing avoidance of this drug. The potential effectiveness of these drugs in the treatment of alcoholism is therefore primarily based upon the patient's fear of becoming sick if he or she drinks ethanol. Indeed, Gordis and Peterson (1977) wrote that, "It is probable that it is the patient's belief that he is taking disulfiram (whether or not he actually is) that is therapeutic and not the action itself."[28]. In general, it is clear that the use of these drugs as adjunct therapies for alcoholism does not aim to pharmacologically attenuate the intrinsic reinforcing effects of alcohol *per se*.

During the past decade there has been substantially increased interest and research on the use of various other pharmacologic agents to alter alcohol use. Many of these studies used agents which exhibit a primary biochemical effect on serotonergic neurotransmission. In general, clinical research in this area indicates that serotonin uptake inhibitors reduce ethanol consumption. These studies suggest that at least part of the mechanism of reinforcement related to alcoholism involves serotonergic neurotransmission, especially when taken together with the literature describing serotonergic influences on ethanol self-administration in animals.

To date, however, no variables have been found which can effectively predict treatment response to serotonergic pharmacotherapies across subjects, and the mechanism by which these medications produce decreases in ethanol intake is unclear. Three hypotheses are: (1) since there is a high comorbidity for depression and alcoholism, the antidepressant properties of these drugs may decrease depression, thus reducing the need for self-medication (2) these drugs, including fluvoxamine, zimelidine and fluoxetine, appear to reverse or inhibit some of the cognitive and memory impairments associated with ethanol consumption and thus, may facilitate behavioral modifications in education-oriented rehabilitation programs; or (3) serotonin uptake inhibitors may attenuate the positive reinforcing effects of ethanol. However, since serotonin uptake inhibitors have been shown to decrease ethanol drinking in patients who exhibit little depression or cognitive dysfunction, the latter hypothesis is most probable, especially since these compounds also decrease consumption of ethanol drinking in animal models which are less confounded by issues of comorbidity with other psychiatric disorders and by chronic effects of ethanol on cognitive dysfunction.

Research is also currently providing information suggesting that opiate antagonists may be useful in treating alcohol abuse. However, the mechanism by which opiate pharmacotherapies may be effective remains unclear.

Chapter 6 discusses mechanisms underlying the effect of opiate antagonists in alcohol self-administration. The clinical implication of opiate antagonists in the treatment of alcohol dependence is described in Chapter 9 Section 3.2..

5.2 PSYCHOSTIMULANTS

Clinical studies of the effectiveness of various treatment regimens for psychostimulant abuse have been based primarily upon three major hypotheses. The first hypothesis is that a pharmacological antagonism of dopaminergic neurotransmission will decrease or inhibit the reinforcing effects of these compounds. Research has illustrated decreases in craving for cocaine and increases in the length of time patients were retained in treatment in open trials with low doses of the dopamine receptor blocker, flupenthixol decanoate. It has also been reported that euphoria induced by amphetamine appears to be sensitive to dopamine receptor blockers. In contrast to these results, haloperidol was reported to have no effect on cocaine-induced rush, and only a limited effect on some of the subjective effects associated with drug "liking". In yet another study, haloperidol or chlorpromazine did not block drug-induced euphoria or increase abstinence, although an attenuation of psychotic symptoms was observed.

In view of these previous inconsistent findings, there has been some interest in developing a direct competitive cocaine antagonist at the dopamine transporter. Such a compound could be useful if it were able to block cocaine binding at the dopamine transporter while leaving the dopamine reuptake process relatively unaffected. Indeed, recent studies involving site directed mutagenesis of the genetic material encoding the dopamine transporter suggest that cocaine binding can be altered independently of dopamine uptake.

Another hypothesis which has guided the search for effective pharmacotherapeutic strategies for cocaine abuse suggests that the chronic use of cocaine results in depletion of brain dopamine levels, which in turn leads to reduction of the euphoric effects associated with cocaine ingestion and, ultimately, "craving" for continued use of the drug. Several clinical tests have suggested that compounds potentiating dopaminergic transmission could decrease craving for cocaine, and these have been recently reviewed in detail[39,56]. For example, studies have evaluated the effectiveness of the dopamine agonist amantadine as an adjunct treatment for cocaine abuse, suggesting that this treatment may be associated with decreased craving, decreased cocaine use, and decreases in several indices of psychiatric adjustment, although these effects were not significantly greater than those observed for placebo treatment groups. In addition, while the dopamine agonist bromocriptine has been shown to antagonize decreases in cerebral glucose metabolism associated with abstinence from chronic cocaine treatment in rats, the results of another recent study utilizing cocaine abusing subjects has shown that this potential pharmacotherapy has no significant effects on either physiological or subjective effects of cocaine which may be predictive of its therapeutic benefit. Thus, studies in this area have yielded inconsistent results and have generally failed to show that drugs acting to facilitate dopaminergic neurotransmission are more clinically effective than placebos in reducing craving or further cocaine intake, and may in fact actually increase the desire to self-administer cocaine.

The other hypothesis upon which studies of the effectiveness of treatments for cocaine abuse are based is that one of the primary aspects of cocaine withdrawal is anhedonia, a symptom which is also common to depressed patients. In a sense, treatment strategies developed with this hypothesis as a basis actually assume that cocaine offers abusers a dual reward, its intrinsic reinforcing effects and its therapeutic value as an antidepressant. From this perspective, the treatment provides an alternative medication for the symptoms of the psychiatric comorbidity. Treatments with the relatively nonselective norepinephrine uptake inhibitor, desimipramine, have yielded variable results. It appears that desipramine does not reduce the reinforcing or craving effects of cocaine, although it may alter other subjective effects associated with its use, and the cardiovascular effects of desipramine appear to enhance the potential for toxicity when this antidepressant is administered in conjunction with cocaine. In general, it appears that a number of adverse side effects of desipramine may contribute to treatment noncompliance by primary cocaine abusers, and, further, that this drug appears to have few benefits with regard to controlling cocaine use in opiate abusers maintained on methadone.

Interest in the efficacy of serotonin uptake blockers as adjunct pharmacotherapeutic agents in addiction treatment programs has also been increasing. Several lines of evidence already discussed above suggest that serotonergic neurotransmission may attenuate self-administration of psychostimulant drugs. Thus, serotonergically mediated antidepressant pharmacotherapies for clients with psychostimulant addiction and concomitant depression may benefit from both a decrease in psychiatric symptoms and a decrease in either the rewarding effects of the drug or craving for repeated administrations. Other antidepressants have also been studied for their potential clinical efficacy in treating psychostimulant abuse, especially cocaine addiction. Unfortunately, there is growing evidence that antidepressant drugs should be used with caution in the treatment of cocaine abuse, especially for patients with histories of cardiovascular or convulsant disorders, since these medications have been shown to increase toxic effects associated with ingestion of high doses of the drug[22].

There is also evidence that opiate compounds could be effective as adjunct pharmacotherapeutic agents for the treatment of cocaine addiction, although the current data are equivocal, and caution must be taken given the likelihood that potential treatment compounds may themselves be reinforcing. For example, buprenorphine, a mixed opiate agonist-antagonist, and naltrexone, an opiate antagonist, have been shown to selectively decrease responding for intravenously administered cocaine in rhesus monkeys. In addition, buprenorphine produces downward shifts in cocaine dose effect curves in rhesus monkeys suggesting that its effects are due to agonist actions, and chronic buprenorphine treatment appears to inhibit cocaine conditioned place preference. Buprenorphine has also been shown significantly to decrease cocaine use in heroin abusers.

However, buprenorphine will itself serve as a reinforcer. Buprenorphine alone produces conditioned place preference in a dose-related manner, and subthreshold doses of cocaine and buprenorphine, given in combination, also produced conditioned place preference. Further, cocaine and buprenorphine both increased extracellular levels of dopamine in the nucleus accumbens as measured by microdialysis techniques. Taken together, these studies suggest that the reinforcing effects of buprenorphine may substitute for, not inhibit, the reinforcing effects of cocaine, and in this way may produce decreases in operant responding for cocaine in experimental studies.

5.3 OPIATES

The etiology of opiate addiction, like addiction to ethanol and psychostimulants, appears to involve the direct or indirect action of opiates on brain reward pathways. Mechanistically, this action involves opiate receptors, interactions with endogenous opiate molecules, and facilitation of dopaminergic neurotransmission. In these ways, opiates produce positive reinforcement.

Theories associated with our understanding of opiate abuse, even more than those associated with alcohol or psychostimulant abuse, must realistically include a recognition of the severe negative reinforcing effects intrinsic to physical dependence on chronic opiate use. At some point in the opiate addiction process, most abusers must seek drug to avoid severe withdrawal symptoms. Indeed, tolerance to the reinforcing and euphoric effects of opiate drugs may have occurred by this stage of the addiction, thus the negative reinforcing effects of the opiate compound may be the primary driving force behind the drug-seeking behavior. At the very least, negative reinforcing effects of drug abstinence are likely additive with the positive reinforcing effects of the drug to produce continued drug consumption.

Pharmacological treatment of opiate addiction have generally involved one of two approaches. First, patients, having completed detoxification procedures, maintain drug-free abstinence or are maintained on an opiate antagonist. Alternatively, patients are maintained on the opiate agonist methadone with the goal of eventual detoxification to a state of abstinence. Both of these approaches are aimed at decreasing the reinforcing or euphoric effects related to opiate administration. Methadone, though it is a opiate agonist and functions as a positive reinforcer, can be taken orally and is associated with less of an acute, highly positively reinforcing "rush" effect than other opiates (eg. morphine and heroin) taken intravenously. This, in addition to its relatively long duration of action, has made it an effective means of treating opiate addicted patients. Opiate antagonists, on the other hand, block opiate receptors directly so that opiate or agonist administration does not result in a rewarding effect.

A related and more recent pharmacotherapeutic strategy involves the use of the partial agonist buprenorphine to facilitate the transition from heroin or methadone dependence to the abstinence phase of recovery. This treatment agent attenuates the opiate withdrawal syndrome and results in minimal withdrawal symptoms as its use is terminated, especially if dosage is gradually decreased.

6 Summary

Over the past several years, our understanding of the various factors which contribute to initiation and maintenance of sustained drug-seeking behavior, the underlying mechanisms of these factors, and their importance in treatment strategies has increased substantially. Motivation and reinforcement are known to be important components of behavior influencing if, to what extent, and under what conditions sustained drug taking will occur. Evidence suggests that a distinction can be made between reinforcing value and incentive value of drugs. Reinforcement appears to be a unique effect of certain psychoactive substances, and does not appear to be significantly related to drug preference, the depressant effects of drugs, psychostimulant drug effects, or mechanisms that result in physical dependence and withdrawal.

Stimulation or enhancement of dopamine transmission in the mesocorticolimbic system appears to be a fundamental property of drugs that are abused. This action on dopaminergic neurotransmission appears to occur by blockade of dopamine reuptake in the case of cocaine, by stimulation of dopamine release in the case of ethanol, and by facilitation of dopamine effects on target neurons in the case of opiates. However, serotonergic neurotransmission appears to modulate the reinforcing properties of drugs. At least some of the scientific evidence to date suggests that serotonin mediates motivational aspects of drug self-administration, especially since drug taking under high work loads can be modulated via changes in serotonergic function. Finally, drug reinforcement appears to be influenced by GABA-ergic neurotransmission and by endogenous opiate systems interacting with brain reward pathways.

The complexity of the neuronal mechanisms associated with drug-seeking behavior suggests that pharmacological treatment interventions need not provide a "magic bullet" to antagonize the drug at its "addiction" site of action. Pharmacotherapeutic treatment approaches can target distinct neuronal mechanisms associated with any of several components of drug-seeking behaviors. Some pharmacological manipulations may decrease the "rush" or positive reinforcing effects of a drug. Another compound may enhance more aversive drug stimuli so that the drug is no longer as desirable. Still another drug therapy may produce agonist actions similar to those of the drug of abuse (e.g., methadone), thus serving as a substitute which hopefully allows for a greater quality of life. An effective pharmacological treatment may decrease the motivation or desire to "work" for the drug. Thus, even though an individual may recognize that the drug would produce rewarding effects if it were immediately available, the decision not to go to great lengths to obtain the drug is made. Finally, a drug may effectively treat another concurrent dysfunction such as depression in order to allow the addicted individual greater clarity in participating and responding to available addiction treatment.

Further more, through an increasingly integrated approach to treatment combining pharmacotherapy and psychosocial intervention, we may better treat the difficult biological, behavioral, and social problems associated with alcoholism and substance abuse disorders. Using these combined therapeutic strategies will enable treatment programs to target multiple aspects of drug taking behavior, thus decrease the likelihood of relapse and continued drug abuse.

References

1. Akil, J., Bronstein, D., Mansour,, A., "Overview of the endogenous opioid systems: anatomical, biochemical and functional issues." In: Rodgers, R.J. and Cooper, S.J. (Eds.), *Endorphins, Opiates and Behavioral Processes.* John Wiley & Sons, London, 1988, pp 1–23.
2. Benloucif, S. and Galloway, M.P., "Facilitation of dopamine release in vivo by serotonin agonists: studies with microdialysis," *Eur. J. Pharmacol.* 1991, **200:** pp 1–8.
3. Benloucif, S., Keegan, M.J., Galloway, M.P., "Serotonin-facilitated dopamine release in vivo: pharmacological characterization," *J. Pharmacol. Exp. Therapeut.* 1993, **265:** pp 373–377.

4. Bergman, J., Madras, B.K., Johnson, S.E., Spealman, R.D., "Effects of cocaine and related drugs in nonhuman primates. III. Self-administration by squirrel monkeys," *J. Pharmacol. Exp. Therapeut.* 1989, **251:** pp 150–155.
5. Bergman, J., Kamien, J.B., Spealman, R.D., "Antagonism of cocaine self-administration by selective dopamine D1 and D2 antagonists," *Behav. Pharmacol.* 1990, **1:** pp 355–363.
6. Bozarth, M.A., "Neuroanatomical boundaries of the reward-relevant opiate-receptor field in the ventral tegmental area as mapped by the conditioned place preference method in rats," *Brain Res.* 1987, **414:** pp 77–84.
7. Bozarth, M.A. and Wise, R.A., "Heroin reward is dependent on dopaamergic substrate," *Life Sci.* 1981, **29:** pp 1881–1886.
8. Bozarth, M.A. and Wise, R.A., "Anatomically distinct opiate receptor fields mediate reinforcement and dependence," *Science.* 1984, **224:** pp 516–517.
9. Brodie, M.S., Shefner, S.A., Dunwiddie, T.V., "Ethanol increases the firing rate of dopamine neurons of the rat ventral tegmental area in vitro," *Brain Res.* 1990, **508:** pp 65–69.
10. Brown, E.E., Finlay, J.M., Wong, J.T.F., Damsma, G., Fibiger, H.C., "Behavioral and neurochemical interactions between cocaine and buprenorphine: Implications for the pharmacotherapy of cocaine abuse," *J. Pharmacol. Exp. Therapeut.* 1991, **256:** pp 119–126.
11. Carboni, E., Imperato, A., Perezzani, L., Di Chiara, G., "Amphetamine, cocaine, phencyclidine and nomifensine increase extracellular dopamine concentrations preferentially in the nucleus accumbens of freely moving rats," *Neuroscience.* 1989, **28:** pp 653–661.
12. Carboni, E., Acquas, E., Frau, R., Di Chiara, G., "Differential inhibitory effects of a 5-HT3 antagonist on drug-induced stimulation of dopamine release," *Eur. J. Pharmacol.* 1989, **164:** pp 515–519.
13. Carroll, M.E., Lac, S.T., Asencio, M., Kragh, R., "Intravenous cocaine self-administration in rats is rduced by L-tryptophan," *Psychopharmacology.* 1990, **100:** pp 293–300.
14. Costall, B., Domeney, A.M., Naylor, R.J., Tyers, M.B., "Effects of the 5-HT3 receptor antagonist, GR38032F, on raised dopaminergic activity in the mesolimbic system of the rat and marmoset brain," *Br. J. Pharmacol.* 1987, pp 881–894.
15. Daoust, M., Chretien, P., Moore, N., Saligaut, C., Lhuintre, J.P., Boismare, F., "Isolation and striatal (^{3}H) seotonin uptake: role in the voluntary intake of ethanol by rats," *Pharmacol. Biochem. Behav.* 1985, **22:** pp 205–208.
16. Davis, W.M. and Smith, S.G., "Effect of haloperidol on (+)amphetamine self-administration," *J. Pharm. Pharmacol.* 1975, **27:** pp 540–542.
17. De Vry, J., Donselaar, I., Van Ree, J.M. "Food deprivation and acquisition of cocaine self-administration in rats: effect of naltrexone and haloperidol," *J. Pharmacol. Exp. Ther.* 1989, **251:** pp 735–740.
18. deWit, H. and Wise, R.A., "Blockade of cocaine reinforcement in rats with the dopamine receptor blocker pimozide but not with the noradrenergic blockers phentolaamine or phenoxybenzamine," *Canad. J. Psychol.* 1977, **31:** pp 195–203.
19. Di Chiara, G. and Imperato, A., "Drugs abused by humans preferentially increase synaptic dopamine concentrations in the mesolimbic system of freely moving rats," *Proc. Nat. Acad. Sci. USA.* 1988, **85:** pp 5274–5278.
20. Ettenberg, A., Pettit, H.O., Bloom, F.E., Koob, G.F., "Heroin and cocaine intravenous self-administration in rats: mediation by separate neural systems," *Psychopharmacology.* 1982, **78:** pp 204–209.
21. Fowler, J.S., Volkow, N.D., Wolf, A.P., Dewey, S.L., Schyler, D.J., MacGregor, R.R., Hitzemann, R., Logan, J., Bendriem, B., Gatley, S.J., and Christman, D., "Mapping cocaine binding sites in human and baboon brain sites in vivo," *Synapse.* 1989, **4:** pp 371–377.
22. George, F.R. and Ritz, M.C., "A psychopharmacology of motivation and reward related to substance abuse treatment," *Exp. Clin. Psychopharmacol.* 1993, **1:** pp 7–26.
23. Gerber, G.J. and Wise, R.A., "Pharmacological regulation of intravenous cocaine and heroin self-administration in rats: a variable dose paradigm," *Pharmacol. Biochem. Behav.* 1989, **32:** pp 527-531.
24. Gerrits, M.A.F.M., Patkina N., Zvartau, E.E., Van Ree, J.M., "Opioid blockade attenuates acquisition and expression of cocaine-induced place preference conditioning in rats," *Psychopharmacology.* 1995, **119:** pp 92–98.

25. Gill, K. and Amit, Z., "Effects of serotinin uptake blockade on food, water, and ethanol consumption in rats," *Alcohol. Clin. Exp. Res.* 1987, **11:** pp 444–449.
26. Goeders, N.E., Dworkin, S.I., Smith, J.E., "Neuropharmacological assessment of cocaine self-administration into the medial prefrontal cortex," *Pharmacol. Biochem. Behav.* 1986, **24:** pp 1429–1440.
27. Goeders, N.E. and Smith, J.E., "Reinforcing properties of cocaine in the medial prefrontal cortex: primary action on presynaptic dopaminergic terminals," *Pharmacol. Biochem. Behav.* 1986, **25:** pp 191–199.
28. Gordis, E. and Peterson, K., (1977) "Disulfiram therapy in alcoholism: patient compliance studied with a urine detection procedure," *Alcohol. Clin. Exp. Res.* 1977, **1:** pp 213–216.
29. Goudreau, J.L., Manzanares, J., Lookingland, K.J., Moore, K.E., "$5HT_2$ receptors mediate the effects of stress on the activity of periventricular hypophysial dopaminergic neurons and the secretion of a-melanocyte-stimulating hormone," *J. Pharmacol. Exp. Therap.* 1993, **265:** pp 303–307.
30. Griffiths, R.R., Bigelow, G.E., Henningfield, J.E., "Similarities in animal and human drug-taking behavior." In: Mello, N.K. (Ed.), *Advances in Substance Abuse,* Vol. 1. 1980, JAI Press, pp 1–90.
31. Gulley, J.M., McNamara, C., Barbera, T.J., Ritz, M.C., George, F.R., "Selective serotonin reuptake inhibitors: effects of chronic treatment on ethanol-reinforced behavior in mice," *Alcohol.* 1995, **12:** pp 177–181.
32. Haraguchi, M., Samson, H.H., Tolliver, G.A., "Reduction in oral ethanol self-administration in the rat by the 5-HT uptake blocker fluoxetine," *Pharmacol. Biochem. Behav.* 1990, **35:** pp 259–262.
33. Kelland, M.D., Freeman, A.S., Chiodo, L.A., "Serotonergic afferent regulation of the basic physiology and pharmacological responsiveness of nigrostriatal dopamine neurons," *J. Pharmacol. Exp. Therap.* 1990, **253:** pp 803–811.
34. Kleven, M.S. and Woolverton, W.L., "Effects of continuous infusions of SCH 23390 on cocaine- or food-maintained behavior in rhesus monkeys," *Behav. Pharmacol.* 1990, **1:** pp 365–373.
35. Koob, G.F., Le, H.T., Creese, I., "The D1 dopamine antagonist SCH 23390 increases cocaine self-administration in the rat," *Neurosci. Lett.* 1987, **79:** pp 315–320.
36. Koob, G.F., "Drugs of abuse: anatomy, pharmacology and function of reward pathways," *Trends Pharmacol. Sci.* 1992, **13:** pp 177–184.
37. Kornet, M., Goosen, C., Van Ree, J.M., "Analysis of spontaneous alcohol drinking in rhesus monkeys," *Physiol. Behav.* 1990, **47:** pp 679–684.
38. Kornet, M., Goosen, C., Van Ree, J.M., "The effect of naltrexone on alcohol consumption during chronic alcohol drinking and after a period of imposed abstinence in free-choice drinking rhesus monkeys," *Psychopharmacology.* 1991, **104:** pp 367–376.
39. Kuhar, M.J., Ritz, M.C., Boja, J.W., "The dopamine hypothesis of the reinforcing properties of cocaine," *Trends Neurosci.* 1991, **14:** pp 299–302.
40. Lecesse, A.P. and Lyness, W.H., "The effects of 5-hydroxytryptamine receptor active agents on D-amphetmaine self-administration in controls and rats with 5,7-dihydroxytryptamine median forebrain bundle lesions," *Brain Res.* 1984, **303:** pp 153–162.
41. Liebman, J.M. and Cooper, S.J. (Eds.), *The Neuropharmacological Basis of Reward.* Clarendon Press, 1989.
42. Lyness, W.H., "Effect of L-tryptophan pretreatment on d-amphetamine self-administration," *Substance Alcohol Actions/Misuse.* 1983, **4:** pp 305–312.
43. Lyness, W.H., Friedle, N.M., Moore., K.E., "Destruction of dopaminergic nerve terminals in nucleus accumbens: effect on d-amphetamine self-administration," *Pharmacol. Biochem. Behav.* 1979, **11:** pp 553–556.
44. Mello, N.K., Mendelson, J.H., Bree, M.P., Lukas, S.E., "Buprenorphine suppresses cocaine self-administration by rhesus monkeys," *Science.* 1989, **245:** pp 859–861.
45. Mello, N.K., Lukas, S.E., Mendelson, J.H., Drieze, J., "Naltrexone-buprenorphine interactions: effects on cocaine self-administration," *Neuropsychopharmacology.* 1993, **9:** pp 211–224.
46. Murphy, J.M. et al., "Effects of acute ethanol administration on monoamine and metabolite content in forebrain regions of ethanol-tolerant and -nontolerant alcohol-preferring (P) rats," *Pharmacol. Biochem. Behav.* 1988, **29:** pp 169–174.

47. Myers, R.D. and Martin. G.E., "The role of cerebral serotonin in the ethanol preference of animals," *Ann. N.Y. Acad. Sci.* 1973, **215:** pp 135–144.
48. Naranjo, C.A., Sellers, E.M., Sullivan, J.T., Woodley, D.V., Kadlec, K.E., Sykora, K., "The serotonin uptake inhibitor citalopram attenuates ethanol intake," *Clin Pharmacol. Ther.* 1987, **41:** pp 266–274.
49. Olds, J. and Milner, P., "Positive reinforcement produced by electrical stimulation of septal area and other regions of rat brain," *J. Comp. Physiol. Psychol.* 1954, **47:** pp 419–427.
50. Pert, C.B. and Snyder, S.H., "Opiate receptor: demonstration in nervous tissue," *Science.* 1973, **179:** pp 1011–1014.
51. Phillips, A.G. and LePiane, F.G., (1980) "Reinforcing effects of morphine microinjection into the ventral tegmental area," *Pharmacol. Biochem. Behav.* 1980, **12:** pp 965–968.
52. Phillips, A.G. and LePiane, F.G., "Reward produced by microinjection of (D-Ala), Met-enkephalinamide into the ventral tegmental area,"*Behav. Brain Res.* 1982, **5:** pp 225–229.
53. Porrino, L.J., Ritz, M.C., Sharpe, L.G., Goodman, N.L., Kuhar, M.J., Goldberg, S.R., "Differential effects of pharmacological manipulation of serotonin systems on cocaine and amphetamine self-administration in rats," *Life Sci.* 1989, **45:** pp 1529–1535.
54. Richter, C.P. and Campbell, K.H., "Taste thresholds and concentrations of solution preferred by rats," *Science.* 1940, **91:** pp 507–508.
55. Risner, M.E. and Jones, B.E., "Role of noradrenergic and dopaminergic processes in amphetamine self-administration," *Pharmacol. Biochem. Behav.* 1976, **5:** pp 477–482.
56. Ritz, M.C., Lamb, R.J., Goldberg, S.R., Kuhar, M.J., "Cocaine receptors on dopamine transporters are related to self-administration of cocaine," *Science.* 1987, **23:** pp 1219–1223.
57. Ritz, M.C. and Kuhar, M.J., "Monoamine uptake inhibition mediates amphetamine self-administration: comparison with cocaine," *J. Pharmacol. Exp. Therap.* 1989, **248:** pp 1010–1017.
58. Ritz, M.C., George, F.R., Kuhar, M.J., "Molecular mechanisms associated with cocaine effects: possible relationships with effects of ethanol." In: Galanter, M. (Ed.), *Recent Developments in Alcoholism: Alcohol and Cocaine: Similarities and Differences*, Vol. X. Plenum Press, New York, 1992, pp 273–302.
59. Roberts, D.C.S., Koob, G.F., Klonoff, P., Fibiger, H.C., "Extinction and recovery of cocaine self-administration following 6-hydroxydopamine lesions of the nucleus accumbens," *Pharmacol. Biochem. Behav.* 1980, **12:** pp 781–787.
60. Roberts, D.C.S. and Koob, G.F., "Disruption of cocaine self-administration following 6-hydroxydopamine lesions of the ventral tegmental area in rats," *Pharmacol. Biochem. Behav.* 1982, **17:** pp 901–904.
61. Rockman, E., Amit, Z., Carr, G., Brown, Z.W., Ogren, S.O., "Attenuation of ethanol intake by 5-hydroxytryptamine uptake blockade in laboratory rats. I. Involvement of brain 5-hydroxytryptamine in the mediation of the positive reinforcing properties of ethanol," *Arch. Int. Pharmacodynamics Therap.* 1979, **241:** pp 245–259.
62. Scheel-Kruger, J., "Dopamine-GABA interactions: evidence athat GABA transmits, modulates and mediates dopaminergic functions in the basal ganglia and the limbic system," *J. Acta Neurol. Scand.* 1986, **73(Suppl. 107):** pp 1–49.
63. Signs, S.A. and Schecter, M.D., "The role of dopamine and serotonin receptors in the mediation of the ethanol interoceptive cue," *Pharmacol. Biochem. Behav.*, 1988, **30:** pp 55–64.
64. Simon, E.J., Hiller, J.M., Edelman, I., "Stereospecific binding of the potent narcotic analgesic ^{3}H-etorphine to rat brain homogenate," *Proc. Nat. Acad. Sci. USA.* 1973, **70:** pp 1947–1949.
65. Smith, F.L., Yu, D.S.L., Smith, D.G., Lecesse, A.P., Lyness, W.H., "Dietary tryptophan supplements attenuate amphetamine self-administration in the rat," *Pharmacol. Biochem. Behav.* 1986, **25:** pp 849–855.
66. Snyder, S.H. and Childers, S .R., "Opiate receptors and opioid peptides," *Ann. Rev. Neurosci.* 1979, **2:** pp 35–64.
67. Spyraki, C., Fibiger, H.C., Phillips, A.G., "Attenuation of heroin reward in rats by disruption of the mesolimbic dopamine system," *Psychopharmacology.* 1983, **79:** pp 278–283.
68. Stolerman, I., "Drugs of abuse: behavioral principles, methods and terms," *Trends Pharmacol. Sci.* 1992, **13:** pp 170–176.

69. Székely, J.I., *Opioid Peprtides in Substance Abuse.* CRC Press, Boca Raton, FL, 1994, pp 1–277.
70. Terenius, L ., "Characteristics of the "receptor" for narcotic analgesics in synaptic plasma membrane fractions from rat brain," *Acta Pharmacol. Toxicol.* 1973, **33:** pp 377–384.
71. Van Ree, J.M. and Ramsey, N.F., "The dopamine hypothesis of reward challenged," *Eur. J. Pharmacol.* 1987, **134:** pp 239–243.
72. Volkow, N.D., Fowler, J.S., Wolf, A.P., Schyler, D., Shiue, CY., Alpert R., Dewey, S.L. Logan, J., Bendriem, B., Christman, D., Hitzemann, R., Henn, F., "Effects of chronic cocaine abuse on postsynaptic dopamine receptors," *Am. J. Psychiatry.* 1990, **147:** pp 719–724.
73. Volkow, N.D., Hitzemann, R., Wang, G.-J., Fowler, J.S., Wolf, A.P., Dewey, S.L., "Long-term frontal brain metabolic changes in cocaine abusers," *Synapse.* 1992, **11:** pp 184–190.
74. Volpicelli, J.R., Davis, M.A., Olgin, J.E., "Naltrexone blocks the post-shock increase of ethanol consumption," *Life Sci.* 1986, **38:** pp 841–847.
75. Wise, R.A., "The role of reward pathways in the development of drug dependence," *Pharmacol. Ther.* 1987, **35:** pp 227–263.
76. Woolverton, W.L., Goldberg, L.I., Ginos, J.Z., "Intravenous self-administration of dopamine receptor agonists by rhesus monkeys," *J. Pharmacol. Exp. Ther.* 1984, **230:** pp 678–683.
77. Woolverton, W.L. and Virus, R.M., "The effects of a D1 and a D2 dopamine antagonist on behavior maintained by cocaine or food," *Pharmacol. Biochem. Behav.* 1989, **32:** pp 691–697.
78. Woolverton, W.L., "Effects of a D1 and a D2 dopamine antagonist on self-administration of cocaine and piribedil by rhesus monkeys," *Pharmacol. Biochem. Behav.* 1986, **24:** pp 531–535.
79. Yi, S.-J., Gifford, A.N., Johnson, K.M., (1991) "Effect of cocaine and 5-HT3 receptor antagonists on 5-HT-induced [^{3}H]dopamine release from rat striatal synaptosomes," *Eur. J. Pharmacol.* 11991, **99:** pp 185–189.
80. Yokel, R.A. and Wise, R.A., "Attenuation of intravenous amphetamine reinforcement by central dopamine blockade in rats," *Psychopharmacology.* 1976, **48:** pp 311–318.
81. Yokel, R.A. and Wise, R.A., "Amphetamine-type reinforcement by dopaminergic agonists in the rat," *Psychopharmacology.* 1978, **58:** pp 289–296.
82. Yoshimoto, K. and McBride, W.J., "Regulation of nucleus accumbens dopamine release by the dorsal raphe nucleus in the rat," *Neurochem. Res.* 1992, **17:** pp 401–407.

Contents Chapter 6

Molecular mechanisms of addictive substances

0-8493-7803-6/99/$0.00+$.50

Chapter 6

Molecular mechanisms of addictive substances

Mary C. Ritz

1 Drug binding sites in the brain

1.1 DRUG INTERACTIONS WITH NEURONS

Drugs which are useful for medicinal purposes perturb biological systems in a variety of predictable ways. For example, compounds such as antacids are effective due to the ability of their basic chemical properties to neutralize endogenous gastrointestinal acidic products. However, most drugs are effective due to their interactions with specific endogenous proteins, cell membrane components, or genetic material enclosed in the cell nucleus. Psychoactive drugs, in particular, produce their effects due to such interactions with brain neurons.

Brain neurons are linked in organized patterns throughout the brain; they communicate with one another in order to process incoming information, store memories, form new associations, prepare outputs to the motor neurons, and numerous other tasks. Although some signals which pass between one neuron and another are direct electrical stimuli, most messages passing between neurons are of a chemical nature. When a neuron "fires", an action potential flows down the length of the axon. When the action potential reaches the axon terminal, neurotransmitter molecules which are stored in vesicles are released into the synapse, or synaptic cleft, that is a tiny space between the axon terminal of one neuron and the dendritic branches of the receiving neuron. These chemical messengers diffuse across the synapse to interact with specific membrane-bound proteins commonly known as receptors. Endogenously produced neurotransmitters are conceptualized to bind to these receptors in a manner analogous to that with which a key fits into a lock. Thus, this "lock and key" model of neurotransmitter/receptor complexes predicts that a particular neurotransmitter is able to fit the receptor lock and "open" or elicit a particular cellular response.
The function of many neurotransmitters, including dopamine, norepinephrine, serotonin, and acetylcholine, is dependent upon the specific brain regions in which each is released and the specific neurons to which it communicates. However, some are almost always excitatory, and some are almost always inhibitory. For example, GABA carries an inhibitory message to neurons surrounding GABAergic neurons. Excitatory amino acids (EAAs) communicate excitatory messages.

Most neurons are thought to synthesize and release only one primary neurotransmitter from their axon terminals. However, it has been shown that some neurons also produce a neuropeptide molecule which is released from the neuron and appears to serve as a neurotransmitter. For example, the synthesis of a neuropeptide called neurotensin appears to be colocalized with dopamine synthesis in at least some dopaminergic neurons. However, brain neurons typically incorporate receptors of many types into the membranes of dendritic surfaces of the cell. Thus, neurons simultaneously receive messages carried by numerous neurotransmitters from surrounding neurons. In addition, neurons utilize receptors for still other endogenous products in order to receive messages related to various physiological systems. For example, receptors for insulin and the endorphins allow neurons to modulate their activity in response to environmental stimuli related to stressors and gustatory behaviors.

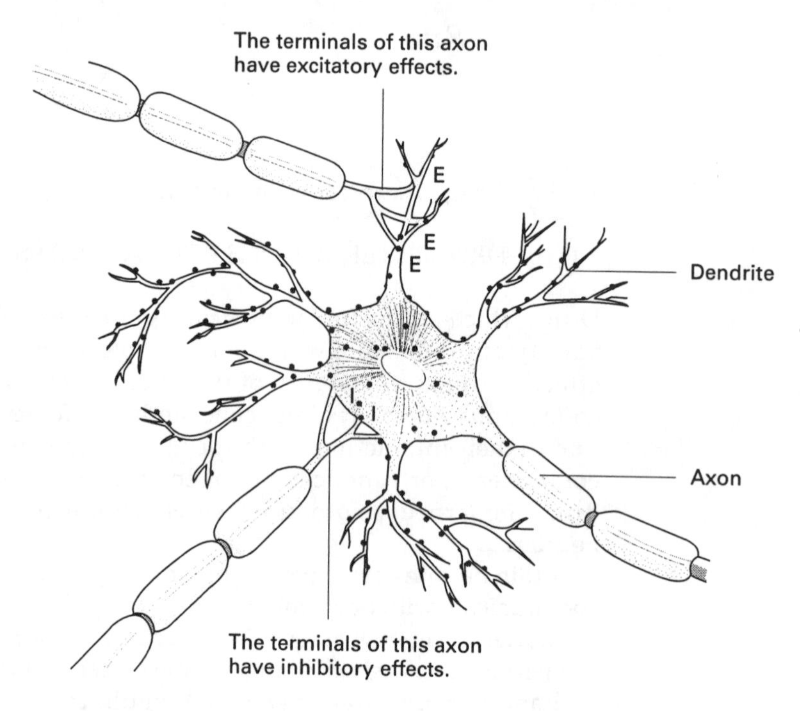

FIGURE 6.1
Excitatory and inhibitory synapses

The particular cellular response may be either to stimulate or to inhibit the initiation of a subsequent action potential in the receiving neuron (Figure 6.1). In many cases this inhibition or excitation is mediated by a cascade of biochemical events occurring within the cell body of the receiving neuron. Indeed, the major biochemical products of these series of biochemical processes are often referred to as second messengers. Some common second messenger systems include the generation of either cyclic adenosine monophosphate (cAMP) or cyclic guanosine monophosphate (cGMP), phosphatidylinositol (PI) turnover, the production of various prostaglandins, thromboxanes and leukotrienes from arachidonic acid metabolism, and the production of nitric oxide (NO) molecules. It is the summation of all such biochemical events, initiated by neurotransmitters binding to a receptors, which determines whether or not a neuron will fire in response to the various stimuli that it has received. A neuron will respond to the composite of stimuli that it receives at any point in time in what is known as an *all* or *none* response. If it fires, it will fire with a characteristic pattern and magnitude of response. It will not moderate or exaggerate each action potential in association with stimuli which are less than or greater than its threshold for such a response.

Neurotransmitter binding to receptors is a transient phenomenon. These molecules bind with receptors, then break away to migrate through the synaptic fluid to collide with enzymatic proteins or with other neuronal membrane components. Over a very short period of time, the neurotransmitter in the synapse is diminished by several processes. First, these chemical messenger molecules may be taken up into the neuron from which they were released via active transporter mechanisms. These "transporters" or "reuptake sites" are specific for the neurotransmitter molecule produced by the releasing neuron, or at least the first stage metabolites. Such reuptake mechanisms allow neurons effectively to recycle neurotransmitter molecules to be stored again in vesicles and released again with the advent of subsequent action potentials. Secondly, neurotransmitter molecules may migrate from their immediate site of action to interact with neighboring neurons. Thus, a single neuron is able to signal many neurons, with the receptors located in more proximal neurons being bound by the greatest number of neurotransmitter molecules and more distant neurons receiving relatively fewer signal molecules. Finally, neurotransmitters may be diminished in the synapse by the action of metabolic enzymes.

Drugs exert their effects on the brain by binding to neuronal sites associated with neurotransmitter receptors or with other sites of action generally related to neuronal firing. Thus, drug molecules may bind to synaptic receptors to form drug-receptor complexes (Figure 6.2). Drug molecules may also bind to sites located on or near ion channels which allow the flow of ions through axonal membranes, thus the propagation of an action potential down the length of the axon. In addition, drugs may block the reuptake of neurotransmitters through presynaptic transporters, leading to higher concentrations of the neurotransmitter in the synapse for longer periods of time and increased innervation of surrounding neurons. Some drugs may cause perturbations in the neuronal cell membrane itself. Such perturbations result in changes in cellular functions of many types, especially those mediated by membrane-bound proteins. Finally, drugs may interact with cytosolic proteins within the neuronal cell. In this way, drugs may produce a variety of effects on neurotransmitter synthesis and release, second messenger system activity, and various intracellular metabolic processes.

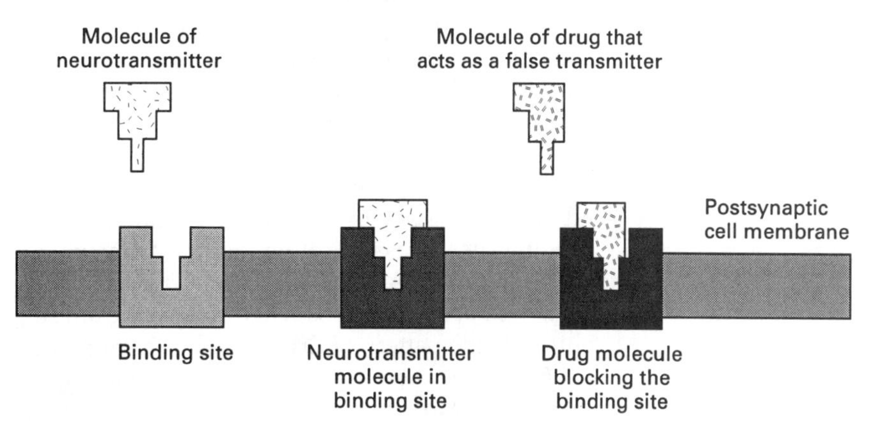

FIGURE 6.2
Lock-and-key view of drug effects in the synapse

Chemical structure determines whether or not drugs will interact with specific neurotransmitter receptors or other binding sites. A drug-receptor complex may be thought of using the lock and key model of receptor binding on a larger scale. Drugs are often larger and more complex than neurotransmitters. Perhaps more importantly, drugs do not often bind specifically to a single neurotransmitter receptor. Different portions of the drug molecules may interact with different receptors. Thus, multiple drug-receptor interactions are associated with particular effects of some drugs. In addition, multiple receptor sites for a drug mean that more than one type of drug effect may be produced. It is common knowledge that most drugs used for medicinal purposes exhibit the "desired effect" as well as some array of potential '"side effects". At the very least, it is likely that such a drug may be effective for a given medicinal purpose only across a limited range of dosage or blood concentration.

Agonism/antagonism

The function of drugs at receptors to which they bind may be either similar or opposed to that of endogenous neurotransmitters that bind to these sites. First, a drug may serve as an *agonist*. In this case, the drug binds to the neurotransmitter receptor and produces an effect which mimics that of the neurotransmitter. Thus, it is able to elicit cellular responses which are similar to that produced by the neurotransmitter. Secondly, a drug may function as an *inverse agonist*, binding to the neurotransmitter receptor but producing an effect opposite to that of the neurotransmitter itself. Third, the drug may act as an *antagonist*. More specifically, if the drug serves to bind to the neurotransmitter receptor but produces no concomitant cellular response, it is known as a *competitive antagonist*. Alternatively, if the drug binds to, not the receptor as recognized by the neurotransmitter, but to a site adjacent to or associated with that receptor, it may effectively block the action of the neurotransmitter (or other agonists) at the receptor site. This type of antagonist action is commonly referred to as a *noncompetitive antagonism* since the antagonist does not actually bind to the neurotransmitter recognition site. Such a compound, in binding to the associated (allosteric) site, might effectively change the neurotransmitter recognition site so that the receptor is "masked". Alternatively, such a compound might block binding to a site for a necessary cofactor.

1.2 RECEPTOR BINDING TECHNIQUES — *IN VIVO* AND *IN VITRO*

Drug binding sites in the brain are identified by means of several receptor binding techniques[175]. Receptor binding assays may be performed under *in vitro* conditions utilizing tissue samples dissected from animals which have been killed for the experiments. Assays may also be performed under *in vivo* conditions involving the injection of the radiolabeled compound into the blood stream of a living organism and the subsequent determination of the pattern of binding using some method of scanning or imaging the brain. In each of these techniques a radioactive label which is known to bind relatively specifically to the receptor of interest is used to identify the receptor. Drug molecules may be added to the assay in order to determine whether they compete for the labeled binding sites, thus indicating that the drug exhibits a measurable affinity for the site in question. Alternatively, the drug itself may be labeled with a radioactive moiety in order to assess whether specific binding sites may be identified or to elucidate the pattern of drug binding throughout the brain.

For *in vitro* assays, the interaction of the ligand and the receptors occurs in test tubes or in other laboratory glassware. Brain slices or tissue homogenates, placed into a buffered solution, are exposed to the radioactive ligand. In a small number of control tubes, a high concentration of another compound which is specific for the receptor of interest, but is not radioactive, is added to the mixture of tissue and radioactive ligand. This compound binds to all of the receptors of interest so that any residual binding of the radioactive ligand to the tissue sample is due to various nonspecific interactions with cellular components. Specific receptor binding sites are then operationally defined as the difference between the total number of radioactive molecules bound by the tissue in the absence of the nonradioactive ligand, and residual binding to nonspecific binding sites in the presence of the nonradioactive compound.

For *in vivo* assays of brain receptors, living tissues are exposed to radiolabeled compounds in an analogous manner. However, these experiments require that the radiolabeled compounds are administered by injection methods and enter the brain and other tissues via the bloodstream. Non-radiolabeled ligands are also used in *in vivo* experiments to elucidate nonspecific binding of the labeled ligand to receptors of interest. Of course *in vivo* assay methods must take into account the effects of metabolic and other physiological functions. In humans and in nonhuman primates, positron emission topography (PET) methodology allows the imaging of patterns of drug binding in living tissues including brain. Smaller animals may be killed after administration of the radiolabeled agent in order to slice tissues so that binding of the ligand to specific brain regions may be studied.

1.3 INFORMATION GAINED FROM RECEPTOR BINDING STUDIES

Rceptor binding specificity

Several types of information about the interactions of drugs with brain receptors may be gained from receptor binding experiments. First, a drug may be shown to bind to specific brain binding sites. In addition, these binding sites may be identified as particular neurotransmitter receptors, binding sites on ion channels or other transport mechanisms, or as various cellular components. For example, a specific drug binding site may be characterized as a serotonergic or GABAergic receptor. Alternatively, the drug may be shown to specific binding sites which are not associated with known neurotransmitters or other endogenous neurochemicals. Moreover, factors which may facilitate or inhibit drug binding may be also be discovered. For example, drug binding may be dependent upon the concentration of various ions or upon the presence of a particular endogenous molecule.

Receptor localization

Drug receptors may also be precisely localized in the brain using either *in vitro* assay conditions involving tissue slices or *in vivo* assay conditions. Indeed, these techniques not only allow the association of drug binding sites with discrete brain formations but also with particular neurons or groups of neurons. Thus, drug binding sites may be observed to be located in specific layers of the cortex or to be colocalized with dopaminergic tracts running through the mesolimbic brain regions. Using PET imaging techniques, changes in the localizaton of drug binding at receptors in various regions of the brain can also be observed, facilitating an understanding of the relationship between these changes and the time course of behavioral or physiological effects.

Receptor function

Finally, the function of a drug at a receptor may be predicted from receptor binding assays. The drug may be identified as agonist or antagonist, for example, on the basis of its binding characteristics in the presence of guanosine triphosphate (GTP), the precursor to the production of the cyclic nucleotide cGMP. An agonist will exhibit decreases in its affinity for the receptor in the presence of GTP. Indeed, research has shown that the greater the decreases in affinity of the drug for the receptor under these conditions, the greater the observed efficacy of the drug is likely to be. Alternatively, the affinity of an antagonist does not change in the presence of GTP.

1.4 ASSOCIATION OF DRUG RECEPTOR WITH A SPECIFIC DRUG EFFECT: CORRELATIONAL ANALYSIS

Pharmacological correlations have been used extensively to identify drug binding sites associated with specific effects of drugs. The underlying premise behind the use of pharmacological correlations is that the identification of a functionally relevant receptor depends upon a significant correlation between the potencies of chemically and pharmacologically related drugs in producing a response, and the potencies of these agents at specific drug binding sites (Figure 6.3). Thus, pharmacological correlations allow the determination of "sites of pharmacological initiation of drug effects". If a significant positive correlation is found, this result suggests that the greater the affinity of each drug for the receptor the more potent the drug in eliciting the drug response. Conversely, if a significant negative correlation is found, this result suggests that the greater the affinity of each drug for the receptor the less potent the drug in eliciting the drug response.

Opiate receptors were identified in part by the use of pharmacological correlations. Brain binding sites were identified which exhibited specificity for the active stereoisomers of opiate agonists[127,146,155]. Subsequently, the correlation between the potencies of various opiate agonists at this opiate receptor site and their potency to produce important pharmacological effects was determined to be quite significant[28,167].

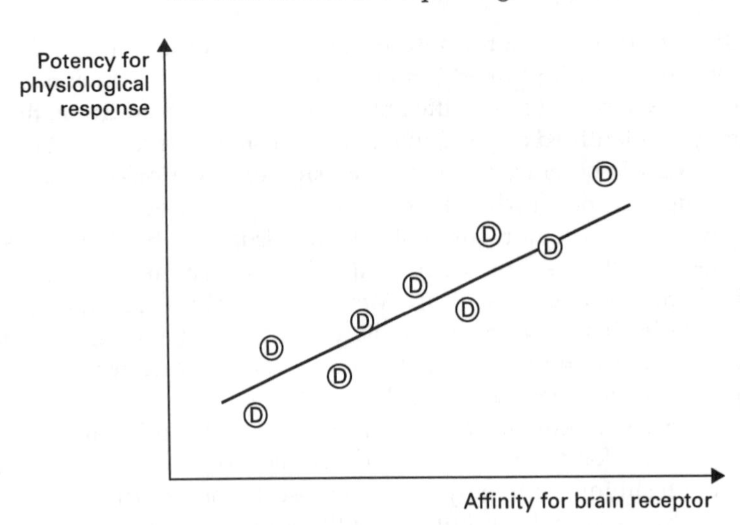

FIGURE 6.3
Pharmacological correlation

Similarly, the hallucinogenic effects of drugs has been shown to be associated with brain $5HT_2$ receptors using correlational methods[51]. The potency of over 20 hallucinogens is highly correlated with their potencies in stimulus generalization tests in animals using the potent hallucinogen DOM (1-(2,5-dimethoxy-4-methylphenyl)-2-aminopropane) as the training drug. Moreover, there is a very strong association between the potency of binding to the $5HT_2$ receptor and reported subjective hallucinatory effects of these drugs in humans[141].

Brain receptors associated with nearly any phenotypic measure of drug effect may be studied using this experimental paradigm. If there is evidence that brain mechanisms mediate the drug response, the affinity of drug binding to specific brain receptors may be associated with many physiologic, behavioral, and neurochemical effects of drugs.

Indeed, the influence of multiple receptors on specific drug effects may be investigated using multiple site analyses. Using multiple regression methods, the influence of several receptor sites on drug response may be examined simultaneously, while accounting for covariance among related transmitter systems. It is possible that the affinities of the drugs studied may be positively correlated with one receptor and negatively correlated with another binding site. This result would suggest that while drug binding to the first site is associated with potency of the drug to *elicit* the effect, while drug binding at the other site would appear to *attenuate or mask* the effect of interest.

Furthermore, multiple regression methods offer at least two advantages over sequential linear analyses. First, the probabilities of both spurious relationships due to covariance between independent variables and false positive relationships are decreased in comparison to linear correlational analyses. If factors other than the brain binding sites assessed in a particular study significantly impact upon the potencies of related drugs to produce a particular effect, then the resultant multiple regression will indicate that only a small portion of the variance in drug response may be accounted for by these brain binding sites. Using this method, it is also possible that none of the receptors studied would be positively or negatively correlated with drug potency to produce seizures. Second, the simultaneous assessment of the influence of several drug binding sites on a specific drug effect allows the single determination of the proportion of the variance in the potency of drugs to produce the specific effect (r^2). In contrast, successive linear correlations would require that the variance accounted for in each analysis would be summed to give and estimate of total variance ($r^2 + r^2 + r^2 + r^2 =$ Total Variance Accounted). In this case, it becomes increasingly likely that the resulting sum will suggest that more than 100% of the total variance is associated with particular drug bindings. Of course, this is impossible.

Thus, pharmacological correlations serve as one of the most effective research strategies to elucidate which of multiple drug binding sites are associated with specific effects of a drug. For example, pharmacological correlations have recently been used as a research strategy helpful for the determination of brain receptors associated with the reinforcing, seizurgenic, and lethal effects of cocaine. In the case of cocaine, a simple correlational approach is immediately made more complex by the fact that this drug produces a variety of pharmacological effects through its interactions with several central nervous system binding sites. In addition to its interaction with binding sites on at least two neuronal ion channels, cocaine binds to a number of synaptically localized binding sites. Therefore, an appropriate analysis must include information about neuronal reuptake sites for dopamine, norepinephrine, and serotonin as well as sigma and muscarinic cholinergic receptors in brain. This method of multiple site analysis, which simultaneously examines the influence of several receptor sites on drug response while accounting for covariance among related transmitter systems, has been a powerful tool which has facilitated the determination of the influence of specific cocaine binding sites on specific cocaine effects.

Similarly, genetic correlational analysis is a powerful research strategy for the determination of biochemical genetic mechanisms of action associated with specific drug effects. Whereas pharmacological correlations involve the study of the relationship between the potencies of drugs in producing a response and their potencies at particular binding sites to determine "sites of pharmacological *initiation* of drug action", genetic correlations involve the determination of the relationship or correlation between biochemical and behavioral phenotypes across distinct genetic strains of animals to determine "sites of genetic *variation* in drug action" (Figure 6.4). For example, research using this strategy has illustrated which of several receptors bound by nicotine are associated with several acute behavioral effects of the drug as well as those which appear to be associated with the development of tolerance to the drug in response to chronic exposure[98,99].

1.5 CORROBORATING EVIDENCE FOR DRUG RECEPTOR IDENTITY AND PHARMACOLOGY

The identity and function of drug receptors in the brain can be corroborated using a number of methods. Neuronal mechanisms associated with behavioral or physiological phenomena may be identified by using an array of compounds that are known to interact with specific receptor subtypes in the attempt to shift dose-response curves to the left or to the right. For example, if the neurotransmission of dopamine is hypothesized to mediate a particular drug effect, then dopamine agonists should shift the dose-response curve for the drug to the left, while a dopaminergic antagonist should shift the dose-response curve to the right.

The release of specific neurotransmitters into synaptic spaces may also be measured by utilizing brain microdialysis methods. These methods allow the direct measurement (via a small probe implanted into the brain) of endogenous chemicals which have been released into extracellular spaces in response to minute quantities of drug which have been applied to neurons. As has been discussed in Chapter 5, brain microdialysis has shown that many drugs which are abused by humans preferentially increase dopamine efflux in limbic areas of the brain, especially in nucleus accumbens, while drugs which are not abused do not have this effect. Interestingly, other drugs exhibiting aversive properties appear to reduce dopamine release in the accumbens and in the caudate. Drugs having neither reinforcing nor aversive properties do not alter dopamine concentrations in these experiments.

Finally, electrophysiological studies allow the measurement of neuronal axon potentials as they are generated in response to a particular drug. For this method, brain slices are oxygenated in buffers approximating natural physiological conditions to allow the study of neuronal signaling in specific neurons or small groups of neurons in response to the administration of drug onto the tissue. As for behavioral studies, a number of pharmacological agents specific for receptors hypothesized to produce particular drug effects may be utilized in these experiments.

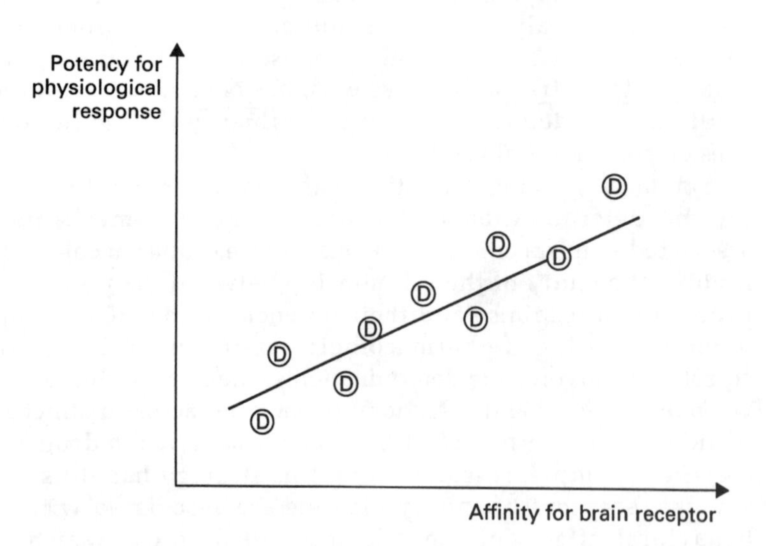

FIGURE 6.4
Example of a genetic correlation analysis

2 Molecular mechanisms of specific drug classes

2.1 COCAINE

Cocaine is a powerfully addicting drug of abuse. However, cocaine use and abuse has increasingly been associated with toxic consequences such as seizures and death. Most of the research strategies which have been discussed here have been applied to the task of determining the neuronal mechanisms associated with various effects of cocaine, including reinforcing properties, effects on cardiac function, and toxic effects. Indeed, the social and economic implications of cocaine abuse have provided a strong rationale for research related to this drug of abuse. The information which has been gained substantially increases our understanding of the mechanisms of action of cocaine, and should aid in the development of pharmacological treatments for cocaine addiction and acute toxicity .

Reinforcement

The pharmacological correlation approach has been utilized to determine the brain binding sites for cocaine which are associated with its reinforcing or addictive effects[135]. Using these methods, the relationship between the potencies of cocaine and several pharmacologically and chemically related compounds in operant studies of drug self-administration behavior and their binding potencies at monoaminergic uptake sites has been determined. This model utilizes the operant paradigm in which an animal performs a particular behavioral response in order to obtain access to the test drug. As discussed in Chapters 3, such drug-reinforced behavior in primates, rodents, and other species is considered to be a useful model of human addiction liability since most drugs that humans abuse have been found to be reinforcing in animals of various species. A fundamental requirement for the identification of receptors related to cocaine self-administration, then, is a significant relationship between the potency of cocaine and related compounds in producing reinforcing effects in operant self-administration studies and the potency of these compounds at particular binding sites in brain.

Cocaine serves to inhibit the reuptake of monoamines into presynaptic terminals. However, the results of this research indicate that the dopamine transporter appears to be the primary cocaine binding site associated with its reinforcing properties. The binding of cocaine and related drugs to the dopamine transporter is highly associated with their potency to produce reinforcing or euphoric effects[135]. Subsequent studies have also provided consistent results indicating the potencies of cocaine and related compounds in self-administration studies correlate positively and significantly with inhibition of the binding of other ligands to the dopamine transporter[11,97]. In humans, positron emission tomography (PET) techniques have indicated a similarity between the time course of cocaine's occupancy of the binding site at the transporter and the time course of associated psychological effects[41]. Thus, cocaine inhibition of dopamine uptake has been shown to be the primary biochemical influence on the reinforcing effects of cocaine.

Many behavioral studies have utilized lesion techniques and pharmacological manipulations to elucidate the biochemical mechanisms associated with the reinforcing effects of cocaine[91,92,137]. Dopaminergic neuronal systems, especially in brain mesolimbic pathways, have been implicated in the reinforcing effects of cocaine. Selective dopaminergic receptor blockade attenuates the reinforcing properties of (-)cocaine in animals, whereas dopaminergic agonists substitute for intravenous self-administration of cocaine.

Several lines of evidence indicate that cocaine, unlike amphetamine, does not promote the release of dopamine. Mathematical modeling studies suggest that the metabolism of dopamine in the synapse is consistent with cocaine inhibition of uptake, but not with influences on the neuronal release of dopamine. Cocaine does little to affect release of dopamine from nerve terminals, rather only inhibits the reuptake of the neurotransmitter subsequent to its release.

The increased dopamine in the synaptic cleft has several effects[172]. First, it decreases the synthesis of dopamine by a feedback mechanism. Second, dopamine in the synaptic cleft acts via D_2 autoreceptors to inhibit the subsequent firing of the presynaptic dopaminergic neuron. Third, the facilitation of dopaminergic neurotransmission results in decreases in postsynaptic neuronal firing, especially in the caudate, nucleus accumbens, and medial prefrontal cortex. This postsynaptic inhibition may be associated with both D_1 and D_2 receptors. Overall, the action of cocaine on these neurons is self-limiting across time because less dopamine will be synthesized and released, decreasing the impact of cocaine as a reuptake inhibitor.

Though substantial effort has been applied to the determination of which dopaminergic receptor subtypes are specifically involved in the reinforcing effects of cocaine, it remains unclear whether this holds true for either D_1 or D_2 receptors[172]. Currently, several studies have illustrated apparent cooperation between D_1 and D_2 receptors in the nucleus accumbens in particular with respect to the induction of adenylate cyclase, and synergism at the cellular level which may be associated with particular dopamine related behavioral phenomena. Overall, however, it may be likely that D_1 antagonists may be more useful as clinical pharmacotherapies for psychostimulant abuse because they typically exhibit fewer of the extrapyramidal side effects associated with D_2 receptor antagonists[172].

Pharmacological correlations have indicated that cocaine binding to serotonin transporters is not associated with the reinforcing effects of cocaine under identical operant conditions utilizing short test sessions and requiring fixed numbers of responses (Fixed Ratio 10) in order to obtain the drug[135]. Consistent with these biochemical results, another study of the reinforcing effects of psychostimulants using similar methods has shown that the potent serotonin uptake inhibitor fluoxetine does not appear to attenuate rates of responding for cocaine, while again decreasing those associated with amphetamine[130]. Nevertheless, some evidence suggests that serotonergic systems may modulate cocaine-reinforced responding under quite different operant conditions. Specifically, if animals have continuous access to the drug over extended periods of time, rather than for only one or a few hours, fluoxetine has been shown to decrease overall intake of cocaine[18]. Further, under operant conditions requiring the completion of an increasingly greater number of responses prior to administration of cocaine (rather than a fixed number of responses) manipulation of serotonergic neurotransmission appears to attenuate responding for cocaine[138]. These results suggest that, under some conditions, factors other than the reinforcing effects of cocaine *per se* may influence cocaine self-administration. The nature of these influences is as yet unknown, although they may be related to some situational and environmental influences that facilitate motivational states or efficacy of the drug. If this were true, these same motivational factors might influence or, perhaps, be analogous to human craving for drugs. In this case, continued chronic abuse of a drug might require not only specific reinforcing effects of a drug per se, but also motivational factors contributed by the subject[47]. Indeed, interest in the efficacy of serotonin uptake blockers as adjunct pharmacotherapeutic agents in addiction treatment programs has been increasing (see also Chapter 9).

Another series of experiments has indicated that the benzodiazepine receptor agonist, chlordiazepoxide, also decreases cocaine-reinforced responding in rats; this effect may be due to specific interactions between benzodiazepine receptors with dopaminergic neurons[52]. Alternatively, it might be speculated that carbamazepine may attenuate the reinforcing effects of cocaine by means of interactions between GABAergic and serotonergic neurons. As mentioned above, although the reinforcing effects of cocaine appear to be mediated primarily through dopaminergic systems, these effects appear to involve serotonergic factors under certain experimental conditions. It is possible that GABAergic inhibition of serotonergic neurons in dorsal raphe and mesolimbic brain regions may lead to decreases in cocaine reinforcement. If further research is consistent with these findings, benzodiazepine agonists may prove to be clinically effective in the treatment of cocaine abuse.

There is growing evidence that opiate compounds may also be effective as adjunct pharmacotherapeutic agents for the treatment of cocaine addiction. Buprenorphine, a mixed opiate agonist-antagonist, and naltrexone, an opiate antagonist, have been shown to selectively decrease responding for intravenously administered cocaine in rhesus monkeys[104]. Further, chronic buprenorphine treatment inhibits cocaine conditioned place preference, suggesting that the reinforcing effects of the drug have been blocked. However, it has been shown that buprenorphine itself produces conditioned place preference in a dose related manner, and that subthreshold doses of cocaine and buprenorphine, given in combination, produced conditioned place preference[13]. These results indicate that buprenorphine may produce reinforcing effects of its own. Further, the same report indicated that cocaine and buprenorphine both increase extracellular levels of dopamine in the nucleus accumbens as measured by microdialysis techniques. Thus, it appears that the reinforcing effects of buprenorphine may substitute for, not inhibit, the reinforcing effects of cocaine, and in this way may produce decreases in operant responding for cocaine in experimental studies.

Cardiotoxicity

In recent years, there has been a marked increase in reports of cardiovascular morbidity associated with cocaine use[57,61]. It is well established that cocaine causes coronary arteries to constrict, reducing blood flow to the heart. Cocaine can also cause coronary arteries to spasm, leading to myocardial infarctions. The mechanism by which this occurs is not yet completely understood.

It is clear, however, that cocaine produces profound effects on the cardiovascular system through its interactions both with the central nervous system and with peripheral neurons associated with the autonomic neural system. Cocaine inhibition of monoamine uptake, in particular, is responsible for increasing the concentrations of these neurotransmitters at synaptic and neuroeffector junctions. Enhanced neurotransmission of norepinephrine and the action of this neurotransmitter at alpha and beta adrenergic receptors mediates the induction of increases in heart rate and blood pressure. At plasma concentrations associated with euphoria, cocaine typically produces a 30–50% increase in heart and blood pressure[40].

Cocaine binds to muscarinic cholinergic receptors at blood concentrations associated with many post mortem cases[144]. Indeed, it appears to serve as an antagonist at muscarinic M_2 receptors, the predominant subtype in cardiac tissue. Since parasympathetic innervation of the heart, via cholinergic receptors, plays an important homeostatic role in modulating and opposing stimulation of heart rate, cocaine may influence cardiac function in yet another way. Specifically, while stimulation of heart rate induced by low doses of cocaine may be modulated by reflex vagal inhibition, the antagonist action of cocaine could prevent vagal inhibition of increased heart rate at high blood concentrations. However, research suggests that cardiac dysfunction, although it is frequently associated with cocaine related death, may be an indirect effect of other physiological responses to cocaine. It is possible, for example, that respiratory failure may precede and lead to subsequent cardiac dysfunction as has been frequently observed in cocaine-related human deaths.

Seizures

Another series of studies utilizing multiple site analysis methods has recently shown that serotonin transporters appear to be the primary initial sites of action mediating seizures induced by cocaine and related drugs in C57BL/6J male mice[136]. In addition, binding of cocaine-like drugs at sigma receptors, or muscarinic M_1 cholinergic receptors appears to be inversely related to seizures produced by cocaine and related drugs, suggesting that drug binding at these receptor sites may attenuate the seizurgenic properties of these compounds. Pharmacological manipulations of all these neurotransmitter systems provide further evidence for the direct involvement of $5HT_2$ receptors in cocaine seizurgenesis. While seizure initiation appears to depend primarily on affinity of cocaine and related compounds for binding sites associated with the serotonin transporter, the seizure-inducing properties of cocaine may ultimately depend on a final summation of its effects not only on serotonergic systems, but on muscarinic and sigma neuronal systems as well.

The observation that serotonin systems are the primary mediator of cocaine-induced seizures is consistent with other lines of evidence suggesting serotonergic involvement in seizures (see ref. 136 for review). For example, infantile spasms have been associated with 5-hydroxytryptophan treatment and with abnormal serotonin metabolism, including high plasma serotonin levels. Other studies also suggest that heightened activity of serotonergic systems, similar to that produced by cocaine binding to serotonin transporters, may result in a seizure-prone state.

The specific brain regions in which sigma and cholinergic receptor systems might interact with serotonergic or serotonergic-innervating neurons, are as yet unclear. Early studies demonstrated that cocaine-induced seizures originate in the limbic system, and, in particular, the amygdala[35]. Receptor binding and autoradiography studies have illustrated the presence of serotonergic, muscarinic, and sigma receptors in this brain region. Limited evidence for interactions between serotonergic and cholinergic systems comes from studies showing that serotonergic neurons projecting to cortical regions from the dorsal raphe nucleus stain for acetylcholinesterase, suggesting that these neurons are responsive to cholinergic stimulation. Other research has shown that neocortical and hippocampal formations receive both cholinergic and serotonergic inputs.

Other neuronal mechanisms such as the gammaaminobutyric acid (GABA)/benzodiazepine (BZ) receptor complex and the excitatory amino acid receptor complex may play roles in modulating seizures induced by acute cocaine injections. Results from these studies suggest that inhibition of cocaine-induced seizures through manipulation of pathways with which cocaine does not directly interact may ultimately be due to an effect of these pathways on serotonergic neuronal activity. It is likely that neurohormones such as GABA, glycine, and glutamate, which have influences on many other types of seizures, whether naturally occurring or induced by chemical or electrical stimuli, may also influence cocaine-induced seizures at some point in the cascade of events subsequent to the binding of cocaine to its initial sites of action.

Lethality

Receptor binding studies have suggested that dopamine transporters are significantly associated with lethality induced by cocaine and related drugs[136]. However, binding of cocaine-like drugs at muscarinic M_1 and sigma receptors also appears to be significantly related to cocaine-induced lethality, and the binding of cocaine and related drugs to these brain sites appears to attenuate subsequent lethal effects. Pharmacological manipulations of these predicted neurotransmitter systems suggest direct involvement of dopaminergic D_1, muscarinic M_1, and sigma receptors in cocaine lethality. It is unclear whether or not and to what extent specific neuroanatomical structures may mediate the lethal effects of cocaine. Further research may illustrate the mechanisms as well as other brain regions in which dopaminergic and sigma receptors are colocalized.

2.2 AMPHETAMINE AND DERIVATIVES

Amphetamine and cocaine have often been viewed as belonging to the same pharmacological class of drugs, the psychostimulants. Research has shown that they can produced similar physiologic, behavioral, and subjective effects. Animal studies also have shown that both drugs exhibit potent reinforcing effects, induce locomotor activation, and increase heart rate and blood pressure. Furthermore, it has been shown that, under certain conditions, neither humans nor animals can discriminate between the effects of the more active stereoisomers of these drugs, *d*-amphetamine and *l*-cocaine.

However, increasing information indicates that amphetamine and cocaine exhibit important differences in both pharmacology and biochemical mechanisms. For example, cocaine is an effective local anesthetic actions while amphetamine is not. Amphetamine is a monoamine oxidase inhibitor, whereas cocaine is not. In general, psychostimulants are now divided into two classes based upon their biochemical effects on catecholaminergic neurons. The amphetamine-like class of compounds inhibit neurotransmitter reuptake and potently stimulate the release of neurotransmitter from these neurons. The nonamphetamine-like (cocaine, mazindol, and methylphenidate) class of drugs inhibit reuptake but are not effective as releasers.

The mechanism by which amphetamine elicits the release of catecholamines has been studied for many years. Two mechanisms have been suggested. First, amphetamine may interact with monoamine transport mechanisms, facilitating monoamine release through heteroexchange[5,6,39]. At the same time, of course, amphetamine molecules would inhibit the further reuptake of monoamines which were previously released into the synapse. Second, amphetamine and its derivatives may migrate through neuronal membranes by means of passive diffusion, displacing monoamines from intracellular binding sites (Liang and Rutledge, 1982). In this case, monoamines would also be released from neurons through active carrier transport. Consistent with both of these hypotheses, it has in fact been shown that the release of dopamine and norepinephrine stimulated by amphetamine is calcium-independent, suggesting that these monoamines are released from cytosolic pools of neurotransmitter not stored in vesicles[5]. More recent evidence also suggests that amphetamine molecules are indeed sequestered into synaptic terminals[178]. However, this research also indicates that passage of the drug molecules across neuronal membranes probably occurs by means of passive diffusion, and that intracellular pH gradients may help to maintain the drug inside the neuron.

Reinforcement

Several lines of evidence indicate that the reinforcing effects of amphetamine are associated with dopaminergic function (see ref. 133 for review). Lesions of dopaminergic neurons in mesolimbic regions of the reward pathway (Chapter 7) decrease amphetamine self-administration in animal models. The dopamine receptor blocker pimozide decreases the euphoria experienced by humans following i.v. injections of amphetamine. Likewise, dopamine receptor antagonists attenuate the reinforcing effects of amphetamine in animals. Dopamine agonists, in contrast, substitute for amphetamine in i.v. self-administration paradigms.

However, amphetamine does not appear to produce its reinforcing effects primarily via binding to the dopamine transporter as does cocaine[133]. Indeed, amphetamine is a much greater reinforcer in self-administration studies than would be predicted by its potency at the dopamine transporter. However, it is possible that the releasing properties of amphetamine are relatively more important than its ability to inhibit dopamine uptake. Alternatively, these biochemical effects of amphetamine together produce the euphoric and reinforcing effects of the drug.

The reinforcing effects of amphetamine, unlike cocaine, are also mediated by serotonergic neurotransmission. Pharmacological correlations indicate that the potency of amphetamine and related drugs at the serotonin transporter is inversely correlated with their potency as reinforcers[133]. Thus, the greater the potency of each of these compounds at the serotonin transporter, the less potent the compound as a reinforcer in self-administration studies. These results indicate that serotonin attenuates the reinforcing effects of amphetamine related compounds. In addition, lesions of serotonergic neurons lead to an increase in amphetamine self-administration. In contrast, pretreatments with the serotonin precursor L-tryptophan or with serotonin uptake inhibitors decrease amphetamine self-administration.

Neurotoxicity of amphetamine derivatives

It is widely known that amphetamine derivatives produce significant neurotoxicity[50]. It is well documented that methamphetamine, the *N*-methylated form of amphetamine, is quite toxic to both dopaminergic and serotonergic neurons. Various experimental paradigms have been used to show that methamphetamine administration produces decreases in neurotransmitter synthesis and release across a number of species.

More recently, there has been a great deal of scientific interest in 3,4-methylenedioxymethamphetamine (MDMA) and methylenedioxyamphetamine (MDA), the ring-substituted derivatives of methamphetamine and amphetamine, respectively. Though it was first synthesized and patented in Europe in 1914, MDMA has only recently gained significant attention as a "designer drug", synthesized in secret by chemists attempting to circumvent laws prohibiting the production, sale, or use of drugs designated as illicit. Similar to other hallucinogens, both MDA and MDMA exhibit significant hallucinogenic properties due to their interaction with serotonergic receptors[96]. Significant scientific debate related to medical ethics also arose in the early 1980s as MDMA gained some popularity as an adjunct to psychotherapy, reportedly producing increased self-confidence, introspection, and empathy while diminishing anxiety and depression. Research confirmed that MDMA is self-administered by primates, suggesting that it might have a high abuse potential in humans.

The use of this compound for either recreational or therapeutic purposes became particularly disturbing, however, as research indicated that MDA and MDMA are potent neurotoxins[124]. These compounds have now been shown to produce selective degeneration of serotonergic neurons, sparing catecholaminergic neurons. This degeneration appears to depend upon the serotonin transporter mechanism, and is prevented by the concurrent administration of serotonin uptake inhibitors. The axons of serotonergic neurons appear to be ablated, while cell bodies are spared and portions of axons remaining are thickened due to the blocked flow of neurotransmitter and other cellular products down the length of the axon. Serotonergic neurons in some brain regions appear to be more resistant to the neurotoxic effects of MDA and MDMA. In addition, thicker axons appear to be generally more resistant to these neurotoxic effects relative to fine, arborized axon terminals.

Regeneration of serotonergic neurons appears to occur over a relatively long period of time, using serotonin transporter binding sites as markers of the regeneration of axons and determining serotonin levels[9]. It has been shown that repeated systemic administrations of MDMA are associated with approximately 75–80% depletions of serotonin transporters and serotonin levels for up to 6 months. These studies indicate that the density or number of serotonin transporters appears to return to normal levels by 12 months following drug administration. However, serotonin synthesis remained 50% below control levels. This latter finding suggests that, although serotonergic fibers may regenerate, the function of these neurons may not be entirely recovered.

Fenfluramine is an amphetamine derivative which has been used as an anorectic agent for the treatment of obesity. It was originally thought to provide a significant advantage over amphetamine for this purpose since it does not produce the psychosis induced by chronic amphetamine treatment. However, like MDA and MDMA, fenfluramine has now been shown to produce neurotoxicity associated with serotonergic neurons, but not dopaminergic neurons[4]. Research has also shown that particular types of neurons and distinct brain regions are more vulnerable to this neurotoxic effect than are others.

2.3 OPIATES

In the 1940s and 1950s, the development of nalorphine and naloxone, drugs which antagonized the physiological and behavioral effects of acute morphine administration, suggested the existence of specific opiate receptors. It was hypothesized that morphine served as an agonist at such a binding site and that nalorphine and naloxone acted as antagonists at the same neuronal site. The localization and function of these neuronal binding sites was unknown.

The discovery of specific opiate receptors in 1973 suggested the existence of endogenous opiates[127,146,155]. Subsequent research lead to the discovery of such compounds, including endorphins and enkephalins[149]. In addition, distinct classes of opiate receptors have been discovered and cloned, including μ, δ, and κ receptors[132]. Opiate binding to each of these receptors is associated with the inhibition of cyclic adenosine monophosphate (cAMP) synthesis and subsequent decreases in levels of phosphoproteins[95]. The mechanism of action of endogenous opiates at the synapse is not as yet completely understood. However, it is generally agreed that endorphins are extremely potent and that their overall effect is to inhibit neuronal firing. They may function as neurotransmitters themselves, as neuromodulators, or as both.

Few opiate receptors are located on dopamine neurons in reward pathways, and opiates generally do not regulate dopaminergic neurons directly. There is some evidence that opiates may act in part through μ receptors localized on GABAergic interneurons in the ventral tegmental area[31]. When these receptors are stimulated by opiate agonists, GABAergic neurons would be hyperpolarized. Thus the firing of these neurons and the release of GABA would be inhibited, resulting in disinhibition of dopaminergic neurons. Alternatively, enkaphalins and opioid agonists acting through δ opioid receptors appear to regulate postsynaptic targets of the mesolimbic dopamine system[34]. Chronic administration of opiate agonists appears to produce a supersensitivity in cultured dopaminergic neurons upon termination of this treatment, such that these neurons are more sensitive to the depolarizing effects of several ions[141].

Dopaminergic systems in turn appear to modulate the function of opiate receptors under some circumstances. Lesions of dopaminergic neurons in the nucleus accumbens result in a decrease in μ opiate receptors, not on dopaminergic neurons themselves, but on the dendrites and cell bodies of postsynaptic neurons[159]. It has also been shown that repeated administrations of cocaine produces a dose-dependent decrease in opiate receptors in several brain regions, nearly all of which are associated with brain reward systems[66]. In addition, cocaine leads to an increase in the synthesis of the opiate peptide dynorphin in the rat striatum, and this effect is dependent on D_1 and D_2 dopaminergic receptors[147].

Though few studies have been undertaken to date, there is some evidence that opiate receptors also interact with serotonergic neurons. The highly potent opiate agonists fentanyl and sufentanyl inhibit agonist binding to serotonergic $5HT_1$ receptors in rat brain, and presumably serve as antagonists at this site since no changes in receptor coupled second messenger systems occur in response to the binding. Chronic administration of morphine results in an increase in the density of $5HT_2$ receptors, referred to as an "upregulation", in several regions of rat brain which are closely associated with reward pathways including amygdala, midbrain, and pons-medulla sections[64]. These phenomena may have important implications for the influence of serotonergic mechanisms in the etiology of opiate addiction, dependence and withdrawal. A substantial literature also indicates that serotonergic neurotransmission influences opiate-induced analgesia in a variety of experimental paradigms[10].

Analgesia

Opioid analgesics function as agonists at opioid receptors in both spinal cord and in brain to produce a decrease in the perception of pain. Opiate receptors in spinal cord may mediate various nociceptive reflexes and agonists acting at these sites produce striking analgesic responses. Opiate agonists administered into the midbrain and medulla, especially the periaqueductal gray matter and the nucleus raphe magnus, produce pronounced analgesia. There is some evidence that descending monaminergic pathways projecting from these brain areas have inhibitory effects on the processing of painful stimuli at the level of the spinal cord.

Indeed, a variety of pharmacological, neurochemical, and behavioral data support the influence of monoamines on opiate-induced analgesia[128]. Dopamine D_2 receptor agonists have been shown to increase pain tolerance using a variety of experimental conditions[106,158]. Serotonergic neurotransmission in both spinal cord and in brain has been linked to the analgesic effects of opiate compounds[73,97,156]. Consistent with this, autoradiographic methods have also provided evidence that μ receptor density in the dorsal raphe nucleus is a primary mechanism associated with genetic differences in mouse strains selectively bred for high (HAR) and low (LAR) analgesic response to levorphanol.

A preponderance of evidence indicates that analgesic effects of opiates are mediated in part by μ receptors. However, there is a growing body of evidence that κ and δ receptors also mediate analgesia. Indeed, current scientific theory suggests that agonists at these receptors may offer a critical therapeutic advantage in that these compounds are not associated either with the abuse potential or the pronounced tolerance and physical dependence typical of agonists such as morphine[108].

Reinforcement

Research has shown that the reinforcing effects of opiates are associated with dopaminergic neurons in brain reward pathways[47,88,89] Electrophysiological evidence, for example, indicates that morphine stimulates neuronal firing of dopaminergic neurons in the ventral tegmental area[100]. Lesions of dopaminergic tracts in the nucleus accumbens or ventral tegmental area decrease or extinguish opiate-reinforced behaviors and place preference[31,151]. In addition, microinjections of opiate agonists and antagonists into these brain areas have been shown to modulate the reinforcing effects of opiates. Nevertheless, under some experimental conditions, modulation of opiate self-administration may not be effectively achieved using agents which inhibit dopaminergic neurotransmission[37,49,160].

Endogenous opiates exhibit reinforcing effects that are similar to those associated with morphine, heroin, and other narcotic compounds. In general, local administration of these compounds into the ventral tegmental area facilitates self-stimulation behaviors and is associated with place preference[31,145]. Intraventricular administration of the endogenous opioid peptide β-endorphin serves as a positive reinforcer in operant studies and produces place preference[3,161]. Intraventricular self-administration of enkephalins has also been illustrated[53,153].

The specific class of opioid receptors which mediate the reinforcing effects of opiate drugs is as yet undetermined. It has traditionally been assumed that μ opioid receptors mediate the reinforcing effects of opiate drugs. Indeed, a number of scientific studies have produced data which is consistent with this hypothesis[89,120]. These studies have shown that activation of μ receptors is sufficient to produce reinforcing effects. However, δ opioid receptors also appear to mediate the reinforcing effects of opiate drugs[121]. In contrast, compounds which have specificity at κ receptors appear to elicit primarily aversive effects, rather than reinforcing effects[121]. The effects of specific subtypes of receptors (e.g., μ_1, μ_2) within these classes is unknown.

Moreover, since morphine, heroin, and other opiates which are readily abused interact with more than one, if not all, of these receptor classes, the reinforcing effects of these compounds are the summation of the effects of drug binding at each of these sites. Studies have shown that μ receptors have a important role in mediating the self-administration of opiates, but that δ receptors may also have a secondary role. Another study utilized two genetically distinct mouse strains which differ substantially in the density of brain μ receptors to study the influence of these receptors on the self-administration of the potent opiate agonist etonetizine[36]. The results of the research indicate that etonetizine served as a reinforcer in both genotypes, although mice having greater numbers of μ receptors ingested more drug. These date suggest that μ receptors may influence the magnitude of the reinforcing effects (i.e., the drug efficacy) of this or other opiate reinforcers, but that still other neuronal sites are likely to be involved in determining the characteristics of these drugs as reinforcing agents. Thus, the abuse liability of each opiate compound depends upon the number of opiate receptors bound and the affinity of the compound at each of the opiate sites to which it binds. In addition, there is evidence that some μ and δ receptors exist together in receptor complexes[75,85,157]. The effects of opiate agonist binding to these receptor complexes on drug reinforcement is unclear.

2.4 BENZODIAZEPINES AND BARBITURATES

The rate of primary sedative abuse in drug abuse treatment programs compared with those associated with the abuse of alcohol, opiates, or psychostimulants is usually less than 10% of drug abusers[168]. However, the incidence of polydrug abuse involving benzodiazepines or barbiturates is frequently reported to be in the range of 40–50% of drug abusers in treatment programs. Indeed, these sedative/hypnotic drugs are frequently used by a psychostimulant abuser who wants to sleep after a prolonged period of stimulation, by a drug abuser experiencing tremors associated with chronic drug intake, or by an opiate abuser who may not have ready access to the next administration of heroin.

Barbiturates are preferred by the addicted population, but benzodiazepines are available more frequently by prescription. Indeed, benzodiazepines are among the most prescribed drugs in the world for the treatment of anxiety and insomnia due to large dose ranges associated with their anxiolytic effects and due to a larger margin of safety relative to barbiturate compounds. While barbiturates have been commonly related to accidental deaths and suicides, benzodiazepines used alone are rarely fatal.

Benzodiazepines and barbiturates interact with binding sites functionally related to GABA receptors[24]. These GABA receptors are associated with neuronal chloride ion channels, which are intrinsically involved in the passage of action potentials down the length of an axon when a neuron fires (Figure 6.5). Since this effect on neuronal systems is not specific to a particular neurotransmitter system or neuronal type, benzodiazepines and barbiturates have significant influence on a number of neurotransmitter systems.

This specific effect is associated with the acute anticonvulsant effects of benzodiazepines and neuronal tolerance to this effect is responsible for observed decreases in the facility with which these compounds can inhibit seizures over time.

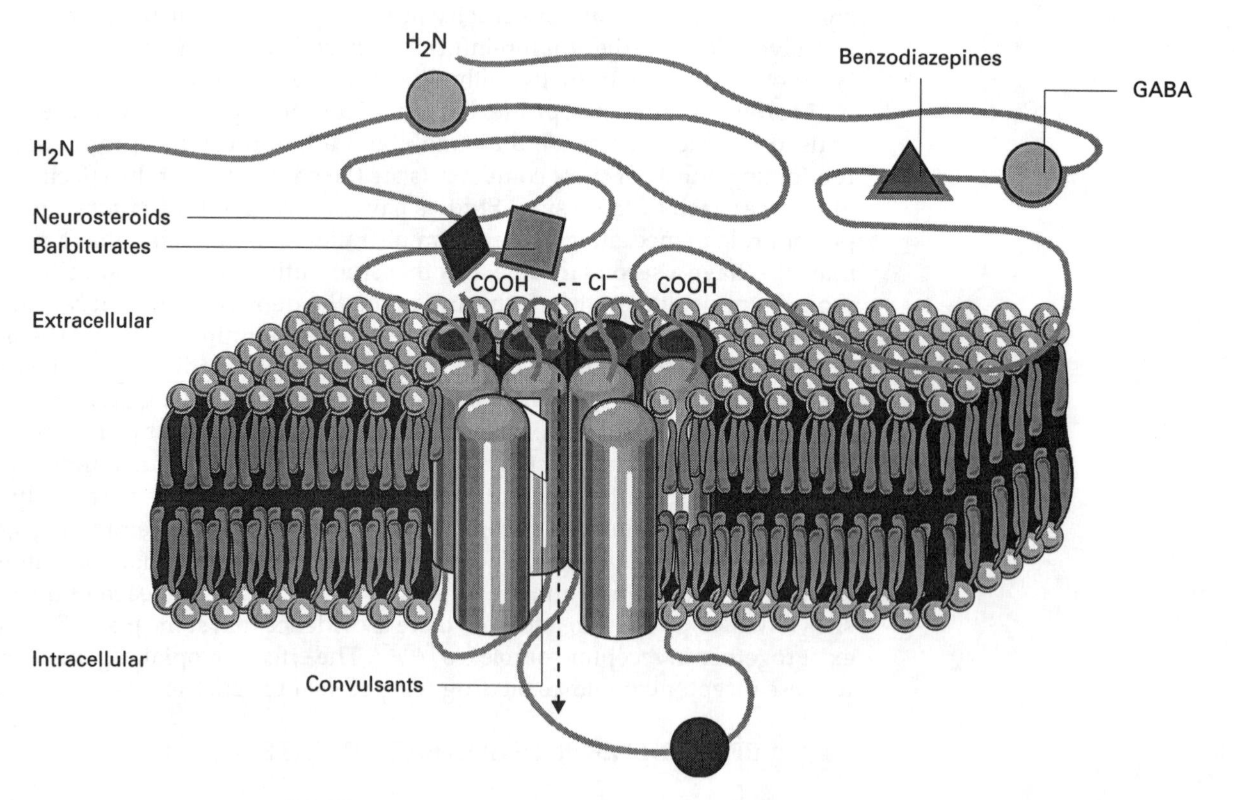

FIGURE 6.5

Schematic representation of the GABA ionotropic receptor channel and imaginary positions of binding sites for ligands

Indeed, the anticonvulsant effects of benzodiazepines are the primary reason that these compounds are included in drug detoxification treatment programs[168]. The withdrawal symptoms associated with abstinence from alcohol or opiates after a long history of chronic use involve powerful rebound effects. Whereas these substances of abuse are behavioral depressants, convulsions and tremors frequently occur during withdrawal from these substances of abuse unless prevented by pharmacological adjuncts to treatment. Thus, benzodiazepines are commonly administered to patients undergoing detoxification in order to inhibit the potentially harmful convulsant effects of these drugs. Barbiturates are similarly effective as anticonvulsants, but are not as commonly utilized in detoxification programs due to their own addictive potential.

The specific influence of GABA on dopaminergic neurons in brain reward pathways and the influence of GABAergic neurotransmission on the effects of ethanol, cocaine, and heroin are well documented[47,89]. The binding of benzodiazepines and barbiturates to other sites on the chloride ion channel are presumed to mediate similar influences on this mesolimbic neuronal system and to be similarly associated with the reinforcing effects of these sedative compounds[177]. Indeed, electrophysiological evidence suggests that benzodiazepines probably potentiate the inhibitory effects of GABA on interneurons which normally act to inhibit dopaminergic neurons in the ventral tegmental area[60]. Thus, by a process of disinhibition, benzodiazepines would have the overall effect of increasing the activity of dopaminergic neurons. Research evidence from behavioral studies also indicates that self-administration of diazepam is decreased by the dopamine receptor antagonist haloperidol[44,129]. Place preference associated with the administration of diazepam is also attenuated by haloperidol and by lesions of the dopamine terminals in nucleus accumbens[150]. These results suggest that dopaminergic mechanisms mediate, at least in part, the reinforcing effects of these benzodiazepines.

However, in contrast to other drugs of abuse, benzodiazepines appear to decrease the release of dopamine in the nucleus accumbens[38]. This paradox raises the possibility that benzodiazepines may not be self-administered for positive reinforcing effects mediated by dopamine neurotransmission in mesolimbic brain regions. Instead, given the important anxiolytic effects of these compounds, it seems possible that these compounds are self-administered for negative reinforcing effects. In behavioral terms, the drug seeking behavior would be likely to continue in order to decrease the negative ramifications of anxiety and stress. However, systematic studies designed to test this hypothesis are lacking.

2.5 ALCOHOL

Ethanol, in contrast to other drugs such as cocaine, exhibits a relatively simple chemical structure. Thus, while other drugs exhibit relatively specific interactions with brain receptors and drug binding sites are localized in distinct brain areas, ethanol appears to have less specific interactions with brain receptors and other endogenous proteins.

Most early studies of the biological mechanisms associated with various ethanol effects centered on its fluidizing effects on cell membranes[58]. Scientific expertise in the area of lipid biochemistry has been a valuable research tool for these endeavors. Thus, though all drugs may have effects on neurotransmission, at least some ethanol effects may involve more generalized influences on cell function. Such a generalized effect might confound linear correlations between particular biochemical and/or behavioral traits. Indeed, this generalized effect may be observed as covariance between the influences of several independent variables on a particular ethanol effect. Current scientific theories concerning the neuronal bases for ethanol effects are consistent with this idea, and generally view most ethanol-induced behaviors as mediated by multiple neuronal components.

Nevertheless, some specific effects of ethanol on receptor function have been recently observed. Serotonergic, dopaminergic, benzodiazepine and opiate receptor systems, and cyclooxygenase enzymes have been shown to modulate ethanol drinking in both operant and preference models of ethanol drinking[134,142]. Similarly, several laboratories have illustrated the influences of NMDA (*N*-methyl-*d*-aspartate) receptors, calcium channels, GABA receptors, cyclooxygenase enzymes, and guanine nucleotide binding proteins on sensitivity to ethanol and on tolerance and physical dependence produced by chronic ethanol treatment.

Brain dialysis has recently been used to show that alcohol, like other drugs which are abused by humans, preferentially increase dopamine efflux in limbic areas of the brain, especially in nucleus accumbens[16,32]. This finding is consistent with evidence from brain slice preparations that ethanol produced a dose-dependent excitation of dopaminergic neurons in the ventral tegmentum, a brain region which has been implicated in the reinforcing properties of many drugs of abuse[12]. Behavioral research has indicated that the dopamine agonist bromocriptine decreased ethanol intake and increased water intake producing a significant decrease in ethanol preference[90]. In the operant ethanol drinking paradigm, the novel dopamine receptor agonist SDZ-205,152 selectively reduced ethanol self-administration[131]. Taken together, these results are consistent with the hypothesis that the reinforcing effects of ethanol may involve a dopaminergic component.

The influence of serotonergic systems on ethanol drinking has been investigated for nearly two decades since initial studies indicated that lesions or pharmacological manipulations that resulted in depletions of brain serotonin were associated with decreased ethanol preference[113,115]. Both preference and operant ethanol-reinforced behavior paradigms have been used to study the effects of serotonin uptake inhibition on ethanol drinking. For example, inhibition of serotonin reuptake or administration of serotonin synthesis precursors has been consistently shown to decrease voluntary consumption of ethanol solutions in rodent preference paradigms[94,101,114,139]. Only a limited number of studies have examined the effects of serotonin uptake inhibitor treatment on ethanol-reinforced responding in an operant paradigm, but have indicated that these compounds may produce a significant decrease in ethanol-reinforced responding beginning on the first day of treatment[65,68]. Similar suppression of ethanol drinking has been found in human subjects treated with a variety of serotonin uptake inhibitors[116,117,118,119]. Taken together, these studies suggest that facilitation of serotonergic neurotransmission may attenuate the reinforcing effects of ethanol.

Recent studies have begun to illustrate the involvement of specific serotonergic receptor subtypes in mediating ethanol drinking behaviors. Both acute and chronic administration of the $5HT_2$ antagonist amperozide have reduced ethanol drinking in the preference paradigm[111,112]. Similarly, the $5HT_2$ antagonist ritanserin also reduces ethanol preference, and this suppression of ethanol drinking has been observed for nearly 3 weeks following the end of treatment[103,126]. Further, $5HT_3$ antagonists have been shown to reduce ethanol preference[87] and to block its discriminative stimulus effects[63]. These latter findings may be related to the findings that ethanol stimulates the release of both serotonin and dopamine in the nucleus accumbens, and that the $5HT_3$ receptor antagonist ICS 205-930 significantly reduces this effect[176].

However, the manner in which serotonergic systems ultimately reduce ethanol intake is unclear[134,143]. A growing body of data also suggests that serotonergic influences on ethanol drinking may result from generalized effects on ingestion of both foods and liquids. Indeed, many serotonin uptake blockers which effectively reduce ethanol intake also are anorexic in nature. Related evidence suggests that these pharmacological manipulations may in fact mediate ethanol preference via the renin-angiotensin system.

Alternatively, it is also possible that serotonergic neurotransmission has a more direct effect on dopaminergic neurons in reward pathways. As mentioned, recent studies have shown that serotonergic $5HT_3$ receptors appear to modulate ethanol-induced dopamine release in the nucleus accumbens[17,173,176]. Specifically, $5HT_3$ receptor antagonists appear to inhibit the release of dopamine in this brain area and to reduce ethanol ingestion using the preference model of ethanol drinking. Preliminary findings also suggest that $5HT_3$ receptor antagonists may decrease the subjective effects of low doses of ethanol in humans, including craving for the drug[86]. Nevertheless, the role of serotonergic systems in mediating the reinforcing effects of ethanol, and the potential for serotonergic compounds to be used in the treatment of alcoholism remains unclear.

Although specific interactions of ethanol with binding sites on the chloride ion channel have not been illustrated (see Figure 6.5), $GABA_A$ receptor function has been consistently associated with both behavioral and biochemical effects of ethanol[62,89,142]. In general, low concentrations of ethanol were consistently observed to potentiate chloride ion flux mediated by $GABA_A$ receptors. Via this mechanism, ethanol potentiates GABAergic inhibition of neuronal firing. Behavioral studies have provided evidence that rats which have been genetically selected for high ethanol preference have greater densities of GABAergic nerve terminals in nucleus accumbens than do rats which have been genetically selected for low preference[81].

The highly publicized finding that the benzodiazepine antagonist RO15-4513 appeared to serve as a specific ethanol antagonist in several behavioral and cellular assays provided convincing evidence that the chloride ion channel was critical in mediating the pharmacological effects of ethanol[154]. It has since been reported that i.p. administration of RO15-4513 significantly and specifically decreased ethanol intake in ethanol-preferring rats without affecting food or water intake[101]. Thus, these data suggest that the reinforcing effects of ethanol may be modulated in part by receptors on the GABA/benzodiazepine/chloride ion channel complex in nucleus accumbens.

However, recent research findings have indicated that ethanol inhibition of neuronal firing in nucleus accumbens slice preparations appears to be mediated by glutamatergic function, and not by GABAergic function[123]. Related to this, ethanol decreases Ca^{2+} conductance at NMDA receptor complexes (Figure 6.6), and blocks NMDA-stimulated neurotransmitter release[33,59,74,171]. Some studies have suggested that ethanol interacts specifically with PCP or glycine binding sites on the channel to produce its effects. In addition, behavioral studies utilizing drug discrimination procedures suggest that ethanol acts as a noncompetitive NMDA antagonist, serving to effectively block the channel to ion conductance. Nevertheless, the mechanism by which ethanol produces these effects at the ion channel formed by the NMDA receptor complex has not been elucidated to date (for review see ref. 62).

In fact, much less is known about the manner in which ethanol interacts with NMDA receptors. Since the molecular characterization of the individual subunits of this ion channel has been more recent, our understanding of the relationship between specific subunit compositions and the functional effects of ethanol on ion conductance is as yet unknown. For the same reasons, we have little information about elements of the protein structure of the channel which may vary across brain region or neuronal type to confer sensitivity to the administration of ethanol.

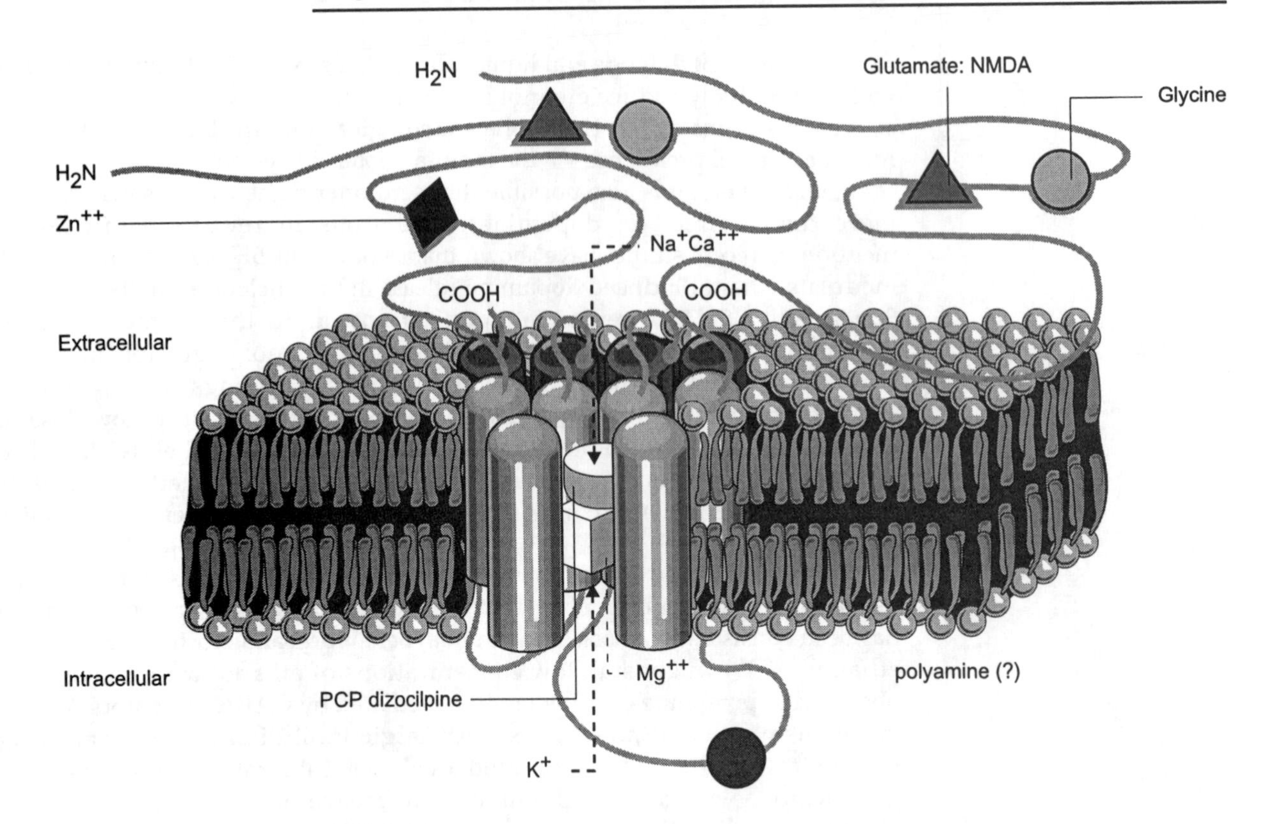

FIGURE 6.6
Model of the NMDA ionotropic receptor channel and imaginary positions of binding sites for ligands

Ethanol inhibition of neuronal firing in nucleus accumbens is reversed by naloxone. Taken together with the finding that this effect is mediated at least in part by glutamatergic receptors, this research suggests that opiate receptors in some way mediate ethanol effects on the function of the ion channel associated with NMDA receptors[123]. Furthermore, these results are consistent with the possibility that endogenous opioid systems influence the reinforcing effects of ethanol.

Some investigators have put forth hypotheses concerning the brain mechanisms associated with the reinforcing effects of ethanol which suggest that ethanol stimulation of endogenous opiates is a key to these rewarding properties. Indeed, a number of studies provide evidence that ethanol administration leads to the expression of POMC mRNA, and ultimately, to increases in brain enkephalin and β-endorphin concentrations. Still other studies provide considerable support for the suggestion that the sensitivity of the POMC system to ethanol stimulation may in part mediate genetic differences in ethanol self-administration.

Morphine, the potent opiate agonist, produces increases in ethanol drinking in rats when administered in low doses. Based on these findings it is theorized by several investigators that the drug may mimic initial ethanol stimulation of endogenous opiates leading to increased subsequent drinking. In contrast, comparable low doses of morphine appeared to reduce alcohol consumption in chronic drinking rhesus monkeys under conditions of unrestricted alcohol supply as well as after several days of alcohol abstinence (see Figure 5.8). These results suggest that in experienced monkeys low doses of morphine substituted for the effects of ethanol drinking. High doses of morphine, in general, decreased ethanol drinking, which could be due to a strong substitution effect of ethanol's reinforcement. On the other hand, high opiate dosing leads to a sedation that might impair behavior in general. Together, it can be concluded that opioid agonists interfere with ethanol consumption, but the exact mechanism is not clear yet.

Ethanol decreases the threshold for brain self-stimulation behavior, and naloxone readily attenuates this phenomenon[95]. Behavioral studies have also shown that the opiate antagonists naloxone and naltrexone will reduce ethanol consumption in preference, free access, and operant models of ethanol self-administration[2,42,43,142,164]. Results are not always unanimous as to what extent these findings are due to specific inhibition of the reinforcing effects of ethanol as mediated in part by specific brain regions associated with the reward pathway. It is known that these opiate antagonists also can reduce food and water intake, even in deprived rats[13,152]. The results of the rhesus monkey study, as shown in Figure 6.7, illustrates an alcohol specific effect of naltrexone, especially when monkeys had been deprived of alcohol but not of water. It is not amazing that alcohol as well as water or food consumption are affected by opiate antagonists, being respectively an unnatural reinforcer (a drug) or natural reinforcers. It is interesting that when craving for a particular reinforcer is high (reduced by deprivation) naltrexones effect seems not specific. Another argument is that naloxone can produce conditioned taste and place aversions to substances with which it is paired[67]. Nevertheless, other studies have illustrated that under certain conditions pairing opiates antagonists with ethanol produce aversions to ethanol.

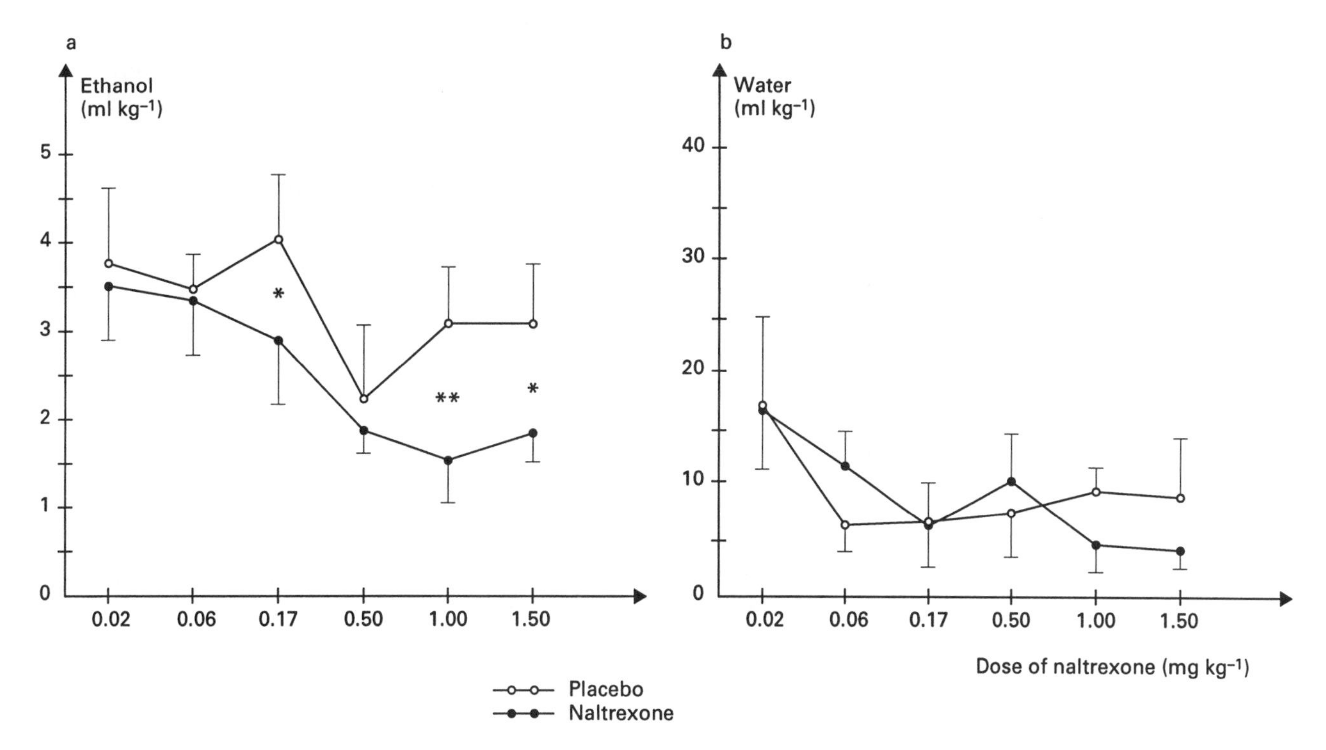

FIGURE 6.7

Mean intake (± SEM) of total net alcohol (a) and drinking water (b) after imposed alcohol abstinence in rhesus monkeys for paired injections of placebo and naltrexone

2.6 CANNABINOIDS

Marijuana is unique among drugs of abuse in that there is little evidence that it serves as a reinforcer in animal models of self-administration. Although tolerance develops to the effects of marijuana, it has also been difficult to illustrate physical dependence. However, although there is also only limited evidence that it stimulates dopaminergic neurons in reward pathways, ^{9}D-tetrahydrocannibinol (THC) has been shown to increase intracranial self-stimulation into the median forebrain and to produce dopamine efflux in the nucleus accumbens and the prefrontal cortex[19,21,46]. Further more, one study has suggested that differences in the facility with which THC induces dopamine efflux from the nucleus accumbens may be associated with genetically determined vulnerability, or sensitivity, to the reinforcing effects of the drug[20]. The neuronal mechanisms underlying these phenomena are as yet unclear.

In 1990, the brain receptor for THC, the active ingredient in marijuana, was isolated and cloned[48]. The gene for the THC-receptor appears to be expressed only in brain. However, another gene with considerable sequence homology has been found in mouse spleen, lymph nodes, and in leukocyte-enriched samples of blood[29]. It is possible that some of the immunosuppressive effects of THC are mediated by these sites in the immune system. Brain receptors for THC have been observed throughout the brain, though most are located in the basal ganglia, hippocampus, cerebellum, striatum, and cerebral cortex. The distribution of these receptors does not parallel that of any other known neurotransmitter or neuromodulator. THC is characterized as an agonist at these receptors; its function appears to involve the inhibition of cAMP production and the reduction of transmembrane ion currents through specific types of calcium channels[77,78].

Since the discovery of a brain receptor implies an endogenous ligand for that binding site, the search for such a compound began immediately thereafter. In 1992 a compound that occurs naturally in the brain called arachidonylethanolamide, belonging to the class of endogenous compounds called eicosanoids, was shown to bind to the THC receptor and to produce pharmacological effects similar to those of THC itself[29,30]. The compound has been given the name anandamide by its Hebrew discoverers after the Sanskrit word *ananda,* meaning bliss. Current research efforts are focused on the determination of the structure-activity relationship for THC, anandamide, and other apparently potent ligands at the THC receptor. In addition, significant research efforts are designed to better understand the biological interactions between eicosanoids and cannabinoids.

The development of several novel cannabinoid-related compounds has facilitated the study of the cannabinoid receptor (Figure 6.8). Although these compounds have proven to be useful tools in the study of both neurochemical and behavioral effects of $^9\Delta$-THC, caution is important in the interpretation of the results of studies utilizing these ligands. $^9\Delta$-THC exhibits nearly equal potency in producing its various effects, including hypoactivity, antinociception, hypothermia, and catalepsy. However, the novel compounds that have been developed are much more potent in producing some of these effects as opposed to others. These differences in response suggest that these ligands may be interacting in a different manner with the cannabinoid receptor, or that they may also interact with other brain binding sites to produce their effects.

Cannabinoids may share at least some common neuronal mechanisms with opioid compounds. As already discussed, studies of intracellular events associated with ligand binding to either cannabinoid or opiate receptors indicate that these receptors are linked via G-proteins to the production of cAMP. However, evidence for a more direct interaction between these receptors is provided by the finding that $^9\Delta$-THC inhibits ligand binding to and opiate receptors in a noncompetitive manner unrelated to its facility to perturb the neuronal membrane bilayer[162]. More recently, microdialysis studies have shown that $^9\Delta$-THC enhancement of dopamine efflux in nucleus accumbens can

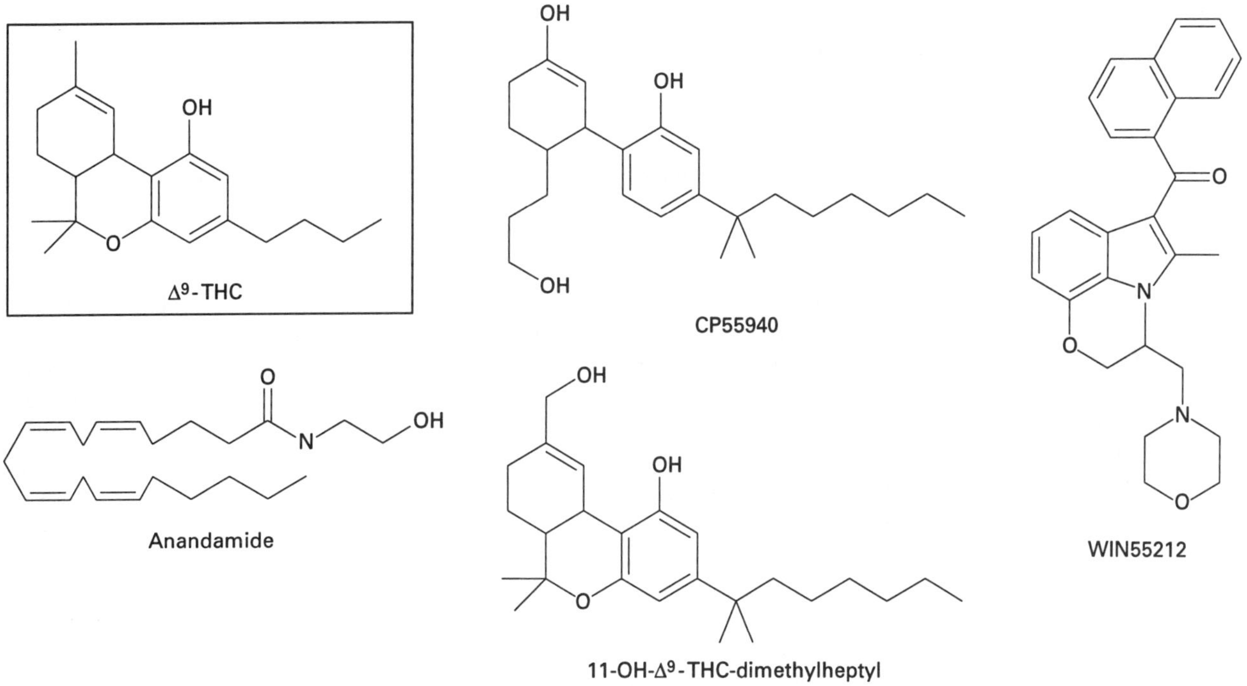

Δ9-THC

CP55940

WIN55212

Anandamide

11-OH-Δ9-THC-dimethylheptyl

FIGURE 6.8
Cannabinoid related compounds

be inhibited by naloxone[22]. Thus, there may be some interaction between cannabinoid binding sites and opiate receptors in this region of the reward pathway. In addition, there is increasing evidence that cannabinoids interact with opiate systems involved in the perception of pain. Indeed, cannabinoids clearly produce analgesic effects in both animals and humans and, of all the potential clinical uses of cannabinoids, the mediation of analgesia has received the most attention. One study has illustrated that the potency with which a number of bicyclic cannabinoid analogs inhibits ^{3}H-CP55940 to cannabinoid receptors is highly correlated with the potency of each of these compounds to produce analgesia[105]. Another study has shown that morphine and THC produce synergistic analgesic effects when administered together[163].

Research designed to utilize opiate antagonists to inhibit the antinociceptive effects of cannabinoid compounds have yielded some inconsistent results. For example, the irreversible opiate antagonist chlornaltrexamine inhibits several effects of THC, including analgesia, hypothermia, and tolerance[158]. Several other studies of the influence of the opiate antagonist naloxone on the antinociceptive effects of several cannabinoids, however, have lead to varying conclusions[23,166,174]. Nevertheless, the most recent studies have begun to suggest that specific opiate receptor subtypes mediate the influence of opiate compounds on cannabinoid-induced analgesia. In particular, the *kappa* opioid antagonist nor-binaltorphimine (nor-BNI) blocks the antinociceptive effects of cannabinoids, including $^{8}\Delta$-THC, $^{9}\Delta$-THC and CP55940, while the *delta* opiate receptor antagonist ICI 174,864 and naloxone, administered in low doses specific for opiate receptors, were not effective[148,165].

Recent reports have suggested that chronic exposure to THC leads to neurodegeneration in hippocampal regions of the brain. Indeed, it has been shown that the neuronal aging induced normally by glucocorticoids is exacerbated by THC. It has been proposed that the mechanism of this neurotoxicity may be related to the chemical structural similarity between cannabinoids and glucocorticoid steroids, and that the effect of cannabinoids is mediated by the direct interaction of THC with glucocorticoid receptors. Alternatively, the interaction of THC with glucocorticoid-receptor-mediated events may be indirect, involving the antagonism of the normal glucocorticoid negative feedback suppression of ACTH and the resultant increase in the release of corticosterone. There are in fact a number of reports that THC administration stimulates pituitary-adrenal secretions, including corticosterone.

2.7 PHENCYCLIDINE

Phencyclidine (PCP) was originally introduced as an anesthetic exhibiting minimal cardiovascular and respiratory depression. However, the advent of its recreational use and abuse in the 1960s has lead to a great deal of investigation of brain mechanisms mediating its psychoactive effects. PCP has now been shown to interact with several brain sites. It binds to potassium channels, biogenic amine transporter molecules, NMDA receptors and sigma receptors with pharmacologically relevant affinities, and to cholinergic receptors and opiate receptors with only relatively low affinity. PCP has also been shown to interact with its own specific binding site, to which no other known neurotransmitters and neuromodulators bind with high affinity[163,179]. It is this distinct PCP binding site that appears to mediate the discriminative stimulus properties and the reinforcing properties of PCP, as differentiated from those of other sympathomimetics or hallucinogens[7,8].

PCP receptors appear to be localized on ion channels associated with the passage of chloride and sodium through neuronal membranes during the propagation of action potentials down the length of an axon. These channels are regulated by the endogenous neurotransmitter glutamate. Several subtypes of glutamate receptors confer specificity of action throughout the brain. PCP interacts with one of these receptor subtypes, the NMDA receptor. This particular glutamate receptor takes its name from *N*-methyl-D-aspartate, the amino acid which has been shown selectively to activate this receptor to mediate that opening of the associated channel. The opening of these ion channels is blocked in the presence of PCP, thus PCP is functionally a pharmacological antagonist of ion channel activation induced by glutamate binding to NMDA receptors.

Many studies have addressed the influence of PCP on dopaminergic function. The results indicate that at least some PCP receptors are located on dopaminergic neurons. Further research has shown that PCP serves as an indirect dopamine agonist, apparently increasing the release of dopamine as well as inhibiting its reuptake. Brain dialysis experiments have shown that PCP, like other drugs of abuse, increases extracellular concentrations of dopamine in nucleus accumbens[17]. These influences on dopaminergic function are presumed to mediate PCP effects such as reinforcing effects, hypothermia and elevation of prolactin levels. In addition, PCP inhibits *N*-methyl-D-aspartate (NMDA) stimulation of dopamine release via a binding site on the NMDA receptor/ion channel protein complex. It appears to act as a noncompetitive voltage-dependent inhibitor of the NMDA receptor, binding to a site inside the ion channel.

Chronic administration of PCP produces a desensitization of dopaminergic neurons. Dopamine release is decreased due to a mechanism which is not associated with a depletion of synthesized and stored dopamine. Instead, the mechanism involves some adaptive response of the dopaminergic neuron to PCP stimulation.

Finally, PCP is also known to inhibit the reuptake of serotonin. Consistent with this, high affinity serotonin uptake inhibitors act as pseudo-irreversible inhibitors of PCP binding to these sites due to their much greater affinity at this macromolecular structure.

2.8 NICOTINE

In classical pharmacology, nicotine has been used to describe the effects of nicotinic receptors in both sympathetic and parasympathetic autonomic ganglia. Nicotine, at low to moderate doses, mimics the interaction of the endogenous neurotransmitter acetylcholine at these receptor sites. However, at high doses, it may produce a pharmacological antagonism of the passage of ions through the associated channel. Though these peripheral binding sites for nicotine are not likely to be associated with the psychoactive properties of nicotine, they have provided a good model for understanding the function of nicotinic receptors in brain.

Nicotine is believed to be the major component of tobacco associated with its addictive potential, or abuse liability. Administered intravenously, nicotine serves as a positive reinforcer in both animals and humans[26,54,55]. In addition, nicotine administration has been associated with conditioned place preference[45], and increases in operant responding maintained by brain stimulation reinforcement[80,169]. However, under some experimental conditions, nicotine can also be shown to exhibit noxious or punishing effects[56,83]. These aversive effects appear to depend largely on the dose of nicotine and the schedule of drug administration. Both the reinforcing and aversive effects of nicotine have been shown to be inhibited by the nicotinic antagonist mecamylamine.

Genetic correlational analysis

The use of genetic correlational analyses has been elegantly applied to studies designed to assess the relationship between various nicotine-induced behavioral phenotypes and several brain cholinergic nicotinic receptor binding parameters[98,99,110]. Utilizing 19 inbred strains of mice in the genetic analysis, it has been illustrated that significant differences in receptor number and regional distribution within brain can be consistently observed across strains in *in vitro* receptor binding techniques. Most importantly, these studies provide information about the association of these genetically determined differences in nicotinic receptor number and regional distribution across inbred mouse strains with specific behavioral phenotypes that have also been measured in each strain. Using these multivariate statistical analyses, observed differences in receptor binding have been correlated with specific behavioral effects of nicotine. For example, these methods have indicated that the greater the density of nicotinic binding sites labeled by ^{3}H *l*-nicotine, the greater the inherited sensitivity to the low dose effects of acutely administered nicotine. High densities of nicotinic binding sites labeled by ^{125}I-α-bungarotoxin appear to be associated with the occurrence of seizures in response to high doses of nicotine. Thus, there appear to be distinct nicotinic receptor mechanisms associated with specific nicotine-induced phenotypes. Moreover, this suggests that it is possible that individuals who smoke may be sensitive to the reinforcing effects and resistant to the toxic or aversive effects of nicotine.

The greatest densities of nicotinic receptors labeled by ^{3}H *l*-nicotine are in the colliculi, midbrain, hypothalamus and striatum, while the greatest densities of nicotinic receptors labeled by ^{125}I-α-bungarotoxin are localized in the colliculi, hypothalamus, hippocampus and striatum. Nicotine is thought to produce some of its reinforcing or addictive effects through excitatory actions on mesolimbic dopaminergic cells[25]. Nicotine, acting at presynaptic nicotinic receptors, increases dopaminergic neuronal firing and enhances dopamine release in striatum and nucleus accumbens[82,107,170]. Nicotinic agonists elicit a dose-dependent increase in dopamine release, while nicotinic antagonists inhibit neurotransmitter release. For example, dopamine release is very sensitive to inhibition by α-bungarotoxin.

2.9 POLYDRUG ABUSE: DRUG INTERACTIONS

The search for effective clinical interventions for human drug abuse must take into account the likely biological mechanisms, which may be predicted from preclinical research involving both animals and humans. In addition, it must be recognized that human drug self-administration often occurs in the context of polydrug use and abuse.

Drug interactions may be considered both at the level of the organism and at the level of the neuronal mechanism of action. Of course, the behavioral effects of one drug may counter the behavioral effects of another. For example, a stimulant may oppose the behavioral effects of a depressant.

At the cellular level, the effects of one drug on neuronal components may modify the manner in which another drug interacts with the neuron. To illustrate this, ethanol and general anesthetics such as nitrous oxide and halothane influence the opiate binding to opioid receptors[125]. Nitrous oxide decreases ligand affinity for μ-receptors without influencing receptor density. In the presence of halothane, decreases in ligand affinity and receptor density have been observed for receptors. Indeed, μ-receptor activity has been shown to depend upon surrounding neuronal membrane lipids[69]. Thus, it is not unexpected that ligand binding to this receptor is influenced by fluidizing effects of ethanol on synaptic membranes.

In addition, the combined effects of two drugs of abuse may be due to some novel product of their interaction. For example, cocaine is commonly used concurrently with ethanol , and it has been reported that ethanol significantly enhances the euphoric effects of cocaine. However, ethanol may significantly increase the toxic effects of cocaine. A recent Drug Abuse Warning Network (DAWN) report indicates that concurrent administrations of cocaine and ethanol was associated most commonly with emergency room admissions related to substance abuse in 27 metropolitan areas surveyed in 1987. Recent evidence suggests that the effects of combined administration of these two substances may be due to the transesterification of cocaine and ethanol to form cocaethylene[71,84]. It has been shown that cocaethylene is found frequently in postmortem brain, blood, and liver from cocaine-related lethal cases, and in concentrations which are often greater than those of cocaine or ethanol themselves. In addition, cocaethylene has an affinity for the dopamine transporter, associated with the euphoric effects of cocaine, which is similar to that of cocaine. Thus, this data suggests that the reinforcing effects of cocaine may indeed be enhanced by ethanol, and ultimately, cocaethylene. However, cocaethylene is also characterized by an LD_{50} which is approximately half that of cocaine[72]. Thus, as blood concentrations of cocaethylene rise, lethality is increasingly likely.

3 Homeostasis and chronic changes in neuronal mechanisms

Chronic drug ingestion is commonly associated with behavioral and physiological tolerance to the drug and with the development of physical dependence. As these processes occur, increasing doses of a drug are required to produce the effects of a drug. Alternatively, an organism may become more sensitive to specific effects of a drug, exhibiting related drug responses at lower doses of the drug. In general, such changes in sensitivity or resistance to drug effects occurs due to homeostasis, a major key to the ultimate survival of most organisms. Like other organisms, in order to function in ever changing environments, animals, and humans must be able to adapt and maintain internal equilibrium. At the organismic level, homeostasis relies on the interaction between the nervous system, including the brain, spinal cord and peripheral nerves, and the endocrine system, which produces hormones essential for the concurrent regulation and modulation of many physiological processes. By means of numerous homeostatic functions, an organism might gradually develop tolerance, sensitization, and physical dependence in response to chronic exposure to a drug. It should be noted, however, that behavioral tolerance is the observable product and summation of numerous neuronal events.

3.1 TOLERANCE

At the level of the neuron, tolerance suggests a decreased neuronal response to a drug. In general, tolerance can be described in terms of two distinct categories. First, observed tolerance may be due to dispositional tolerance, involving enhanced or facilitated metabolic function. For example, metabolic systems may adapt to chronic presence of a drug with the induction of specific enzymes which are relatively inactive under normal baseline conditions. By doing so, organisms may metabolically reduce the blood concentration of the drug more quickly, thus reduce the effects of a particular drug dose on the organism.

Alternatively, neurons may develop pharmacodynamic tolerance to the chronic presence of a drug. Like other tissues in the body, brain neurons also typically respond to the chronic presence of a drug by gradually limiting the effect of the drug on their functional responses. The neuronal mechanisms associated with these effects of frequent and repeated drug administration involve various cellularly based homeostatic processes. These processes combine to neutralize the effect of the drug on the neuron, allowing it to function as closely to its genetically coded baseline as possible. Changes in neuronal membrane fluidity, increases or decreases in receptor number or density, and the inactivation of receptors are all possible neuronal adaptations to consistent bombardment of neurons by a drug.

Chronic nicotine administration, for example, leads to dose-dependent and time-dependent tolerance[98,99]. However, although chronic treatment with most other agonists results in a down-regulation or decrease in receptor density, tolerance to the effects of nicotine are associated with a paradoxical up-regulation of receptors[170]. Based upon this evidence, a related hypothesis has been put forth that the up-regulation may occur as chronic treatment results in desensitization or inactivation of nicotinic receptors[98,99]. Indeed, tolerance to nicotine has been associated with decreases in nicotine-induced dopamine release and with decreases in ion efflux from synaptosomal tissue preparations.

It is well established that chronic ethanol ingestion is associated with profound changes in membrane fluidity characteristics of cellular membranes throughout the body[55]. The effect of acute administration of ethanol is to cause swelling of cellular membranes and greater movement of lipids within membrane bilayers. In response, neuronal membranes modify their membrane composition to incorporate more rigid lipid molecules, especially more highly saturated lipids and cholesterol. This makes the neuronal membrane more resistant to the intercalation of ethanol into the membrane, and maximizes the function of various membrane-bound receptors and other proteins which are dependent upon the lipid bilayer for specific conformational support.

Recent research illustrates that second messenger systems play an important role in regulating drug self-administration, and that changes in the activity and function of these systems are associated with the development of tolerance[122]. It has been shown that pharmacological inhibition of G proteins and elements of the cAMP synthesis pathway in the nucleus accumbens reduces the self-administration of cocaine and opiates. Chronic administration of these drugs leads to decreased expression of G proteins in this brain region, and to changes in the levels of cAMP produced by drug administration. Such changes in intracellular biochemistry may be at the heart of homeostatic neuronal adaption to chronic drug exposure in that they are associated with changes in gene expression. Elements of second messenger systems appear to alter the function of transcription factors which bind to regulatory regions of DNA to increase or decrease the rate at which certain genes are transcribed.

3.2 SENSITIZATION

At the level of the behavior of the organism, sensitization to a drug effect may occur. Such sensitization, often called reverse tolerance, is a paradoxical effect since it is realistic to assume that most individual neurons and other cells will tend to limit the effect of the drug on their function. However, when sensitization occurs, lower doses of a drug are required to elicit effects on the behavior of an organism.

The relationship between adaptive mechanisms of neurons in response to the chronic presence of a drug and the observed resistance or sensitization to a drug at the level of the organism is not well understood. However, it is most reasonable to hypothesize that individual neuronal systems with which a drug interacts adapt to chronic treatment by decreasing their response to a drug. However, since action potentials are generated as summary responses to multiple inputs to the cell, responses of a neuron to external stimuli are difficult to predict following their adaptation to chronic drug treatments. In addition, the time course required for each of these neuronal systems to adapt to the drug may be different, complicating our understanding of neuronal responses following chronic drug intake. This is perhaps especially true when a specific behavioral effect is mediated by a number of neurotransmitters or interactions between several types of neurons in several brain regions.

One of the best examples of sensitization is that repeated administration of cocaine elicits a phenomenon which is commonly called kindling. Though it is known that high doses of cocaine may produce seizures, it is also widely known that repeated administration of a subthreshold dose of this psychostimulant leads to the "kindling" of seizures. After repeated treatments, seizures eventually occur in response to a dose of cocaine which does not produce convulsions following the first administration. In these experiments, dose-response curves for cocaine-induced seizures have been shifted to the left and subjects are said to be sensitized to the seizurgenic effects of cocaine.

The neuronal mechanism associated with the kindling of seizures by cocaine is not well understood. Likewise, the relationship between acute cocaine-induced seizures and seizures kindled by repeated cocaine administration is not yet known. Recent hypotheses suggest that cocaine-induced kindling is a discontinuous process involving discrete, stepwise transitions from one state of neuronal organization to another, with each transitional stage mediated by different neuronal mechanisms in specific brain regions[15]. Current hypotheses suggest that the kindling process is best described in terms of a summation mechanism involving noradrenergic, cholinergic, GABAergic, and excitatory neurotransmitter systems[102]. It is not yet known whether serotonergic pathways are involved in this process.

3.3 CROSS TOLERANCE

Cross tolerance occurs when the development of tolerance to one drug confers tolerance to another drug. This phenomena is based on similarities in the mechanism of action for each drug. Drugs having similar chemical structure interact with the same neuronal binding sites, produce similar neuronal changes and elicit similar behavioral or physiological effects. However, if drugs act at the same or related sites, regardless of their structure, then adaptation to one drug will affect the action of another.

Examples of cross tolerance are common, especially within a particular pharmacological class. For sedative-hypnotic compounds, chronic treatment with alcohol, a benzodiazepine, or a barbiturate produces cross tolerance to the other two classes of compounds. This is presumably due to the fact that barbiturates, benzodiazepines and alcohol elicit their pharmacological effects through a common mechanism of action associated with the chloride ion channel complex[1,14]. Likewise, psychostimulants often exhibit cross tolerance with each other, most likely due common mechanisms of action associated with biogenic amine receptors. Finally, alcohol also exhibits cross tolerance with inhaled general anesthetics. This phenomenon is presumably due to similar mechanisms of action associated with fluidization of neuronal membranes. This particular example of cross tolerance has special practical implications for the provision of anesthesiology services to alcoholic patients since these patients tend to require higher doses of general anesthetics for necessary medical purposes.

3.4 PHYSICAL DEPENDENCE AND WITHDRAWAL

Physical dependence is a state of an organism which is operationally defined by the onset of withdrawal symptoms associated with the termination of drug administration. In this case, neuronal systems which have adapted to the chronic presence of a drug must readapt to the absence of these same compounds. Each neuronal component which was modified in response to the presence of drug must begin to reapproach baseline function levels. These readaption processes follow varying time courses for completion. However, the initial impact of this reverse homeostatic process on neuronal function and on the behavior and physiology of the organism cannot be underestimated, and withdrawal effects are characterized by massive rebound effects.

Indeed, the anticonvulsant effects of benzodiazepines are the primary reason that these compounds are included in drug detoxification treatment programs. The withdrawal symptoms associated with abstinence from alcohol or opiates after a long history of chronic use involve powerful rebound effects. Whereas these substances of abuse are behavioral depressants, convulsions and tremors frequently occur during withdrawal from these substances of abuse unless prevented by pharmacological adjuncts to treatment. Thus, benzodiazepines are commonly administered to patients undergoing detoxification in order to inhibit the potentially harmful convulsant effects of these drugs. Barbiturates are similarly effective as anticonvulsants, but are not as commonly utilized in detoxification programs due to their own addictive potential.

4 Summary

For a number of addictive drugs it is known that they show specific receptor binding in the central nervous system. Biochemical, behavioral, and pharmacological studies are used to understand the interactions between a specific drug and its effects on CNS processes. These include receptor binding and localization studies. In this chapter the results of molecular mechanisms of various classes of drugs are presented. Aspects of interactions with more than one neurotransmitter system are discussed, and the molecular mechanisms of tolerance, sensitization, and withdrawal have been dealt with in this chapter.

References

1. Allan, A.M. and Harris, R.A., "Involvement of the neuronal chloride channels in ethanol intoxication , tolerance, and dependence." In: Galanter, D.M. (Ed.), *Recent Developments in Alcoholism*, Vol. 5. 1987, pp 313–326.
2. Altshuler, H.L., Phillips, P.E., Feinhandler, D.A., "Alteration of ethanol self-administration by naltrexone," *Life Sci.* 1980, **26:** pp 679–688.
3. Amalric, M., Cline, E.J., Martinez, J.L., Bloom, F.E., Koob, G.F., "Rewarding properties of endorphin as measured by conditioned place preference," *Psychopharmacology.* 1987, **91:** pp 14–19.
4. Appel, N., Contrera, J.F., DeSouza, E.B., "Fenfluramine selectively and differentially decreases the density of serotonergic nerve terminals in rat brain: evidence from immunocytochemical studies," *J. Pharmacol. Exp. Ther.* 1989, **249:** pp 928–942.
5. Arnold, E.B., Molinoff, P.B., Rutledge, C.O., "The release of endogenous norepinephrine and dopamine from cerebral cortex by amphetamine," *J. Pharmacol. Exp. Ther.* 1977, 202, pp 544–557.
6. Azzaro, A.J., Ziance, R.J., Rutledge, C.O., "The importance of neuronal uptake of amines for amphetamine-induced release of ^{3}H-norepinephrine from isolated brain tissue," *J. Pharmacol. Exp. Ther.* 1974, **189:** pp 110–118.
7. Balster, R.L. and Willets, J., "Receptor mediation of the discriminative stimulus properties of phencyclidine and sigma-opioid agonists." In: Balster, R.L. (Ed.), *Psychopharmacology Series 4: Transduction Mechanisms of Drug Stimuli.* Springer-Verlag, Berlin, 1988, pp 122–135.
8. Balster, R.L., "The behavioral pharmacology of phencyclidine." In: Meltzer, H. (Ed.), *Psychopharmacology: The Third Generation of Progress.* Raven Press, New York, 1987, p 1573.
9. Battaglia, G., Yeh, S.Y., DeSouza, E.B., "MDMA-induced neurotocicity: parameters of degeneration and recovery of brain serotonin neurons," *Pharmacol. Biochem. Behav.* 1988, **29:** pp 269–274.
10. Berge, O-G., Post, C., Archer, T., "The behavioural pharmacology of serotonin in pain processes." In: Bevan, P., Cools, A.R., Archer, T. (Eds.), *Behavioural Pharmacology of Serotonin.* Lawrence Erlbaum Associates, Hillside, NJ, 1989, pp 301–319.
11. Bergman, J., Madras, B.K., Johnson, S.E., Spealman, R.D., "Effects of cocaine and related drugs in nonhuman primates. III. Self-administration by squirrel monkeys," *JPET.* 1989, **51:** pp 150–155.
12. Brodie, M.S., Shefner, S.A., Dunwiddie, T.V., "Ethanol increases the firing rate of dopamine neurons of the rat ventral tegmental area in vitro," *Brain Res.* 1990, **508:** pp 65–69.
13. Brown, R.D. and Holtzman, S.G., (1979) "Suppression of deprivation-induced food and water intake in rats and mice by naloxone," *Pharmacol. Biochem. Behav.* 1979, **11:** pp 567–573.
14. Buck, K.J. and Harris, R.A., "Benzodiazepine agonist and inverse agonist actions on $GABA_A$ receptor-operated chloride channels II. Chronic effects of ethanol," *J. Pharmacol. Exp. Ther.* 1990, **253:** pp 713–719.
15. Burchfiel, J.L. and Applegate, C.D., "Stepwise progression of kindling: perspectives from the kindling antagonism model," *Neurosci. Biobehav. Rev.* 1989, **13:** pp 289–300.
16. Carboni, E., Imperato, A., Perezzani, L., DiChiara, G., "Amphetamine, cocaine, phencyclidine and nomifensine increase extracellular dopamine concentrations preferentially in the nucleus accumbens of freely moving rats," *Neuroscience.* 1989, **28:** pp 653–661.
17. Carboni, E., Acquas, E., Frau, R., DiChiara, G., "Differential inhibitory effects of a $5HT_3$ antagonist on drug-induced stimulation of dopamine release," *Eur. J. Pharmacol.* 1989, **164:** pp 515–519.
18. Carroll, M.E., Lac, S.T., Asencio, M., Kragh, R., "Intravenous cocaine self-administration in rats is reduced by L-tryptophan," *Psychopharmacology.* 1990, 100, pp 293–300.
19. Chen, J., Paredes, W., Lowinson, J.H., Gardner, E.L., "$^9\Delta$-Tetrahydrocannabinol enhances presynaptic dopamine efflux in medial prefrontal cortex," *Eur. J. Pharmacol.* 1990, **190:** pp 259–262.
20. Chen, J., Paredes, W., Lowinson, J.H., Gardner, E.L., "Strain-specific facilitation of dopamine efflux by delta^9-tetrahydrocannabinol in the nucleus accumbens of rat: an in vivo microdialysis study," *Neurosci. Lett.* 1990, **129:** pp 136–180.

21. Chen, J., Marmur, R., Pulles, A., Paredes, W., Gardner, E.L., "Ventral tegmental microinjection of delta-9-tetrahydrocannibinol enhances ventral tegmental somatodentritic dopamine levels but not forebrain dopamine levels: evidence for local neural action by marijuana's psychoactive ingredient," *Brain Res.* 1992, **621:** pp 65–70.
22. Chen, J., Paredes, W., Li, J., Smith, D., Lowinson, J.H., Gardner, E.L., "$^{9}\Delta$-tetrahydrocannibinol produces naloxone-blockable enhancement of presynaptic basal dopamine efflux in nucleus accumbens of conscious freely-moving rats as measured by intracerebral microdialysis," *Psychopharmacology.* 1990, **102:** pp 156–162.
23. Chesher, G.B., Dahl, C.J., Everingham, M., Jackson, D.M., Marchant-Williams, H., Starmer, G.A., "The effects of cannabinoids on intestinal motility and their antinociceptive effects in mice," *Br J Pharmacol.* 1973, **49:** pp 588–594.
24. Cooper, J.R., Bloom, F.E., Roth, R.H., *The Biochemical Basis of Neuropharmacology.* Oxford University Press, New York, 1982.
25. Corrigal, W.A., Franklin, K.B.J., Coen, K.M., Clarke, P., "The mesolimbic dopaminergic system is implicated in the reinforcing effects of nicotine," *Psychopharmacology.* 1992, **107:** pp 285–289.
26. Cox, B.M., Goldstein, A., Neslon, W.T., "Nicotine self-administration in rats," *Br. J. Pharmacol.* 1984, **83:** pp 49–55.
27. Coyle, J.T. and Snyder, S.H., *Science.* 1969, **166:** pp 899–901.
28. Creese, I., Snyder, S.H., "Receptor binding and pharmacological activity of opiates in the guinea pig intestine," *J. Pharmacol. Exp. Ther.* 1975, **194:** pp 205–219.
29. Devane, W.A., "New dawn of cannabinoid pharmacology," *TIPS.* 1994, **15:** pp 40–41.
30. Devane, W.A. et al., "Isolation and structure of a brain constituent that binds to the cannabinoid receptor," *Science.* 1992, **258:** pp 1946–1949.
31. Di Chiara, G. and North, R.A., "Neurobiology of opiate abuse," *TIPS* 1992, **13:** pp 185–193.
32. Di Chiara, G. and Imperato, A., "Drugs abused by humans preferentially increase synaptic dopamine concentrations in the mesolimbic system of freely moving rats," *Proc. Nat. Acad. Sci. USA.* 1988, **85:** pp 5274–5278.
33. Dildy-Mayfield, J.E. and Leslie, S.W., "Mechanism of inhibition of *N*-methyl-D-aspartate-stimulated increases in free intracellular Ca^{2+} concentrations by ethanol," *J. Neurochem.* 1991, **56:** pp 1536–1543.
34. Dilts, R.P. and Kalivas, P.W., "Autoradiographic localization of delta opioid receptors within the mesocorticolimbic dopamine system using radioiodinated [2-D-penicillamine 5-D-penicillamine] enkaphalin (^{125}I-DPDPE)," *Synapse.* 1990, **6:** pp 121–132.
35. Eidelberg, E., Lesse, E., Gault, F.P., "An experimental model of temporal lope epilipsy: studies of the convulsant properties of cocaine." In: Glasser, G.H. (Ed.), *EEG and Behavior.* Basic Books, New York, 1963, pp 277–283.
36. Elmer, G.I., Pieper, J.O., Goldberg, S.R., George, F.R., "Opioid operant self-administration analgesia stimulation and respiratory depression in μ-deficient mice," *Psychopharmacology.* 1994.
37. Ettenberg, A., Pettit, H.O., Bloom, F.E., Koob, G.F., "Heroin and cocaine intravenous self-administration in rats: mediation by separate neural systems," *Psychopharmacology.* 1982, **78:** pp 204–209.
38. Finlay, J.M., Damsma, G., Fibiger, H.C., "Benzodizepine-induced decreases in extracellular concentrations of dopamine in the nucleus accumbens after acute and repeated administration," *Psychopharmacology.* 1992, **106:** pp 202–208.
39. Fischer, J.F., Cho, A.K., "Chemical release of dopamine from striatal homogenates: Evidence for an exchange diffusion model," *J. Pharmacol. Exp. Ther.* 1979, **208:** pp 203–209.
40. Foltin, R.W. and Fischman, M.W., "Smoked and intravenous cocaine in humans: acute tolerance cardiovascular and subjective effects," *J. Pharmacol. Exp. Ther.* 1991, **257:** pp 247–261.
41. Fowler, J.S., Volkow, N.D., Wolf, A.P., Dewey, S.L., Schyler, D.J., MacGregor, R.R., Hitzemann, R., Logan, J., Bendriem, B., Gatley, S.J., Christman, D., "Mapping cocaine binding sites in human and baboon brain sites *in vivo. Synapse.* 1989, **4:** pp 371–377.
42. Froelich, J.C., Harts, J., Lumeng, L. et al., "Naloxone attenuation of voluntary alcohol consumption," *Alcohol Alchohol Suppl.* 1987, **1:** pp 333–337.

43. Froelich, J.C., Harts, J., Lumeng, L., Li, T.-K., "Naloxone attenuates voluntary ethanol intake in rats selectively bred for high ethanol preference," *Pharmacol. Biochem. Behav.* 1990, **35:** pp 385–390.
44. Fuchs, V., Burbes, E., Coper, H., "The influence of haloperidol and aminooxyacetic acid on etonitaxene alcohol diazepam and barbital consumption," *Drug Alcohol Depend.* 1984, **14:** pp 179–186.
45. Fudala, P.J., Teoh, K.W., Iwamoto, E.T., "Pharmacologic characterization of nicotine-induced conditioned place preference," *Pharmacol. Biochem. Behav.* 1985, **22:** pp 237–241.
46. Gardner, E.L., Paredes, W., Smith, D., Donner, A., Milling, C., Cohen, D., Morrison, D., "Facilitation of brain stimulation reward by delta 9-tetrahydrocannabinol," *Psychopharmacology.* 1988, **96:** pp 142–144.
47. George, F.R. and Ritz, M.C., "A psychopharmacology of motivation and reward related to substance abuse treatment," *Exp. Clin. Psychopharmacol.* 1993, **1:** pp 7–26.
48. Gerard, C.M., Mollereau, C., Vassart, G., Parmentier, M., Molecular cloning of a human cannabinoid receptor which is also expressed in testis. *Biochem J.* 1991, **279:** pp 129–134.
49. Gerber, G.J. and Wise, R.A., "Pharmacological regulation of intravenous cocaine and heroin self-administration in rats: a variable dose paradigm," *Pharmacol. Biochem. Behav.* 1989, **32:** pp 527–531.
50. Gibb, J.W., Johnson, M., Stone, D.M., Hanson, G.R., "Mechanisms mediating biogenic amine deficits induced by amphetamine and its congeners." In: Arinoff, L. (Ed.), *Assessing Neurotoxicity of Drugs of Abuse.* NIDA Monograph 136, National Institute on Drug Abuse, 1993, pp 226–241.
51. Glennon, R.A., Titeler, M., McKenney, J.D., "Evidence for 5-HT_2 involvement in the mechanism of action of hallucinogenic agents," *Life Sci.* 1984, **35:** pp 2505–2511.
52. Goeders, N.E., "Cocaine differentially affects benzodiazepine receptors in discrete regions of the rat brain: persistence and potential mechanisms mediating these effects," *J. Pharmacol. Exp. Ther.* 1991, **259:** pp 574–581.
53. Goeders, N.E., Lane, J.D., Smith, J.E., "Self-administration of methionine enkephalin into the nucleus accumbens," *Pharmacol. Biochem. Behav.* 1984, **20:** pp 451–455.
54. Goldberg, S,R. and Henningfield, J.E., "Reinforcing effects of nicotine in humans and experimental animals responding under intermittent schedules of IV drug injection," *Pharmacol. Biochem. Behav.* 1988, **30:** pp 227–234.
55. Goldberg, S.R., Spealman, R.D., Goldberg, D.M., "Persistent behavior at high rates maintained by intravenous self-administration of nicotine," *Science.* 1981, **214:** pp 573–575.
56. Goldberg, S.R. and Spealman, R.D., "Maintenance and suppression of behavior by intravenous nicotine injections in squirrel monkeys," *Fed. Proc.* 1982, **41:** pp 216–220.
57. Goldfrank, L.R. and Hoffman, R.S., "Cardiovascular effects of cocaine," *Ann. Emerg. Med.* 1991, **20:** pp 165–175.
58. Goldstein, D.B., *The Pharmacology of Alcohol.* Oxford University Press, New York, 1983.
59. Gonzales, R.A. and Woodward, J.J., "Ethanol inhibits *N*-methyl-D-aspartate-stimulated [^{3}H]norepinephrine from rat cortical slices," *J. Pharmacol. Exp. Ther.* 1990, **253:** pp 1138–1144.
60. Grace, A.A. and Bunney, B.S., "Paradocical GABA excitation of nigral dopaminegic cells: indirect mediation through reticulata inhibitory neurons," *Eur. J. Pharmacol.* 1979, **59:** pp 211–218.
61. Gradman, A.H., "Cardiac effects of cocaine: a review," *Yale J. Biol. Med.* 1988, **61:** pp 137–147.
62. Grant, K., "Emerging neurochemical concepts in the actions of ethanol at ligand-gated ion channels," *Behav. Pharmacol.* 1994, **5:** pp 383–404.
63. Grant, K.A. and Barrett, J.E., "Blockade of the discriminative stimulus effects of ethanol with 5-HT_3 receptor antagonists *Psychopharmacology.* 1991, **104:** pp 451–456.
64. Gulati, A. and Bhargava, H.N., "Brain and spinal cord 5-HT_2 receptors of morphine-tolerant-dependent and -abstinent rats," *Eur. J. Pharmacol.* 1989, **167:** pp 185–192.
65. Gulley, J.M., McNamara, C., Barbera, T.J., Ritz, M.C., George, F.R., "Selective serotonin reuptake inhibitors: effects of chronic treatment on ethanol-reinforced behavior in mice," *Alcohol.* 1985, **12:** pp 177–181.

66. Hammer, R.P., "Cocaine alters opiate receptor binding in critical brain reward regions," *Synapse*, 1989, **3:** pp 55–60.
67. Hand, T.H., Koob, G.F., Stinus, L. et al., "Aversive properties of opioid receptor blockade are centrally mediated and are potentiated by previous exposure to opiates," *Brain Res.* 1988, **474:** pp 364–468.
68. Haraguchi, M., Samson, H.H., Tolliver, G.A., "Reduction in oral ethanol self-administration in the rat by the 5-HT uptake blocker fluoxetine," *Pharmacol Biochem Behav.* 1990, **35:** pp 259–262.
69. Hasegawa, J.-I., Loh, H.H., Lee, N.M., "Lipid requirement for μ opioid receptor binding," *J. Neurochem.* 1987, **49:** pp 1007-1012.
70. Hatsukami, D., Keenan, R., Halikas, J., Pentel, P.R., Hartman-Brauer, L., "Effects of carbamazepine on acute responses to smoked cocaine-base in human cocaine users," *Psychopharmacology.* 1991, **104:** pp 120–124.
71. Hearn, W.L., Flynn, D.D., Hime, G.W., Rose, S., Cofino, J.C., Mantero-Atienza, E., Wetli, C.V., Mash, D.C., "Cocaethylene: a unique cocaine metabolite displays high affinity for he dopamine transporter," *J. Neurochem.* 1991, **56:** pp 698–701.
72. Hearn, W.L., Rose, S., Wagner, J., Ciarleglios, A., Mash, D.C., "Cocaethylene is more potent than cocaine in mediating lethality," *Pharmacol. Biochem. Behav.* 1991, **39:** pp 531–533.
73. Ho, B.Y. and Takemori, A.E., "Serotonergic involvement in the antinociceptive action of and the development of tolerance to the kappa-opioid receptor agonist U-50 488H," *J. Pharmacol. Exp. Ther.* 1989, **250:** pp 508–514.
74. Hoffman, P.L., Rabe, C.S., Moses, F., Tabakoff, B., "*N*-methyl-D-aspartate receptors and ethanol: inhibition of calcium flux and cyclic GMP production," *J. Neuorchem.* 1989, **52:** pp 1937–1940.
75. Holaday, J.W., Long, J.B., Tortella, F.C., "Evidence for κ, μ, and δ opioid binding site interactions in vivo, *Fed. Proc.* 1985, **44:** pp 2860–2862.
76. Horn, A.S., Cuello, C., Miller, R.J., *J Neurochem.* 1974, **22:** pp 265–270.
77. Howlett, A.C., Qualy, J.M., Khachatrian, L.L., "Involvement of Gi in the inhibition of adenylate cyclase by cannabimimetic drugs," *Mol. Pharmacol.* 1986, **29:** pp 307–313.
78. Howlett, A.C., Qualy, J.M., Khachatrian, L.L., "The cannabinoid receptor: biochemical anatomical and behavioral characterization," *TINS* 1990, **13:** pp 420–423.
79. Hurd, Y.L. and Ungerstadt, U., "In vivo neurochemical profile of dopamine uptake inhibitors and releasers in rat caudate-putamen," *Eur. J. Pharmacol.* 1989, **166:** pp 251–260.
80. Huston-Lyons, D. and Kornetsky, C., "Effects of nicotine on thethreshold for rewarding brain stimulation in rats," *Pharmacol Biochem Behav.* 1992, **41:** pp 755–759.
81. Hwang, B.H., Lumeng, L., Wu, J.-Y. et al., "GABAergic neurons in nucleus accumbens: A possible role in alcohol preference," *Alcohol. Clin. Exp. Res.* 1988, **12:** p 306.
82. Imperato, A., Mulas, A., Di Chiara, G., "Nicotine preferentially stimulates dopamine release in the limbic system of freely moving rats," *Eur. J. Pharmacol.* 1986, **132:** pp 337–338.
83. Iwamoto, E.T. and Williamson, E.C., "Nicotine-induced taste aversion: characterization and preexposure effects in rats," *Pharmacol. Biochem. Behav.* 1984, **21:** pp 527–532.
84. Jatlow, P., Elsworth, J.D., Bradberry, C.W., Winger, G., Taylor, J.R., Russell, R., Roth, R.H., "Cocaethylene: a neuropharmacologically active metabolite associated with concurrent cocaine-ethanol ingestion," *Life Sci.* 1991, **48:** pp 1787–1794.
85. Jiang, Q., Mosberg, H.I., Porreca, F., "Modulation of the potency and efficacy of mu-mediated antinoception by delta agonists in the mouse," *J. Pharmacol. Exp. Ther.* 1990, **254:** pp 683–689.
86. Johnson, B.A., Campling, G.M., Griffiths, P., Cowen, P.J., "Attenuation of some alcohol-induced mood changes and the desire to drink by $5HT_3$ receptor blockade. A preliminary study in healthy male volunteers," *Psychopharmacology.* 1993, **112:** pp 142–144.
87. Knapp, D.J. and Pohorecky, L.A., "Zacopride a $5\text{-}HT_3$ receptor antagonist reduces voluntary ethanol consumption in rats," *Pharmacol. Biochem. Behav.* 1992, **41:** pp 847–850.
88. Koob, G.F. and Bloom, F.E., "Cellular and molecular mechanisms of drug dependence," *Science.* 1988, **242:** pp 715–723.
89. Koob, G.F., "Drugs of abuse: anatomy pharmacology and function of reward pathways," *TIPS,* 1992, **13:** pp 177–184.

90. Koob, G.F. and Weiss, F., "Pharmacology of drug self-administration," *Alcohol.* 1990, **7:** pp 193–197.

91. Kosten, T.R., Neurobiology of abused drugs: opioids and stimulants. *J. Nerv. Ment. Dis.* 1990, **178**: pp 217–227.

92. Kuhar, M., Ritz, M.C., Boja, J.W., The dopamine hypothesis of the reinforcing properties of cocaine. *TINS.* 1991, **14**: pp 299–302.

93. Larson, A.A. and Takemori, A.E., Effect of fluoxetine hydrochloride (Lilly 110140), a specific inhibitor of serotonin uptake on morphine analgesia and the development of tolerance. *Life Sci.* 1978, **21:** pp 1807–1812.

94. Lawrin, M.O., Naranjo, C.A., Sellers, E.M., "Identification of new drugs for modulating alcohol consumption," *Psychopharmacol. Bull.* 1986, **22:** pp 1020–1025.

95. Lorens, A.A. and Sainati, S.M., "Naloxone blocks the excitatory efect of ethanol and chlordiazepoxide on lateral hypothalamic self-stimulation behavior," *Life Sci.* 1978, **23:** pp 1359–1364.

96. Lyon, R.A., Glennon, R.A., Titeler, M., "3,4-Methylenedioxymethamphetamine (MDMA): stereoselective interactions at brain 5-HT_1 and 5-HT_2 receptors," *Psychopharmacology.* 1986, **88:** pp 525–526.

97. Madras, B.K., Fahey, M.A., Bergman, J., Canfield, D.R., Spealman, R.D., "Effects of cocaine and related drugs in nonhuman primates I [^{3}H]. Cocaine binding sites in caudate-putamen," *J. Pharmacol. Exp. Ther.* 1989, **251:** pp 131–141.

98. Marks, M.J., Grady, S.R., Collins, A.C., "Down regulation of nicotinic receptor function after chronic nicotine infusion," *J. Pharmacol. Exp. Ther.* 1993, **266:** pp 1268–1276.

99. Marks, M.J., Romm, E., Campbell, S.M., Collins, A.C., "Genotype influences the development of tolerance to nicotine in the mouse," *J. Pharmacol. Exp. Ther.* 1991, 259, pp 392–402.

100. Matthews, R.T. and German, D.C., "Electrophysiological evidence for excitation of rat ventral tegmental area dopamine neurons by morphine," *Neuroscience.* 1984, **11:** pp 617–625.

101. McBride, W.J., Murphy, L., Lumeng, L., Li, T.-K., "Effects of RO 15-4513 fluoxetine and desipramine on the intake of ethanol water and food by the alcohol-preferring (P) and nonpreferring (NP) lines of rats," *Pharmacol. Biochem. Behav.* 1988, **30:** pp 1045–1050.

102. McIntyre, D.C. and Racine, R.J., "Kindling mechanisms: current progress on an experimental epilepsy model," *Prog. Neurobiol.* 1986, **27:** pp 1–12.

103. Meert, T.F., Awouters, F., Niemegeers, C.J.E., Schellekens, K.H.L., Janssen, P.A.J., "Ritanserin reduces abuse of alcohol cocaine and fentanyl in rats," *Pharmacopsychiatry.* 1991, **24:** pp 159–163.

104. Mello, N.K., Lukas, S.E., Mendelson, J.H., Drieze, J., "Naltrexone-buprenorphine interactions: effects on cocaine self-administration," *Neuropsychopharmacology.* 1993, **9:** pp 211–224.

105. Melvin, L.S., Milne, G.M., Johnson, M.R., Subramaniam, B., Wilken, G.H., Howlett, A.C., "Structure-activity relationships for cannabinoid receptor-binding and analgesic activity: studies of bicyclic cannabinoid analogs," *Mol. Pharmacol.* 1993, **44:** pp 1008–1015.

106. Michael-Titus, A., Bousselmame, R., Costentin, J., "Stimulation of dopamine D_2 receptors induces an analgesia involving an opioidergic but enkaphalinergic link," *Eur. J. Pharmacol.* 1990, **187:** pp 201–207.

107. Mifsud, J.C., Hernandez, L., Hoebel, B.G., "Nicotine infused into the nuclesu accumbens increases synaptic dopamine as measured by in vivo microdialysis," *Brain Res.* 1989, **478:** pp 5–367.

108. Millan, M., "Kappa-opioid receptors and analgesia," *TIPS.* 1990, **11:** pp 70–76.

109. Milligan, G. (Ed.), *Signal Transduction: A Practical Approach.* IRL Press, Oxford, 1992.

110. Miner, L.L., Marks, M.J., Collins, A.C., "Classical genetic analysis of nicotine-induced seizures and nicotinic receptors," *J. Pharmacol. Exp. Ther.* 1984, **231:** pp 545–554.

111. Myers, R.D., Lankford, M., Björk, A., "Selective reduction by the 5-HT antagonist amperozide of alcohol preference induced in rats by systemic cyanamide," *Pharmacol. Biochem. Behav.* 1992, **43:** pp 661–667.

112. Myers, R.D., Lankford, M., Björk, A., "Irreversible suppression of alcohol drinking in cyanamid-treated rats after sustained delivery of the 5-HT_2 antagonist amperozide," *Alcohol.* 1993, **10:** pp 117–125.

113. Myers, R.D. and Veale, W.L., "Alcohol preference in the rat: reduction following depletion of brain serotonin," *Science.* 1968, **160:** pp 1469–1471.

114. Myers, R.D. and Martin, G.E., "The role of cerebral serotonin in the ethanol preference of animals," *Ann. N.Y. Acad. Sci.* 1973, **215:** pp 135–144.

115. Nackman, M., Lester, D., Le Magnen, J., "Alcohol aversion in the rat: behavioral assessment of noxious drug effects," *Science.* 1970, **168:** pp 1224–1226.

116. Naranjo, C.A., Sellers, E.M., Sullivan, J.T., Woodley, D.V., Kadlec, K., Sykoro, K., "The serotonin uptake inhibitor citalopram attenuates ethanol intake," *Clin. Pharmacol. Ther.* 1987, 41, pp 266–274.

117. Naranjo, C.A., Sellers, E.M., Roach, C., Woodley, D.V., Sanchez-Craig, M., Sykoro, K., "Zimeldine-induced variations in alcohol intake by nondepressed heavy drinkers," *Clin. Pharmacol. Ther.* 1984, **35:** pp 374–381.

118. Naranjo, C.A., Sullivan, J.T., Kadlec, K., Woodley-Remus, D.V., Kennedy, G., Sellers, E.M., "Differential effects of viqualine on alcohol intake and other consummatory behaviors," *Clin. Pharmacol. Ther.* 1989, **46:** pp 301–309.

119. Naranjo, C.A., Kadlec, K., Sanhueza, P., Woodley-Remus, D.V., Sellers, E.M., "Fluoxetine differentially alters alcohol intake and other consummatory behaviors in problem drinkers," *Clin. Pharmacol. Ther.* 1990, **47:** pp 490–498.

120. Negus, S.S. and Dykstra, L., "Neural substrates mediating the reinforcing properties of opioid analgesics." In: Watson, R.R. (Ed.), *Focus on Biochemistry and Physiology of Substance Abuse*, Vol. 1. 1989, pp 211–242.

121. Negus, S.S. et al., "Effect of antagonists selective for mu, delta, and kappa opioid receptors on the reinforcing effects of heroin in rats," *J. Pharmacol. Exp. Ther.* 1993, **265:** pp 1245–1252.

122. Nestler, E.J., "Molecular neurobiology of drug addiction," *Neuropsychopharmacology.* 1994, **11:** pp 77–87.

123. Nie, Z., Yuan, X., Madamba, S.G., Siggins, G.R., "Ethanol decreases glutamatergic synaptic transmission in rat nucleus accumbens in vitro: naloxone reversal," *J. Pharmacol. Exp. Ther.* 1993, **266:** pp 1705–1712.

124. O'Hearn, E., Battaglia, G., DeSouza, E.B., Kuhar, M.J., Molliver, M.E., "Methylenedioxyamphetamine (MDA) and methylenedioxymethamphetamine (MDMA) cause selective ablation of serotonergic axon terminals in forebrain: immunocytochemical evidence for neurotoxicity," *J. Neurosci.* 1988, **8:** pp 2788–2803.

125. Ori, C., Ford-Rice, F., London, E.D., "Effects of nitrous oxide and halothane on μ and κ opioid receptors in guinea-pig brain," *Anesthesiology.* 1989, **70:** pp 541–544.

126. Panocka, I. and Massi, M., "Long-lasting suppression of alcohol preference in rats following serotonin receptor blockade by rintanserin," *Brain Res. Bull.* 1992, **28:** pp 493–496.

127. Pert, C.B. and Snyder, S.H., "Opiate receptor: demonstration in nervous tissue," *Science.* 1973, **179:** pp 1011–1014.

128. Pert, A., "Cholinergic and catecholaminergic modulation of nociceptive reactions: interactions with opiates," *Pain Headache.* 1987, **9:** p 1.

129. Pilotto, R., Singer, G., Overstreet, D., "Self-injection of diazepam in naive rats: Effects of dose schedule and blockade of different receptors," *Psychopharmacology.* 1984, **84:** pp 174–177.

130. Porrino, L.J., Ritz, M.C., Sharpe, L.G., Goodman, N.L., Kuhar, M.J., Goldberg, S.R., "Differential effects of pharmacological manipulation of serotonin systems on cocaine and amphetamine self-administration in rats," *Life Sci.* 1989, **45:** pp 1529–1535.

131. Rassnick, S., Pulvirenti, L., Koob, G.F., "SDZ-205152 a novel dopamine receptor agonist reduces oral ethanol self-administation in rats," *Alcohol,* 1993, **10:** pp 127–132.

132. Reisine, T. and Bell, G.I., "Molecular biology of opioid receptors," *TINS.* 1993, **16:** pp 506–510.

133. Ritz, M.C. and Kuhar, M.J., "Monoamine uptake inhibition mediates amphetamine self-administration: comparison with cocaine," *J. Pharmacol. Exp. Ther.* 1989, **248:** pp 1010–1017.

134. Ritz, M.C., George, F.R., Kuhar, M.J., "Molecular mechanisms associated with cocaine effects: possible relationships with effects of ethanol." In: Galanter, M. (Ed.), *Recent Developments in Alcoholism: Alcohol and Cocaine: Similarities and Differences*, Vol. X. 1992, Plenum Press, New York, pp 273–302.

135. Ritz, M.C., Lamb, R.J., Goldberg, S.R., Kuhar, M.J., "Cocaine receptors on dopamine transporters are related to cocaine self-administration," *Science.* 1987, **237:** pp 1219–1223.
136. Ritz, M.C. and George, F.R., "Cocaine-induced seizures and lethality appear to be mediated by distinct CNS receptors," *J. Pharmacol. Exp. Ther.* 1993, **264:** pp 1333–1343.
137. Ritz, M.C. and Kuhar, M.J., "Psychostimulant drugs and a dopamine hypothesis regarding addiction: update on recent research." In: Wonnacott, S. and Lunt, G.G. (Eds.), *Biochemical Society Monographs: Neurochemistry of Drug Dependence.* Portland Press, London, 1993, **59:** pp 51–64.
138. Roberts, D.C.S., "Neural substrates mediating cocaine reinforcement: the role of monoamine systems." In: Lakoski, J.M., Galloway, M.P., Galloway, F.J. (Eds.), *Cocaine: Pharmacology, Physiology and Clinical Strategies.* The Telford Press, 1991.
139. Rockman, E., Amit, Z., Carr, G. et al., "Attenuation of ethanol intake by 5-hydroxytryptamine uptake blockade in laboratory rats I Involvement of brain 5-hydroxytryptamine in the mediation of the positive reinforcing properties of ethanol," *Arch. Int. Pharmacodyn. Ther.* 1979, **241:** pp 245-259.
140. Ronken, E., Mulder, A.H., Schoffelmeer, A.N.M., "Chronic activation of mu- and kappa-opioid receptors in cultured catecholaminergic neurons from rat brain causes neuronal supersensitivity without receptor desensitization," *J. Pharmacol. Exp. Ther.* 1994, **268:** pp 595–599.
141. Sadzot, B., Baranban, J.M., Glennon, R.A., Lyon, R.A., Leonhardt, S., Jan, C-R., Titeler, M., "Hallucinogenic drug interactions at human brain 5-HT_2 receptors: Implications for treating LSD-induced hallucinogenesis," *Psychopharmacology.* 1989, **98:** pp 495–499.
142. Samson. H,H, and Harris, R.A., "Neurobiology of alcohol abuse," *TIPS.* 1992, **13:** pp 206–211.
143. Sellers, E.M., Higgins, G.A., Sobell, M.B., "5-HT and alcohol abuse," *TIPS.* 1992, 13, pp 69–75.
144. Sharkey, J., Ritz, M.C., Schenden, J.A., Hanson, R.C., Kuhar, M.J., "Cocaine inhibits muscarinic cholinergic receptors in heart and brain," *J. Pharmacol. Exp. Ther.* 1988, **246:** pp 1048–1052.
145. Shaw, S.G., Vives, F., Mora, F., "Opioid peptides and self-stimulation of the medial prefrontal cortex," *Psychopharmacology.* 1984, **83:** pp 288–292.
146. Simon, E.J., Hiller, J.M., Edelman, I., "Stereospecific binding of the potent narcotic analgesic ^{3}H-etorphine to rat brain homogenate," *Proc. Nat. Acad. Sci. USA,* 1973, **70:** pp 1947–1949.
147. Sivam, S.P., "Cocaine selectively increases striatonigral dynorphin levels by a dopaminergic mechanism," *J. Pharmacol. Exp. Ther.* 1989, **250:** pp 818–824.
148. Smith, P.B., Welch, S.P., Martin, B.R., "Interactions between $^9\Delta$-tetrahydrocannibinol and kappa opioids in mice," *J. Pharmacol. Exp. Ther.* 1994, **268:** pp 1381–1387.
149. Snyder, S.H. and Childers, S.R., "Opiate receptors and opioid peptides," *Ann. Rev. Neurosci.* 1979, **2:** 35–64.
150. Spyraki, C. and Fibiger, H.C., "A role for the mesolimbic dopmine system in the reinforcing properties of diazepam," *Psychopharmacology.* 1988, **94:** pp 133–137.
151. Spyraki, C., Fibiger, H.C., Phillips, A.G., (1983) "Attenuation of heroin reward in rats by disruption of the mesolimbic dopamine system," *Psychopharmacology.* 1983, **79:** pp 278–283.
152. Stapleton, J.M., Ostrowski, N.L., Merriman, V.O. et al., "Naloxone reduces fluid consumption in deprived and nondeprived rats," *Bull. Psychol. Soc.* 1979, **13:** pp 237–239.
153. Stein, L. and Belluzzi, J.D., "Brain endorphins: possible role in reward and memory formation," *Fed. Proc.* 1979, **38:** pp 2468–2472.
154. Suzdak, P.D., Glowa, J.R., Crawley, J.N., Schwartz, R.D., Skolnick, P., Paul, S.M., "A selective imidazobenzodiazepine antagonist of ethanol in the rat," *Science,* 1986, **234:** pp 1243–1247.
155. Terenius, L., "Characteristics of the 'receptor' for narcotic analgesics in synaptic plasma membrane fractions from rat brain," *Acta Pharmacol. Toxicol.* 1973, **33:** pp 377–384.
156. Tjolsen, A., Lund, A., Hole, K., "Atinociceptive effect of paracetomol in rats is partly dependent on spinal serotonergic systems," *Eur. J. Pharmacol.* 1991, **193:** pp 193–201.
157. Traynor, J.R. and Elliott, J., "*d*-Opioid receptor subtypes and cross-talk with m-receptors," *TIPS.* 1993, **14:** pp 84–86.
158. Tulunay, F.C., Ayhan, I.H., Portoghese, P.S., Takemori, A.E., "Antagonism by chlornaltrexamine of some effects of Δ^9-tetrahydrocannibinol in rats," *Eur. J. Pharmacol.* 1981, **70:** pp 219–224.

159. Unterwald, E.M., Tempel, A,, Koob, G.F., Zukin, R.S., "Characterization of opioid receptors in rat nucleus accumbens following mesolimbic dopaminergic lesions," *Brain Res.* 1989, **505:** pp 111–118.

160. Van Ree, J.M. and Ramsey, N.F., "The dopamine hypothesis of reward challenge," *Eur. J. Pharmacol.* 1987, **134:** pp 239–243.

161. Van Ree, J.M., Smyth, D.G., Colpaert, F., "Dependence creating properties of lipotropin C-fragment (-endorphin): evidence for its internal control of behavior," *Life Sci.* 1979, **24:** pp 495-502.

162. Vaysse, P.J., Gardner, E.L., Zukin, R.S., "Modulation of rat brain opioid receptors by cannabinoids," *J. Pharmacol. Exp. Ther.* 1987, **241:** pp 534–539.

163. Vincent, J.P., Kartalovski, B., Ganeste, P., Kamenka, J.M., Lazdunski, M., "Interaction of phencyclidine ('angel dust') with specific receptor in rat brain membranes," *Proc. Nat. Acad. Sci. USA,* 1979, **76:** pp 4678–4682.

164. Volpicelli, J.R., Davis, M.A., Olgin, J.E., "Naltrexone blocks the post-shock increase of ethanol consumption," *Life Sci.* 1986, **38:** pp 841–847.

165. Welch, S.P., "Blockade of cannabinoid-induced antinociception by nor-binaltorphimine but not *N,N*-diallyl-tyrosine-aib-phenylananine-leucine, ICI 174864, in mice," *J. Pharmacol. Exp. Ther.* 1993, **256:** pp 633–640.

166. Welch, S.P. and Stevens, D.L., "Antinociceptive activity of intrathecally administered cannabinoids alone and in combination with morphine in mice," *J Pharmacol Exp Ther.* 1992, **262:** pp 10–18.

167. Wilson, R.S., Rogers, M.E., Pert, C.B., "Homologous *N*-alkylnorketobemidones correlation of receptor binding with analgesic potency," *J. Med. Chem.* 1975, **18:** pp 240–242.

168. Winger, G., Hofmann, F.G., Woods, J.H., *A Handbook on Drug and Alcohol Abuse: The Biomedical Aspects.* Oxford University Press, Oxford, 1992.

169. Wise, R.A., Bauco, P., Carlezon, W.A., Trojniar, W., "Self-administration and drug reward mechanisms. *Ann. N.Y. Acad. Sci.* 1992, **654:** pp 192–198.

170. Wonnacott, S., "The paradox of nicotinic acetylcholine receptor upregulation by nicotine," *Trends Pharmacol. Sci.* 1990, **11:** pp 216–219.

171. Woodward, J.J. and Gonzales, R.A., "Ethanol inhibition of *N*-methyl-D-aspartate-stimulated endogenous dopamine release from rat striatal slices: reversal by glycine," *J. Neurochem.* 1990, **54:** 712–715.

172. Woolverton, W.L. and Johnson, K.M., "Neurobiology of cocaine abuse," *TIPS.* 1992, **13:** pp 193–200.

173. Wozniak, K.M., Pert, A., Linnoila, M., "Antagonism of 5-HT_3 receptors attenuates the effects of ethanol on extracellular dopamine," *Eur. J. Pharmacol.* 1990, **187:** pp 287–289.

174. Yaksh, T.L., "The antinociceptive effects of intrathecally administered levonantradol and desacetyllevonantradol in the rat," *J. Clin. Pharmacol.* 1981, **21:** pp 334s–340s.

175. Yamamura, H.I., Enna, S.J., Kuhar, M.J., *Neurotransmitter Receptor Binding.* Raven Press, New York, 1985.

176. Yoshimoto, K., McBride, W.J., Lumeng, L., Li, T.-K., "Alcohol stimulates the release of dopamine and serotonin in the nucleus accumbens," *Alcohol.* 1991, **9:** pp 17–22.

177. Yu, S. and Ho, I.K., "Effects of acute barbiturate administration tolerance and dependence on brain GABA system: comparison to alcohol and benzodiazepines," *Alcohol.* 1990, **7:** pp 261–272.

178. Zaczek, R., Culp, S., De Souza, E.B., "Intrasynaptosomal sequestration of [^{3}H]amphetamine and [^{3}H]methylenedioxyamphetamine: characterization suggests the presence of a factor responsible for maintaining sequestration," *J. Neurochem.* 1990, **54:** pp 195–204.

179. Zukin, S.R. and Zukin, R.S., "Specific [^{3}H]phencyclidine binding in rat central nervous system." *Proc. Nat. Acad. Sci. USA,* 1979, **76:** pp 5372–5376.

Contents Chapter 7

Drug use and addiction: human research

0-8493-7803-6/99/$0.00+$.50

Chapter 7

Drug use and addiction: human research

L. Marleen W. Kornet, Raymond J.M. Niesink, and Jan M. van Ree

INTRODUCTION

Human substance use and related problems of abuse and addiction, are considered to be determined by a complex interaction of biological, psychological, and social factors. Therefore, the ideal approach of research on human addiction should be based on the interaction between endogenous neurochemical, physiological, and psychological systems with exogenous systems of social, ecological, and life historical (career) factors. Practically, this is very hard to realize, but multidisciplinary approaches are certainly developed. On the other hand, unidisciplinary studies still remain important to unravel specific problems and to develop or test new hypotheses, requiring highly specialized methodology and professionals. The discovery of the existence of neurobiological systems for behavioral reinforcement has changed the whole field of drug research drastically. Equally, the evidence concerning the influence of social factors on drug taking behavior is unanimously accepted.

Human versus animal research

Animal research on drug use and addiction offers the advantages of optimal experimental control, more possibilities for invasive research and of providing relatively simple experimental designs. It is evident, however, that not all questions can be answered by animal studies and that human research is necessary either to test the validity of the animal results or to address specific human-related issues.

Aim of this chapter

The aim of the present chapter is to show what scientific research approaches are available to perform with human subjects. A global distinction is made between epidemiological, experimental laboratory, and clinical research approaches. Although (techniques of) these approaches might be integrated in one design, for example in multidisciplinary studies, operational hypotheses, methodology, and selected subject sample as well as data acquired can be distinguished for the diverse approaches.

In short, *epidemiological research* investigates general or specific subject populations to quantify drug use and drug-related problems, to estimate high risk groups, to identify risk factors, and to assess the effectiveness of health services (e.g., prevention programs). *Experimental laboratory approach* is used to study specific relationships between *a priori* determined independent and dependent variables in a strictly controlled environment using experimental and control groups of critically selected subjects. *Clinical research* specifically addresses clinical populations, i.e., subjects with (an increased risk for) a medical/psychiatric diagnosis related to drug use. Here, clinical relevance, e.g., individual variables for etiology and treatment, play a more significant role compared to the other two approaches.

Analyses in the field of human research can move from the level of the population to the level of the person, each having its specific contribution. A general distinction can be made between cross-sectional studies, generating data of individuals/groups of interest at one specific point of time, and longitudinal studies, generating follow-up data of identified individuals/groups by repeated measuring.

Data sampled can vary from heterogeneous material, such as in unstructured individual reports or case descriptions, to semi-structured as in interviews with standard questions but open answers, to structured questionnaires with standard questions and standard answer items, to standardized psychometric test results, quantified operant response behavior or measures of physiological and chemical parameters.

1 Epidemiological approaches of drug use and addiction

1.1 GENERAL POPULATION SURVEYS

General population surveys are considered as epidemiological research at a descriptive level. They provide estimates of the rate of some characteristic, e.g., cigarette smoking in a country, by drawing a sample from the population[9]. There are several techniques to take a sample. The simplest way of sampling subjects from a population is called the *simple random sample,* which consists of using random numbers to draw subjects from a list of the entire study population. A more complicated design is known as *stratified random sampling*. This technique provides a mean to correct for the different proportions in a population. It includes drawing samples randomly from within different subgroups of the study population in order to create more representative samples. For example, if there are more men than women in the sample, subjects can be drawn separately from the male and female population. A third sampling design called *cluster sampling* is convenient whenever individual cases cannot be easily rostered. Combinations of stratified and clustered sampling are often used for epidemiological surveys in the community[9].

General population surveys concerning drugs estimate quantitative information on drug use in the community, on use patterns and trends, and can reveal subpopulations at increased risk. Obtained data can relate to quantity frequency and occasion of drug use, related morbidity and mortality, adverse social consequences, cessation rates, and prevalence of drug-related problems.

Drug-related morbidity and mortality are frequently derived from annual data of hospitals, and of death certificates. However, this is very likely an underestimation due to reporting bias, lack of information on decedents' drug-use histories, or both. In addition, drug involvement in for example, traffic accidents, falls, drowning, violence, or suicides is usually underestimated. To give an impression on drug-related mortality: more than 400,000 deaths occur annually in the U.S.A directly attributed to cigarette smoking; 12,500–15,000 due to alcohol; 4000 due to alcohol plus another drug; 4000 due to heroin, 2000–4000 from cocaine, and 75 from marijuana[15].

Epidemiological household surveys on drug consumption and drug use patterns have been collected in almost every country of the world. For legal psychoactive substances, such as nicotine and alcohol, official sales data can be used in estimating consumption by certain sociogeographic areas. Of course, no such source is available for illicit drugs.

In a general survey on the amount of alcohol consumed in the United States, amounts have been estimated on the basis of alcohol sales in stores and/or reports from beverage industry sources[16]. Apparent per capita consumption is determined by dividing total alcohol by the total population aged 14 or older.

Apparent per capita consumption of pure alcohol in the U.S.A from 1977 through 1987 was shown to peak in 1980 and 1981 and subsequently to decline (see Figure 7.1). The term "apparent" is used because these estimates artificially attribute average consumption to all persons in this population, regardless of their actual consumption.

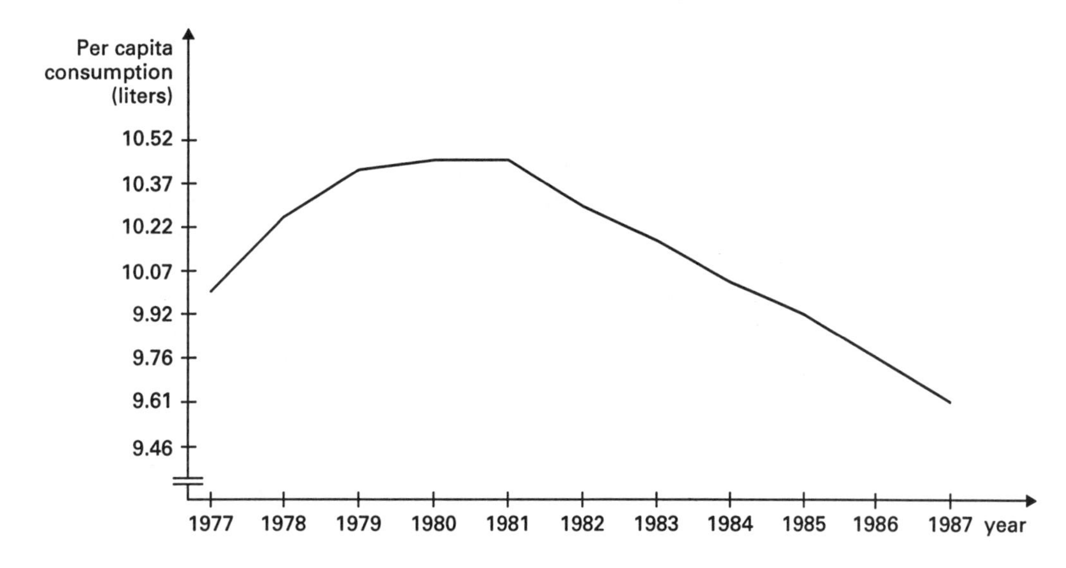

FIGURE 7.1

Apparent U.S. per capita consumption of pure alcohol, 1977-1987
(Based on original data from Surveillance Report 13 of the National Institute on Alcohol Abuse.[63])

Of 25 countries surveyed in this period, two thirds experienced the same trend in apparent per capita alcohol consumption. Although general surveys give valuable descriptions of population profiles, they are not adequate to analyze causal relationships between drug use and specific subpopulations. This kind of analyses can be performed by conducting specific epidemiological field studies, that can determine the influence of environmental factors on drug use and dependence.

1.2 FIELD STUDIES

Field studies can contribute in identification of risk factors, can test hypotheses originated in clinical and experimental studies, or can be used to monitor the effects of treatment and prevention programs. They produce quantified differences in distribution of drug-related items among different groups and reveal associations between specific population characteristics with types of drug taking behavior. When looking at the distribution of drug use and abuse in the population, rates are commonly associated with various factors such as age, sex, occupation, state of health, level of education, ethnicity, urbanism-ruralism, socio-economic status, and geographic site.

Field studies on alcohol use patterns across the same time period as the consumption survey example in the previous paragraph (1977–1987), revealed for instance an increasing proportion of heavy drinkers among young people in their twenties and a small increase in the prevalence of dependence problems[77]. Patterns were defined using classification of number of drinks consumed per month, frequency of drinks per occasion and getting drunk at least once a week. These results contrasted with the general survey data of a decline in alcohol consumption in the population (see U.S., 1990). A similar general trend has been reported for smoking which declined from 50% of all adult men in 1965 to

33.5% in 1985[8]. This trend appeared in striking contrast to that observed in surveyed alcoholics, of which smoking rates remained rather consistent (about 90%!).

Using field studies, six distinct developmental patterns of cocaine use have been defined[39]:

1. Much cocaine use at first, then slowly decreasing use
2. A slow and steady increase in cocaine use
3. A stable level from first to current use
4. "Up-top-down", i.e., a relatively low cocaine use at first that rises to a peak and then drops to the early levels
5. Intermittent use at the same level, but with significant breaks in time
6. Varying cocaine use with peaks and troughs in an irregular trajectory

In an Amsterdam study the most prevalent patterns were Pattern 4 (39% "up-top-down") and Pattern 6 (33% irregular trajectory). Only 3% of the sample presented Pattern 2 (slow and steady increase), the developmental profile that would characterize classical opiate dependence[13].

The given examples make clear that to determine particular characteristics of people with or at higher risk for drug-related problems, highly targeted field studies are necessary in specific population subgroups. Examples of specific subpopulations under study are adolescents, individuals with socioeconomical disadvantages (no job, single, cultural minority, low income, little education), children with behavioral disorders, children from drug using parent(s), individuals with mental disorders, recreational drug users not in treatment, or drug users in treatment.

1.3 HIDDEN POPULATIONS

There has become increasing awareness of methodological problems derived from selection biases in the field of drug use studies, especially with respect to illicit drugs. Illicit drug users often will refuse to participate in surveys because of fear of losing anonymity and getting arrested. These problems are tied to the fact that drug use and addiction are socially undesirable behaviors and are characterized by social stigma. In addition, many users stay out of professional supervision. As a consequence, patterns, use scales and risks of recreational (non-clinical) illicit drug use are hard to trace since they must be extracted from so-called "hidden" populations. To get insight in the determinants of the drug use by this group of users, the behavior of the users in his/her natural environment is a core element of epidemiological field research.

Alternative sampling techniques for hidden populations have therefore been developed, such as *"snowball sampling"*. In this methodology, sample selection begins by making contact with members of a specific population that may be characterized as "rare", "elusive", or "hidden"[43]. The knowledge accrued in this initial fieldwork (a simple or stratified random sample may also be used) is used to find the "starters", composing the "zero stage" of a snowball sample. These starters, in turn, "nominate" other members of the population. Interviews with the starters and nominees are routinely conducted. The snowball procedure "ascends" through these chains of referrals into more general levels of the population. The disadvantage is the uncertain generalizability of the sample. If it is not known how representative the sample is, valid results may be difficult to obtain. However, through a strategy of selecting multiple snowball samples that are started at independent field sites more precision and valid results can be obtained.

Experience Sampling Methodology

Another example is the *Experience Sampling Methodology* (ESM) which collects data on the person in his or her natural setting (for an example of the use of ESM see section 1.8). ESM employs a specially designed technology that signals the subject with a random beep to self-report on experience within 10 minutes. Beeps are preprogrammed and vary from five to ten per day depending on the conditions of specific subpopulations under study. Neither the subject nor the researcher can control the initiation of an observation thereby eliminating one source of persistent bias found in non-randomized observational studies.
ESM has been employed in study of the daily lives of over 40 active heroin addicts in and out of clinical settings[45].

1.4 RISK FACTORS

To identify risk factors, a first step is to examine subgroups in the population to determine where the rates of drug use and abuse are higher than usual. Table 7.1 shows the frequency of crack use in various survey populations and high-risk groups[78].

TABLE 7.1

Frequency of crack use in various survey population and high-risk groups
(Based on data from: Smart, R.G. *Am. J. Drug Alcohol Abuse.* 1991, **17(1):** pp 13–26. With permission.)

Authors	*N*	*Sample*	*Place*	*Year*	*Rate of use*
Smart and Adlaf (1987)	1,040	Adults 18 and over household sample	Ontario	1987	0.7% (ever)
Smart and Adlaf (1987)	4,267	Students (aged 13-19), Grades 7-13 classroom sample, stratified	Ontario	1987	1.4% (past year)
Chamberlayne et al. (1988)	14,750	Students (aged 14-18), Grades 8-12	British Comulbia	1988	1.6
Johnston et al. (1987)		Senior students in high schools	U.S.A. national sample	1987 1988	5.6 1.4 (lifetime)
				1986 1987 1988	4.1 4.0 3.1 (past year)
				1987 1988	1.5 1.6 (past 30 days)
Johnston et al. (1987)		College students, 1-4 years beyond high school	U.S.A. national sample	1986 1987 1988	1.3 2.0 1.4 (past year)
Washton et al. (1986)	458	Cocaine abusers calling cocaine hotline	U.S.A	1986	32.0% (ever)
Inciardi (1987)	254	Street youths involved in drugs and crime	Miami	1986	95.5 (tried) 87.3 (regular use)
Wish et al. (1986)	250	Intensive supervision of probationers	Brooklyn	1986	38.0% (ever)

It can be deduced from the table that crack is used by a small minority of adult and student populations, but by a large proportion of cocaine users and heavy drug-using groups. Street youths involved in drugs and crime in Miami seem to be especially at high risk for crack use. Further knowledge on high-risk factors is of importance since crack has many serious adverse effects, including those for offspring and for children living in crack smoke-filled rooms. Characteristics of most crack users were adolescent or young adult age; heavy use of alcohol, tobacco and many illicit drugs — many have switched to crack after intranasal cocaine[85]. In treatment seeking crack users, the following characteristics were found: unemployment, domestic violence or alcoholism; in 39% of the patients, a major depression or dysthymic disorder was diagnosed. Based on this analysis of high-risk groups of crack users, it can be expected that treatment of crack dependence will be more difficult compared to cocaine because of the polydrug use pattern and life historical variables.

Risk factors can vary from biological, genetic, environmental, infectious, or other characteristics of the group. As already mentioned, the association with life-historical variables, such as domestic violence or neglect, can be included as well.

Risk factors that might determine the genesis of drug-related problems can be best examined by comparing rates of new cases, *incidence rate*, in the population. *Prevalence rates* represent the proportion of a population who have drug-problems at a given point of time (i.e., incidence and the long-standing cases together). Incidence studies are more complex than prevalence studies because the development of new cases usually requires passage of time. Incidence studies are for instance *prospective longitudinal studies* in which a risk group is sampled and characterized without *a priori* incidents of problems. Subjects are followed longitudinally and repeatedly reexamined for incidence. For example, the relationship between childhood behavioral disorders and later drug use and dependence has been studied in children that were repeatedly examined during their development and adulthood[46].

Lifetime prevalences of different drug dependencies can have a wide variation among various cultures and across different time points. However, risk factors such as age of onset, gender, antisocial personality, and major depression appeared to be quite general variables in the development of abuse and addiction. The strongest risk factor to be specified in current research is age. The average risk of becoming a new case of drug abuse/dependency in the United States for adults of all ages is 1.09% per year, for 18–29 years it is 2.84%, for 30–44 years 0.69%, and for 45 years and older 0.05%[2]. If the age of onset of drug use and manifestation of first problems occur before the age of 18 years, the risk of later addiction problems is substantially higher. Other risk factors associated with becoming a new case of drug abuse/dependency in the United States includes being without a high school diploma, being a worker in a low prestige job, never being married or being separated/divorced, being black, and having an antecedent mental disorder[2].

1.5 PROTECTIVE FACTORS

The determination of risk factors for drug use and later addiction still will require considerable research efforts in the future. Important is the distinction between use and abuse. According to international treaty regulations various substances are classified in terms of their perceived risks. Within The Netherlands, for example, drugs are scheduled in terms of being of "acceptable" and "unacceptable" risk. Currently, alcohol, tobacco, prescription drugs and cannabis are classified as acceptable risk drugs while all others are unacceptable. Drugs used without medical indication or "designer drugs" are usually soon classified as having an unacceptable risk. This has more to do with a policy of "prophylactics" concerning all new drugs rather than any firmly established abuse liability. As with all governmental regulations, the scheduling of drugs is a time-consuming process and the "unscheduling" is even more difficult. Alcohol is an example of a drug that has been unscheduled after being scheduled in the United States. In the Netherlands, personal use of marijuana is not considered an illegal act anymore or as bearing unacceptable risks. The pros and cons of legislation of marijuana and of other illegal drugs is an international issue of conflict and debate.

Most risk factors become noticeable only after drugs are used. Therefore, it is important to be able to identify the protective factors which intervene or prevent drug use and addiction. The little known suggests that peer pressure and peer support are important. Peer pressure on persons who have never used drugs have proven effective if vulnerability is otherwise relatively low. Once drug use begins, other strategies must be introduced. Peer support is especially relevant in this situation. This approach to prevention has been termed "harm reduction"[66]. For example, the improvement of drug treatment services to be more responsive to early case detection and to work with social networks of peer support clients can function as a protective factor stabilizing an otherwise escalating addiction syndrome. The "low threshold" methadone services widely used in The Netherlands is one concrete example of how addict life is reorganized to provide more time for health supporting activities[44]. Other prevention efforts that aim at teaching 'safe' and regulated drug use are examples of efforts to promote the strengthening of existing protective factors that have emerged within drug subcultures themselves to minimize risk. And finally, the policies of controlling availability through a flexible market strategy can be seen as yet another sort of protective factor. The controlled availability of heroin, for example, has been proposed as one social experiment that could help to prevent the spread of AIDS (injecting drug users are currently the group most at-risk for contracting the virus). It would alter the current monopolistic illegal drug market which has come to be characterized by its aggressivity in recruiting new users and stop the corruption of existing official institutions[34]. Changes in public perception of the drug user away from a stigmatized monster image will also help in facilitating peer support and self-help.

1.6 EXPERIMENTAL EPIDEMIOLOGICAL STUDIES

Experimental epidemiology can give evidence that an association observed in field studies between a risk factor and a drug-related problem may be a real cause-and-effect relationship[9].This can be achieved by designing (preventive) interventions and testing how effective they are in reducing the problem of interest. For example, community-based interventions to reduce cigarette smoking may lead to a reduction in subsequent development of lung cancer. The methods for assessing a preventive intervention trial are analogous to those for determining therapeutic interventions. However, preventive studies are mostly more extensive. The central feature of the experimental method is *random assignment* of subjects to either the experimental intervention group or the control group without intervention. This process of assigning subjects at random is expected to prevent selection bias because any unknown underlying causes would be equally distributed between both groups. Observer bias is minimized by having examiners and subjects 'blind' to their assignment in experimental or control group. This methodology can be used for clinical trials (to measure treatment success) and for testing the effectiveness of preventive programs in high-risk groups. One problem for the latter application is to determine how many subjects are needed to show that any differences between the control and experimental group are "true" (i.e., drug-related) differences. Advanced statistical techniques have been developed to estimate this and other related parameters.

By following specifically defined cohorts in the development of antecedents (e.g., childhood behavioral disorders), early onset and disease course of clinically relevant addiction syndromes and patterns, it is also possible to study causal relationship between antecedent variables and later development of drug-related problems. The experimental epidemiological studies can also concern the assessment of the effect of preventive or therapeutic interventions designed to alter the development or outcome of illnesses.

1.7 META-ANALYSIS

Methods of targeting specifically defined subpopulations to obtain meaningful and ecologically valid results do challenge the generalizability of these findings for other populations or the discovery of general causalities. By conducting meta-analysis the extent can be assessed to which cross-study, longitudinal findings from different cultural and ecological context replicate. Meta-analysis involves the collecting together of a set of similar studies which are then submitted to a conjoint analysis[17]. In this way, it becomes possible to differentiate between cohort-specific and generalized effects. It is evident that these are complex studies, which require the construction of a new and common coding frame. A promising project is a collaborative alcohol-related longitudinal project by Middleton and many others[57]. Studies included were 41 general population longitudinal studies and two adoption studies. Various broad research agendas are addressed:

1. The consistency across data sets of prevalence, incidence, and chronicity of drinking patterns and problems by age and sex
2. The influence of biological, sociological and psychological factors on initiation and alteration in drinking careers for several age strata, across the cohorts;
3. The determination to what extent antisocial youthful behavior is a general predictor of adult drinking problems
4. The determination of the (biological) influence of parental drinking

Table 7.2 gives an overview of the wide variety of data samples included in this project.

1.8 CASE STUDIES

An advanced design is the *case control study*, which is especially useful in an exploratory stage before deciding to conduct a costly, time consuming and complex prospective longitudinal study. In a case control design known cases are identified, e.g., from hospitals or treatment centers, and individuals who are similar in sex, age, socioeconomic status, or other relevant characteristics and without drug-related problems are matched to the cases in the comparison group.

TABLE 7.2 (right)
Brief description of samples in the collaborative alcohol-related longitudinal archive
T1 = first time; T2 = second time; N = sample size. (Based on original data from: Middleton Fillmore, K. et al., *Br. J. Addict.* 1991, **86**: pp 1203–1210.)

TABLE 7.2

Study	*Collaborator*	*Date T1*	*Date T2*	*N waves*	*Interval*	*T1 age*	*N T1*	*Sampling frame*
Canada-g	Giesbrecht	1984	1986	2	2	15-70	1700	Males in 2 Ontario communities
Canada-m	Schlegel/ Manske	1978	1986	12	8	19-22	1076	Schools in 2 Ontario school districts
Czech	Kubicka	1983	1989	2	6	21-34	1074	Males in Prague
Denmark-sa	Ahlstrom	1960	1964	2	4	14	185	Males in Copenhagen
Finland-sa	Ahlstrom	1960	1964	2	4	14	186	Males in Helsinki
FRG-g	Guether	1975	1983	2	1	24-53	9000	National sample/Germany
FRG-s	Silbereisen	1982	1985	3	3	10-19	1432	Multiple Berlin schools
FRG-s	Silbereisen	1982	1985	3	3	19-70	805	Representative sample of parents of FRG-s
Ireland	Grube/ Morgan	1984	1985	2	1	13-17	2927	Schools in greater Dublin
Israel	Teichman	1982	1983	2	1	15-19	1900	Schools in central region of Israel
N.Z.	Casswell	1981	1985	2	4	9	743	Born in 1 hospital, Dunedin
Norway-a	Amundsen	1951	1987	1	36	19	17340	Male conscripts in Norway
Norway-sa	Ahlstrom	1960	1964	2	4	14	202	Males in Oslo
Poland	Silbereisen	1985	1986	2	1	6-24	1200	Multiple Warsaw schools
Scotland-p	Plant	1979	1983	3	4	15	1036	Schools in Lothian region
Scotland-r	Peck	1978	1982	2	4	17-79	676	Citizens in Lothion region
Shetlands	Rosen	1975	1978	2	3	15-65	533	Citizens in 2 communities one rapidly industrializing
Sweden-s	Sigvardsson	1930	1972	1	42	birth	1990	Adoptees born in Stockholm
Sweden-sa	Ahlstrom	1960	1964	2	4	14	199	Males in Stockholm
Sweden-o	Ojesjo	1957	1972	2	15	0-96	2297	Citizens of Lundby
Sweden-r	Allebeck/ Romelsjo	1969	1984	2	15	18-22	50465	Male conscripts/Sweden
Switzer.	Sieber	1971	1983	2	12	18-21	6315	Males conscripts in Canton of Zurich
UK-b	Bagnall	1986	1988	2	2	13-15	1586	Nine schools from Highland area, Scotland, Berkshire area England & Dyfed area, S. Wales
UK-p	Power	1974	1981	2	7	16	1036	Infants born March 3-9, 1958 in U.K.
USA-b	Brunswick	1975	1983	3	9	18-23	535	Central Harlem housing units, Non-Latin Blacks
USA-c	Cadoret	1938	1982	1	44	birth	84	Adoptees in Iowa
USA-d	deLabry	1973	1982	2	9	28-84	1897	Male volunteers in Boston without cirrhosis, pancreatitis or elevated SGOT
USA-g	Greenfield	1978	1980	2	2	18-48	975	University in Washington State
USA-l	Leino	1928	1984	2	56	birth	248	Every 3rd birth in Berkeley, CA
						birth	74	Infants born to White, English speaking Berkeley parents
						11	212	Whites in Oakland, Ca schools
USA-m	Mulford	1979	1980	3	1	17-93	1535	Citizens of Iowa
USA-NLSY	Harford/ Grant	1982	1986	5	4	17-25	12686	National sample of US with supplements of Blacks, Hispanics, economically disadvantaged non-Black & Hispanic, & young men in military
USA-r	Robins	1972	1974	2	2	19-31	920	Male enlisted men departing Vietnam in Sept., 1971
USA-o	Temple	1964	1979	12	15	16	1227	White males of one county in Oregon
USA-w	Wilsnack	1981	1986	2	5	21-84	917	National sample of women stratified by alcohol consumption
USA-2/5	Public domain	1967	1974	2	7	21-70+	1359	Male national sample/USA
USA-3/4	Public domain	1969	1973	2	4	21-69	978	White male national sample/USA
USA-05	Temple/Stall	1964	1984	2	20	23-88	970	Citizens of San Francisco
USA-07	Temple/Stall	1967	1984	2	17	21-59	786	White male citizens/San Francisco

Intensive case findings and descriptions without matched controls are essential as well in addiction research because of the hidden population issue. The *a priori* control of these studies have been improved through the use of observational protocols and multisite comparisons. A recent ethnographic study of the American crack users employed a design that applied a common protocol across a number of cities[32]. Further refinements in qualitative field studies can be seen in the use of complementary observational techniques such as participant observation, narrative interviews, and focus groups.

The case presented below was guided by the hypothesis that the use of cocaine would influence the daily craving patterns of heroin addicts and require extraordinary self-management behaviors above and beyond the daily methadone maintenance regime employed to treat heroin addiction.

Example: The experience of craving in daily life of Bob, a heroin addicted person

This example is adapted and slightly modified from a manuscript by Drs. Kaplan and De Vries (see also ref. 45).

Method

In this case, craving is measured by use of a methodology of the self-monitoring of the subjective experience (ESM) (see Section 1.3), which is considered as a diagnostic tool to measure the experience of psychopathology in patients in their natural settings. Craving experience was evaluated over the period of a week. The craving module consisted of five items. The items were composed of individual Likert scales ranging from "very" to "not". These scales were self-administered by the subject within 10 minutes of the signal of the beep:

Are you thinking about using?
Do you have yourself under control?
Do you feel yourself restless?
Do you need dope quickly?
Do you need money badly?

Complementing the ESM data collection was ethnographic fieldwork in which the researchers accompanied the subject when possible in the daily routines. Daily "debriefing" interviews were conducted with the subject. Interviews and observations with the others related to the subject also were conducted. These qualitative data were stored as fieldnotes and referred to in subsequent analysis.

Subject

Bob is a fairly representative case of heroin addicts of Rotterdam, the city where the fieldwork has been conducted. In terms of background characteristics Bob is typical. He is a 30-year-old, unemployed white male from a working class family of eight children. Along with three of his brothers, he spent much of his growing up in a children's home as his family could not support so many children. He had been using heroin for 10 years. He was registered as a client in a "low threshold" methadone program. He was provided with daily doses of 20 cc by the program to help him gradually in detoxification from heroin. In the last years, he had added cocaine to his daily heroin taking regime to form a polydrug pattern of use. This characteristic has become representative of the large majority Rotterdam cocaine users in the city's methadone programs.

Subjective report

In order to help finance his polydrug use Bob had done some small-scale dealing in the employment of larger dealers. Drug dealing had become a significant symbol in his life and was directly associated with the frequency and amounts of his own drug taking as evidenced by his own story:

"I've used more the last days. I had some money. Last Tuesday a couple of old customers came by from Belgium. They wanted me to help them to buy 8 grams of heroin. I hadn't

seen them for about six months. Normally, I don't go out looking for other people to find dope. But for 8 grams I'll do it and it brought something for me, too."

Bob's main psychopathological complaints are loneliness and addiction. He recognizes the association of these complaints in himself:

"I'm living alone in this house — my girlfriend lives in another town. The only thing I got outside is the methadone program. It breaks my day; I meet people down there. For the rest, I'm mostly at home watching television. And I really feel bad about that. You can see it in the diary [the ESM self-administered instrument] that when somebody is coming by I feel better right away. I would like to have my girlfriend around more. When I'm lonely I use more, too".

Drug dealing is the behavior that has become the self-recognized source of Bob's complaints while his relationship with his girlfriend has come to symbolize a way out of the complex of addiction and loneliness. But dealing and the presence of suppliers still exert their influence to counter the effect of Bob's girlfriend. The dealing life has its material benefits and Bob's suppliers still drop in to his home offering friendly new opportunities:

"Last year on my birthday I stopped dealing. I couldn't let my hands off the coke. Well you know how it is when you are in a coke binge, you can hardly stop then. One moment I really spit on it and said to myself, 'Knock it off, quit with that stuff'. My girlfriend didn't want to see me anymore because I was dealing. Not for the dealing itself, but because she can't accept that I earned money like that on other people. I offered her a trip to Paris and she said she did not want to go with me the way I was now. She then went with a friend and paid me one ticket back."[45]

Objective data

Analysis proceeds by the inspection of plots of the Bob's beep scores on two items that load highly on the craving factors, acquisition (need dope), and control (in control yourself). Bob's scores are plotted on a y-axis representing the Likert items and the x-axis of time intervals marked by each beep. The x-axis plotting is further refined by aggregating the subject-beeps by each day and breaking down each subject-beep by the drug used. The analysis provides a profile of the dynamics between craving, drugs, and intimacy on three successive days in the life of Bob.

Figures 7.2 and 7.3 subsequently plot Bob's scores on the craving items — need dope quickly and have yourself under control — across the first 3 successive days in the week. An overview of the three days indicates a set of complex inverse relationships between the two craving factors. When the need for dope is high, control is low. Inversely, when control is high, the need for dope is low.

On inspection Bob's first day (Figure 7.2) the continuous use of cocaine is progressively associated with heightening need for dope. This heightening need puts pressure on control. Control is maintained at a relatively high level through the first two time intervals. Suddenly control noticeably drops as the need for drugs continues to accelerate. At t3 Bob is still using only cocaine, but he self-reports the thought of "smack" (heroin). Bob has learned that heroin has a new function. It comes to mind when a sense of control is lost from escalating cocaine use. The thought of heroin alone has an effect in lowering the need for dope and heightening the control. At interval t4, although cocaine is still the only drug used, the mere anticipation of heroin invokes an equilibrium. At interval t5 anticipation becomes action. Bob substitutes heroin for cocaine and achieves a state of calm with a very low need for dope and a high sense of control. This state continues to be maintained by the heroin at t6 as the day ends.

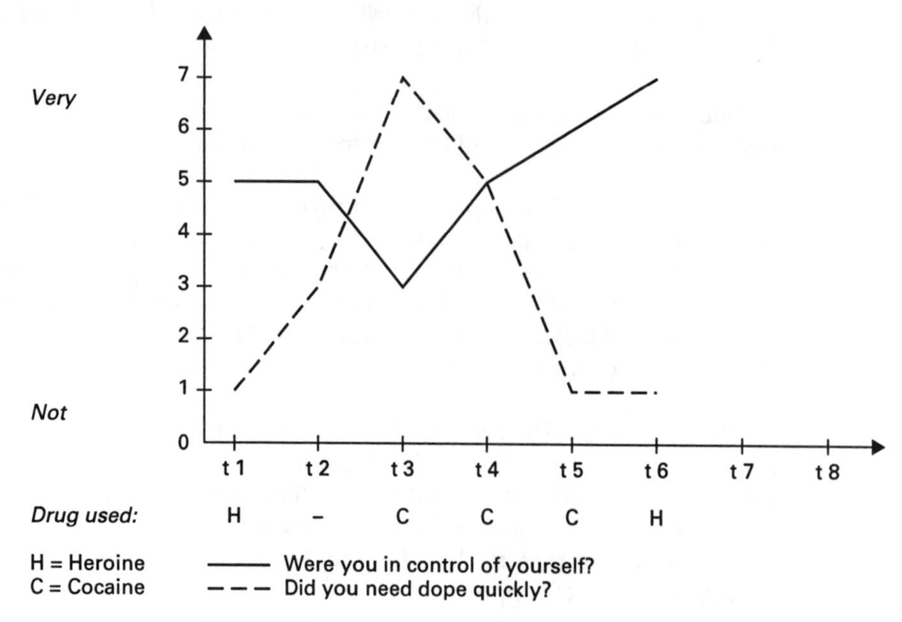

FIGURE 7.2

Bob's responses on the craving items during day 1

(Data from Kaplan, C.D., In: de Vries, M.W. (Ed.), *The Experience of Psychopathology: Investigating Mental Disorders in Their Natural Settings*. Cambridge University Press, Cambridge, 1992, pp 193–219. With permission.)

Bob's second day is characterized by relatively high control and low need of dope. Heroin is taken throughout the day having the effect of keeping control at a constant level while lowering the need for dope. As the need for dope stabilizes at a low level, control increases to a high level. This optimal state is maintained throughout the day. In addition to this drug effect, there is a setting effect. During the day Bob is visited by his brother and his girlfriend. This human contact seems to reinforce the stable sate brought about by the heroin "self-medication".

On the third day the destabilizing effect of cocaine on Bob can be observed. He starts the day in a stable way with heroin and no drugs at t2. At t3, when alone, he uses cocaine. The need for drugs immediately increases and the control drops. He uses cocaine at t4 in the company of others. The need for dope goes down, but so does the control. At this moment, the need for drugs is dampened by company. However, this company is not intimates of the same importance as his brother and girlfriend. By taking cocaine with them and experiencing the drug's socializing effect, the need for drugs does go down. But this benefit is experienced without an accompanying sense of gain in control. In fact, this company actually produces a lowering sense of control given that they are all mutually reinforcing each other's cocaine taking behavior. At t5 Bob uses cocaine again alone. The need for dope rises to a daily high. Control also rises to a high level. This rise in control is yet another case of anticipation effect of using heroin. As the day ends at t6, Bob reports he "was fucked up from the coke." He then takes heroin without the cocaine achieving his optimal state of a low need for dope and high control.

Discussion

The case of Bob provides an intensive and microscopic view of the variations in addiction symptoms and associated conditions at the level of a person. This person-level data is not idiosyncratic but representative of broader epidemiological patterns and changes in the heroin user population in the last decade. Drug patterns are constantly shifting and with these shifts the drug-using populations change their drug preferences. Drugs also change their function in this dynamic situation. One such noticeable shift has been in the

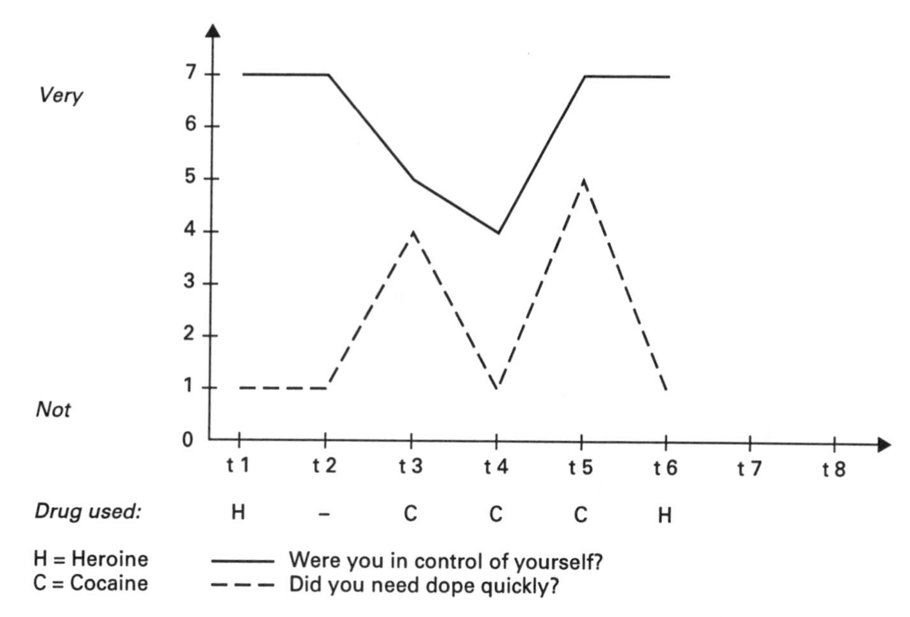

FIGURE 7.3

Bob's responses on the craving items during day 3

(Data from Kaplan, C.D., In: de Vries, M.W. (Ed.), *The Experience of Psychopathology: Investigating Mental Disorders in Their Natural Settings.* Cambridge University Press, Cambridge, 1992, pp 193–219. With permission.)

heroin-using population where cocaine use is having increasing importance. In Rotterdam, the drug-taking rituals have transformed themselves into new varieties of polydrug use involving cocaine occurs on a daily basis[32]. This pattern of high cocaine use in the heroin addict population has been documented in other European cities as well[34]. The case of Bob reveals in great detail the behaviors related to this broad epidemiological trend.

1.9 USE AND DEPENDENCE

In most western societies a large percentage of the population has tried illegal drugs. For example, in the United States, it is officially estimated in 1991 that over 55% of the population of 21–34 year olds have used illegal substances[29]. In addition, about 60 million Americans smoke tobacco.

From American epidemiological studies, the percentage of the adult population that can be estimated as having a clinical diagnosis of drug abuse or dependency is only 2.67% which is comprised of 4.0% of the adult men and 1.37% of the adult women[2]. With respect to benzodiazepines, which are frequently prescribed as medications, only a minority (16%) of users abuse sedatives[24]. Even for cocaine, that has been demonstrated in animal studies as an extremely potent positive reinforcer[75], the epidemiological evidence shows that the majority of persons who have used cocaine (crack excluded) encountered no serious clinical problems[24]. This suggests that subtle patterns such as "casual use" exist beyond experimental use and dependency and are also quite prevalent[42]. Thus, one of the research issues is to describe, explain and classify patterns, trends and problems that are related to "casual" use and to find ways of preventing and/or reducing possible harms and hazards of "casual" drug use. In addition, by comparing epidemiological results found in specific subgroups such as non-users, casual users, dependent users without seeking treatment and clinically treated users, more knowledge can be gathered on the critical factors determining abstinence, no-problem use or the onset of dependence.

The issue of use and abuse must therefore eventually rely on specific and detailed field studies. In addition, attention is increasing for subpopulations with non-drug dependence syndromes. Behavioral symptoms of compulsive gambling and eating share many commonalties with clinically defined substance disorders [50].

2 Experimental laboratory approaches

Compared to epidemiological studies, the researcher in a laboratory study has more *a priori* control over independent variables and the context in which dependent variables are measured, and includes more control data to compare the experimental data with. Determination of baseline values (pre-experimental level of the dependent variable), placebo control measurement (effect of experimental situation using a neutral variable), randomization (balanced groups of subjects), and "blind" administration procedures (subject and sometimes investigator do not know to which experimental condition the subject is exposed) are typical characteristics of laboratory designs.

Human experimentation thus shares many formal conditions with animal experimentation, but meets more restrictions due to the ethical aspects of exposing subjects (especially naive or abstinent) to drugs or drug stimuli. Nevertheless, human laboratory studies are important because they generate new hypotheses that can lead field research in new directions; otherwise they can test patterns and hypotheses generated by field studies under conditions of high control and with less influence of confounding variables. A laboratory paradigm for a human model of addiction involves three dimensions:

1. Subjective reports of drug effects
2. Drug identification or drug discrimination
3. Self-administration behavior

2.1 MEASURING SUBJECTIVE EFFECTS

The Addiction Research Center Inventory (ARCI)

Early interest in measurement of subjective effects of drugs stemmed from the need to identify which of the pharmaceutical products being tested prior to release had a high potential for abuse[87]. Extensive testing was carried out with simple yes/no questions concerning feelings and moods and perceived bodily changes in order to find out which changes were produced by the various types of drugs. A question might be concerned with whether the subjects feel light-headed, whether they have a metallic taste in their mouth, whether they find it hard to move around, or whether they have a sentimental feeling. By testing hundreds of such questions in subjects affected by a variety of drugs, as well as in the non-drugged state, it was found that a number of different types of subjective drug effects could be distinguished.

The result of this work was The Addiction Research Center Inventory (ARCI)[33]. The complete questionnaire consists of 550 items, but a short form consists of only 49 items which can be completed in less than 5 minutes[23]. The responses on the questions produce a particular pattern for various types of drugs. Thus, if a drug of unknown effects is given, ARCI results can be used to categorize the drug on the basis of its subjective effects. Results have shown that the classification of drugs according to their responses on the inventory is much the same as one would expect on the basis of their pharmacological activity.

The ARCI provides scores on five scales. One scale of the ARCI is sometimes referred to as the "euphoria" scale (indicated as the morphine-benzedrine group scale (MBG)) and has questions related to feelings of popularity, efficiency, social effectiveness, pleasant feelings, absence of worry, good self-image, and feelings of insight and satisfaction. One scale is based on the dose-dependent assessment of the effects of amphetamine (A-scale). One scale consists of questions related to intellectual efficiency and energy (benzedrine group (BG) scale). One scale focuses on a measure of sedation (pentobarbital-chlorpromazine alcohol group (PCAG)), and one scale provides questions on items assessing dysphoria and somatic complaints (LSD-scale).

Table 7.3 gives a summary of studies on subjective reports for stimulants. First, typical profiles of responses are shown on the five scales for amphetamine and cocaine, both drugs with high abuse liability[23]. Amphetamine and cocaine administration generated positive answers in the A (amphetamine-like), MBG ("euphoria"), and BG ("efficiency", and "energy") scales, occasionally negative answers for the PCAG-scale ("sedation") and positive answers for the LSD-scale ("dysphoria", "somatic complaints") following higher doses. Peak effects for amphetamine were evident 1–3 hours after administration; for cocaine peak effects occurred 2–15 minutes after administration. Most scores returned to baseline, depending on dose, by 6 hours for amphetamine and by 30–90 minutes for cocaine.

TABLE 7.3

Changes in the Addiction Research Center Inventory as a function of drug and route of administration

(Based on data from Foltin, R.W. and Fischman, M.W., *Drug Alcohol Depend.* 1991, **28:** pp 3–48.)

Drug	*Route*[a]	*Dose (mg)*	*Amphetamine (A)*	*Benzedrine Group (BG)*	*Morphine-Benzedrine Group (MBG)*	*Pentobarbital-Chlorpromazine-ETOH group (PCAG)*	*LSD*
d-Amphetamine	s.c.	7.5-30	+ 1,2	+ 1,2	+ 1,2	– 1	+ 1
	Oral	5-40	+ 2-10,22,25,27, 28	+ 2,5,6,8-10, 22,25,27,28	+ 2,3,5-10, 22,25,27,28	– 1,5,6,8,-10,22, 27,27	+ 1,5,10, 22,26
	i.v.	10	+ 12	+ 12	+ 12	– 12	
dl-Amphetamine	Oral	5	+ 11,26	+ 11,26	+ 11,26	– 11	
Methylamphetamine	s.c.	15-30	+ 1	+ 1	+ 1	– 1	+ 1
Cocaine	i.n.	96	+ 13	+ 13	+ 13		
	i.v.	8-48	+ 12,14-17,21, 24	+ 12,16-18,24	+ 12,14-21, 24,29	– 12,16-18	+ 12,14- 20,24,29
	Smoke	50	+ 21,24	+ 24	+ 21,24		+ 24
Phenmetrazine	Oral	25-70	+ 1,22,23	+ 1,22,23	+ 1,22,23	– 1	– 1,22
Diethylpropion	s.c.	150-600	+ 2	+ 2	+ 2		
	Oral	100-400	+ 2	+ 2	+ 2		
Mazindol	Oral	0.5-2.0					+ 9,22
Methylphenidate	Oral	16-60	+ 1	+ 1	+ 1	– 1	+ 1
dl-Fenfluramine	Oral	60-240	+ 3		+ 3		+ 3
Phenylpropanolamine	Oral	25-75					+ 7,22
Caffeine	Oral	100-800	+ 6,30	+ 30	+ 6,30	– 10,30	+ 6,10

+, Increase; –, decrease.
[a] s.c., subcutaneous; i.v., intravenous; i.n., intranasal powder.

This profile was compared with ARCI responses for other stimulants (see Table 7.3). Phenmetrazine produced the most similar profile of subjective effects compared to amphetamine and cocaine. Dietylproprion and methylphenidate produced some similar changes. Mazindol and fenfluramine had a small number of subjective effects and showed little resemblance with amphetamine and cocaine. Phenylpropanolamine, the over the counter medication, produced no changes at all in ARCI-scores. Caffeine had some subjective effects that are qualitatively similar to amphetamine and cocaine, but only in very high doses.

With respect to sedative drugs, ARCI results are less robust. Increases in positive MBG scores for pentobarbital, a barbiturate with abuse potential, appear not to be very reliable[14]. In normal healthy volunteers, a single dose of pentobarbital (160 mg) or placebo was administered and subjects completed the

ARCI repeatedly every 30 minutes for 4 hours following drug administration. Relative to placebo, pentobarbital significantly increased MBG scores at a single time point, 1.5 hours after drug administration. The elevation in MBG scores was short-lived. In contrast, scores on the PCAG scale increased within 30 minutes and remained elevated for the entire 4-hour session. Overall, several versions of the MBG or other "euphoria" scales have been used to assess the subjective effects of pentobarbital. Of the ten studies, six have shown significant increases in MBG or "euphoria" ratings.

The effects of diazepam, a benzodiazepine, on MBG scores are even more inconsistent. Of the 23 studies that have measured MBG or "euphoria", less than half of them have demonstrated increases[14]. Of the nine studies that have specifically tested diazepam in individuals with sedative or alcohol abuse histories, five have reported significant increases. Of the nine studies that tested other benzodiazepines, including oxazepam, lorazepam, trialozam, alprazolam and flurazepam, only two reported increases on ARCI euphoria scales. The fact that fewer studies found significant MBG increases with diazepam than with pentobarbital is consistent with the widely recognized differences between these drugs with respect to their risk to cause addiction.

Several non-barbiturate, non-benzodiazepine sedative drugs have been tested as well[14]. Methaqualone, a drug which has been widely abused, reliably increased MBG scores in subjects with histories of sedative abuse. PCAG scores were increased in two of the three studies. Buspirone, an anxiolytic with purportedly less sedative effects and no clinical evidence of abuse, failed consistently to increase measures of "euphoria".

The Profile of Mood States (POMS)

An instrument that orginally was validated for clinical evaluation of mood states, but also used quite frequently to assess subjective effects of drugs is the profile of mood states (POMS)[23]. The POMS consists of 72 adjectives (short form 65) commonly used to describe mood and feelings. Following drug or placebo administration, the subjects are asked to indicate on a scale from zero to four whether or not that adjective accurately describes how they feel. From these answers, levels can be assessed of anxiety, anger, depression, fatigue, vigor, confusion, friendliness, arousal, elation, and positive mood.

Whereas the ARCI is frequently used in drug dependent populations, the POMS seems especially sensitive in normal non-drug or infrequent using subjects. The greater sensitivity may be related to the fact that it was developed and validated with normal (college students) and symptomatic (e.g., anxious) volunteers rather than with drug abusers.

Using the POMS, amphetamine and cocaine increase vigor, elation, positive mood, arousal, and friendliness while decreasing confusion and depression and increasing Anxiety only following higher doses[23]. Phenmetrazine and methylphenidate (given in large doses), can produce a similar profile of subjective effects. The other stimulants mentioned in Table 7.3 predominantly increased in normal volunteers arousal and anxiety (mazindol, dietylproprion, caffeine), or had only dysphoric effects at higher doses (fenfluramine). However, in clinically depressed patients fenfluramine increased vigor and decreased fatigue, anger, confusion, and depression. This example indicates the importance of subject population in determining the subjective effects of a drug. In non-abusing volunteers, benzodiazepines generally fail to produce positive mood effects. It has been suggested that the insensitivity of the POMS in drug abusing subjects might limit its usefulness as predictor of abuse liability[14].

Visual Analog Scales (VAS)

Other tests, such as Visual Analog Scales (VAS) are also used in the assessment of drug-induced momentary changes in e.g., affect[22]. In a visual analog scale, a 100 mm line is presented with at the ends showing either opposing adjectives (e.g., happy-sad) or labels "not at all" and "an awful lot". In the latter case, the adjective which is being rated is centrally located above or below the analog line. Subjects are instructed to rate how they feel along that dimension by making a mark anywhere along the line. Although various VAS's are used, items should include "stimulated", "anxious", "liking", "sedated", "high", "down", and "hungry"[23].

Estimates of "drug liking" are usually assessed using a VAS because ratings of "liking a drug" have high face validity (as a measurement of "euphoria") for being predictive of abuse liability.

Figure 7.4 illustrates the mean "liking" scores for diverse drugs, measured by the Single Dose Questionnaire 5-point Liking Scale in subjects that have previously used or abused psychoactive substances[40]. As shown in the figure, drugs known to produce widespread use, such as morphine, amphetamine, and pentobarbital, produced dose-related increases in "liking" scores. Other substances not known for abuse (e.g., zomepirac, chlorpromazine, placebo) do not significantly elevate "liking"-scale scores.

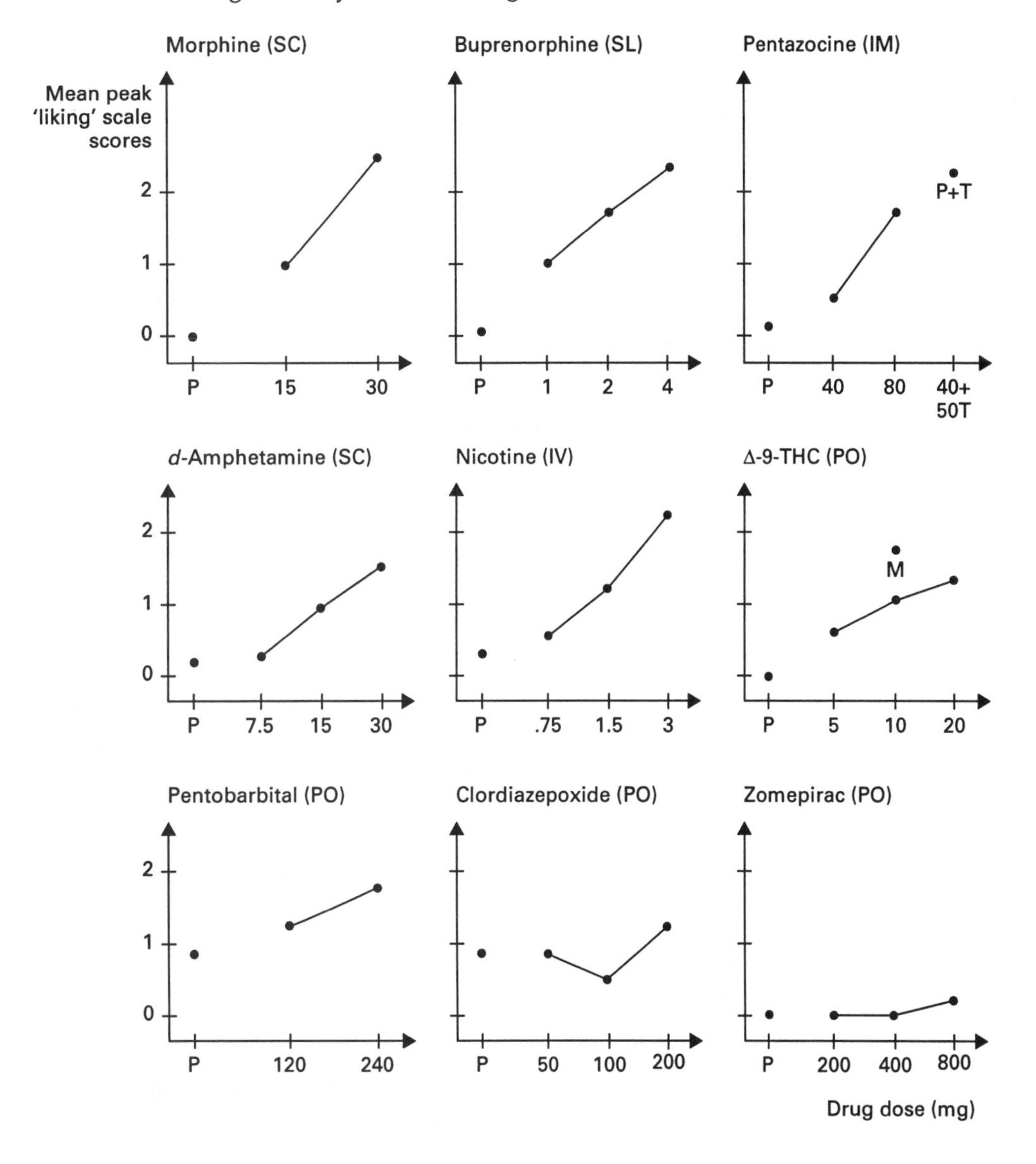

FIGURE 7.4

Mean scores on the "liking" scale of the Single Dose Questionnaire from subjects at an Addiction Research Center

P+T = combination with tripelennamine, a street drug called 'T's and Blues'; M = after smoking a marihuana cigarette. (Based on original data from Jasinski, D.R. et al., *TIPS*. May, 1984, pp 196–200.)

Diazepam and other benzodiazepines have also been found to increase ratings of liking in almost every study using subjects with histories of drug abuse. Thus, "liking" rates seem to be more sensitive indicators for benzodiazepines than MBG scores. In contrast, three studies with normal, non-drug abusing volunteers reported no increases in liking after pentobarbital, nor after benzodiazepines[14].

Different "liking"-scales have been found to lead to different outcomes and in addition measured rates can vary dependent on time points of measuring, i.e., shortly after drug administration versus retrospective ratings. It has been advised therefore to use more than one scale of subjective effects, and to measure at multiple time points.

Estimates of value

Estimates of value are useful within a single study, or group of subjects, as an index of the relationship between doses tested in the laboratory and those used on the street. Questions asked are e.g., "How much would you pay for the dose you received today?", or "What is the street value of the dose you received today?".

The variability in estimates of value is huge and it is difficult, if not impossible, to limit individual differences in this regard. In addition, estimates of value are tied to the current street values, popularity, quality, and availability of drugs which vary from location to location. Thus, in contrast to changes in ARCI, "liking" and POMS scores, it is more difficult to compare absolute estimates of value across studies[23]. The usefulness of estimates of street value as an indicator of abuse liability may be enhanced in the future as more data are obtained with a range of drugs, doses, and subject populations[18].

2.2 DRUG IDENTIFICATION

A different approach of assessing a drug's effects is to ask which drugs give subjects similar subjective experiences and which drugs are experienced as e.g., placebo. This kind of drug identification is performed using standardized procedures and structured questionnaires for subjects with prior drug experience.

One of the earliest standardized questionnaires for drug identification is the Single Dose Questionnaire[25]. It contains four scales: the first asks whether the drug was felt, and thereby determines whether the drug is "psychoactive"; the second is a 14-item list of substances from which the subject is asked to choose which the administered compound is most similar to (includes "blank" or "other" items), thereby permitting classification; the third is a 14-item list of "sensations" (including "normal" and "high") which characterizes and quantifies symptoms; the fourth is a 5-point "liking" scale which measures euphoria[40]. Another method is to ask subjects to classify the drug as a stimulant, tranquilizer (sedative), or placebo. In both methods, accuracy of identification is dependent either upon history of drug use, or expectations, or instructions about the effects of stimulants or tranquilizers. For example, when asked to classify the dose of a drug as a placebo, tranquilizer, or stimulant, experienced subjects always labeled cocaine as a stimulant, but non-using volunteers labeling showed much more variability. By stimulant users, amphetamine is mostly recognized as a stimulant, occasionally as placebo, and rarely as a tranquilizer. Of the other stimulant drugs, phenmetrazine, diethylproprion, and caffeine were labeled stimulants. Results for mazindol (stimulant or tranquilizer), methylphenidate (stimulant or placebo), fenfluramine (placebo or tranquilizer) were not unanimous. Phenylpropanolamine was labeled as a placebo[23].

2.3 DRUG DISCRIMINATION

In the drug discrimination design subjects receive, prior to administration of a test drug, a drug discrimination training. After training subjects are asked to identify the test dose with one of the initial discrimination options. Drug discrimination studies generally consist of three phases: sampling, training/assessment, and testing. Subjects are instructed that their job is to learn to discriminate between two different drugs, Drug A and Drug B, based on the effects produced by each.

In most studies one drug is placebo while the other is a psychoactive substance. On different days during the initial sampling phase subjects are given capsules containing Drug A and capsules containing Drug B and are then instructed which they were given. This phase consists of one or two exposures to each drug. Each day during the training/assessment phase subjects are given one of the drugs without identification. Subjects identify which drug they believed they received at several time points after drug administration and they are paid if their last identification of the day is correct. This phase lasts 6–7 days. Only those subjects who reliably discriminate drug A from drug B continue in the next testing phase. Test phases consist of additional training days intermixed with test days. On test days subjects are given a different drug or a different drug dose than they have experienced under sampling and training conditions. Subjects proceed as they normally would by identifying the test dose as Drug A or Drug B throughout the day. After their last identification of the day, the subjects are instructed that it was a test day and are paid regardless of their identification. Training days are the same during this phase as in the previous phase, i.e., subjects are paid only for correct identifications[23].

The drug discrimination procedure has many advantages:

1. All the subjects are provided with similar drug histories in close proximity to testing.
2. Even subjects with limited drug use histories may participate.
3. Reinforcing correct choices with monetary gain increases the experimental control of the discrimination, limiting day-to-day variability in the subjects efforts to accurately identify the drug.
4. Subjective-effects questionnaires are completed throughout the procedure so that information on both drug discrimination and subjective effects are obtained concurrently.

Disadvantages of the procedure are that:

1. Many more experimental days are required, e.g., 25–35, than have been used in most studies of the subjective effects of drugs;
2. Not all subjects can learn the required discrimination;
3. Motivational factors of drug taking (initiation, maintenance, cessation) are not addressed.

Table 7.4[23] gives an overview of human drug discrimination studies on stimulants and diazepam. There have been no drug discrimination studies with cocaine as a test drug. Administration of high caffeine doses, medium fenfluramine doses, or medium mazindol doses resulted in responding that was not appropriate for placebo or amphetamine, while administration of high phenylpropanolamine and mazindol produced amphetamine-appropriate responding. Thus, dissimilar results have been found between identification and discrimination studies.

In addition, there are several discrepancies in subjective effects and discriminative effects. For instance, while caffeine, phenmetrazine, and mazindol produce some similar subjective effects, mazindol engenders

TABLE 7.4
Amphetamine appropriate responding following oral drug administration in subjects trained to discriminate 10 mg *d*-amphetamine from placebo
(Based on data from Foltin, R.W. and Fischman, M.W., *Drug Alcohol Depend.* 1991, 28, pp 3–48.)

Drug	*Dose (mg)*	*Amphetamine-appropriate responding (%)*
d-Amphetamine	2.5[1,2]	10-20
	5[1,2]	40-50
	10[1-5]	83-93
Phenmetrazine	25[4]	80
	50[4]	85
Mazindol	0.5[3]	40
	2.0[3]	75
Fenfluramine	20[4]	25
	40[4]	50
Phenylpropanolamine	25[3]	15
	75[3]	80
Caffeine capsules	100[5]	42
	300[5]	68
Diazepam	10[1,2]	30-40

Studies: [1]Chait et al., 1984; [2]Chait et al., 1985; [3]Chait et al., 1986a; [4]Chait et al., 1986b; [5]Chait et al., 1988.

more amphetamine-like responses than does caffeine in subjects trained to discriminate amphetamine from placebo. Phenmetrazine was consistently responded to as if the subjects had been given amphetamine.

With respect to barbiturates and benzodiazepines, one drug discrimination study is known[70]. In experienced drug users accuracy of identification of pentobarbital (100–600 mg) as a barbiturate increased in a dose-dependent manner, reaching a peak of 78% correct at the highest dose and accuracy of triazolam (0.5–3.0 mg) as a benzodiazepine increased dose-dependently, reaching a peak of 89% correct.

It seems probable that drugs that cannot be discriminated from placebo are less likely to be consumed again by a subject. Presumably those drugs that are most highly addictive drugs in subjective as well as in discriminative effects will be those drugs that have high abuse liability too.

2.4 SELF-ADMINISTRATION

The experimental design that directly measures drug-taking behavior is that of operant drug self-administration, in which determinants of human drug taking are quantitatively studied in laboratory settings. The self-administration paradigm is considered as the primary method to determine directly whether or not a drug produces reinforcing effects and thus has potential for abuse[36].

Procedures used are derived from animal drug self-administration models, which were first developed in the 1960s (see Chapter 4). In general, the similarities in drug taking behavior between humans and animals are greater than the differences. Various examples of these similarities are given in figures presented below. Of course, the human variant can yield data not obtainable with animals, such as self-reported drug effects and comparisons of population subgroups[36]. Examples of applications for the human drug self-administration paradigm are assessment of the reinforcing effects of drugs, analysis of behavioral and pharmacological mechanisms of drug self-administration, evaluating a drug's potential of abuse liability, monitoring behavioral toxicity and aversive effects of drugs[36].

Human subjects have complex personal life histories. The results of self-administration studies depend also critically upon the histories and current state of the subjects. Subjects' prior drug use, as well as demographic, medical, and psychiatric characteristics should be documented as completely as possible[14]. Depending upon the purpose of the study, certain subject populations may be more appropriate than others. Furthermore, drug reinforced responding is also under influence of the prevailing social, psychological, or experimental circumstances[36].

The strengths of the human operant self-administration paradigm are that it can be readily integrated with other methodologies, such as collecting subjective and physiological effects and effects on human performance. Furthermore, it has face-validity of modeling phenomena of natural drug-taking and it can be used to determine empirically whether or not a medication selectively can reduce drug taking, while leaving other desirable behaviors intact. Limitations are that more test sessions are required than can be conducted practically (mostly not more than 20 sessions). For some drugs or routes of administration (e.g., i.v.), a considerable amount of professional (medical, technical) staffing is required. For safety reasons, intakes have to be limited sometimes. Finally, ethical arguments are involved when exposing individuals to addictive compounds[36].

Models

Models used for human self-administration are; (1) free-access, (2) choice procedures, and (3) operant conditioning techniques. Common elements of these models are experimental control of drug taking conditions, the control of associated contexts, and the quantification of behavioral parameters. Table 7.5 lists the common elements in more detail.

TABLE 7.5

General elements of human self-administration models
(Based on data from Henningfield, J.E. et al., *Pharmacol Biochem Behav.* 1983, 19, pp 887–890.)

General elements
Optimal control of experimental environment
Control of contingencies related to drug self-administration
Objective measurement of drug self-administration
Free frequency of self-administration
Control of drug dose
Recording of temporal response patterns and drug delivery

2.4.1 *Free-access procedures*

In a free-access procedure, subjects are given essentially unlimited access to standard doses per session (day) and are allowed to take the drug, assuming it is medically safe to do so, according to their own patterns[23]. If active drug maintains responding above levels observed during placebo sessions then that drug has positively reinforcing properties. These procedures are most often accomplished with subjects residing on a Clinical Research Center. Free-access procedures are frequently applied for observing use patterns, such as "binges" or "runs" and total drug intakes when no limitations are present, e.g., financial restrictions.

One of the earliest laboratory studies of human drug self-administration was by Wikler[88], who provided a volunteer with a history of morphine dependence with continuous access to morphine and other drugs. This subject consistently chose to inject morphine, and self-administration was marked by a gradually escalating daily intake during the first 3 months, followed by a fairly stable rate of morphine intake for the next month.

Figure 7.5 illustrates how similar this pattern of self-administration behavior was to that of rhesus monkeys[30].

Experienced cocaine users will ingest available cocaine doses in a very regular tempo. In one study, subjects were given access to either 96 mg or 4 mg (a placebo dose) for intranasal cocaine self-administration. When 96 mg were available subjects requested the drug as soon as possible and inhaled about five doses. For safety reasons, sessions were always terminated by the researchers. Sessions with 4 mg were always terminated by the subjects, suggesting little reinforcing effects of this dose[23].

Different drugs of abuse, i.e., morphine and alcohol, produce different characteristic use patterns under conditions of unrestricted drug availability. This is illustrated by Figure 7.5. Whereas morphine intake by a volunteer drug abuser gradually increased over time with only small daily fluctuations, alcohol drinking by volunteer alcoholics (as well as by monkeys) showed wide fluctuations over time and was interspersed by self-inflicted withdrawal periods (dots).

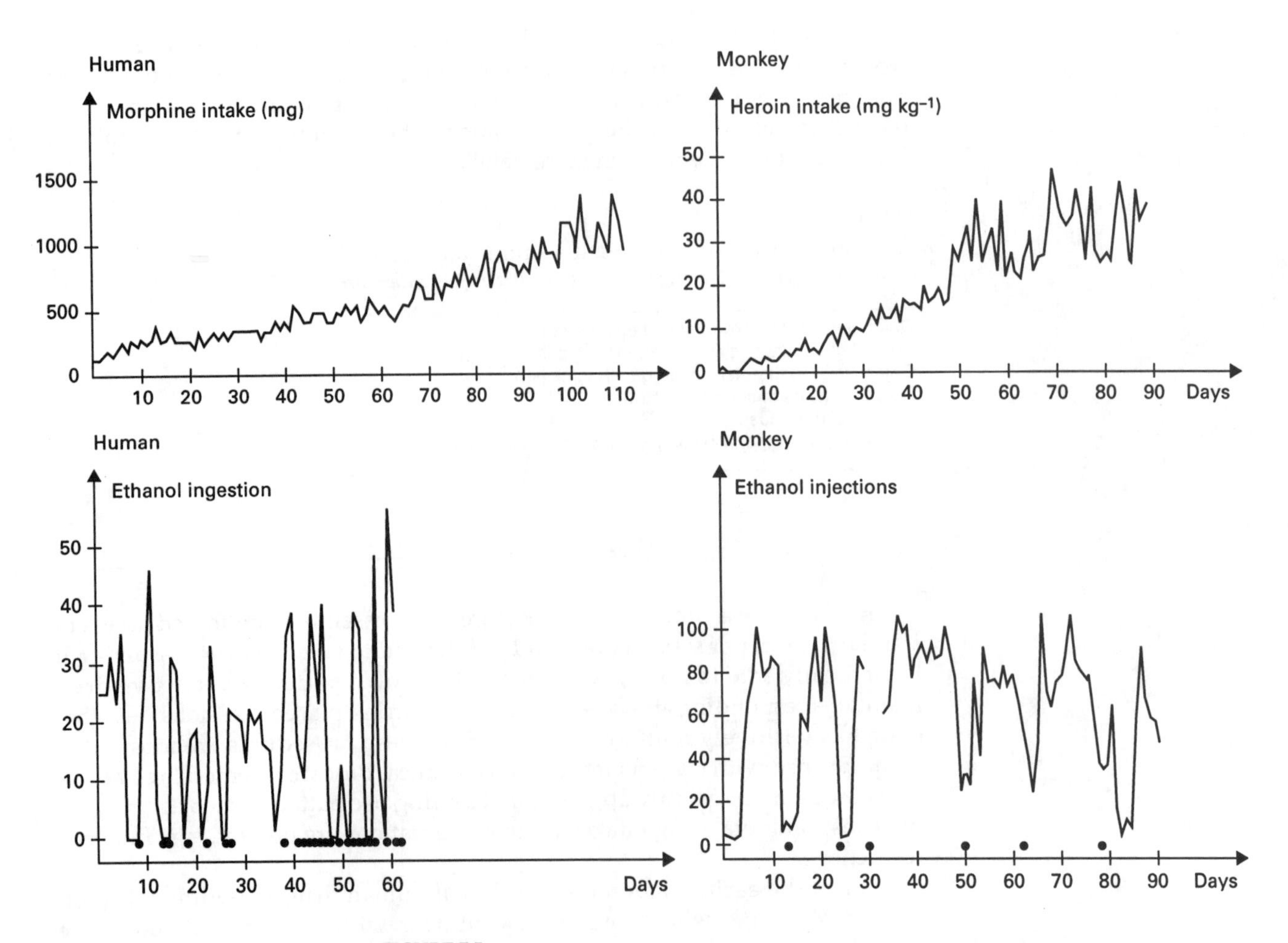

FIGURE 7.5

Patterns of opioid and ethanol intake in humans and monkeys

(Based on data from Griffiths, R.R., In: Mello N.K. and Mendelson, J.H. (chairmen), "Behavioral pharmacology of substance abuse: recent advances." *Psychopharmacol. Bull.*, 1980, **16:** 1, pp 45–47.)

There have been no free-access studies of intravenous amphetamine self-administration and only one study that has provided relevant data to free-access oral amphetamine[23]. In that study, overweight female hospital employees participated in a weight-loss program with *d*-amphetamine (5 mg per capsule) as pharmacological adjunct. They could self-regulate their intake between zero and six capsules per day. Under all conditions, subjects consumed about 1.5 capsule per day. This small amount of amphetamine ingestion in patients seeking to lose weight suggests that amphetamine used for medical reasons has only moderate potential to be addictive.

2.4.2 *Choice designs*

A self-administration design that readily lends itself to research with human subjects is the discrete-trial choice or preference procedure. In such procedures, when drugs are available, subjects make a discrete choice between, typically, either drug "A" or drug "B". Drug "A" and drug "B" may be active drug and placebo, two different drugs, or two doses of the same drug[23,32]. Choice paradigms provide information about the relative reinforcing effects of drugs. If active drug is chosen about equally as often as placebo, the drug has minimal potential to be addictive. If chosen above placebo levels then it has potential to be addictive. If below placebo levels the drug is considered to have aversive effects.

Intravenous cocaine is the most reliably chosen drug of stimulants. Every subject tested chose cocaine over placebo, and most exclusively chose higher over lower doses of cocaine[23]. Oral amphetamine, phenmetrazine, and diethylproprion are chosen over placebo about 60–80% of the time with approximately 40–60% of subjects exclusively choosing active drug. Noteworthy is that diethylproprion is rarely abused in natural ecology. Although caffeine is readily consumed in caffeinated beverages, an equally robust choice of caffeine capsules has not been demonstrated.

In all studies with subjects having histories of sedative abuse[14], pentobarbital is preferred over placebo. When allowed to take multiple doses per day (range 30–200 mg), subjects with histories of sedative abuse preferred intermediate doses of pentobarbital over low and high doses, whereas for diazepam (range 2–40 mg) no consistent dose preferences have been demonstrated. In a study in which subjects with known sedative abuse were allowed one dose per day, higher doses of pentobarbital were chosen over lower ones (range 200–900 mg) and pentobarbital was preferred over diazepam (50–400 mg)[14].

An interesting finding is that a high dose of diazepam (160 mg) was preferred over a comparable effective dose of oxazepam (480 mg), produced significantly higher "liking" scores and MBG scores, and was more often identified as a sedative. However, the doses of both compounds were equivalent in therapeutic respect and produced comparable psychomotor impairment.

In contrast to the above results, it has been found that neither barbiturates nor benzodiazepines were self-administered (neither in choice nor operant design) by normal volunteers without histories of abuse[14]. Most volunteers, and even populations thought to be at higher than average risk (e.g., with high anxiety levels), found the effects of sedatives aversive and preferred a placebo to drug. The only exception is a study that used a group of non-drug abusing individuals who were moderate alcohol consumers (about 10 drinks per week). These subjects consistently chose diazepam over placebo and ingested an average of 25 mg diazepam per choice session.

The general contrast between results using sedative users versus non-users imply that studies with normal volunteers do not provide a sensitive indicator of relative abuse liabilities of novel sedative compounds. Thus, abuse liability should first be assessed in the highest risk population (i.e., drug abusers)[14].

2.4.3 *Operant designs*

Whereas in most choice procedures the experimenter controls the total number of doses and timing of administration, in operant studies this is determined by the subjects. Operant studies provide measures of frequency, use patterns, and quantity of drug ingestion. Moreover, variables that might affect drug taking behavior can be systematically studied[36]. It defines the stimulus conditions in which the behavior might occur, the operant behavior itself, and describes the consequences of emitting the behavior. This model thus permits translation of non-laboratory phenomena into rudimentary elements that can be experimentally manipulated and measured.

Independent variables commonly studied include behavioral manipulations such as the conditions of drug availability (i.e., reinforcement schedule), the subject population, and the consequences of drug self-administration. Pharmacological manipulations involve presence or absence of drug, type of drug, drug dosage, duration of drug deprivation, and pretreatment with the drug under study or another drug[36].

Dependent variables include rate and pattern of responding in order to obtain drug delivery (drug seeking behavior), drug intake itself, and persistence of responding when the drug is not available (extinction). Other behavioral drug effects can be assessed, such as responding maintained by other reinforcers (e.g., food, companions, money). Additionally, subjective-report data, such as liking of the drug, urge to use drug, pharmacological identification of the drug, and antecedent or consequent symptoms, can be measured. Table 7.6 gives diverse examples of the various applications of operant designs[36].

Use Patterns

Examples of self-administered drug intake across daily sessions are shown in Figure 7.6[31] for pentobarbital, chlorpromazine, and placebo, using double-blind conditions. Subjects with histories of sedative abuse had to ride on an exercise bicycle for 15 minutes to earn one dose of drug. It is evident that chlorpromazine, nor placebo had positive reinforcing effects, in contrast to pentobarbital of which intake was reliably maintained.

In a study with intravenous nicotine self-injection, habitual smokers had to press a lever ten times (FR10) to obtain an injection with 1.5 mg of nicotine or saline (see Figure 7.7). The number of injections increased over the consecutive days (7 days), confirming the positive reinforcing effects of nicotine without interference of non-drug variables associated with smoking. The figure also shows a typical extinction pattern in responding when only saline became available during days 7–14, characterized by an initial increase in responding followed by a sharp decline.

In a study in which alcoholics could work 24 hours per day (pressing a response key) to obtain alcohol or money, the subjects preferred alcohol to monetary reinforcers and worked for several days until they had earned enough to drink for a prolonged period without working. During working periods subjects (almost) abstained completely, even though they had sufficient earnings and were experiencing partial withdrawal symptoms. As a result daily alcohol intakes fluctuated widely.

TABLE 7.6

Studies illustrating applications of human drug self-administration paradigm

(Based on data from Henningfield, J.E. et al., *Br. J. Addict.* 1991, **86:** pp 1571–1577.)

Authors	*Study design*	*Conclusions*
Mendelson & Mello, 1966	4 subjects with histories of alcoholism could work 24 hrs/day on an operant task (key pressing) for money or alcohol. The number of required responses per reinforcement from 60 to 360.	All subjects achieved high blood levels (150-300 mg/100 ml). They tended to maintain stable blood levels although their daily alcohol intake fluctuated widely.
Griffiths et al., 1974	6 subjects with histories of alcoholism were given daily access to 17 alcoholic drinks (1 oz 95-proof ethanol in orange juice) with a minimum of 40 min between drinks. During social time-out phase, subjects were not allowed to socialize for the 40 min period following each drink.	Contingent social time-out suppressed drinking.
Bigelow et al., 1976	5 subjects with histories of sedative drug abuse could self-administer up to 20 oral doses per day of 10 mg diazepam or 30 mg sodium pentobarbital. Each dose was purchased with 1 to 10 tokens earned by exercising on a stationary exercise bicycle.	Drug intake decreased as a function of response requirement for purchasing the drug.
Babor et al., 1978	34 subjects with histories of casual or heavy drinking could purchase alcohol under a single-price condition (50¢/drink) or with a 25¢ price reduction between 2 and 5 pm ('happy hour').	Price reduction increased alcohol consumption and reinstatement of price returned drinking to previous levels.
Mello et al., 1981	12 heroin addicts could work on an operant task for money or heroin (up to 40 mg/day) for 10 days. Subjects were maintained on placebo or naltrexone, 50 mg/day.	Naltrexone-maintained subjects took 2 to 7.5% of total available heroin, whereas placebo-maintained subjects took 58 to 100% of available heroin.
Mello et al., 1982	10 subjects with histories of heroin abuse could acquire heroin (21 to 40.5 mg/day, i.v.) for 10 days by working on an operant task. Subjects were maintained on buprenorphine (4 or 8 mg/day s.c.) or on placebo,	Buprenorphine-maintained subjects took 2 to 31% of total available heroin, whereas placebo-maintained subjects took 93 to 100% of available heroin.
Nemeth-Coslett & Henningfield, 1986	5 smokers were free to smoke their usual brand of cigarettes using portable puff monitors during 12 hr sessions. Either placebo gum or 2 or 4 mg nicotine gum was administered every 2 hours.	Nicotine 2 and 4 mg reduced the total number of puffs per day; 4 mg dose decreased the total number of cigarettes smoked per day compared to placebo.
Griffiths et al., 1979	19 subjects with histories of sedative abuse could acquire up to 10 doses per day of either pentobarbital (30-90 mg), diazepam (10-20 mg), chlorpromazine (25-50 mg) or placebo by exercising on a bicycle.	Diazepam and pentobarbital maintained self-administration with the higher dose associated with higher average levels of self-administration than the lower dose. Chlorpromazine and placebo did not maintain self-administration.
Henningfield et al., 1983	6 smokers could work on an operant task to receive i.v. injection of saline or nicotine 0.75-3 mg for 3 hr.	Nicotine injections were taken in orderly patterns related to unit dose.
Benowitz & Jacob, 1984	11 smokers were required to smoke ad libitum for 1 day their usual brand or commercial high-yield or low-yield cigarettes. The yields were taken from Federal Trade Commission (FTC) smoking machine tests.	Subjects obtained 60% more nicotine per low-yield cigarettes than predicted by FTC yield. Intake from high-yield cigarettes was 76% of that predicted.
Fischman & Schuster, 1982	4 cocaine users could work on an operant task to receive an i.v. injection of either cocaine (16-32 mg) or saline. A maximum of 10 injections per day (one every 15 min) could be requested for a 2-week period.	Cocaine was consistently chosen over saline. The initial injections of cocaine increased heart rate and self-report of stimulant effects, with subsequent repeated injections having smaller effects.

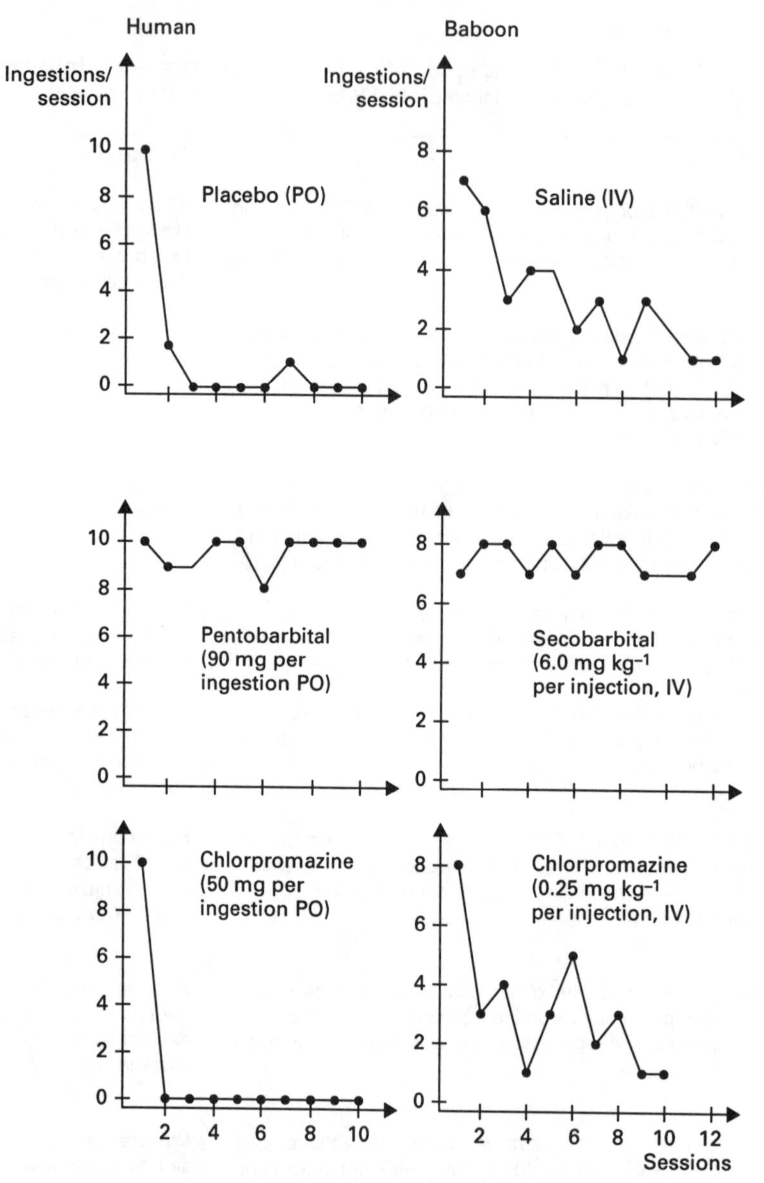

FIGURE 7.6

Effect of type of drug on drug self-administration in humans and monkeys
(Based on data from Griffiths, R.R. and Bigelow, G.E., In: Fishman, J. (Ed.), *The Bases of Addiction*. Dahlem Konferenzen, Berlin, 1978, pp 157–174.)

In contrast nicotine , being a short-acting stimulant, produced a regular spaced pattern, as is illustrated in Figure 7.8[35]. Subjects were permitted three puffs (see vertical hatch mark) from a cigarette each time they rode an exercise bicycle for a 1-minute period for 4–5 hours per day.

2.4.3.1 Experimental manipulation

Non-drug reinforcers

Self-administration behavior can be experimentally changed by manipulation of immediate contingent variables, e.g., non-drug reinforcers. Alcoholics for instance could get 17 alcoholic drinks per day, but they had to wait minimal 40 minutes between each drink. When during this 40 minute period subjects were not allowed to socialize or to participate in recreational activities, or both, alcohol ingestion decreased accordingly[31].

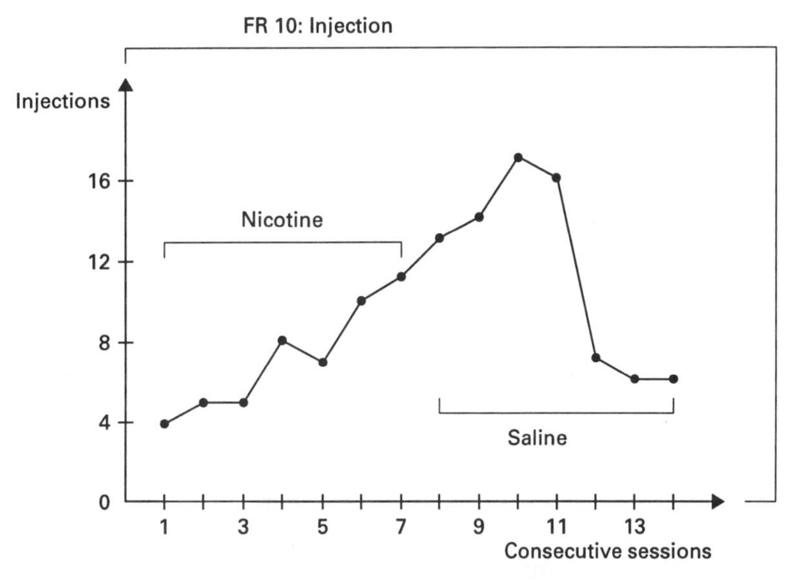

FIGURE 7.7

Intravenous nicotine self-injection by a cigarette smoker
(Based on data from Henningfield, J.E. and Goldberg, S.R., *Pharmacol Biochem Behav.* 1983, 19, pp 1021–1026.)

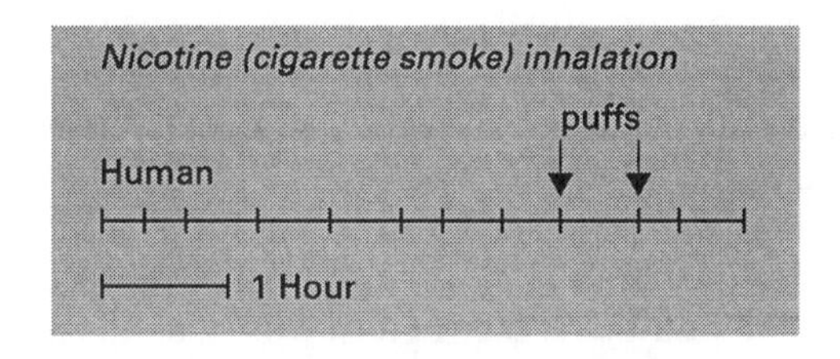

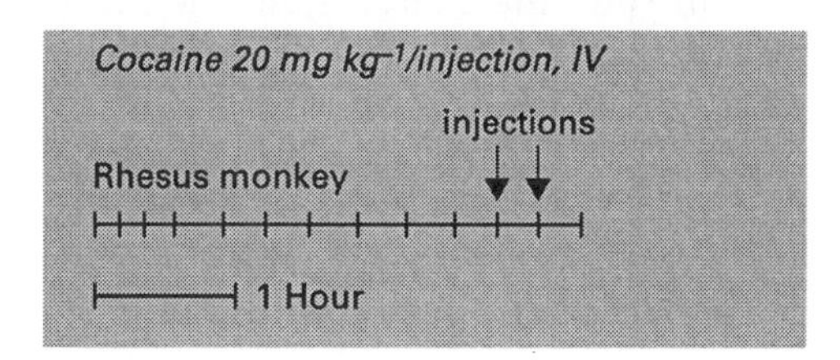

FIGURE 7.8

Patterns of nicotine and cocaine self-administration in humans and animals under conditions of relatively unrestricted drug-availability
(Based on data from Henningfield, J.E. and Goldberg, S.R., *Pharmacol Biochem Behav.* 1983, 19, pp 1021–1026.)

Cost-benefit

Another way of manipulating self-administration behavior is by varying the effort needed to obtain the drug. In one study, subjects could earn up to 20 oral doses per day of diazepam or pentobarbital by riding an exercise bicycle, earning 1 token per 15 minutes. Required number of tokens were varied from day to day. When the response requirement was low, subjects ingested every possible dose, but as response requirement increased, the amount of drug decreased[31]. Similar findings have been found with price reductions and increased drinking rates[3].

For cocaine however it has been reported that cocaine users will be hardly influenced by increasing response cost (button presses) when high doses (> 8 mg) are available[23].

Pretreatment

The utility of operant self-administration to determine the efficacy of pharmacological versus placebo pretreatment has been demonstrated with heroin users[36]. Subjects receiving naltrexone (an opioid antagonist) or buprenorphine (a mixed opioid agonist/antagonist) reduced the daily self-administration of heroin dramatically compared to placebo[53,54]. Another example is when cigarette smokers were given nicotine polacrilex (gum), the number of cigarettes smoked per day was decreased compared to placebo[64].

For intravenous cocaine-taking behavior, the effect of desipramine was evaluated[23]. Cocaine intakes were recorded before and during desipramine treatment. Desipramine had many significant influences on the subjective effects of cocaine including decreasing "I want cocaine" scores, decreasing the magnitude of many stimulant scores on POMS and ARCI and increasing the magnitude of negative subjective scores, e.g., anger and confusion. However, there was no change in cocaine intake during desipramine maintenance compared to baseline!

Dose-reponse

Orderly relationships have been found to exist between dose and operant behavior. Figure 7.9 illustrates the general finding for addictive drugs that higher doses result in more *total drug* (here: alcohol) *intake*[31]. With respect to *response rates*, increasing doses produce an inverted U-form dose-response function, so that greatest number of operant responses are made for intermediate doses. This dose-response behavior is demonstrated for intravenous nicotine self-injection behavior in which subjects had to make 10 responses (FR10) for one injection with nicotine. As shown in Figure 7.10, the intermediate dose of 0.75 mg resulted in the highest response rate.

The reinforcing effects of different doses of pentobarbital have also been assessed using a progressive ratio procedure[14] in which subjects with histories of sedative abuse were required to either button press or ride an exercise bicycle to obtain placebo or pentobarbital (200–600 mg). Each day the response requirement was increased and the point at which subjects chose not to work for a single day dose (i.e., break point) was recorded. Higher doses yielded higher break points. At the highest doses, subjects pressed a button as many as 90,000 times, or rode the exercise bicycle for as long as 6 hours each day to obtain a dose of drug!

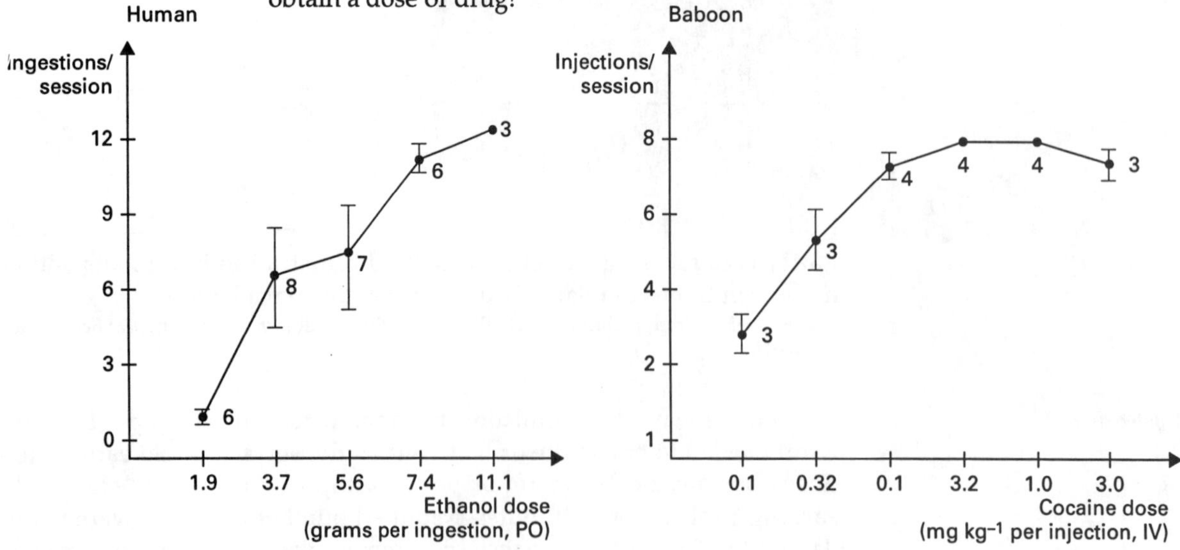

FIGURE 7.9

Dose-effect relationship in self-administration of ethanol in humans and cocaine in baboons

Numbers indicate total number of observations at each dose. (Based on data from Griffiths, R.R. and Bigelow, G.E., In: Fishman, J. (Ed.), *The Bases of Addiction*. Dahlem Konferenzen, Berlin, 1978, pp 157–174.)

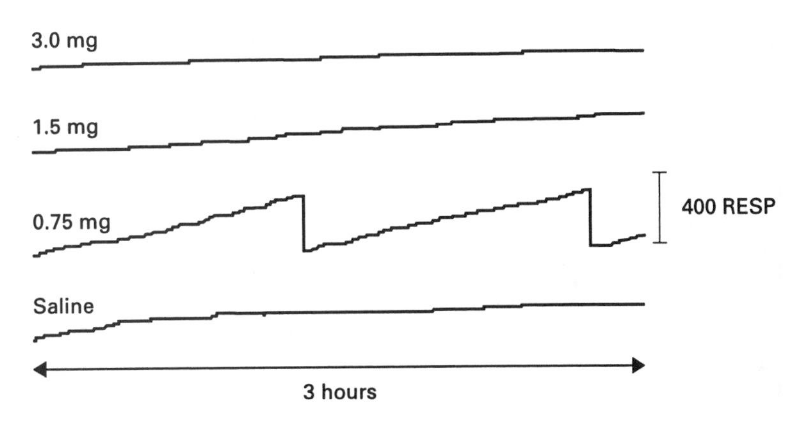

FIGURE 7.10

Dose-effect relationship of iv nicotine self-administration in a human subject

Figures represent cumulative records from one subject. Lever presses are indicated by vertical increments. (Based on data from Henningfield, J.E. et al., *Pharmacol Biochem Behav.* 1983, **19:** pp 887–890.

2.4.4 *Multimeasure studies*

An interesting approach is to study the relationship between subjective effects after drug administration in combination with other drug-related parameters. For example, Figure 7.11[40] shows the mean different time action curves of 12 subjects for orally (20 mg/kg) and subcutaneously (20 mg/kg) administered methadone and for subcutaneously (20 mg/kg) administered morphine with respect to subjective effects (MBG scale and Single Dose Questionnaire scale of "liking"), clinical observation of symptoms, and physiological reaction (pupillary diameter). It demonstrates that during the first hours after administration pupillary responses were quite parallel for the various administrations, whereas the physiological reactions did not correspond neatly with the other measures. Furthermore, it shows that duration of action differed for the various drug administrations. "Symptoms" and "MBG" corresponded reasonably with each other. It is noteworthy that subjective effects measured with "Liking" did not correspond reliably with those measured with MBG-scale: e.g., morphine was much more "liked" than that it produced "euphoria" in MBG.

In a multimeasure study with alcohol, nonalcoholic volunteers experienced both "euphoria" as well as mildly pleasant states. These pleasant states were found during the ascending portion of the blood alcohol curve and were associated with increases in the production of plasma ACTH, as well as the increase in brain electrical alpha activity measured by EEG. Neither these subjective nor physical effects were associated with the descending portion of the curve.

It has been found that after experimental aversive stress induction, both subjective ("anxiety" ratings) as stress responses (heart rate) can be reduced by alcohol. In another study, subjective stress experience and heart rate were both reduced in subjects with increased alcoholism risk (due to an alcoholic parent or a risk-related personality pattern) compared to no-risk controls[16]. Alcohol is indeed believed to be also reinforcing because of the anxiety-reducing properties. Sober alcoholics report tension and anxiety reduction as a desirable consequence of their drinking although they often report feelings of depression and aggressiveness after drinking[61]. An explanation for this discrepancy might be that initially (low doses of) drinking produces pleasant states, but continued drinking (increasing the dose) which alcoholics often do, produces dysphoric states[55].

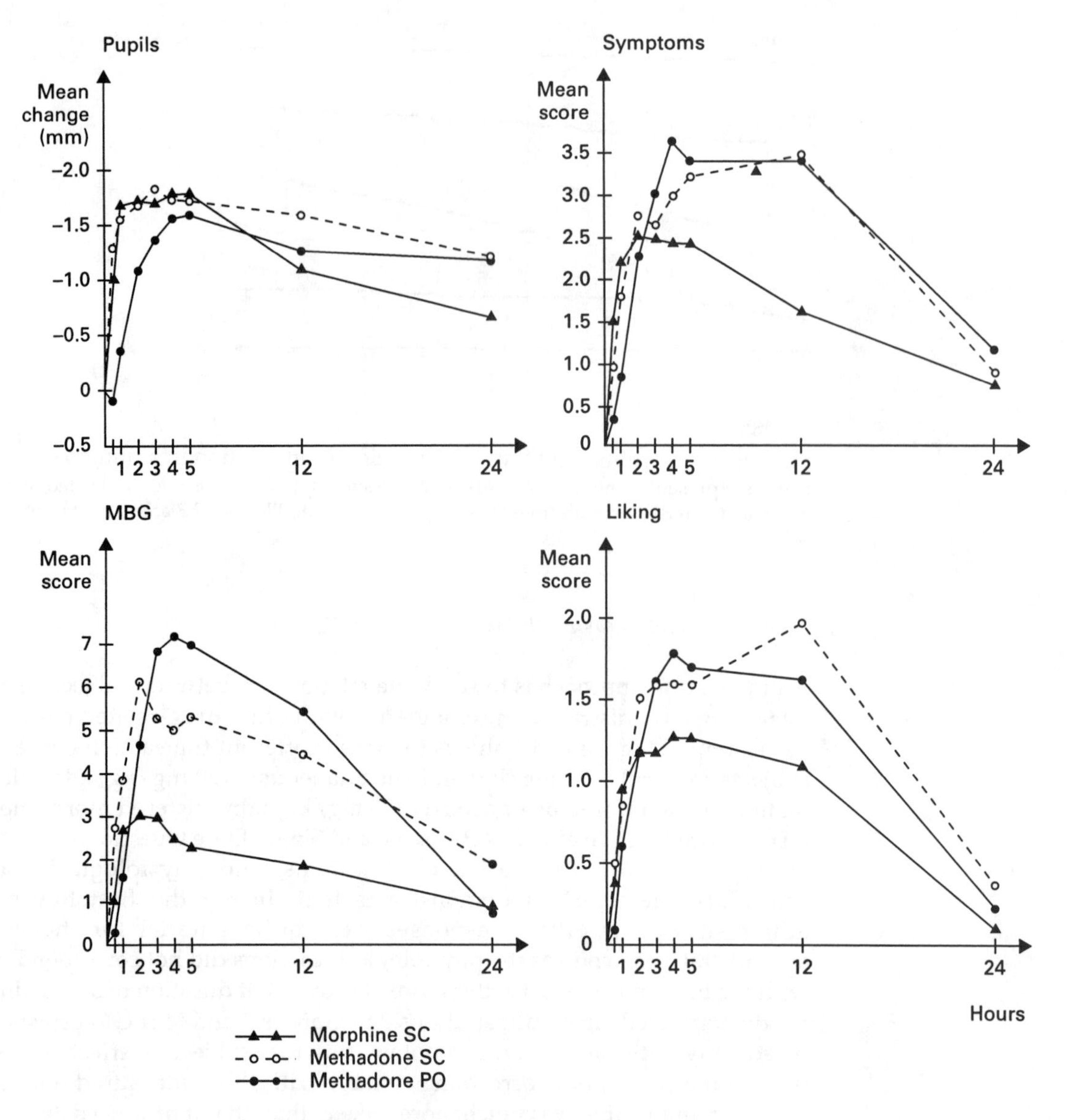

FIGURE 7.11

Time action curves of subcutaneous morphine sulfate, and oral and subcutaneous methadone hydrochloride

(Based on data from Jasinski, D.R. et al., *TIPS*. May, 1984, pp 196–200.)

Subjective measures have also been correlated with operant self-administration designs. Increases in the MBG scale, assumed to be related to euphoria, are observed across a wide range of drugs of abuse and several researchers have argued that production of euphoria is the common link across drug classes such that euphoria is the subjective state responsible for illicit drug self-administration[40]. Increases in the MBG scale indeed frequently correspond with the extent of drug self-administration. Nevertheless, there are sufficient cases of dissociations between MBG indications of euphoria and the extent of drug self-administration to doubt the predictive value of subjective effects for abuse liability[23]. Apparently, the role of situational variables in determining self-administration, such as social, setting, historical, and personality variables must be taken into account as well.

Morphine self-administration, for example, can be maintained at doses that do not produce reliable subjective effects and minimal miosis, whereas placebo does not maintain responding[36]. Similar results were found in subjects with histories of sedative abuse who preferred a dose of pentobarbital in a choice design without elevated liking of the drug[14]. These findings suggest that assessment of reinforcing effects might be a more sensitive way than measurement of subjective effects[36].

3 Clinical research approaches

Clinical studies can be distinguished from epidemiological and experimental laboratory studies in that clinical research focuses specifically on (groups) of individuals that are under medical and/or therapeutical supervision with respect to drug dependence or to other forms of addictive behaviors, such as excessive eating or gambling.

The aim of clinical research is to identify clinical characteristics and syndromes of one or a group of (drug-)dependent individual(s) and to develop and evaluate various treatments. A global distinction can be made between studies on etiology, on changes in mental and physical functions induced by drug use (pathology), on the course of the addiction (prognosis), and on treatment outcome.

Methodologies include various levels of experimental control, with single case descriptions at the lowest level, followed by comparisons with matched control subjects and, at the highest experimental level, randomized, placebo-controlled, double blind designs.

Dependent persons frequently differ in biological susceptibility, late versus early onset, psychiatric comorbidity, differences in personality, life experiences, family background, and in social-economic status. These multiple factors are supposed to contribute to the etiology of addiction as well as to influence prognosis and treatment results.

Identifying interindividual differences with respect to the manifestation and maintenance of dependence can lead to better insight in individual etiology and in developing more effective patient-matched treatments.

For example the identification of two clinical subtypes of alcoholics revealed quite distinct etiological loading in terms of genetic and environmental influences[12]. In Type 2 alcoholics, with a strong inheritance factor, drinking patterns are characterized by persistent consumption accompanied by aggressive behavior and involvement with the police. Type 1 alcoholics, with weak genetic factor, can be characterized by guilt and periodically uncontrolled drinking. Consequently, training social coping skills in the treatment program has been found more successful for Type 2 sociopathic alcoholics, whereas treatment aimed at exploring personal coping styles appeared more effective for alcoholics with low self-esteem and high feelings of guilt about their dependence[16].

3.1 ETIOLOGY

To date, the etiology or etiologies of addiction appear too complex to be easily understood. No single category of variables can explain the wide variety of individual developments of addiction. Although, for instance, children from alcoholic parents have a significant higher chance to develop alcoholism (one third of alcoholics have at least one alcoholic parent), less than half of high risk children develop drinking problems and only a portion of these develop alcohol dependence[16].

Thus, even if facilitative genes are inherited, they may not be expressed in the absence of provocative environmental factors. Of importance can be, for instance, whether the parental alcoholism also involves a general impairment in parental role, thus impairing the quality of family environment. Or, how the relationship between parental alcoholism and offspring alcohol involvement is influenced by alcohol expectancies of the child, which can be present even before direct pharmacological experience.

Hence, the conclusion can be drawn that the *interaction of genetic factors with psychological and social factors* is of fundamental importance in understanding drug dependence. Zucker and Gomberg[90] stated that the development of alcoholism is the result of interaction between four distinct classes of factors: physiological factors and other individual influences, sociocultural and environmental factors, familial factors, and peer influences. These researchers proposed an integrated theory on etiology of alcohol dependence that involves biological, psychological and social processes operating within a developmental framework. Given a genetic causal basis, alcoholism would result from a continuous process in which childhood behavior problems and childhood environment play significant roles. Childhood factors that have been identified through research include antisocial behavior, achievement deficits, heightened activity level, and interpersonal difficulties of the child, as well as marital friction, parental deviance, and deficient rearing practices in the family environment.

TABLE 7.7

Characteristic features distinguishing two types of alcoholism

(From Cloninger, C.R., *Science.* 1987, **236:** pp 410–416. Copyright 1987 by the AAAS.)

Characteristic features	*Type of alcoholism*	
	Type 1	*Type 2*
Alcohol-related problems		
Usual age of onset (years)	After 25	Before 25
Spontaneous alcohol-seeking (inability to abstain)	Infrequent	Frequent
Fighting and arrests when drinking	Infrequent	Frequent
Psychological dependence (loss of control)	Frequent	Infrequent
Guilt and fear about alcohol dependence	Frequent	Infrequent
Personality traits		
Novelty seeking	Low	High
Harm avoidance	High	Low
Reward dependence	High	Low

Another frequently cited, integrated theory comes from Cloninger[12]. Based on adoption data, he distinguished two clinical alcoholic subtypes, that differ in genetic loading, age of onset, drinking behavior, and personality traits. Type 1 was found in 13% of the adopted male alcoholics, Type 2 in 4%. Table 7.7 indicates the main characteristics of the types. Type 1 involves strong environmental influences whereas Type 2 (male-limited) is highly heritable with limited environmental involvement and is associated with both parental alcoholism and parental antisocial behavior. Key symptoms are psychological dependence (i.e., loss of control) and guilt about drinking in Type 1, and spontaneous alcohol-seeking (i.e., inability to achieve complete abstinence), drinking-involved aggressivity and law involvement in Type 2. Furthermore, Cloninger[12] proposed three personality traits — novelty seeking, harm avoidance, and reward dependence — that together describe subtype personalities. Type 1 involves behaviors consistent with a "passive dependent" personality, including inflexibility and contemplative behavior (low novelty seeking), careful and inhibited behavior (high harm avoidance), and concern about the feelings and thoughts of others (high reward dependence). Type 2 involves behaviors consistent with antisocial personality, including impulsivity and excitability (high novelty seeking), brash and uninhibited behavior (low harm avoidance), and distant social relations (low reward dependence). Low sensation-seeking scores have been found as well for opioid-dependent subjects, whereas high sensation-seeking individuals were found to use cocaine[24].

The predictive value of the characteristics of novelty-seeking, harm avoidance and reward dependence, identified in clinical patients, have been investigated in children 11 years old. This group was followed and was investigated again at age 27 for alcohol-related problems[11]. The most significant finding was that a high (i.e., beyond the normal range) score on novelty seeking and a low score on harm avoidance during childhood predicted alcoholism later on.

Psychopathology in children

Behavioral pathology in children, such as in children with Conduct Disorder or Attention-Deficit Hyperactivity Disorder (ADHD) (DSM-IV) has been found to increase the risk for later drug dependence[72]. Follow-up studies have shown that e.g., specific ADHD symptoms decline when children grow up, but as adults still have more frequent behavioral problems, such as antisocial behavior and drug dependence, compared to controls[46,86]. These findings are summarized in Table 7.8.

TABLE 7.8

Percentages of ADHD, antisocial personality disorder and drug-dependency
(Based on data from Klein, R.G. and Mannuzza, S., *J. Am. Acad. Child. Adolescent Psychiatry.* 1991, **30:** pp 383–387.)

	18 year			*26 year*		
	ADHD	anti-soc.	drug dep.	ADHD	anti-soc.	drug dep.
ADHD group	31%	27%	16%	8%	18%	16%
Controls	3%	8%	3%	1%	2%	4%

Other mental disturbances, such as *mood and affect disorders*, have been suggested as causal factors by increasing the vulnerability of a person for subsequent drug addiction[24]. A strong interaction has been mentioned between dominant dysphoric feelings and drug preference. The so-called self-medication hypothesis postulates that individuals self-select drugs on the basis of personality and ego impairments. It was found in opioid-addicted persons that they had difficulty in maintaining intimate relationships, partly related to their social withdrawal and avoidance of sexual and aggressive conflicts. In cocaine addicts who sought clinical treatment, 55% had a current psychiatric diagnosis. Anxiety disorder, ADHD- and anti-social personality disorder were found to be frequently preceding cocaine dependence. In cocaine addicts, dysthymic patients used cocaine to avoid depressive affect, whereas cyclothymic patients used cocaine to heighten or maintain elevated mood. In patients with ADHD disorder, self-medication with cocaine, paradoxically, sedated, decreased stimulation, and improved concentration.

Marijuana is sometimes used by anxious individuals due to its sedative effects.

Environmental circumstances

Acute environmental circumstances can become a causal factor in opiate dependence, as shown in United States soldiers in Vietnam[71], who faced the major stressors of loneliness, fear and social disruption along with heroin availability (and restricted alcohol availability). Although they had high rates of opioid dependency (one out of five of all army enlisted men from Vietnam), in most instances this appeared reversible after returning from the war. (Only 10% of the veterans reported use of opiates at 8–10 months after their return home; less than 1% reported addiction).

Although nicotine addiction, together with alcohol, represents the largest contribution of all addictions, very little research on etiology has been performed. Recently, it has become quite evident that there is a strong association between *alcohol drinking and smoking* and that nicotine addiction shows great similarities with opioid and alcohol addiction[24].

Summarizing, it is evident that certain characteristics prevail in addicted persons compared with non-addicted populations, but that individual etiologies for specific drug dependencies can be divergent. Using clinical patients as subjects has the serious drawback that consequences and antecedents are hard to distinguish, and that life histories are difficult to analyze due to omissions and subjective recall from the addict and his relatives. Therefore, prospective studies in possible high-risk, no dependence groups are an absolute must in confirming or rejecting the clinically apparent etiology.

3.2 PATHOPHYSIOLOGY

Addicted subjects have been studied to find relevant data on potential behavioral, physiological, and biochemical markers of (1) individual susceptibilities for addiction (trait markers), and (2) characteristic consequences of drug use (state markers). The general approach in searching for markers is to compare dependent individuals with non-dependent subjects to see if they differ in particular measurements. If differences are found it must be specified whether they concern trait or state markers.

3.2.1 *State markers*

State markers indicate drug-induced changes and pathology. Well-known measures are drug concentrations in the body, elevations in liver enzymes for alcohol (ASAT, ALAT and gamma-GT), and mean corpuscular volume. The marker for recent heavy alcohol use, carbohydrate-deficient transferrin[24], is very useful, because alcoholic patients frequently deny or lie about recent drinking. This state marker is 86% sensitive for daily intake of at least 80 g for a minimum of 3 weeks and 98% specific. A recent developed technique concerns measuring specific antibodies to morphine by means of radioimmunological screening to indicate chronic use of morphine especially in periods long after the last drug intake[27].

3.2.2 *Trait markers*

Trait markers can indicate etiological or high-risk factors. Criteria for biological trait markers have been defined[16] as follows: in patients it must be shown that the trait is common, persists during symptom remission, occurs among first-degree relatives at a higher frequency than in the general population and tends to accompany the illness among relatives who have the same condition as the patients.

The distinction between trait and state markers can be demonstrated by studies on alterations in electrical brain wave activity using the technique of *event-related potentials (ERP)*. Electrical potentials are recorded by EEG in response to discrete visual, auditory or sensory stimuli. These potentials are presumed to reflect stages of the brain's information processing. When subjects are required to discriminate between stimuli, there is usually a large positive brain wave that occurs approximately 300 milliseconds after the stimulus presentation and is known as the *P3 component* or *P3 wave* It has been found that latency and amplitude of the P3 are highly similar between identical twins and siblings, which suggests a genetic basis for the P3 response[16].

Begleiter and coworkers[5] did a number of studies on P3 responses and alcohol[16]. In abstinent alcoholics attempting to discriminate between visual stimuli, delays in the P3 response as well a deficit in the amplitude of the P3 response were found[69]. Reduction in peak amplitude appeared to exist also in subjects at high risk for alcoholism, i.e., young boys with alcoholic Type 2 fathers, as illustrated in Figure 7.12[5]. Matched low risk sons of non-alcoholic fathers did not show this effect. These results show that the P3-trait precedes the development of alcoholism, thus might be a marker of predisposition. The reduced P-3 response has been interpreted as a deficit in cortical brain areas involved in motivational and cognitive systems implicated in information processing. Another electrical brain wave, the auditory brainstem potential (ABP) appeared also to be significantly delayed in abstinent alcoholics[4,69]. In contrast to the P3 studies, however, no differences were found between matched sons of alcoholics and non-alcoholic, indicating that ABP deficits in alcoholism are consequences of alcoholism (state markers) rather than a marker of susceptibility (trait marker).

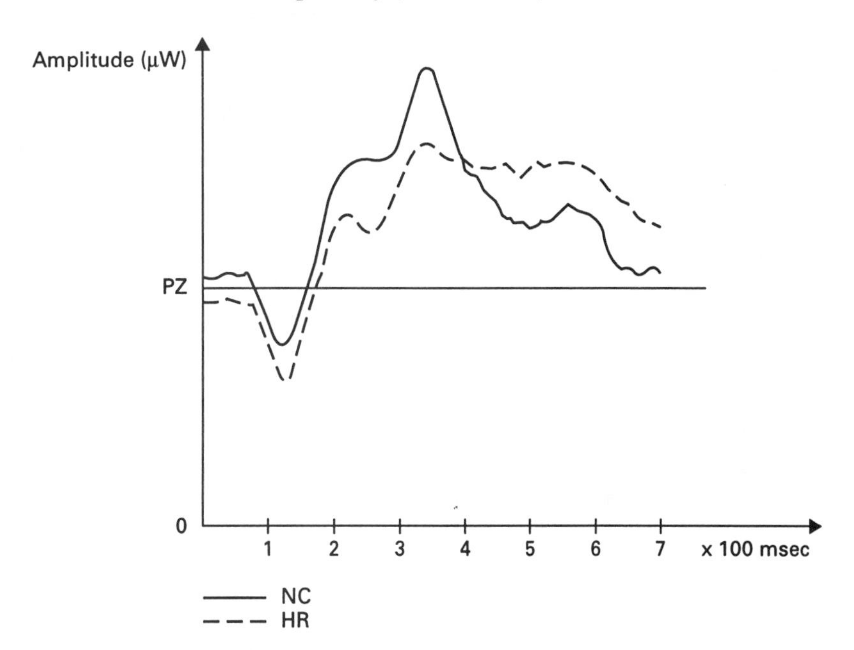

FIGURE 7.12

Event-related brain potentials from a subject at high-risk for alcoholism and from a control subject

(Based on an figure from Begleiter, H. et al., *Science.* 1984, **225:** pp 1493–1496.)

The endocrine system

The endocrine system has also received attention in search of potential trait markers. Although differences between alcoholics and controls have been reported by Schuckit[73,74] with respect to plasma levels of hormones such as prolactin, cortisol, and ACTH, other's studies could not confirm these findings[16].

Interesting findings have been reported on the interaction between alcohol drinking and *beta-endorphin* levels in abstinent alcoholics, in high and in low risk individuals[28]. Stress as well as exposure to diverse addictive drugs stimulate the endocrine system indicated as the hypothalamic-pituitary-adrenal (HPA)-axis.

This stimulation results in the release of CRF from the hypothalamus, which in turn stimulates the release of ACTH and beta-endorphin from the pituitary gland. ACTH in turn stimulates the release of cortisol from the adrenal cortex.

Example

Since it has been demonstrated that animals can be bred selectively for high endocrine response to stress and high preference for ethanol, it has been investigated whether or not predisposed human individuals might have inherited an increased sensitivity in the activity of the HPA-axis to alcohol leading to an increased sensitivity to the reinforcing effects of ethanol.

Table 7.9 shows the group characteristics of subjects that participated in such a study. Subjects of high-risk (HR) groups were individuals from families with a strong positive history of alcoholism across three generations, but not showing alcohol abuse themselves. Low-risk (LR) subjects came from families with absence of alcoholism across three generations. Abstinent alcoholics were selected from the same or similar families as the non-affected HR-individuals and were matched in age and sex with the unaffected HR and LR subjects. A first finding was that plasma baseline levels of immunoreactive beta-endorphin levels, but not of cortisol, in HR-subjects and abstinent alcoholics were significant different from LR-subjects, as is illustrated in Figure 7.13. Administration of placebo did not change these levels.

TABLE 7.9

Characteristics of subjects in a study on plasma levels of beta-endorphin

(Based on data from Gianoulakis, C. et al., In: Reid, L.D. (Ed.), *Opioids, Bulimia, and Alcohol Abuse and Alcoholism*. Springer-Verlag, New York, 1990, pp 229–246.)

	High risk	*Low risk*	*Abstinent*
Number of males & females	33	20	13
Number of males	15	11	7
Number of females	18	9	6
Age	34±1.86	31±2.3	38±2.89
Race	Caucasian	Caucasian	Caucasian
Alcoholism in family	8.6±0.7	0	12.62±1.42
Scores on MAST	1.94±0.54	0.89±0.30	27.62±5.47
Scores on DBI	6.15±1.66	6.64±1.9	Abstinent

Note: A score on MAST (Michigan Alcoholism Screening Test) less than 3 and a DBI (Drinking Behavior Interview) score less than 20 indicate that the individual was not alcoholic at the time of testing.

Ingestion of a low dose of 0.5 g/kg ethanol induced an increase of about 70% over baseline within 45 minutes in plasma immunoreactive beta-endorphin levels in the HR-group, bringing the plasma content close to the levels of the LR-subjects. No increase, but a gradual decrease over time was observed for LR-subjects (Figure 7.14). The abstinent subjects were not challenged with ethanol. Plasma cortisol levels in the HR-group approximately increased a 20% at 15 minutes after ethanol ingestion and returned to baseline at 120 min. In the LR-group again a gradual decrease was observed, now for cortisol, so that at 120 minutes cortisol levels were about 30% lower than baseline levels. Since both groups presented similar contents of ethanol in blood plasma (max. 50 mg/dl at 45 minutes), the different results can not be explained by differences in rate of absorption or clearance of ethanol. When controls used much higher doses, similar endocrine increases were found, suggesting a supersensitivity in HR-persons. The impairment of beta-endorphin and cortisol activity during basal non-stress, non-drug conditions, and a supersensitivity for HPA-axis stimulation might thus be a potential trait marker for HR-individuals. Also in the brain and the pituitary gland of chronic alcoholics low beta-endorphin levels have been found.

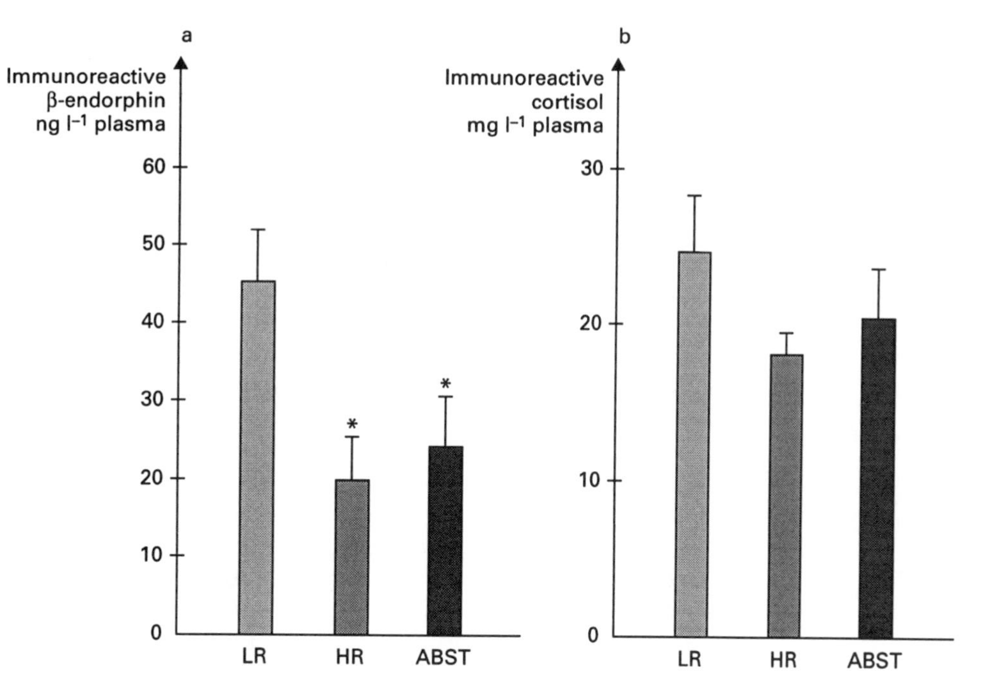

FIGURE 7.13

Mean baseline plasma levels (SEM) of immunoreactive beta-endorphin (A) and cortisol (B) for low risk (LR), high-risk (HR) and abstinent (Abst) individuals
(Based on data from Gianoulakis, C. et al., In: Reid, L.D. (Ed.), *Opioids, Bulimia, and Alcohol Abuse and Alcoholism*. Springer-Verlag, New York, 1990, pp 229–246.)

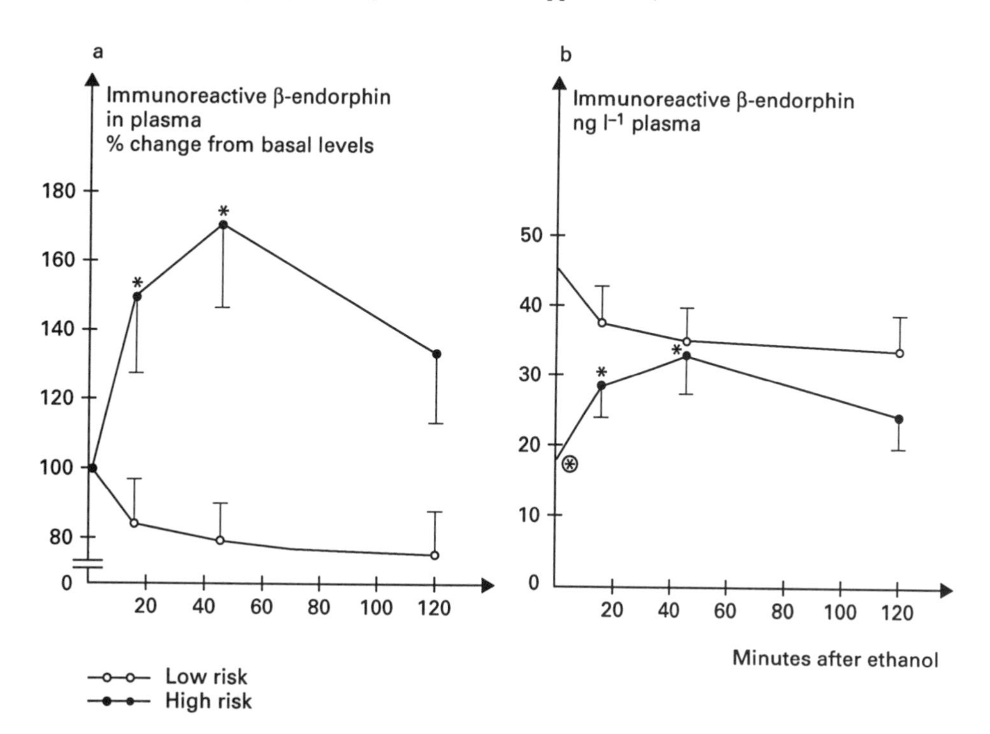

FIGURE 7.14

Plasma immunoreactive beta-endorphin during 120 minutes after intake of ethanol, expressed as percent change from baseline, in individuals at high-risk (HR) and at low risk (LR) for alcoholism
(Based on data from Gianoulakis, C. et al., In: Reid, L.D. (Ed.) *Opioids, Bulimia, and Alcohol Abuse and Alcoholism*. Springer-Verlag, New York, 1990, pp 229–246.)

The presented study appears to support the hypothesis formulated by other researchers that dysfunctions in the endogenous opioid system are involved in the development of alcohol and opiate dependence, as well as in addictive behavior in general[7,82].

The necessity of studying unaffected HR-subjects can be demonstrated by the fact that in a heterogeneous group of drinking alcoholics plasma beta-endorphin were half the levels of matched controls (17.1 ± 5.3 pg/ml), but after 5 weeks of forced abstinence levels rose to those observed in normal controls (30.1 ± 4.9)[83]. These results suggest the finding to be a state marker rather than a trait marker.

Although other markers have been suggested as well, such as *enzyme markers* (e.g., platelet monoamine oxidase (MAO) levels), and *serological markers* (e.g., human leukocyte antigens; HLA), their current clinical usefulness is limited and future research will be needed.

3.2.3 *Tetrahydroisoquinolines (TIQs)*

It was discovered that in chronic alcoholic patients urinary levels of salsolinol were enormously high following hospital admission for detoxification. Also the cerebrospinal fluid (CSF) salsolinol levels were elevated, even 8 days after detoxification. Salsolinol, an alkaloid, is formed via condensation of a catecholamine (e.g., dopamine) with acetaldehyde. Since acetaldehyde is the main intermediate in the metabolism of alcohol, speculation has been risen that an acetaldehyde-amine derived alkaloid is involved in alcoholism. Salsolinol and similar condensation products (e.g., tetrahydropapaveroline, THP) are classified as tetrahydroisoquinolines (TIQs). It has been shown that salsolinol and THP possess opioid-like activity because they bind to opioid receptors and have an analgesic effect. In addition, THP is self-administered by rats. To test TIQs central effects on alcohol ingestion, THP has been administered intracerebroventriculary in rat ventricles. As a result, the voluntary consumption of alcohol rose sharply in these animals, and this effect remained permanent[60]. This TIQ-induced alcohol drinking in animals has been replicated in various studies. It is postulated that the chronic ingestion of alcohol can lead to the biosynthesis of an alkaloid that seems to possess opioid–like and addictive properties.

3.3 TREATMENT

3.3.1 *Clinical trials*

A clinical trial compares the effect(s) of a particular component of a treatment or of a pharmacon with the effects of a treatment without this component or pharmacon. In the case of pharmaca, it is common to use a *placebo* medication for the control group or control situation, which can be unknown for the subject (*single-blind* procedure) as well as for the experimenter (*double-blind* procedure). Ideally, except the experimental intervention, other variables of the treatment would be the same. In practice, this criterion is difficult to realize and is frequently not met.

Furthermore, to obtain comparable results, it is desirable to have homogenous subgroups of selected participants (e.g., age, sex, education level, socioeconomic status, identical symptoms, mental health, etc.). This implicates on the other hand that results are not per definition generalizable to other addicted (sub)populations[48].

van Limbeek[48] gives an example of methodological impairments in a clinical trial in which three kinds of treatment were compared in 227 patients with excessive alcohol use who volunteered for the "Employee Assistance Program": inpatient treatment, self-help group, and a self-selected treatment.

Inclusion criteria for participation were: first time participation in this program (no matter previous treatments), alcohol-related professional problems, no acute hospital indication, and a blood alcohol concentration of 0.2% or less. Exclusion criteria were psychiatric and somatic comorbidity. As a result the researchers obtained a sample with 96% males (90% white and age 30–40 years), 10 women and 23 non-whites. It would have been better if the researchers either had restricted their sample to males only, or had performed a stratification technique. Furthermore, no distinction was made between subjects with first-time problems or with previous alcohol problems. Furthermore, since conclusions are only valid for individuals with the same specific in-and exclusion criteria, these criteria should have reflected relevant parameters with respect to theoretical assumptions.

Another methodological factor in a clinical design is *randomization*. Subjects are randomly distributed over the diverse experimental groups in order to distribute eventual unknown factors equally over the research groups. A problem hereby is that patients frequently *drop out of* the program prematurely and thus make the randomization process invalid. This problem can be countered by performing a pre-experimental *qualification period*[47] in which subjects are screened in more detail for in- and exclusion criteria, for assessing the motivation of the subjects, and specific individual prognoses are made. For example, in a qualification period patients can be exposed to a baseline period (e.g., placebo treatment) after detoxification. Non-compliers can already be excluded and the compliance rate of the others can be classified. For these separate classes treatment effect can be investigated (start of the experiment). The advantage of such a qualification method is that early drop-outs have been excluded and different motivation classes have been determined. Drop-outs during the experimental phase can of course still occur and validity can then be improved by adapting the previously determined in-and exclusion criteria for the remainders as well as reconsidering the representativity of the results. Examples of clinical trials will be presented in Chapters 8 and 9.

3.3.2 *Case studies*

Although clinical trials are a necessary approach in addiction research, statistical results do not always predict individual effects due to the methodological problems discussed in the preceding paragraph. Therefore, it is also useful to document individual experiences in order to discover exceptions and to adapt or develop new research questions or hypotheses. The example given comes from a study on the effects of naltrexone or imipramine on binge eating in bulimia[1]. The clinical trial favored imipramine as a significant effective pharmacon (Table 7.10). The presented case revealed a different effect.

TABLE 7.10

Effect of imipramine or naltrexone on frequency of food binges
(Based on data from Alger, S.A. et al., In: Reid, L.D. (Ed.), *Opioids, Bulimia, and Alcohol Abuse and Alcoholism*. Springer-Verlag, New York, 1990, pp 131–142.)

	Groups		
	Placebo	*Imipramine*	*Naltrexone*
100%	0	56%	29%
75%-99%	44%	11%	14%
50%-74%	22%	22%	29%
25%-50%	22%	11%	29%
< 25%	11%	0	0

Note: Values are % of patients receiving either placebo, imipramine, or naltrexone, who either decreased 100%, or some other amount, in binges across the period of treatment. A score of 100% means no reported binges across Weeks 7 and 8.

Laura had a 25-year history of binge-eating and a family history of obesity in both parents. She reported a history of drug- and/or alcohol-abuse in two of her four siblings. She is 1.80 m. (= 5 foot, 11 inches) tall and weighed 123 kg (= 272 pounds) at the start of the study. Laura related the onset of her bingeing to a serious illness during childhood when she was strongly encouraged by her parents to "eat to keep up her strength." She began to eat large amounts of food and found it provided comfort and support. Her weight steadily increased, and by puberty Laura weighed 118 kg. Laura described a continuous pattern of binge eating, experiencing 3 to 4 binges a week over the past 20 years. She stated, "I felt ashamed that I had no self-control, that I couldn't lose weight despite the ridicule. I was an embarrassment to my family. My distress grew; so did my addiction to food. The hole in my stomach also grew. I remember standing at a bus stop and literally cramming cakes and cookies into my mouth; I was ashamed of my size and still denied reality. I tried dieting, using drugs, and drank to intoxication frequently".

As Laura entered the study she averaged, during the 2 weeks of baseline monitoring, three binges per week, with a typical binge described as two bowls of ice cream, a one pound bag of M&Ms, and six candy bars. Laura received naltrexone (50 mg increasing to 100 mg/day) for 4 weeks. Within one week of starting medication, she completely eliminated her binge eating. She described losing the "anxious knot that builds in my stomach prior to a binge." She also lost her intense craving for sweets. Her "food thought" became less frequent, and depression ratings were lower. When using imipramine, however, the "anxious knot in her stomach" returned, as did her cravings for sweets.

Opioidergic systems were apparently significantly involved in Laura's addictive behavior. A hypothesis could be that the fact that two of her siblings had drug-related problems as well, might indicate that some common susceptibility for addictive behaviors was present in her family specifically linked with opioidergic activity.

3.3.3 *Craving and relapse*

After leaving the treatment center, the treated, drug-free former addict may report occasional episodes, appearing without warning, of a sudden compulsion or drive to obtain the drug, despite his/her intentions never to abuse drugs again. The mechanism of these relapses, that frequently occur in all addictions including alcoholism, cocaine addiction, heroin addiction, nicotine addiction, and obesity[10] is not well understood. It has been proposed that the reasons for relapse after treatment may be totally different from the reasons which caused the patient to begin using drugs[65].

Conditioning factors

Research has been conducted on the role of classical conditioning processes in triggering relapse. Two categories of conditioning processes are relevant: drug-opposite and drug-like conditioned responses.

Drug-opposite conditoned responses

Drug-opposite conditioned responses are responses that are opposite to the effects of the drug itself. For example, opiate injections produce elevations in skin temperature, but stimuli that have repeatedly preceded opiate injections (e.g., sights, sounds, smells, situations) will reliably produce reductions in skin temperature prior to receiving the drug. The drug-opposite responses can mimic drug withdrawal signals (*conditioned withdrawal*) and produce feelings of craving. If drug administration follows at these responses, drug effects are attenuated (*conditioned tolerance*). Furthermore, after as few as seven pairings between mild withdrawal syndromes and a conditioned stimulus (e.g., peppermint odor), narcotic addicts begin to show signs of withdrawal (conditioned response) when exposed to the previously neutral stimulus (odor) alone. This conditioned response appeared to occur long after detoxification when the subjects were exposed to the odor again. This mechanism could explain why a drug-free patient experiences withdrawal symptoms, when he returns to an area where withdrawal symptoms had occurred in the past[65].

Conditioned drug-like responses

Conditioned *drug-like* responses are produced by pairing distinct stimuli with drug administration. After repeated pairing, the stimuli themselves can produce drug-like effects. Such a conditioning mechanism may form a partial explanation for what are known as "placebo" effects of drugs. It has been discovered in animal as well as in human research, that stimulants are more likely to produce drug-like conditioned responses, while opioids in human subjects produce more prominent drug-opposite responses. Drug-like effects are found clinically in patients known as "needle freaks", reporting euphoric effects and showing clinical symptoms, such as pupillary constriction, after injecting saline and occasionally even after naloxone[65].

Cue exposure studies

Of clinical importance is to determine whether modification of conditioned responses can influence the course of addiction. Studies have been performed on opioid and cocaine addicts[65]. These studies using *extinction techniques* are often indicated as cue exposure treatments.

Detoxified opioid addicts that volunteered for receiving additional extinction therapy were compared to opioid addicted patients that did receive normal treatments. After first having selected conditioned "trigger" stimuli in studies with chronic opioid users (e.g., sight of syringe, drug-talk, video showing drug administration, cook-up paraphernalia), the researchers attempted to reduce the conditioned responses to these stimuli in abstinent users through repeated, non-reinforced exposure (extinction).

Target conditioned responses were subjective responses ("craving", feelings of "high", and feelings of drug withdrawal), as well as autonomic responses (pulse rate, blood pressure, skin resistance, and skin temperature) at presentation of the "trigger" stimuli. It was found that conditioned opiate-like responses extinguished rapidly in most patients, but responses which were opposite to the effects of opiates (withdrawal-like physiological responses and craving) were very resistant to extinction. Even in methadone-maintained patients, the drug-related stimuli were reliable elicitors of conditioned craving and conditioned withdrawal. After 20 or more extinction sessions, conditioned craving was significantly reduced in this group, but conditioned withdrawal signs remained present.

Also in *detoxified cocaine-users* responses to cocaine-related stimuli have been studied with respect to subjective measures (10-point scale) of experiencing "high", "craving" or "crash" (withdrawal) as well as to physiological measures, simultaneously recorded on a polygraph. Figure 7.15 illustrates the mean reduction in skin temperature of 30 detoxified cocaine users after diverse stimulus conditions: a neutral video tape (a nature story), neutral activity (video game), a video with buy-sell and cocaine administration, and a drug-related activity (handling drug paraphernalia and performing a simulated cocaine administration). In addition, skin resistance was significantly decreased and heart rate significantly increased. These findings went along with subjective reports of craving and withdrawal/crash. Craving was however reported two to three times as often as either "high" or "withdrawal/crash" responses.

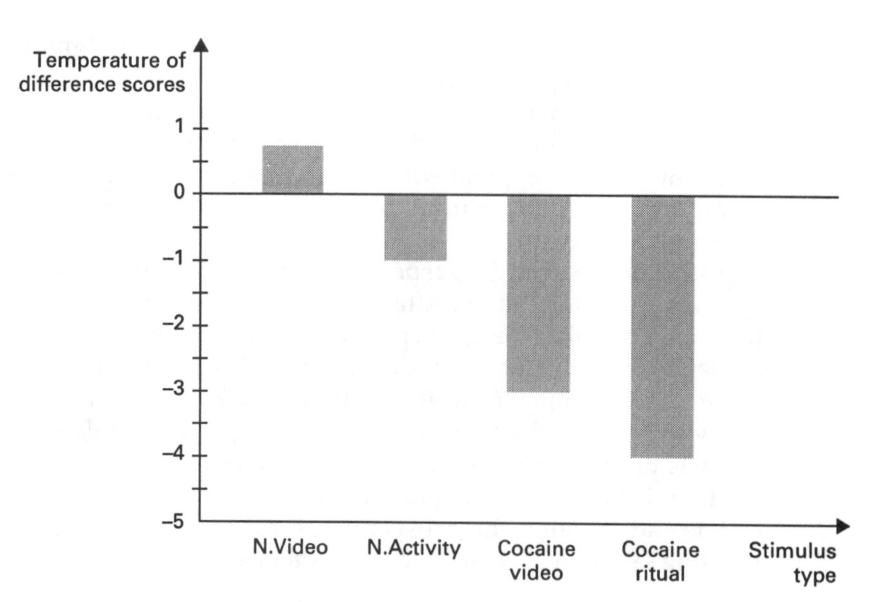

FIGURE 7.15

Effect of neutral versus cocaine-related stimuli on skin temperature in cocaine abusers

(Based on data from O'Brien, C.P. et al., In: O'Brien, C.P. and Jaffe, J.H. (Eds.), *Addictive States*. Raven Press, New York, 1992, pp 157–177.)

These effects did not decline a month after detoxification, in which conventional treatments were applied. These findings implicate that traditional abstinence-oriented treatments do not alter conditioning processes that seem to be (at least partly) responsible for craving and conditioned withdrawal experiences.

The effect of extinction therapy (135 exposures to conditioned stimuli over the course of 15 sessions) in detoxified cocaine addicts is illustrated in Figure 7.16. Of the subjective responses craving was the most prevalent and persistent, reducing gradually over the course of the extinction sessions. Reports of "high" and "withdrawal/crash" were less common and were largely extinguished by the sixth hour of extinction. Skin responses declined over the sessions, but heart rates appeared resistant to extinction. As a result of the extinction therapy, patients showed better retention in outpatient treatment and a higher proportion of clean urines than control groups.

These data on conditioned responses in users have opened new perspectives in understanding relapse and finding new modification techniques and certainly future research will follow.

Relapse prevention

A recently developed treatment approach is based on a *cognitive-behavioral* theoretical perspective in which relapses are not considered as treatment failures or full remissions, but as an inherent phenomenon of being addicted[52]. The addicted individuals are trained to identify high-risk situations and to develop coping strategies to neutralize them. For example, how to deflect the offer to drink, taking up a hobby to minimize boredom, or cognitive techniques to counter mood disturbances. Associated questionnaires have been developed[49], such as the Relapse Precipants Inventory, the Coping Behaviours Inventory, the Effectiveness of Coping Behaviours Inventory, and the Dependence Inventory.

A direct relationship was demonstrated in alcoholics between the alcoholics' perception of self-efficacy (i.e., coping) behaviors and outcome.

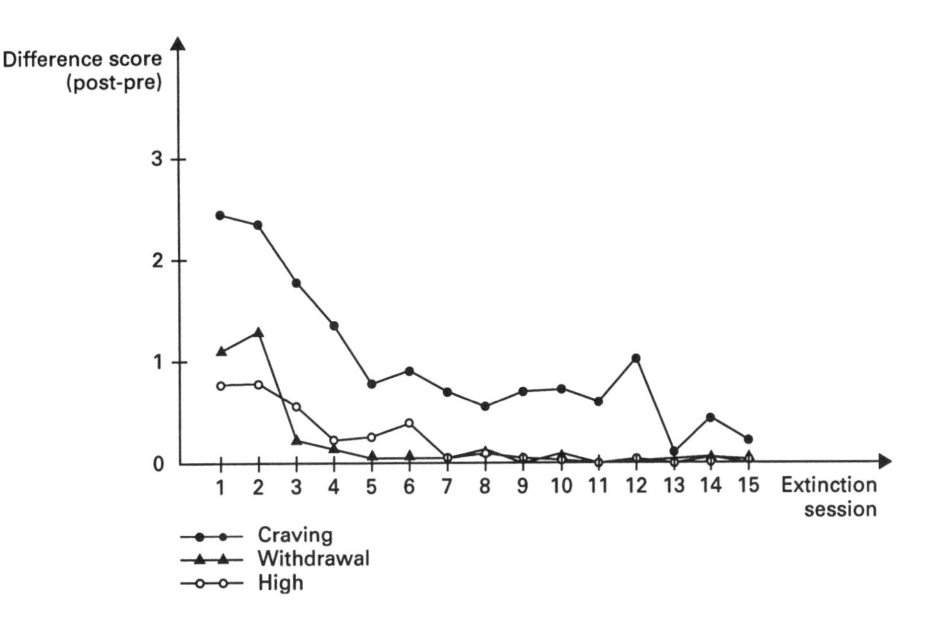

FIGURE 7.16

Reduction in subjective responding to cocaine related stimuli as a function of extinction

(Based on data from O'Brien, C.P. et al., In: O'Brien, C.P. and Jaffe, J.H. (Eds.), *Addictive States*. Raven Press, New York, 1992, pp 157–177.)

Those who did not relapse in the post-treatment period of 15 months reported belief in their coping style and perceived fewer situations as trigger for relapse in contrast to those that did relapse. Alcoholics trained for coping techniques during a treatment program showed marked decreases in consumption rates and a substantially improved ratings of self-efficacy. From a mean of 46 drinks a week at intake, 47% reported total abstinence after 3 months post-treatment and 29% after 6 months[79].

In the field of heroin addiction some research efforts based on this approach have begun. Eleven factors reported by opiate addicts were identified as having caused a lapse to renewed opiate use. The three factors most commonly reported were: negative mood states, cognitive factors, and external situations. Almost two-thirds of those who lapsed indicated that cognitive factors (usually some explicit plan or decision to resume drug use) were implicated in their initial lapse, while more than half indicated that negative moods, such as sadness, boredom, anxiety or tension) preceded the lapse[89].

This approach which takes away the stigma of hopelessness at relapsing and which make addicted persons active in coping with this addiction-inherent phenomenon will need further research, but seems very promising[52].

In conclusion, an important contribution of cue exposure and relapse prevention procedures is that talking about addiction behavior, drugs, and relapse situations is overtly accepted in the therapeutic communities, whereas this frequently is a taboo in conventional treatment programs. Furthermore, individuals are being prepared that after treatment conditioned cues and situations still bear a risk to relapse despite the abstinence during treatment.

3.3.4 *Treatment outcome evaluations*

Drug abuse treatment programs have proliferated during the last 20 years. Enormous amounts of money are spent on treatment programs. A recent estimate, for example, is $3 billion yearly in the U.S.[38]

A wide variety of treatments dealing with drug (and other forms of) dependence has been developed, which can be globally distinguished as social, behavioral, psychotherapeutic or pharmacological treatments. To date, an intense debate is still going on about the effectivity of these treatments and about what can be considered as "success" (i.e., outcome criteria). This is also due to the fact there are no simple tools to define and measure success rates and methodological problems are multiple[38,62]. Several factors have been identified that can affect treatment outcome.

Treatment factors

One of the earliest and most dramatic demonstrations of the relation between treatment content and outcome was reported by Edwards et al.[18]. Comparing the outcome of a state-of-the-art outpatient program with that of telling matched alcoholics that they could not drink without risking serious consequences, he reported no difference whatsoever in outcomes, despite the great differences in professional effort, time, and expense. Later extensive surveys (in total on 649 outcome studies) confirmed that different treatment methods did not significantly differ in effect on long-term outcome[20,21]. More recent research nevertheless indicated some factors that seemed to distinguish if a treatment program will be more or less effective[58]. However, these researchers also discovered that clinical treatment selection in general appeared not to be based on probability of success[68]. Exhaustively surveying the literature on comparative or controlled research on alcoholism treatment, Miller and Hester[58] noted:

"Not only is the volume of research large, but it is gratifyingly consistent. The results of well-controlled studies in this area have seldom contradicted one another...Certain methods have a very good track record, working well across a wide range of populations and settings. Others seem to have little therapeutic value, and are rather consistently found to yield little impact on drinking behavior when subjected to controlled evaluation... As we constructed a list of treatment approaches most clearly supported as effective, based on current research, it was apparent they all had one thing in common... they were very rarely used in American treatment programs. The list of elements that are(!) typically included in alcoholism treatment in the U.S.A likewise evidenced a commonality: virtually all of them lacked adequate scientific evidence of effectiveness." (p. 122)

Miller and Hester constructed a table summarizing effective and much-utilized therapies in Table 7.11. They found aversion therapies and behavioral self-control training generally effective for initial change in drinking; social-skills training and marital/family therapy, and a combination of these elements increased the probability of prolonged sobriety. Training persons' coping skills could reduce anxiety and enhance achieving realistic goals. Noteworthy is the lack of overlap between these empirically supported interventions and the list of treatments that are "standard practice" in alcoholism programs such as AA, alcoholism education, confrontation, group and individual therapy, and disulfiram use.

The "self-help" programs (e.g., Alcoholics Anonymous, Cocanon, Narcotics Anonymous) are quite popular and publicly considered to be successful. Nevertheless, they deal with self-selected samples of clients, have tremendous attrition rates, and provide no statistics about outcome[38]. These programs will thus need careful evaluation for instance of targeting for whom they are and are not effective[76].

Based on three surveys of effective treatments, Peele[68] concludes that effective therapy:

TABLE 7.11

Supported versus standard alcoholism treatment methods
(Based on data from Peele, S., *Int. J. Addict.* 1991, **25(12A)**: pp 1409–1419.)

Treatment methods currently supported by controlled outcome research	*Treatment methods currently employed as standard practice in alcoholism programs*
Aversion therapies	Alcoholics Anonymous
Behavioral self-control training	Alcoholism education
Community reinforcement approach[b]	Confrontation
Marital and family therapy	Disulfiram
Social skills training	Group therapy
Stress management	Individual counseling

[a] Miller and Hester (1986).
[b] The community reinforcement approach combines marital and family therapy, job interventions, and self-control training using a time-out procedure under conditions of high likelihood of relapse.

1. Teaches people real skills for dealing with the world, with other people, with work and with themselves;
2. Confronts without apology the negative value system of addicts and their worlds;
3. Concentrates on broader social units (families, social groups and communities) both as causes of and as resolutions for addiction.

A methodological factor is that evaluations have traditionally been difficult because of the *complexity of treatment methods*, which often include multimodal therapies. In the course of therapy an individual may experience a variety of treatments, used at different stages of the recovery process. For instance, initial treatment starts with detoxification, followed by behavioral modification therapy directed at drug use habits, followed by psychotherapy to cope with psychological problems and/or psychiatric therapy, including pharmacotherapy to treat comorbid psychiatric disorder; in addition social and/or self-help groups are provided to support someone after treatment period. It is difficult to assess the separate contributions of the diverse treatment components.

Investigation of the role of *extra-treatment factors* needs more attention[62]. Recovery without any formal treatment, so-called *natural recovery*, has been documented in alcoholics, cocaine addicted- and heroin addicted people, marijuana abusers, obese individuals, and cigarette smokers[10]. For instance, nicotine addiction is considered by most addicted people to be the most difficult drug to stop using, but 95% of those who stop do it on their own[67]. In a follow-up study of opiate addicts, Vaillant[80] recorded that only 10% had become abstinent at the end of 1 year, 25% at the end of 5 years and 40% at 10 years. It is of significant clinical relevance to discover the ways in which dependent persons rehabilitate themselves and what can be considered as relevant factors. For example, cocaine-addicted people who attained natural recovery appeared to utilize religion, volunteer work, formal education, and recommitment to friends as relapse prevention techniques[10]. Vaillant[81] found that spontaneously remitted alcoholics utilized compulsive work or developed a new love relationship.

It has even been suggested that research protocols per se may obscure the very process of recovery. Forced assignments to a treatment protocol, exclusion criteria, random allocations, and lack of compliances (drop-outs are considered as treatment failures) might lead to unrealistic conclusions[79].

Patients factors

Drop-outs are a general phenomenon in treatment programs. It means that the individuals selected for a study drop out prematurely from the treatment program or cannot be found for subsequent follow-up. Since this is often considered as treatment failure (although outcomes for these individuals are not known), success estimates are unfairly inflated[62].

It has been stated that *psychiatric disorders* have prognostic significance for the outcome of drug treatment[89]. For example, individuals with psychiatric and emotional problems seem not to benefit from "A"-groups. Some persistently depressed alcoholics may find AA discouraging because they fail to experience the improvements described by fellow AA members and can get demoralized even further by demeaning jargon (e.g., "sitting on the pity pot")[10]. These subgroup of dependent individuals probably would have more benefit by adjunctive antidepressant therapy.

Alcoholic and opiate addicts with few additional psychiatric symptoms generally did well in a 6-month post-treatment outcome study, regardless of the type of treatment program or inpatient versus outpatient setting[89]. Patients with high symptom levels did poorly in all forms of programs, although outpatient medication-free treatments were particularly ineffective. Hence, treatments have to be better adapted to specific types of patients and, consequently, validity of treatment success has to be measured in specific subpopulations (patient-matched treatments). It must be mentioned here that some psychiatric disorders have been found to resolve with abstinence. Alcoholics for example show symptoms of depression at the beginning of treatment, but most symptoms disappear after 3–4 weeks of abstinence.

Although little is known about members of "Anonymous" groups, even less is known about those who have not chosen the associated 12-step method of recovery. Nationwide estimates of drug users far exceed the number of "Anonymous" members, suggesting that these methods are not the answer for each dependent individual[10].

The presented findings created the insight that specific kinds of treatments are probably differentially effective with specific kind of drug addicts and have to be investigated with much more sophisticated methods[62].

Outcome assessment factors

Since addiction is in most persons a remitting problem, 3-, or 6-month *post-treatment follow-ups* have been considered as insufficient to enable prediction of the long-term effects of treatment. Follow-up periods of 2 years or more have been advised to determine long-term effectivity[62].

Comparative treatment research typically contrasts a promising new treatment with existing ones. High commitment by the therapists for the new treatment versus lost enthusiasm for the conventional one can result in a false positive judgment for the new treatment due to improved quality of care.

The validity of the very frequently used *self-report procedures*, by interviewing or giving questionnaires to addicted individuals, is another factor of consideration in assessing outcome success. It has been reported that barely half the variance in the alcoholics' self-reports corresponded to assessments of the alcoholics' drinking behavior made by friends or relatives. Comparison of relatives' reports with alcoholics' self-reports suggested that alcoholics may have underestimated their alcohol use about three times as often as they overestimated it[62]. In a study by Fuller et al.[26], self-reported rate of relapse was 59%, whereas relative interviews together with tests of blood and urine alcohol revealed a rate of slightly more than 72%. These investigators reported only 65% probability that patients who claimed continuous abstinence for a year were in fact continuously abstinent. It has been proposed to supplement self-report data with tracking of biochemical markers[16] as a verification measure and as a measure of negative drug-related health consequences, as well as other multiple follow-up measures, such as the reports by significant familial, vocational, and peer relationships[62].

Criteria for treatment "success" can be different in different outcome studies. For some researchers it has to be complete abstinence, for others controlled use, and yet for others social-economic rehabilitation. Outcome dimensions that can be measured can include treatment completion, readmission to treatment, mortality, related health problems, other substance abuse, legal problems, vocational functioning, family/social functioning, emotional functioning, and life stressors.

Future directions

In conclusion it can be stated that treatment outcome research will have to be further developed to become a valid clinical tool. To date, the most agreed consensus is about which treatment is most ineffective to get free from addiction: i.e., detoxification. Detoxification fails as a treatment in itself since it does not modify drug-taking behavior thereafter[79].

Positively, there is also growing consensus that treatment outcome should be evaluated according to multiple variables such as the drug use, vocational adjustment, psychological functioning, interpersonal relations, and criminal behavior at various time points[19]. A promising approach to this end is the *process-oriented framework* of Moos and Finney[59] that focuses on the causal link between specific treatment components and dimensions of outcome. Figure 7.17 illustrates that diverse treatment components and processes have to be identified and assessed for quality. Secondly, it considers life-context or extra-treatment factors as additional determinants of treatment entry, duration, and outcome. Within this framework, an intervention program is only one among many sets of factors that influence subsequent changes in addictive behavior.

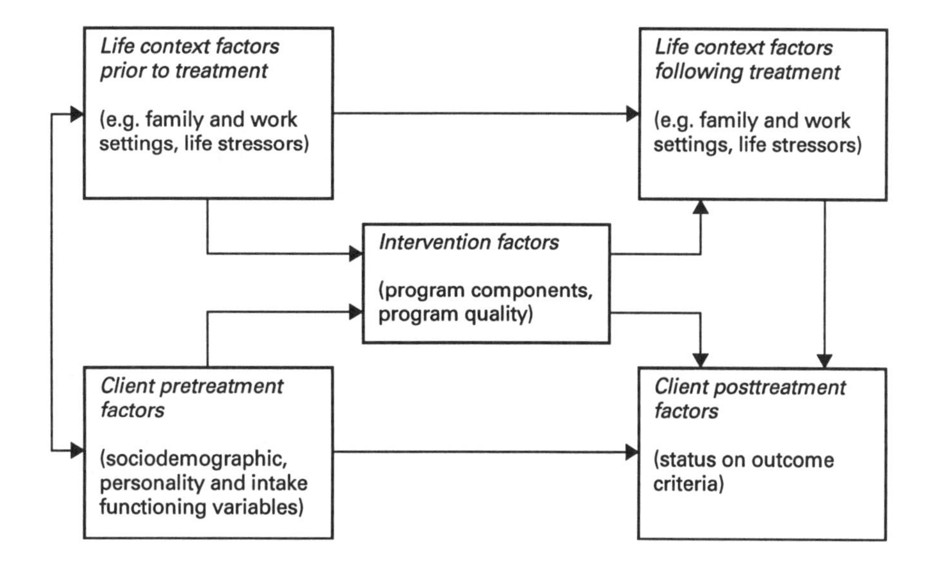

FIGURE 7.17

Process-oriented framework for evaluation of treatment of alcoholism

(Based on data from Peele, S., *Int. J. Addict.* 1991, **25(12A):** pp 1409–1419.)

Another example of a useful research tool, as well as a comprehensive clinical assessment device is the Alcohol Use Inventory[84]. This instrument contains 17 scales designed to capture a multidimensional range of determinants of alcohol use and abuse. Furthermore, longitudinal studies are necessary tools to evaluate the interactions between treatment impacts and lifehistorical events.

4 Summary

The previous chapter provides an overview of the diversity of scientific research on human drug use and addiction. These involve human aspects that cannot be studied in animal models e.g., such as subjective feelings of euphoria or population characteristics, as well as testing the validity of animal results.

A distinction is made between epidemiological, experimental laboratory and clinical research approaches, and related methodologies. In practice, more than one approach or research method is sometimes integrated within one experimental design, due to the multidisciplinary nature of the problem of addiction. Examples of studies originating from the different disciplinary approaches are given and the results are discussed.

Contributions and limitations of the various approaches and of human addiction research in general are evaluated. It is concluded that advanced experimental epidemiological field research will be necessary to define risk and protective factors. Experimental laboratory research is needed to elucidate the impact of the interaction between biological, environmental, and psychological variables. In addition, clinical research is of importance to identify interindividual differences with respect to onset and course of addiction, as well as to develop new treatment strategies that are directed towards specific (sub) groups of patients.

References

1. Alger, S.A., Schwalberg, M.J., Bigaouette, J.M., Howard, L.J., Reid, L.D., "Using drugs to manage binge-eating among obese and normal weight patients." In: Reid, L.D. (Ed.), *Opioids, Bulimia, and Alcohol Abuse and Alcoholism*. Springer-Verlag, New York, 1990, pp 131–142.
2. Anthony, J.C. and Helzer, J.E., "Syndromes of drug abuse and dependence." In: Robins, L.N. and Regier, N.G.D.A. (Eds.), *Psychiatric Disorders in America*. Free Press, New York, 1990, pp 116 –154.
3. Babor, T.F., Mendelson, J.H., Greenberg, I., Kuehnle, J., "Experimental analysis of the 'happy hour': effects of purchase price on alcohol consumption," *Psychopharmacology*. 1978, **58:** pp 35–41.
4. Begleiter, H., Porjesz, B., Bihari, B., "Auditory brainstem potentials in sons of alcoholic fathers," *Alcohol. Clin. Exp. Res.* 1987, **11:** pp 477–480.
5. Begleiter, H., Porjesz, B., Bihari, B., Kissin, B., "Event-related brain potentials in boys at risk for alcoholism," *Science*. 1984, **225:** pp 1493–1496.
6. Benowitz, N.L. and Jacob, III, P., "Nicotine and carbon monoxide intake from high- and low-yield cigarettes," *Clin. Pharmacol. Ther.* 1984, **36:** pp 265–270.
7. Blum, K. and Topel, H., "Opioid peptides and alcoholism: genetic deficiency and chemical management," *Funct. Neurol.* 1986, **1:** pp 71–83.
8. Bobo, J.K., "Nicotine dependence and alcoholism epidemiology and treatment," *J. Psychoact. Drugs*. 1992, **24(2):** pp 123–129.
9. Burke, J.D. and Regier, D.A., "Epidemiology of mental disorders." In: Hales, R.E., Yudofsky, S.C., Talbott, J.A. (Eds.), *Textbook of Psychiatry*, Second edition. The American Psychiatric Press, Washington, D.C., 1994, pp 81–104.
10. Chiauzzi, E.J. and Liljegren, S., "Taboo topics in addiction treatment: an empirical review of clinical folklore," *J. Subst. Abuse Treat.* 1993, **10:** pp 303–316.

11. Cloninger, C.R., Sigvardsson, S., Bohman, M., "Childhood personality predicts alcohol abuse in young adults," *Alcohol. Clin. Exp. Res.* 1988, **12:** pp 494–505.
12. Cloninger, C.R., "Neurogenetic adaptive mechanisms in alcoholism," *Science.* 1987, **236:** pp 410–416..
13. Cohen, P., *Cocaine Use in Amsterdam in Non-Deviant Subcultures.* University of Amsterdam, Amsterdam, 1989.
14. De Wit, H. and Griffiths, R.R., "Testing abuse liability of anxiolytic and hypnotic drugs in humans," *Drug Alcohol. Depend.* 1991, **28:** pp 83–111.
15. Department of Health and Human Services, *The Health Consequences of Smoking: Nicotine Addiction.* DHHS Publication No. CDC 88-8406, U.S. Government Printing Office, Washington D.C., 1988.
16. Department of Health and Human Services, *Seventh Special Report to the U.S. Congress on Alcohol and Health From the Secretary of Health and Human Services.* DHHS Publ. No. (ADM)-281-88-0002, 1990.
17. Edwards, G., "Meta-analysis in aid of alcohol studies," *Br. J. Addict.* 1991, **86:** pp 1189–1190.
18. Edwards, G., Orford, J., Egert, S., Guthrie, S., Hawker, A., Hensman, C., Mitcheson, M., Oppenheimer, E., Taylor, C., "Alcoholism: a controlled trial of 'treatment' and 'advice'." *Q. J. Stud. Alcohol.* 1977, **38:** pp 1004–1031.
19. Emrick, C.D. and Hansen, J., "Assertions regarding effectiveness of treatment of alcoholism," *Am. Psychol.* 1983, **38:** pp 1078–1088.
20. Emrick, C.D., "A review of psychologically oriented treatment for alcoholism: I. The use and interrelationships of outcome criteria and drinking behavior following treatment," *Q. J. Stud. Alcohol.* 1974, **35:** pp 534–549.
21. Emrick, C.D., "A review of psychologically oriented treatment for alcoholism: II. The relative effectiveness of different treatment approaches and the effectiveness of treatment versus no-treatment," *Q. J. Stud. Alcohol.* 1975, **36:** pp 88–108.
22. Folstein, M.F. and Luria, R., "Reliability validity and clinical application of the visual analogue mood scale," *Psychol. Med.* 1973, **3:** pp 479–486.
23. Foltin, R.W. and Fischman, M.W., "Assessment of abuse liability of stimulant drugs in humans: a methodological survey," *Drug Alcohol Depend.* 1991, **28:** pp 3–48.
24. Frances, R.J. and Franklin, J.E., "Alcohol and other psychoactive substance use disorders." In: Hales, R.E., Yudofsky, S.C., Talbott, J.A. (Eds.), *Textbook of Psychiatry,* Second edition. The American Psychiatric Press, Washington, D.C., 1994, pp 355–410.
25. Fraser, H.F., Van Horn, G.D., Martin, W.R., Wolbach, A.B., Isbell, H., "Methods for evaluating addiction liability; (a) 'attitude' of opiate addicts toward opiate-like drugs, (b) a short-term 'direct' addiction test," *J. Pharmacol. Exp. Ther.* 1961, **133:** pp 371–387.
26. Fuller, R.K., Lee, K.K., Gordis, E., "Validity of self-report in alcoholism research: results of a veterans administration cooperative study," *Alcoholism.* 1988, **12(2):** pp 201–205.
27. Gamaleya, N.B., Parshin, A.N., Tronnikov, S.I., Yusupov, D.V., "Induction of antibodies to morphine during chronic morphine treatment in rodents and opiate addicts," *Drug Alcohol Depend.* 1993, **32(1):** pp 59–64.
28. Gianoulakis, C., Angelogianni, P., Meaney, M., Thavundayil, J., Tawar, V., "Endorphins in individuals with high and low risk for development of alcoholism," In: Reid, L.D. (Ed.), *Opioids, Bulimia, and Alcohol Abuse and Alcoholism.* Springer-Verlag, New York, 1990, pp 229–246 .
29. Goode, E., *Drugs in American Society,* Fourth edition. Knopf, New York, 1993.
30. Griffiths, R.R. "Common factors in human and infrahuman drug self-administration." In: Mello, N.K. and Mendelson, J.H. (chairmen), Behavioral pharmacology of substance abuse: recent advances. *Psychopharmacol. Bull.,* 1980, **16(1):** pp 45–47.
31. Griffiths, R.R. and Bigelow, G.E., "Commonalities in human and infrahuman drug self-administration." In: Fishman, J. (Ed.), *The Bases of Addiction.* Dahlem Konferenzen, Berlin, 1978, pp 157–174.
32. Grund, P.C., Stern, L.S., Kaplan, C.D., Adriaans, N.F.P., Drucker, E., "Drug use contexts and HIV-consequences: the effect of drug policy on patterns of every day drug use in Rotterdam and the Bronx," *Br. J. Addict. (Special Edition on HIV/AIDS).* 1992, **87:** pp 381–392.

33. Haertzen, C.A., Hill, H.E., Belleville, R.E., "Development of the Addiction Research Center Inventory (ARCI): selection of items that are sensitive to the effects of various drugs," *Psychopharmacologia.* 1963, **4:** pp 155–166 .
34. Hartnoll, R., *Multi-City Drug Misuse: Trends in 13 European Cities.* Pompidougroup, Strassbourg, 1994.
35. Henningfield, J.E., Miyasato, K., Jasinski, D.R., "Cigarette smokers self-administer intravenous nicotine," *Pharmacol. Biochem. Behav.* 1983, **19:** pp 887–890.
36. Henningfield, J.E., Cohen, C., Heishman, S.J., "Drug self-administration methods in abuse liability evaluation," *Br. J. Addict.* 1991, **86:** pp 1571–1577.
37. Henningfield, J.E. and Goldberg, S.R., "Control of behavior by intravenous nicotine injections in human subjects," *Pharmacol. Biochem. Behav.* 1983, **19:** pp 1021–1026.
38. Hollister, L.E., "Treatment outcome: a neglected area of drug abuse research," *Drug Alcohol Depend.* 1990, **25:** pp 175–177.
39. Inciardi, J.A., "Beyond cocaine: Basuco crack and other cocaine products," *Contemp. Drug Probl.* 1987, **14:** pp 461–492.
40. Jasinski, D.R., Johnson, R.E., Henningfield, J.E., "Abuse liability assessment in human subjects," *TIPS.* May 1984, pp 196–200.
41. Johnston, L.D., O'Malley, P.M., Bachman, J.G., *Drug Use Among American High School Students, College Students, and Other Young Adults: National Trends through 1987.* National Institute on Drug Abuse, Washington, D.C., 1987.
42. Kaplan, C.D., Bieleman, B., TenHouten, W.D., "Are there 'casual users' of cocaine?" In: *Cocaine: Scientific and Social Dimensions.* CIBA Foundation Symposium, 166, John Wiley & Sons, Chicester, England, 1992, pp 57–73.
43. Kaplan, C.D., Korf, D., Sterk, C., "Temporal and social contexts of heroin-using populations. An illustration of the snowball sampling technique." In: de Vries, M.W. (Ed.), *Journal of Nervous and Mental Disorders: Mental Disorders in their Natural Settings.* 1987, pp 566–575.
44. Kaplan, C.D., de Vries, M.W., Grund, J-P. C., Adriaans, N.F.P., "Protective factors: Dutch intervention health determinants and the reorganization of addict life." In: Ghodse, H., Kaplan, C.D., Mann, R.D. (Eds.), *Drug Misuse and Dependence.* Parthenon, London, 1990.
45. Kaplan, C.D., "Drug craving and drug use in the daily life of heroin addicts." In: de Vries, M.W., (Ed.), *The Experience of Psychopathology: Investigating Mental Disorders in their Natural Settings.* Cambridge University Press, Cambridge, 1992, pp 193–219.
46. Klein, R.G. and Mannuzza, S., "Long-term outcome of hyperactive children: a review," *J. Am. Acad. Child Adolescent Psychiatry.* 1991, **30:** pp 383–387.
47. Knipschild, P., Leffers, P., Feinstein, A.R., "The qualification period," *J. Clin. Epidemiol.* 1991, **44(6):** pp 461–464.
48. Limbeek, van J., Alem van V., Wouters, L., "Onderzoek in de verslavingszorg: mogelijkhden en onmogelijkheden van vergelijkend klinisch onderzoek." In: Geerlings, P. (Ed.), *Hersenen en Verslaving.* Van Gorcum, Assen, 1993, pp 94–110.
49. Litman, G.K., Stapleton, J., Oppenheim, A.N., Pelleg, M., "An instrument for measuring coping behaviours in hospitalised alcoholics: implications for relapse prevention and treatment," *Br. J. Addict.* 1983, **78:** pp 269–276.
50. Marks, I., Bradley, B.P., Miele, G.M., Tilly, S.M., First, M., Frances, A., Jaffe, J.H., "Behavioral (non-chemical) addictions," *Br. J. Addiction.* 1990, **85:** pp 1389 –1394.
51. Marlatt, G.A. and George, W.H., "Relapse prevention: introduction and overview of the model," *Br. J. Addict.* 1984, **79:** pp 261–273.
52. Marlatt, G.A., Baer, J.S., Donovan, D.M., Kivlahan, D.R., "Addictive behaviors: etiology and treatment," *Ann. Rev. Psychol.* 1988, **39:** pp 223–252.
53. Mello, N.K., Mendelson, J.H., Kuehnle, J.C., Sellers, M.S., "Operant analysis of human heroin self-administration and the effects of naltrexone," *J. Pharmacol. Exp. Ther.* 1981, 216, pp 45–54.
54. Mello, N.K., Mendelson, J.H., Kuehnle, J.C., "Buprenorphine effects on human self-administration: an operant analysis," *J. Pharmacol. Exp. Ther.* 1982, **223:** pp 30–39.
55. Mendelson, J.H. and Mello, N.K., "One unanswered question about alcoholism," *Br. J. Addict.* 1979, **74:** pp 11–14.

56. Mendelson, J.H. and Mello, N.K., "Experimental analysis of drinking behavior of chronic alcoholics," *Ann. N.Y. Acad. Sci.* 1966, **133:** pp 828–845.
57. Middleton Fillmore, K., Hartka, E., Johnsone, B.M., Leino, E.V., Motoyoshi, M., Temple, M.T., "Preliminary results from a meta-analysis of drinking behavior in multiple longitudinal studies. The collaborative alcohol-related longitudinal project," *Br. J. Addict.* 1991, **86:** pp 1203–1210.
58. Miller, W.R. and Hester, R.K., "The effectiveness of alcoholism treatment: what research reveals." In: Miller, W.R. and Heather, N. (Eds.), *Treating Addictive Behaviors: Processes of Change.* Plenum Press, New York, 1986, pp 121-174.
59. Moos, R.H. and Finney, J.W., "The expanding scope of alcoholism treatment evaluation," *Am. Psychol.* 1983, **38:** pp 1037–1044.
60. Myers, R.D., "Pharmacological effects of amine-aldehyde condensation products," Rigter, H. and Crabbe, J.C. (Eds.), *Alcohol Tolerance and Dependence.* Elsevier Biomedical Press, Amsterdam, The Netherlands, 1980, pp 339–370.
61. Nathan, P.E. and O'Brien, J.S., "An experimental analysis of the behavior of alcoholics and nonalcoholics during prolonged experimental drinking: a necessary precursor of behavior therapy?" *Behav. Ther.* 1971, **2:** pp 455–476.
62. Nathan, P.E. and Skinstead, A., "Outcomes of treatment for alcohol problems: current methods problems and results," *J. Consult. Clin. Psychol.* 1987, **55:** pp 332–340.
63. National Institute on Alcohol Abuse, *Apparent Per Capita Alcohol Consumption: National State and Regional Trends 1977–1987.* Surveillance Report 13, NIAA, Washington, D.C., 1983.
64. Nemeth-Coslett, R. and Henningfield, J.E., "Effects of nicotine chewing gum on human cigarette smoking and subjective psychological effects," *Clin. Pharmacol. Ther.* 1986, **39:** pp 625–630.
65. O'Brien, C.P., Childress, A.R., McLellan, A.T., Ehrman, R., "A learning model of addiction," O'Brien, C.P. and Jaffe, J.H. (Eds.), *Addictive States.* Raven Press, New York, 1992, pp 157–177.
66. O'Hare, P.A., Newcombe, R., Matthews, A., Bunning, E.C., Drucker, E. (Eds.), *The Reduction of Drug-Related Harm.* Routledge, London, 1992.
67. Peele, S., *Diseasing of America: Addiction Treatment Out of Control.* Lexington Books, Lexington, MA, 1989.
68. Peele, S., "What works in addiction treatment and what doesn't. Is the best therapy no therapy?" *Int. J. Addict.* 1991, **25(12A):** pp 1409–1419.
69. Porjesz, B., Begleiter, H., Bihari, B., Kissin, B., "Event-related brain potentials to high incentive stimuli in abstinent alcoholics," *Alcohol Int. Biomed. J.* 1987, **4:** pp 283–287.
70. Roache, J.D. and Griffiths, R.R., "Comparison of triazolam and pentobarbital: performance impairment subjective reports and abuse liability," *J. Pharmacol. Exp. Ther.* 1985, **234:** pp 120–133.
71. Robins, L.N., Davis, D.H., Goodwin, D.W., "Drug users in Vietnam: a follow-up on return to the USA," *Am. J. Epidemiol.* 1974, **99:** pp 235–249.
72. Robins, L.N., "Conduct disorder," *J. Child Psychol. Psychiatr.* 1991, **32:** pp 193–212.
73. Schuckit, M.A., Gold, E.O., Risch, S.C., "Serum prolactin levels in sons of alcoholics and control subjects," *Am. J. Psychiatry.* 1987, **144(7):** pp 854–859.
74. Schuckit, M.A., Risch, S.C., Gold, E.O., "Alcohol consumption ACTH level and family history of alcoholism," *Am. J. Psychiatry.* 1988, **145(11):** pp 1391–1395.
75. Siegel, R.K., *Intoxication. Life in Pursuit of Artificial Paradise.* Pocket Books, New York, 1990.
76. Smart, R.G., Allison, K.R., Cheung, Y., Erickson, P.G., Shain, M., Single, E., "Future Research needs in policy prevention and treatment for drug abuse problems," *Int. J. Addict.* 1990, **25(2A):** pp 117–126.
77. Smart, R.G. and Adlaf, E.M., *Alcohol and Other Drug Use Among Ontario Adults 1977–1987.* Addiction Research Foundation, Toronto, 1987.
78. Smart, R.G., "Crack cocaine use: a review of prevalence and adverse effects," *Am. J. Drug Alcohol Abuse.* 1991, **17(1):** pp 13–26.
79. Strang, J., "The fifth Thomas James Okey Memorial Lecture. Research and practice: the necessary symbiosis," *Br. J. Addict.* 1992, **87:** pp 967–986.
80. Vaillant, G.E., "A twenty-year follow-up of New York narcotic addicts," *Arch. Gen. Psychiatry.* 1973, **29:** pp 237–241.

81. Vaillant, G.E., *The Natural History of Alcoholism*. Harvard University Press, Cambridge, MA, 1983.

82. Van Ree, J.M., "Reward and abuse: opiates and neuropeptides." In: Engel, J., Oreland, L., Ingvar, D.H., Pernow, B., Rössner, S., Pelborn, L.A. (Eds.), *Brain Reward Systems and Abuse*. 7th Int. Berzelius Symp., Göteborg, Sweden. Raven Press, New York, 1987, pp 75–88.

83. Vescovi, P.P., Coiro, V., Volpi, R., Giannini, A., Passeri, M., "Plasma beta-endorphin but not met-enkephalin levels are abnormal in chronic alcoholics," *Alcohol Alcohol*. 1992, **27(5):** pp 471–475.

84. Wanberg, K.W. and Horn, J.L., "Assessment of alcohol use with multidimensional concepts and measures," *Am. Psychol.* 1983, **38:** pp 1055–1069.

85. Washton, A.M., Gold, M.S., Pottash, A.C., " 'Crack': early report on a new drug epidemic," *Postgrad. Med.* 1986, **80:** pp 52–58.

86. Weiss, R.D. and Mirin, S.M., "Subtypes of cocaine users," *Psychiatr. Clin. N. Am.* 1986, **9:** pp 491–501.

87. White, J.M., *Drug Dependence*. Prentice-Hall, Englewood Cliffs, NJ, 1991.

88. Wikler, A., "A psychodynamic study of a patient during self-regulated readdiction to morphine," *Psychiatr. Q.* 1952, **26:** pp 270–293.

89. Woody, G.E., McLellan, A.T., O'Brien, C.P., "Research on psychopathology and addiction: treatment implications," *Drug Alcohol Depend.* 1990, **25:** pp 121–123.

90. Zucker, R.A. and Gomberg, E.S.L., "Etiology of alcoholism reconsidered: the case for a biopsychosocial process," *Am. Psychol.* 1986, **41:** pp 783–793.

References of studies mentioned in the tables

Babor, T.F., Mendelson, J.H., Greenberg, I., Kuehnle, J., " Experimental analysis of the 'happy hour': effects of purchase price on alcohol consumption," *Psychopharmacology*. 1978, **58:** pp 35–41.

Benowitz, N.L. and Jacob, III, P., "Nicotine and carbon monoxide intake from high- and low-yield cigarettes," *Clin. Pharmacol. Therap.*, 1984, **36:** pp 265–270.

Bigelow, G., Griffiths, R.R., Lierson, I.A., "Effects of the response requirement upon human sedative self-administration and drug-seeking behavior," *Pharmacol. Biochem. Behav.* 1976, **5:** pp 681–685.

Chait, L.D., Uhlenhuth, E.H., Johanson, C.E., "An experimental paradigm for studying the discriminative stimulus properties of drugs in humans," *Psychopharmacology*, 1984, **82:** pp 272–274.

Chait, L.D., Uhlenhuth, E.H., Johanson, C.E., "The discriminative stimulus and subjective effects of d-amphetamine in humans," *Psychopharmacology*. 1985, **86:** pp 307–312.

Chait, L.D., Uhlenhuth, E.H., Johanson, C.E., "The discriminative stimulus and subjective effects of phenylpropanolamine, mazindol and d-amphetamine in humans," *Pharmacology Biochem. Behav.* 1986a, **24:** pp 1665–1672.

Chait, L.D., Uhlenhuth, E.H., Johanson, C.E., "The discriminative stimulus and subjective effects of d-amphetamine, phenmetrazine and fenfluramine in humans," *Psychopharmacology*. 1985, **89:** pp 301–306.

Chait, L.D. and Johanson, C.E., "Discriminative stimulus and subjective effects of caffeine and benzphetamine in amphetamine-trained volunteers," *Psychopharmacology*. 1988, **96:** pp 302–308.

Chamberlayne, R., Kierans, W., Fletcher, L., "British Columbia alcohol and drug problems adolescent survey, 1987," British Columbia Ministry of Health, Victoria, 1988.

Cloninger, C.R., "Neurogenetic adaptive mechanisms in alcoholism," *Science*. 1987, **236:** pp 410–416.

Fischman, M.W. and Schuster, C.R., "Cocaine self-administration in humans," *Fed. Proc.* 1982, **41:** pp 241–246.

Griffiths, R.R., Bigelow, G.E., Liebson, I., "Suppression of ethanol self-administration in alcoholics by contingent time-out for social interactions," *Behav. Res. Ther.* 1974, **12:** pp 327–334.

Griffiths, R.R., Bigelow, G.E., Liebson, I. "Human sedative self-administration: double blind comparison of pentobarbital, diazepam, chlorpromazine and placebo," *J. Pharmacol. Exp. Therap.* 1979, **210:** pp 301–310.

Henningfield, J.E., Miyasato, K., Jasinski, D.R., "Cigarette smokers self-administer intravenous nicotine," *Pharmacol. Biochem. Behavior.* 1983, **19:** pp 887–890.

Inciardi, J.A., "Beyond cocaine: Basuco, crack and other cocaine products," *Contemp. Drug Probl.* 1987, **14:** pp 461–492.

Johnston, L.D., O'Malley, P.M., Bachman, J.G., *Drug Use Among American High School Students, College Students, and Other Young Adults: National Trends through 1987,* National Institute on Drug Abuse, Washington DC, 1987.

Mello, N.K., Mendelson, J.H., Kuehnle, J.C., Sellers, M.S., "Operant analysis of human heroin self-administration and the effects of naltrexone," *J. Pharmacol. Exp. Therap.* 1981, **216:** pp 45–54.

Mello, N.K., Mendelson, J.H., Kuehnle, J.C., "Buprenorphine effects on human self-administration: an operant analysis," *J. Pharmacol. Exp. Therap.* 1982, **223:** pp 30–39.

Mendelson, J.H. and Mello, N.K., "Experimental analysis of drinking behavior of chronic alcoholics," *Ann. N.Y. Acad. Sci.* 1966, **133:** pp 828–845.

Miller, W.R. and Hester, R.K., "The effectiveness of alcoholism treatment: what research reveals." In: Miller, W.R. and Heather, N. (Eds.) *Treating Addictive Behaviors: Processes of Change.* Plenum Press, New York, 1986, pp 121–174.

Nemeth-Coslett, R. and Henninfield, J.E., "Effects of nicotine chewing gum on human cigarette smoking and subjective psychological effects," *Clin. Pharmacol. Ther.* 1986, **39:** pp 625–630.

Smart, R.G. and Adlaf , E.M., *Alcohol and Other Drug Use Among Ontario Adults 1977–1987.* Addiction Research Foundation, Toronto, 1987.

Washton, A.M., Gold, M.S., Pottash, A.C., " 'Crack': early report on a new drug epidemic," *Postgrad. Med.* 1986, **80:** pp 52–58.

Wish, E.D., Cuadro, M., Martorana, J.A., "Estimates of drug use in intensive probationers: results from a pilot study," *Fed. Probat.* 1986, **50:** pp 4–16.

Note

The example of the case study in section 1.8 is adapted and slightly modified from a manuscript by Drs. Kaplan and De Vries (see also ref. 45).

Contents Chapter 8

Current clinical assessment and management of drug dependence

0-8493-7803-6/99/$0.00+$.50

Chapter 8

Current clinical assessment and management of drug dependence

Claudio A. Naranjo, Usoa Busto, Vural Özdemir

INTRODUCTION

Once a person has entered a treatment facility, the first step is diagnosis or assessment. Although superficially simple, adequate assessment is a complex task and is important in the design of therapy. The first part of this chapter deals with clinical assessment of substance abuse disorders, while the second part mainly deals with the management of drug dependence.

1 Clinical assessment

The clinical assessment of alcohol and psychotropic substance use disorders is important both for clinicians and researchers. Effective prevention and treatment as well as research on the mechanisms of these disorders depend on having valid and reliable methods of diagnosis.

The clinical assessment of drug dependence is a continuous and dynamic process which determines the various types of substances abused and evaluates associated problems along a spectrum. Ideally, this assessment should employ a battery of diagnostic approaches including screening questionnaires, clinical signs and laboratory tests as well as formal diagnostic techniques such as structured clinical interviews. In this way, clinicians are able systematically to assess substances that are being used and abused as well as the medical and psychosocial consequences associated with their use. Figure 8.1 shows a stepwise approach for the application of these assessment instruments. A brief review of the most commonly used instruments follows: questionnaires, clinical signs, laboratory tests, and structured clinical interviews.

1.1 QUESTIONNAIRES

Several questionnaires have been developed in response to a need for simple, practical and yet accurate methods for rapidly identifying individuals with substance use disorders. The information gathered by questionnaires can be used to corroborate the results of structured clinical interviews.

By and large, these assessment questionnaires have been developed for alcoholism: alcohol has been in existence even in prehistoric times and is socially and legally more acceptable compared to other drugs of abuse. In addition, the cost of alcoholism is substantial: in 1990, in the United States the cost of alcohol-related problems was $136.3 billion resulting from increased morbidity, mortality, diminished productivity at work, and disrupted families. These factors might have prompted the development of assessment instruments mainly in the area of alcoholism. However, there is a large overlap between

alcoholism and other drug addiction: as many as two-thirds of alcoholics are cross-addicted to other drugs, particularly nicotine and sedative-hypnotics such as benzodiazepines and barbiturates[2,25]. It is also quite common for drug abusers to use alcohol concomitantly. Although the assessment questionnaires for alcohol-related problems are mainly validated on alcoholic subjects, the behavioral and social indicators of alcoholism (e.g., impaired psychosocial functioning, disrupted family structures) tapped in these questionnaires appear to be common in other drug abusers. Therefore, detection of alcoholism in a patient should also prompt the physician/therapist to seek for other potential drug abuse.
Based on their purpose, the questionnaires can be divided into two groups:

1. Screening questionnaires
2. Questionnaires to evaluate the severity of substance use disorders.

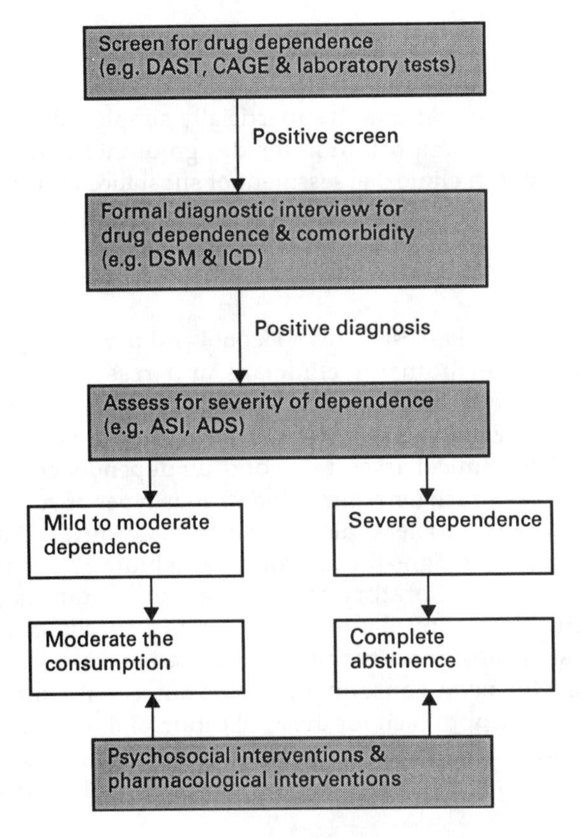

FIGURE 8.1
Clinical assessment and treatment of drug dependence

1.1.1 *Questionnaires for screening drug problems*

The Drug Abuse Screening Test (DAST) is a 28-item questionnaire for clinical screening of drug use problems and evaluation of response to treatment[46]. The DAST does not concentrate on any particular class of drugs and poses questions on drug use in a generic manner (e.g., "do you ever feel bad about your drug abuse?"). A shorter 20-item version highly correlates with the original instrument. In addition to screening of patients for problems associated with non-medical use of drugs, the DAST can potentially be used to provide a baseline for monitoring changes in patients during treatment and follow-up. A positive answer to 8 or 9 items in the DAST detects alcohol-related problems with a sensitivity and specificity of 91% and 89%, respectively.

The Michigan Alcoholism Screening Test (MAST) is a standard screening test for alcohol problems[43]. This is a 24-item self-report questionnaire and takes about 5 minutes to complete. A shorter 13-item version is also available[42]. The MAST has been useful in identifying alcoholic patients in clinical settings and requires the patients' willingness and cooperation to admit their alcohol-related problems. The MAST has a sensitivity and specificity of approximately 90% and 80%, respectively[41].

The Alcohol Use Disorders Identification Test (AUDIT) is designed for the early detection of alcohol-related problems in primary health care settings[5]. AUDIT is based on a WHO collaborative project on identification and treatment of persons with harmful alcohol consumption, conducted in six countries. Therefore, it has a cross-national validity with a sensitivity and specificity of 92% and 94%, respectively.

The CAGE is a frequently used four-item screening test for alcoholism and consists of yes/no questions[27]:

- Need to **C**ut down on drinking?
- **A**nnoyed by criticism about your drinking?
- **G**uilty about drinking?
- Need a morning drink or **E**ye-opener?

A "yes" answer for 2 or more questions is considered to be evidence of alcoholism. In a study of 4203 patients visiting their general practitioners, the CAGE questionnaire was included in a general health assessment questionnaire. Under these conditions, the CAGE showed a sensitivity of 59% in men and 70% in women and a specificity of 91% in men and 99% in women in the detection of alcohol-related problems[13].

An even shorter instrument consisting of two questions has also been proposed to screen alcoholism[14,53]. The two items in this very brief test are "Have you ever had a drinking problem?" and "When was your last drink?". A "yes" answer to the first question and admitting to drink in the 24 hours preceding the interview is considered a positive reply. An affirmative response to either or both questions was reported to have a 91.5% sensitivity to detect alcohol-related problems. Although further experience is needed on the clinical value of this two question test, Cyr and Wartman (1988)[14] propose that detection of alcohol and drug problems may be hampered by not asking direct and straightforward questions.

1.1.2 *Questionnaires for assessing the severity of drug problems*

These instruments determine the degree of dependence and the severity of the associated medical and psychosocial problems. This information can be used for matching patients with appropriate therapeutic interventions. In general, patients with a severe substance dependence are encouraged to enroll in complete abstinence oriented treatment programs while those patients with mild to moderate substance dependence are encouraged to be in programs designed to achieve reduction or moderation of substance consumption (Figure 8.1).

The Addiction Severity Index (ASI) is an interview based questionnaire which collects information on seven problem areas frequently affected in substance abusers: medical condition, employment status, legal status, alcohol use, drug use, family and social functioning and psychological status[28]. The severity ratings in each area is based on the interviewer's estimates of the patients' need for treatment after a detailed evaluation of the patients' history. The severity is marked on a scale of 0 (no treatment necessary) to 9 (treatment necessary to intervene in a life-threatening situation). Although the ASI is a reliable and valid instrument to assess the severity of drug dependence, it takes approximately 30 to 40 minutes to complete, thus has limited application in common clinical practice[19]. However, it is useful in the evaluation of outcome in treatment research studies.

The Alcohol Dependence Scale (ADS) is a validated self-report questionnaire designed to identify and assess problems associated with alcohol use and provides a quantitative index of the severity of alcohol dependence[38,47]. The ADS has a sensitivity and specificity of 91% and 82%, respectively. The ADS measures several features of alcohol dependence such as impaired control over alcohol use, presence of tolerance and withdrawal symptoms, and a compulsive drinking pattern. ADS is particularly recommended to differentiate moderate to severely alcohol dependent patients (ADS score > 9) from those with a low dependence (e.g., problem drinkers) (ADS score < 10).

The Severity of Alcohol Dependence Questionnaire (SADQ) is another instrument to assess the severity of alcohol dependence[16,49]. It contains 20 items tapping problems related to severe alcohol dependence.

1.2 CLINICAL SIGNS

The diagnostic value of clinical signs in substance use disorders has been studied mostly in the field of alcoholism. A clear constellation of clinical findings, specific for each kind of drug abuse, has not been defined other than alcoholism: an alcohol clinical index has been constructed to differentiate outpatients with alcohol-related problems from family practice patients and social drinkers[45]. In this study, patients with alcohol-related problems were excessive drinkers, consuming more than 400 g ethanol/week for a mean of 7.7 years, and had significant consequences related to their drinking habits as assessed by the Michigan Alcoholism Screening Test. The social drinkers consumed fewer than 7 drinks (95 g ethanol) per week and no more than 2 drinks (27 g ethanol) per occasion. The alcohol clinical index incorporates 13 medical history items primarily related to the presence of alcohol withdrawal symptoms, anxiety or depression, and 17 clinical signs such as presence of more than 5 spider naevi (telangiectatic arterioles in the skin with radiating capillary branches simulating the legs of a spider), hand tremor, tandem gait, rhinophyma (hypertrophy of the nose with follicular dilatation), and the signs of past or present injury (bruises, scars or burns). Presence of four or more of either clinical signs or medical history items from the index was associated with a high probability (> 90%) of alcohol-related problems. On the basis of distinguishing patients with alcohol-related problems from social drinkers, this index has a sensitivity of 89% and a specificity of 88%.

Similarly, in a World Health Organization (WHO) collaborative study, the presence of abnormal skin vascularization, conjunctival injection, tongue or hand tremor, or soft hepatomegaly (increase in liver size) has been reported to correlate with the presence of alcohol-related problems[40]. This association was more significant in patients consuming more than 80 g of alcohol per day.

In the case of drugs other than alcohol, the earliest clinical clues consist of behavioral and social indicators as discussed in the previous section. In addition, the following physical examination findings may be helpful in some cases of drug abuse. Heavy users of cocaine (by inhalation through their nose) may have a perforated nasal septum or nasal inflammation. Unexpected cardiovascular (e.g., myocardial infarction) or cerebrovascular (e.g., transient ischemic attacks) events in a young patient (less than 40 years of age) as well as frequent panic attacks or paranoid psychosis followed by acute depression episodes may also suggest an on-going cocaine abuse. Patients who use marijuana may appear to lack ambition and have injected conjunctivae. Multiple scars suggestive of venipuncture may also be a clue for intravenous drug use such as heroin. In addition, indirect clinical findings may assist in the diagnosis of drug abuse: depression, anxiety and antisocial personality disorder frequently coexist with drug abuse and may be the reason for initial clinical presentation (see coexisting psychiatric disorders under Section 1.4).

In summary, diagnosis of substance use disorders requires a high index of clinical suspicion, and elicitation and evaluation of information coming from various clinical sources.

1.3 LABORATORY TESTS

The self-report methods have enjoyed considerable popularity in the clinical assessment as well as in the evaluation of treatment efficacy for substance use disorders. Retrospective methods such as self-reports of substance consumption are practical and cost-effective assessment techniques. In addition, such self-reports are considered to be fairly reliable and accurate provided that they are conducted in a hospital or research setting, patients are given reassurances of confidentiality, and are drug-free at the time of the interview[4]. However, there is still a need for objective measures of substance consumption due to the following reasons:

1. The accuracy of self-reports of substance use is dependent on patients' cooperation and can not expected to be reliable in all patients under all circumstances.
2. Drug consumption, with or without associated abuse or dependence, can put the individual at risk for medical problems, particularly with chronic use and in the presence of concomitant diseases.
3. Accurate measurement of drug consumption is necessary for the evaluation of response to treatment, particularly in treatment programs emphasizing complete abstinence.

Several laboratory tests are currently being used for the detection of substance use such as urinalyses, breath alcohol test, alcohol dipstick, hair sample analysis, and liver function tests.

1.3.1 *Urinalyses*

When physicians are uncertain about the substance use patterns of suspected patients, they often request a "comprehensive drug screen". This screen is usually performed by using urine samples. Some laboratories employ the most inexpensive and relatively insensitive technique — thin-layer chromatography (TLC). This chromatographic technique separates different substances in a biological fluid based on the polarity of each substance which in turn determines the degree of migration of a substance from the origin point where the sample of a biological fluid (e.g., urine) is applied. The final location (migration) of the involved substance(s) on the chromatographic plate is visualized by illumination with ultraviolet or fluorescent lights or by spraying with chemical dyes and is compared with the location of standards for different substances such as amphetamines, cocaine, and opioids. This allows a qualitative determination of the presence of a particular substance(s) in a biological fluid. However, TLC has a low sensitivity and requires the presence of high concentrations of substances in the urine. In addition, this technique is not sensitive enough to detect several substances such as marijuana, phencyclidine, and mescalin. Therefore, false-negative results may occur with TLC and a negative result cannot rule-out substance use. False-positive results are less common with TLC and may occur due to the presence of other molecules in the urine, with chromatographic properties similar to some of the frequently abused substances. Urinalyses requires time for the processing of samples and is unable to provide immediate feedback to clinicians or researchers.

Other more sophisticated and sensitive techniques such as gas chromatography/mass spectrometry, gas-liquid chromatography, and enzyme-immunoassay are also used for drug screening. However, they are more expensive, and are usually applied to detect smaller concentrations of abused substances or when a definitive confirmation of use is necessary (e.g., in forensic cases or for treatment outcome evaluation in complete-abstinence oriented rehabilitation programs).

1.3.2 *The breath alcohol test and the alcohol dipstick*

A small portion of the ingested alcohol is excreted through the lungs and can be measured by portable, inexpensive breath testers (breathalyzers). The breath alcohol test provides a practical and immediate way of assessing current alcohol use. However, the results, at times, can be inconclusive due to the difficulties in calibration of the instrument. As an alternative, an alcohol dipstick test has been developed to measure ethanol concentration in different biological fluids such as urine, saliva, and serum. The dipstick is a cellulose pad impregnated with the enzyme alcohol dehydrogenase and color-developing agents. Upon contact with ethanol, the dipstick changes its color, the intensity of which correlates with the concentrations of ethanol in the biological fluid. Thus, it provides a quantifiable index of ethanol concentration. These two techniques offer clinicians and researchers an objective, reliable and prompt index of ethanol consumption in a variety of settings[21]. The results must be interpreted in light of the kinetic properties of ethanol (e.g., half life ranges from 2 to 6 hours).

1.3.3 *Hair sample analysis*

Hair sample analysis has been used to assess exposure to trace elements. Recently, hair sample analysis has also been applied to assess substance use[11]. The hair follicles are nourished by capillary circulation in hair roots, and drugs in the blood are incorporated into the hair strands from blood as the hair grows. Since the scalp hair grows approximately 1.1 cm/month, analysis of sections of hair strand gives the amounts and patterns of drug use over the course of several months. For example, the different patterns of morphine use in two patients are illustrated in Figure 8.2. Therefore, this new approach may offer the following potential advantages over the currently available alternative techniques such as urinalyses: an ability to obtain an objective history of cumulative drug use dating back over a period of several weeks to years, provision of information on patterns of use, i.e., whether drug use has increased, remained stable or decreased, and clinical convenience of obtaining a hair sample compared to more invasive procedures such as blood sampling. Several drugs of abuse including opiates, cocaine, phencyclidine, methamphetamine, nicotine, and phenobarbital have been reported to be detected in hair samples. This technique can validate self-reports, and thus is very appealing for drug abuse screening in individuals in the work place and in those making critical decisions requiring full cognitive and psychomotor abilities (e.g., pilots, armed force personnel, police).

However, more work has to be done to establish and improve the accuracy and reliability of this technique such as a clear understanding of the dose-response relationships between the administered drugs and their concentrations in hair samples.

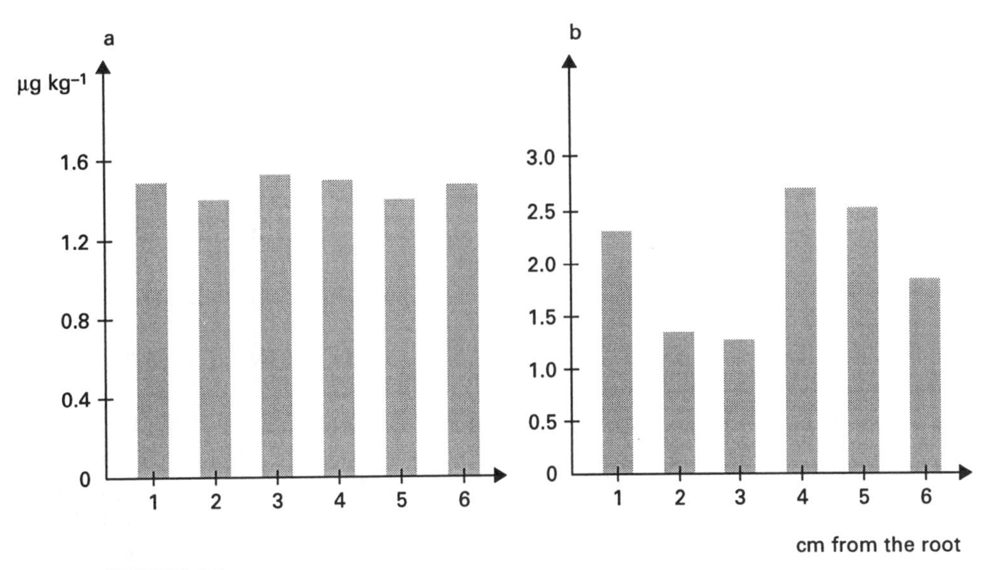

FIGURE 8.2

Patterns of morphine use

A, constant use; B, with variation. (Based on data from Kintz, P. et al., *Int. J. Leg. Med.* 1992, **105:** pp 1–4.)

1.3.4 *Biological markers*

An objective biological marker for substance abuse or dependence would facilitate the screening and diagnosis of these problems. In addition, such markers could prove to be useful in clinical research as a therapeutic endpoint and an index of relapse. Several biological markers have been proposed as possible indicators of excessive substance consumption, mainly for alcohol. However, such markers are also affected by conditions other than substance use such as concomitant liver disease and the biological half-life of the potential markers. In general, they suffer from a low sensitivity and specificity for the detection of substance use problems.

Serum gamma-glutamyltransferase (GGT) activity and mean corpuscular volume (MCV) were proposed as potential markers of alcoholism[51]. Values of GGT above 50 $IU.l^{-1}$ and values of MCV greater than 96 fl are considered to be indicative of heavy alcohol consumption, but lower criteria have also been used (40 $IU.l^{-1}$ and 90 fl, respectively). Although both tests are generally elevated above normal values in heavy drinkers, they have a low sensitivity for the identification of people with less severe alcohol problems.

Recently, more specialized laboratory tests have been proposed as potential indicators of excessive alcohol consumption such as carbohydrate-deficient transferrin and acetaldehyde-protein adducts[1]. Transferrin is a protein produced by liver cells and contains carbohydrate moieties. It binds iron in the blood and transports it to tissues. Upon moderate to heavy consumption of alcohol, liver cells release an abnormal transferrin without the carbohydrate moieties. Therefore, this carbohydrate-deficient transferrin can be used as a marker of potential alcohol abuse.

Acetaldehyde is a highly reactive metabolite of ethanol and binds to several proteins in the blood such as plasma proteins and hemoglobin. The major portion of acetaldehyde in the blood is derived from the metabolism of ethanol. Therefore, the concentration of protein-acetaldehyde complexes (adducts) in the blood can be quantified by specific antibodies as a measure of alcohol consumption. This research line on biological markers of substance use may eventually lead to the development of valid, reliable, and practical laboratory tools for the diagnosis of substance use disorders.

In summary, laboratory tests should be used to corroborate the information obtained by other means such as self-reports, clinical examination findings, and interviews. A positive laboratory test should alert the physician to the possibility of substance abuse or dependence. However, negative laboratory tests, alone, are not enough to exclude the diagnosis of a substance use disorder. False positive results are also possible. The relative values of questionnaires, clinical signs, and laboratory tests are discussed elsewhere[24].

1.4 STRUCTURED CLINICAL INTERVIEWS

Structured clinical interviews are used for a formal and systematic diagnosis of substance use disorders by a trained clinician (e.g., a psychiatrist). However, they are not usually applied in common clinical practice because they take 1 to 2 hours to perform. This structured and standardized approach facilitates communication and allows clinicians and researchers to reduce uncertainty and variability in their diagnosis.

At present, there are two structured diagnostic systems for the clinical assessment of alcohol and psychoactive substance use disorders:

1. *The Diagnostic and Statistical Manual* (*DSM*) diagnostic system, prepared by the American Psychiatric Association. Presently, the fourth edition, (*DSM-IV*) is in use[3].
2. *The International Classification of Diseases* (*ICD*) diagnostic system, prepared by the World Health Organization (WHO). Presently the tenth edition (*ICD-10*) is in use[54].

The diagnostic criteria in these two systems has been further operationalized in the Composite International Diagnostic Interview (CIDI), a standardized instrument for the study of psychiatric epidemiology across different cultures and nations[55].

1.4.1 *Diagnostic and Statistical Manual (DSM) diagnostic system*

The DSM diagnostic system has evolved since the publication of its first edition in 1952. Over the years, substance use disorders received growing emphasis in the DSM. This is reflected by the increasing scope and prominence of the proportion of the manual devoted to these problems in newer editions.

In the first edition of the DSM, alcohol and drug use disorders were represented by three quarters of a page under "Acute Brain Syndromes", 6 lines under "Chronic Brain Syndromes Associated with Intoxication", and 8 lines under "Sociopathic Personality" sections. In the current edition, (*DSM-IV*), substance-related disorders are represented as a separate and prominent section and are divided into two subgroups:

- Substance Use Disorders: Substance Dependence and Substance Abuse
- Substance-Induced Disorders: Substance Intoxication, Substance Withdrawal, Substance-Induced Delirium, Substance-Induced Persisting Dementia, Substance-Induced Psychotic Disorder, Substance-Induced Mood Disorder, Substance-Induced Anxiety Disorder, etc.

In the *DSM-IV* manual, substance dependence and abuse are described by the following criteria.

SUBSTANCE DEPENDENCE

DSM-IV criteria for substance dependence

A maladaptive pattern of substance use leading to clinically significant impairment or distress, as manifested by three (or more) of the following, occurring at any time in the same 12-month period:

1. Tolerance, as defined by either of the following:
   a. A need for markedly increased amounts of the substance to achieve intoxication or desired effect.
   b. Markedly diminished effect with continued use of the same amount of the substance.

2. Withdrawal, as manifested by either of the following:
   a. The characteristic withdrawal syndrome for the substance.
   b. The same or a closely related substance is taken to relieve or avoid withdrawal symptoms.

3. The substance is often taken in larger amounts or over a longer period than was intended.

4. There is a persistent desire or unsuccessful efforts to cut down or control substance use.

5. A great deal of time is spent in activities necessary to obtain the substance (e.g., visiting multiple doctors or driving long distances), use the substance (e.g., chain smoking), or recover from its effects.

6. Important social, occupational, or recreational activities are given up or reduced because of substance use.

7. The substance use is continued despite knowledge of having a persistent or recurrent physical or psychological problem that is likely to have been caused or exacerbated by the substance (e.g., current cocaine-induced depression, or continued drinking despite recognition that an ulcer was made worse by alcohol consumption).

Specify if:

With physiological dependence:	evidence of tolerance or withdrawal (i.e., either Item 1 or 2 is present).
Without physiological dependence:	no evidence of tolerance or withdrawal (i.e., neither Item 1 nor 2 is present).

SUBSTANCE ABUSE

DSM-VI criteria for substance abuse

A A maladaptive pattern of substance use leading to clinically significant impairment or distress, as manifested by one (or more) of the following, occurring within a 12-month period:

1. Recurrent substance use resulting in a failure to fulfil major role obligations at work, school, or home (e.g., repeated absences or poor work performance related to substance use; substance-related absences, suspensions, or expulsions from school; neglect of children or household).

2. Recurrent substance use in situations in which it is physically hazardous (e.g., driving an automobile or operating a machine when impaired by substance use).

3. Recurrent substance-related legal problems (e.g., arrests for substance-related disorderly conduct).

4. Continued substance use despite having persistent or recurrent social or interpersonal problems caused or exacerbated by the effects of the substance (e.g., arguments with spouse about consequences of intoxication, physical fights).

B The symptoms have never met the criteria for Substance Dependence for this class of substance.

The *DSM-IV* is a result of a joint effort between the American Psychiatric Association and the World Health Organization (WHO). The *DSM-IV* asks the clinician to subtype dependence as existing with or without evidence of tolerance and physical withdrawal. The *DSM-IV* conceptualizes substance abuse as a maladaptive pattern of substance use that leads to adverse consequences and that occurs in the absence of dependence. The substance abuse criteria set in the *DSM-IV* have been expanded from two to four (compared to the *DSM-IIIR*) by the addition of "failure to fulfil major role obligations" and "recurrent legal problems". The *DSM-IV* cautions against diagnosing major psychiatric disorders based on symptoms that occur during intoxication or within a month after cessation of acute withdrawal.

1.4.2 *International Classification of Disease (ICD) diagnostic system*

ICD classification

Similar to the *DSM*, amendments have been made in the drug dependence sections of the *ICD* over the years. The *ICD* system was initially based on the alcohol dependence syndrome and has been extended for all drugs of dependence. The *ICD* diagnostic system does not promote the use of the word "drugs" because it is considered to be too broad. Specifically, the *ICD* targets psychoactive drugs that are likely to be self-administered owing to their reinforcing properties and substances that are used non-therapeutically. In the *ICD*, the term "physical dependence" is replaced by "neuroadaptation". The latter term was proposed because it is possible for a person to show signs of neuroadaptation without signs of dependence. Also, the term "abuse" is not used and is replaced by more specific terms such as unsanctioned use, hazardous use or, as in the latest version (*ICD-10*), harmful use. The diagnostic criteria for psychoactive substance use disorders in the latest version (*ICD-10*) are as follows:

HARMFUL USE

ICD criteria for harmful use

A pattern of psychoactive substance use that is causing damage to health. The damage may be physical (as in cases of hepatitis from the self-administration of injected drugs) or mental (e.g., episodes of depressive disorders secondary to heavy consumption of alcohol).

Diagnostic guidelines
The diagnosis requires that actual damage should have been caused to the mental or physical health of the user. Harmful patterns of use are often

criticized by others and frequently associated with adverse social consequences of various kinds. The fact that a pattern of use or a particular substance is disapproved by another person or by the culture, or may have led to socially negative consequences such as arrest or marital arguments is not in itself evidence of harmful use. Acute intoxication or "hangover" is not in itself sufficient evidence of the damage to health required for coding harmful use.

Harmful use should not be diagnosed if dependence syndrome, a psychotic disorder or another specific form of drug- or alcohol-related disorder is present.

DEPENDENCE SYNDROME

ICD criteria for dependence syndrome

Dependence syndrome is a cluster of physiological, behavioral, and cognitive phenomena in which the use of a substance or a class of substances takes on a much higher priority for a given individual than other behaviors that once had greater value. A central descriptive characteristic of the dependence syndrome is the desire (often strong, sometimes overpowering) to take psychoactive drugs (which may or may not have been medically prescribed), alcohol, or tobacco. There may be evidence that return to substance use after a period of abstinence leads to a more rapid reappearance of other features of the syndrome than occurs with nondependent individuals.

Diagnostic guidelines
A definite diagnosis of dependence should usually be made only if three or more of the following have been experienced or exhibited at some time during the previous year:

1. A strong desire or sense of compulsion to take the substance.

2. Difficulties in controlling substance-taking behavior in terms of its onset, termination, or levels of use.

3. A physiological withdrawal state when substance use has ceased or been reduced, as evidenced by the characteristic withdrawal syndrome for the substance or use of the same (or a closely related) substance with the intention of relieving or avoiding withdrawal symptoms.

4. Evidence of tolerance, such that increased doses of the psychoactive substance are required in order to achieve effects originally produced by lower doses (clear examples of this are found in alcohol- and opiate-dependent individuals who may take daily doses sufficient to incapacitate or kill nontolerant users).

5. Progressive neglect of alternative pleasures or interests because of psychoactive substance use, increased amount of time necessary to obtain or take the substance or to recover from its effects.

6. Persisting with substance use despite clear evidence of overtly harmful consequences, such as harm to the liver through excessive drinking, depressive mood states consequent to periods of heavy substance use, or drug-related impairment of cognitive functioning. Efforts should be made to determine that the user was actually, or could be expected to be, aware of the nature and extent of the harm.

1.4.3 *The Composite International Diagnostic Interview (CIDI)*

Although the diagnosis of substance use disorders has been standardized by the *DSM* and the *ICD* diagnostic systems, there is still a need for the rationalization and the standardization of the interview techniques to elicit the pertinent information from the patient. The Composite International Diagnostic Interview (CIDI) is a standardized and structured interview for the assessment of psychiatric disorders, including the substance use disorders, according to the *DSM-IIIR* (the third revised edition) and the *ICD-10* criteria[55]. The CIDI was designed to be used across different countries and cultures and was developed by the WHO and the United States Alcohol, Drug Abuse and Mental Health Administration (ADAMHA) Joint Project on Diagnosis and Classification of Mental Disorders and Alcohol and Drug Related Problems in 18 different locations around the world. The CIDI allows non-clinician interviewers to inquire about substance use disorders by translating the diagnostic criteria into standardized questions. The patients' answers can then be combined by a computer to yield diagnosis of substance abuse or dependence. In a recent field trial to assess the cross-cultural acceptability and reliability of the CIDI questions, the substance use questions were reported to be reliable and well accepted[12]. There were, however, problems in translating the relevant substance use concepts into different languages. The CIDI can be viewed as an important development for the study of substance abuse and dependence across different cultures and can improve the communication regarding these problems among various countries.

1.4.4 *Comparisons between the DSM and the ICD diagnostic systems*

Comparison of DSM-IIIR with ICD

The agreement between the *DSM* and the *ICD* systems for diagnosis of substance use disorders have been assessed in two studies[7,12]. In a field trial[12] evaluating the acceptability of the CIDI questions, nosological comparisons between the DSM-IIIR and the *ICD-10* systems were analyzed for alcohol, marijuana, opioid, and stimulant abuse and dependence (Table 8.1). Although there was a good agreement for alcohol and opioid use disorders, agreement was rather low for marijuana and stimulant use disorders. A study by Bryant and colleagues (1993)[7] found a high cross system agreement for substance dependence between the *DSM-IIIR* and *DSM-IV*. In this study, similar to the findings of Cottler and associates (1991)[12], the agreement between the *DSM-IV* and *ICD-10* for marijuana use disorders was low when compared to alcohol use disorders.

1.4.5 *Coexisting (comorbid) psychiatric disorders in patients with substance use disorders*

Psychiatric disorders frequently coexist with substance use disorders. In the United States, 52% of alcoholics and 75% of drug abusers were reported to have a second mental health disorder[36]. Depression, antisocial personality disorder and anxiety disorders, particularly phobias, are among the most commonly encountered psychiatric diagnosis in substance abusers and may be the reason for initial clinical presentation. The recognition of the coexisting psychiatric disorders is important as they may interfere with the diagnosis and treatment of the substance use disorders. For example, patients with concomitant antisocial personality disorder do not usually admit the problems associated with their substance use and are less likely to seek treatment for such problems. Similarly, presence of major depression in an alcohol dependent patient may increase the consumption of alcohol and the associated problems. Conversely, substance use

TABLE 8.1

Nosological comparisons between the *DSM-IIIR* and *ICD-10* substance use disorders (n = 581)

In this study an intraclass correlation test, Cohen's kappa measure of agreement, was used. A kappa > 0.75 means that the similarity between both tests is excellent; 0.4 < kappa < 0.75 means that the similarity is moderate; and a kappa < 0.40 means poor similarity. (Table based on a study by Cottler et al.[12])

DSM-III-R category	*ICD-10 category: no. of patients*			
	Dependence	*Harmful*	*Neither*	*Total*
Alcohol[1]: kappa = 0.81				
Dependence	85	11	8	104
Abuse	8	11	16	35
Neither	1	26	262	289
Total	94	48	286	428
Marijuana[2]: kappa = 0.48				
Dependence	0	12	5	17
Abuse	0	6	6	12
Neither	0	0	39	39
Total	0	18	50	68
Opioids[2] (heroin and nay opiates): kappa = 0.82				
Dependence	0	0	0	0
Abuse	0	10	2	12
Neither	0	0	8	8
Total	0	10	10	20
Stimulants[2] (includes cocaine, amphetamines): kappa = 0.46				
Dependence	0	5	7	12
Abuse	0	6	4	10
Neither	0	0	16	16
Total	0	11	27	38

1. Excludes 147 persons who never had a drink and six persons with missing data.
2. Includes only persons who had used these drugs at least six times.

can also affect the course of the existing mental illness by altering the effects of the psychiatric medications, decreasing compliance to medications, and precipitating psychiatric syndromes.

In summary, the assessment of a patient with drug dependence involves a multi-staged process (Figure 8.1). The assessment starts with screening tools and the suspected cases are further evaluated by formal diagnostic interviews.

The following case study illustrates several differential diagnoses in substance use disorders:

"Ms. A was a 52 year old married housewife who presented to the outpatient family practice services with complaints of slight tremor, anxiety, difficulty in concentrating, palpitations, feeling down and loss of pleasure and interest in life, difficulty in sleeping at night, and a decreased appetite.

Upon questioning, she admitted to have a depressed mood most of the time for the last few weeks. She has missed her bi-weekly bridge sessions which she used to enjoy very much. She also acknowledged being fearful, worried. She found it difficult to do housework because she felt she did not have the energy anymore. She had no previous history of depression or anxiety. Her anxiety and concentration problems were more recent and had started 2 days ago.

During the routine assessment of health risk factors such as smoking, blood pressure, and cholesterol, she admitted to her physician that she drank 4 to 5 glasses of red wine per day on most days of the week. She also added that she generally needed to have the first drink after she woke up in the morning, in order to make a good start for the day. She often felt annoyed when her daughter expressed concerns about her drinking habits.

Family history: Her mother is still alive and in good health for her age and her father died 12 years ago due to prostate cancer. Grandfather died of cirrhosis. One uncle committed suicide in his twenties. Her marriage has "some problems" and she thinks her husband is having an affair. Her children are all living on their own and generally have a good relationship with her.

Medication history: She was prescribed diazepam (a benzodiazepine sedative) 10 mg at night time for sleep problems 6 months ago and she has been using it regularly for the last 4 months. She had stopped using it 5 days ago since her medication finished. She said she did not have the energy to call her physician to renew the prescription. The only other medications she admitted using were vitamins.

Physical examination: Physical examination was unremarkable except a slight enlargement of the liver and several burn scars on the back of both hands. She explained that they happened when "she was ironing her clothes."

Differential diagnoses:

1. *Major depressive episode?*
2. *Alcohol abuse/dependence?*
3. *Sedative/hypnotic withdrawal?*
4. *Anxiety disorder?*

Follow-up analysis of the patient:

Patient's drinking history such as the need to drink as an eye-opener and feeling annoyed when criticized for her drinking habits (two positive answers according to the CAGE questionnaire) suggested a potential alcohol use disorder. Further assessment of Ms. A for DSM-IV diagnostic criteria for substance use disorders revealed the following findings: she stated (a) that she tried to cut down her drinking several times without any successful results, (b) that she would go to the liquor store to buy her wine even if she was ill and under heavy snow, (c) that she continued drinking even though her physician asked her to stop drinking due to peptic ulcer problems, (d) that she in fact started drinking approximately 8 years ago (only a glass of red wine before dinner) and that she never had the intention to drink 4 or 5 drinks a day at that time. In light of this evidence, Ms. A's clinical condition was diagnosed as alcohol dependence. Similar findings were also present for her diazepam use. For example, she could not sleep or function without taking her diazepam. Her physician tried to cut down her diazepam but she managed to obtain her prescription by going to several different doctors. She also had to take more diazepam in order to have clinical effects compared to 6 months ago when she first started using the medication. Therefore, she was also found to suffer from sedative/hypnotic (benzodiazepine) dependence. This latter diagnosis, the nature of clinical findings (e.g., anxiety, palpitations, difficulty in concentration), time-course of her anxiety (started a few days ago after she stopped taking diazepam) were consistent with a sedative/hypnotic withdrawal syndrome.

In addition, she was also diagnosed as having depression (e.g., depressed mood, loss of pleasure and interest in life, impaired daily functioning). She had no prior history of depression and her depression was found to be secondary to alcohol dependence.

2 Treatment of substance dependence

Traditionally, the focus of treatment for substance dependence has been on individuals demonstrating severe substance dependence with medical and psychosocial complications[44]. However, the spectrum of substance use disorders is not limited to such heavy users. For example, in the case of alcohol, late stage dependent alcoholics constitute only 5–6% of the adult population while approximately 20% of the adult population can be classified as "problem drinkers"[44]. This latter group of individuals differs from the severely dependent people in that they do not have a history of physical dependence or withdrawal symptoms and usually have a shorter alcohol-related problem history, greater social and economical stability and resources[48]. Although these problem drinkers use health care systems frequently, their problems may remain unrecognized and under-treated as a result of the current unwarranted emphasis on severely dependent cases[8,23].

Substance use disorders may affect individuals of all ages and socioeconomic backgrounds, and should be evaluated in all patients as an integral part of a medical assessment. Substance abuse or dependence may involve alcohol and illegal drugs as well as prescription medications. However, this is not usually recognized by physicians. In a study of general practitioners' beliefs about addiction, physicians were reported to be sympathetic to alcoholics, hostile towards opiate users, and unconcerned about anxiolytic abuse in their patients[37].

There is an increasing awareness of the importance of preventing health risk factors and most patients expect their physicians to inquire about their alcohol consumption and other health risks. Therefore, questions related to substance use disorders are more easily posed within the context of a general assessment of life-style and health risk factors following a non-threatening and non-judgmental approach.

The treatment of substance use disorders can be conceptualized within the context of the following four clinical problem areas: intoxication, withdrawal, rehabilitation and relapse prevention, and medical complications. In each area, treatment involves pharmacotherapy and/or psychosocial interventions such as a brief advice, counseling, group therapy, and cognitive therapy.

2.1 INTOXICATION

The clinical presentation of acute intoxication with substances may vary based on type, dose and route of administration of the substance being used, presence of tolerance, and the general medical condition of the individual. The clinical manifestations of substance intoxication and the toxic doses for various drug classes are illustrated in Table 8.2. Differential diagnoses should always be considered since other medical problems such as hypertensive crisis (e.g., cocaine intoxication) can be mistaken for substance intoxication.

Several substances such as alcohol, at intoxicating doses, depress cardiovascular and respiratory functions and can result in death. Substance intoxication can also result in disruptive behaviors. For example, alcohol has been implicated in the 4 leading causes of accidental deaths in the U.S.A (motor vehicle accidents, falls, drownings, and fires and burns). Therefore, the principle objective in the treatment of drug intoxication is the prevention of medical and psychosocial sequela during the intoxication period. In most substance intoxication cases, non-pharmacological measures such as provision of a safe and stable environment and reassurance are sufficient for the treatment of the patient. In severe cases, depending on the type of substance (e.g., alcohol), general medical support such as intravenous fluid treatment and, in some cases, respiratory assistance may be necessary until the blood drug concentrations return to physiologically tolerable levels. More vigorous interventions are rarely necessary and include the following measures:

- Elimination of the substance from the body can be enhanced by peritoneal or hemodialysis or through alterations of the pharmacokinetics of the drug by manipulation of the urinary pH.
- Inhibition of the effects of the drug at the receptor level by administration of receptor antagonists (pharmacological antagonists).
- Administration of agents which activate a physiological system that counteracts the effects of the drug (physiological antagonists).

Examples to pharmacological and physiological antagonism are the reversal of opiate coma by the opiate receptor antagonist naloxone, and stimulant (e.g., caffeine) treatment of hypnotic intoxication, respectively. The clinical value of physiological antagonism in the treatment of substance intoxication is not, however, well established and requires more research evidence before it can be recommended.

TABLE 8.2
Clinical manisfestations of substance intoxication and the associated toxic doses

Substance	*Signs and symptoms of intoxication*	*Toxic doses*
Alcohol	Impaired attention, poor motor coordination, ataxia, slurred speech, respiratory depression and death.	*Alcohol* blood levels > 200 mg/dl result in gross intoxication, and a blood level > 400 mg/dl is potentially lethal.
Sedative-Hypnotics	Findings of acute intoxication is similar to alcohol intoxication but patients do not have the alcoholic smell in their breath.	Toxic doses vary depending on the tolerance and the type of sedative-hypnotic.
Cocaine and other stimulants	Repetitive stereotyped behavior, panic, paranoia, irritability, dilated pupils, increased blood pressure, hyperthermia, seizures, respiratory depression, acute heart failure, stroke and death.	Toxic dose of *cocaine* is variable, more common with intravenous users or in those who smoke free base or crack. *Amphetamine,* at doses > 60 mg/day, may cause intoxication.
LSD (d-Lysergic acid diethylamide)	Panic reactions, extreme emotional lability, visual illusions, increased blood pressure and heart rate.	LSD is a potent hallucinogen and doses as low as 50 μg can cause intoxication.
PCP (phencyclidine)	Dose-related effects ranging from muscular incoordination and nystagmus to seizures, hypertensive crisis, respiratory depression, coma and death.	Doses > 5 mg may cause intoxication and a dose > 20 mg can be lethal.
Marijuana	Panic reactions, delirium, psychosis, flashbacks and emotional lability.	Toxic dose varies among individuals depending on the presence of tolerance. Smoking is more likely to cause acute intoxication compared to oral ingestion.
Opioids	Decreased gastrointestinal motility, constricted pupils, respiratory and central nervous system depression, coma and death.	Toxic dose varies depending on the tolerance.

2.2 WITHDRAWAL

Chronic and excessive drug ingestion leads to compensatory adaptive changes in the organism to counteract the effects of the addictive drug. Upon abrupt discontinuation of drug consumption, drug-dependent individuals suffer from withdrawal reactions which reflect the unbalanced compensatory biological changes that have taken place previously. The main objectives in the treatment of withdrawal reactions are prevention of complications and ensurance of a smooth transition into a long-term rehabilitation program.

Most withdrawal reactions are treatable by non-pharmacological interventions. For example, opioid withdrawal syndrome is characterized by findings such as diarrhea, vomiting, insomnia, runny nose, increased blood pressure and heart rate which last for approximately 2 weeks. These reactions are time-limited and are not life-threatening, thus can easily be controlled by reassurance, personal attention and general nursing care without a need for any pharmacotherapy. Recently, clonidine (a centrally acting alpha-2 receptor agonist) was reported to alleviate some of the opioid withdrawal findings such as increased blood pressure and heart rate and may shorten the duration of opioid withdrawal[35]. In some withdrawal reactions, such as barbiturate withdrawal and in severe alcohol withdrawal syndrome (AWS), pharmacological treatment may be necessary to avoid potentially serious complications (e.g., convulsions). In addition, severe withdrawal reactions, together with the increased craving for the addictive drug, may result in the resumption of drug ingestion. Administration of a pharmacological agent which is cross-tolerant to the addictive drug results in milder withdrawal reactions and enables the patient to successfully withdraw from the involved drug and start a rehabilitation program. This rationale is used in the pharmacological treatment of withdrawal reactions such as the AWS: benzodiazepines, owing to their cross-tolerance with alcohol and wide margin of safety are very effective and are currently the drugs of choice for the treatment of AWS[29]. Similarly, in opioid withdrawal, methadone, a long-acting opioid agonist, can be used to alleviate the initial symptoms of the opioid withdrawal. After several days, methadone dose can be tapered by approximately 10% every 1 or 2 days until the patient is free of methadone.

Recently, there has been important progress in the understanding of the neurobiological mechanisms mediating various withdrawal reactions. This may be expected to result in an expansion of available pharmacotherapies in the near future.

2.3 REHABILITATION AND RELAPSE PREVENTION

Recovery from substance use disorders involves not only the initial withdrawal from the involved substance(s) but also prevention of the resumption of substance use. However, it is not easy to achieve abstinence since one of the key features of drug dependence is a compulsive behavior pattern which maintains uncontrolled drug ingestion[26]. For example, up to 80% of alcohol-dependent individuals may relapse within 6 months of withdrawal from alcohol[32]. Moreover, the confirmation of total abstinence is fraught with difficulty. Relapse to substance use may be due to several factors such as social pressure, interpersonal conflicts, positive or negative mood states and craving for the addictive substance.

Craving is defined as an intense desire to use a drug and may play an important role in the initiation of relapse process. Approximately 4 decades ago,

Wikler (1948) reported an interesting clinical observation on craving in drug-free ex-opioid users: upon talking about their prior drug use, they often demonstrated signs and symptoms of opioid withdrawal such as sniffing, yawning and lacrimation, and a craving for the opioid drug. In recent years, several studies reported the contribution of classically conditioned drug-compensatory responses to craving. Siegel and other investigators have shown that the organism's homeostatic efforts to counteract the effects of the administered drugs may be classically conditioned to the administration procedures and rituals accompanying drug administration which may serve as "cues" elicitating craving (for a review, see ref. 10). The following clinical anecdote provides an example of craving and cues in drug-free former drug addicts:

Mr. Smith has been treated for alcoholism for the last 3 months. He had to go to another city for this treatment because there was no alcoholism rehabilitation facility in his home town. He has been completely sober for the last 2 months and was feeling very confident and happy for staying alcohol-free. He had repeatedly indicated to his therapist that he had no desire to drink alcohol anymore. Mr. Smith was discharged from the rehabilitation unit to return to his home town. Upon arrival to his town, Mr. Smith decided to visit a couple of friends to say hello. It was a dark Monday morning. He never liked this kind of weather as he used to be severely intoxicated with alcohol on days like this in the past. This weather made him feel somewhat shaky and restless. He also felt some desire to have a beer which disturbed him. While walking on the street, all of a sudden he came across to the bar where he used to drink all the time. At this point, he was started feeling extremely restless, irritable, shaky and had a strong urge to have several beers. He could not help entering the bar and started to drink. Within two hours, he drank 9 beers and was severely intoxicated.

As illustrated in the clinical anecdote, drug-compensatory responses, which are clinically similar to withdrawal signs and symptoms from the involved drug (e.g., irritability, restlessness) can create a craving for the drug and may result in relapse to uncontrolled drug use. Desensitization and/or avoidance of conditional stimulus (i.e., cues: the dark weather and the sight of the bar in the anecdote) which may produce such drug-compensatory responses, may assist ex-drug users to remain in control of their drug-seeking behaviors.

In addition to the awareness of potential reasons underlying craving, establishment of appropriate treatment goals is also important for a successful rehabilitation of the dependent individual. For a long time, the traditional goal of treatment for drug dependence has been complete and total abstinence from the addictive drug(s). This stems from the "disease" model of drug dependence which indicates that there is a progressive, and eventually terminal, loss of control over drug consumption. In the past several decades, this view point has been challenged: it has been shown that those individuals with a mild to moderate dependence problem and even some of those with heavy and chronic dependence, are able to moderate their drug intake at non-hazardous levels. For example, in a study of individuals with alcohol problems and mild dependence to alcohol, those who were instructed to moderate their alcohol intake were rehabilitated significantly better than those who were instructed to achieve complete abstinence[39].

In general, patients with mild to moderate drug dependence are encouraged for moderation of their drug intake while those with severe dependence are encouraged to achieve complete abstinence.

Treatment goals can be achieved by psychosocial and/or pharmacological interventions.

2.3.1 *Psychosocial interventions*

Psychosocial interventions can be performed following several approaches such as cognitive therapy, individual, group or family therapy, and self-help groups. A few of the many examples are described below.

Cognitive therapy
Cognitive therapy is based on the premise that habitual errors in perception, information processing, and decision making may lead to psychological disturbances and dysfunctional beliefs such as a self-destructive need to use substances in an attempt to cope with life stressors[6,56]. In the context of substance abuse, cognitive therapy aims to identify, modify, and change dysfunctional beliefs and attitudes towards substance use and build more realistic and functional thinking patterns.

Beliefs are relatively stable thoughts and images that, once formed, are not easily modified by ordinary life experiences. For example, a student can have a belief that, "Unless I use amphetamines, I cannot succeed in any examination," or a competitive executive may think, "If I use cocaine, it will help me to go through this stressful period in my job". Under high risk-circumstances such as social stress and negative emotional states (e.g., depression), these belief systems, in return, result in "automatic thinking patterns" resulting in decisions to seek an immediate relief or escape from the current unpleasant situation by substance use, without considering the long-term adverse consequences. These dysfunctional thinking and behavioral patterns tend to persist despite the presence of objective evidence that there are alternative and more functional ways of coping with these situations.

Cognitive therapy is an active form of psychotherapy and requires patients' cooperation. A significant portion of substance abusers view their problem as "beyond their control" and may not be willing to show initiative to develop insight to review their beliefs and thought patterns. Therefore, in such individuals, initial stages of cognitive therapy consist of developing skills for self-examination. After a certain degree of introspection has developed, the therapist starts challenging and probing the patient by asking questions such as, "Is there any objective evidence for this belief?" or, "Is there really no other way to cope with this situation?" in order to prompt the patient to test her/his belief systems objectively and build up constructive and adaptive behavior patterns. After the patient has reviewed her/his behavior, depending on the receptive capacity, the therapist gives the patient assignments or 'homework' to apply the alternative thinking patterns and coping skills discussed during the therapy to situations in everyday life. As a result, the patients learn to intervene with their destructive automatic thinking patterns leading to substance use. For example, when they feel an urge to use cocaine, they ask, "Is there any other way of looking to my situation?" or "What are the long-term consequences if I use cocaine now?". The active application of new forms of thinking and feedback from the therapist assist the patient to adapt a healthy life-style.

Individual and group counseling
Individual counseling plays a prominent role to educate and to provide support and advice to patients during the rehabilitation process. Cognitive therapy, coping skills training, and various other behavioral modification techniques can be used to counsel and assist the patients in giving up their addiction to drugs.
Contingency contracting is a behavioral treatment commonly employed during counseling. In this technique, a contract is negotiated with the patient in which positive changes in the addictive behaviors are rewarded while absence of progress or resumption of uncontrolled substance use can result in

the application of some penalties. Accordingly, under these positive and negative contingencies, patients are expected to show an initiative to change their dysfunctional behaviors. For example, in the methadone maintenance treatment for opioid addiction, patients are usually required to report to the methadone clinic on a daily basis to provide a urine sample (to confirm their abstinence from opioids) and to receive their daily methadone dose. Using contingency contracting technique, those patients who remain opioid free during the treatment can be rewarded by allowing take home privilege for the methadone, thereby waiving the requirement for daily reporting to the clinic. Conversely, failure to remain opioid free during methadone treatment may result in discontinuation of methadone treatment.

Psychotherapy and counseling can also be performed on a group basis. This may offer several advantages over individual therapy. Patients may be assisted by other members of the group, by role modeling, and by realizing that their situation is not "terminally unique" — similar problems are shared by other members of the society. The commonality of the substance use problems gives a sense of being understood and provides an incentive and tolerance to confront themselves and initiate a change to replace their destructive substance use behaviors. However, the composition and the infra-structure of the group is very important to obtain positive results with group therapy. For example, a severely impaired cocaine-dependent individual can be offended and humiliated in a group of substance abusers with greater functionality and less severe psychosocial problems. This may result in the failure of treatment attempts. Therefore, groups need to be carefully designed based on the characteristics of the individuals and the treatment goals.

Family counseling

Many of the substance users have marital and family problems. Frequently, the family structure is altered to accommodate the patient's destructive substance use problems. This affects the whole family, including the children, which puts them at increased risk to develop similar problems in their adulthood. Conversely, marital and family problems also may result in substance use disorders.

Counseling of the family and improvements in family and marital functioning result in more favorable treatment outcomes. For this purpose, several techniques such as training in communication skills, behavioral contracting, and conjoint hospitalization can be employed in order to restore the family functioning and to treat substance use problems.

Self-help groups

Twelve-step self-help programs such as Alcoholic Anonymous (AA), Narcotic Anonymous (NA), and Cocaine Anonymous (CA) are non-profit fellowship programs run by former substance users. The only requirement for a membership is a desire to give-up substance use. These programs endorse complete abstinence from the abused substance and accept the disease model of substance abuse which proposes that the individual is powerless to control her/his substance use under this disease. Self-help groups aim to rehabilitate the addicted individuals through recognition of the substance use problem (as an illness), promotion of spirituality, self-honesty, self-care, and role modeling with successful (drug-free) members of the group. Empirical evidence regarding the effectiveness of these programs has been difficult to collect. Therefore, a critical assessment is not possible at this moment.

2.3.2 *Pharmacological interventions*

Although psychosocial interventions are very useful in the rehabilitation of addicted individuals, they may have to be complemented with pharmacological interventions, particularly in cases with moderate to severe drug dependence.

Pharmacological interventions for drug dependence can be conceptualized within the context of the following three, mechanistically different, approaches:

1. Creation of an aversion for the drugs
2. Modifications of neurotransmitter function to decrease drug consumption
3. Long-term maintenance (substitution) with a less addictive and cross-tolerant medication

Aversive treatments

This form of pharmacotherapy is only available for the treatment of alcohol dependence. Two examples are disulfiram and calcium carbimide which inhibit aldehyde dehydrogenase, the enzyme responsible for the intermediary metabolism of ethanol. Metronidazole, an antiparasitic drug, also has similar pharmacological properties and has been used to treat alcoholism in the past. These drugs cause an episode of unpleasant reactions such as flushing, nausea, and vomiting upon ethanol consumption. In theory, then, these agents may potentially diminish the incentive for alcohol consumption and its rewarding effects. Although these medications are frequently prescribed for alcoholics, controlled clinical trials failed to show their efficacy in the treatment of alcohol dependence[18,34]. A recent meta-analysis of randomized clinical trials with these drugs also showed negative results[33]. In this study, even though the combined mean drug effects were 19.2%, 16.1%, and 27.5% for disulfiram (DS) (oral form), DS (implant form), and metronidazole (oral), respectively, the combined therapeutic differences (drug effect minus placebo effect) were not significant — 3.9% (–11.1 to 3.3%) (mean; 95% confidence interval) for DS-oral, –0.25% (–11.3 to 10.8%) for DS-implant and –0.5% (–15.4 to 14.4%) for metronidazole-oral. In this analysis, drug effects were expressed as percentage of complete abstainers and successfully treated cases (as indicated by study clinical global ratings), in the disulfiram and metronidazole groups, respectively. Meta-analysis was not possible for calcium carbimide, as there was only one study (with negative results) available[34]. These results indicate that alcohol sensitizing agents are not clinically efficacious in the treatment of alcohol dependence. The reported therapeutic effects in the literature are possibly due to nonpharmacological factors. However, these agents are still being prescribed, probably due to slow dissemination of research results into clinical practice. Critical analysis of the effectiveness of other drugs by meta-analysis is very difficult because of difficulties in data extraction and summary of results from various studies.

Modifications of neurotransmitter availability

Recent progress in the understanding of the molecular mechanisms of neuronal function and the neurobiological substrates of drug dependence have allowed the development of novel pharmacological interventions for the treatment of substance use and associated problems. There is considerable research evidence that drug dependence is mediated through a complex interaction of serotonergic, dopaminergic, and opioidergic neurotransmitter systems. Clinical trials with medications modulating the function of these neurotransmitter functions have yielded promising results[31]: selective serotonin reuptake inhibitors such as zimeldine, viqualine, citalopram, and fluoxetine have decreased alcohol consumption by averages of 14–20% compared to baseline consumption. Reductions of up to 60% were observed in some patients. Similarly, the dopamine agonist bromocriptine was reported to decrease alcohol consumption in a preliminary study while this was not confirmed in a recent multi-center clinial trial[30]. The opioid receptor antagonist naltrexone has been reported to decrease relapse rates and craving for alcohol after withdrawal from alcohol[50]. Although these results are promising, they need further verification to determine whether or not changes in alcohol consumption also translate to improvements in psychosocial functioning and well-being of the addicted individuals.

Future progress in the understanding of the molecular mechanisms of drug dependence should be able to synthesize these findings and guide the design of individually tailored pharmacotherapies matching the neurochemical profiles of drug-dependent patients.

Long-term maintenance treatments

A long-term maintenance treatment with a cross-tolerant medication that has less addictive properties and lower potential for causing medical complications alleviates the need to maintain destructive behaviors to obtain the drug. Counselling and application of behavioral treatments during the maintenance treatment assist the addicted individuals to establish more healthy lifestyles.

A typical example for this kind of pharmacotherapy is the methadone replacement therapy for opiod-dependent patients. Recently, nicotine patches have been introduced to replace smoking in nicotine-dependent individuals. In the latter case, nicotine patches eliminate the need for smoking and prevents the harmful effects of smoke and tobacco combustion products. However, more systematic and well-controlled clinical studies are needed to establish the efficacy of the replacements treatments.

2.4 MEDICAL COMPLICATIONS

Drug dependence can directly result in significant medical complications such as alcoholic cirrhosis, chronic organic brain syndrome, peptic ulcer, and death. However, some of the medical complications may also occur secondary to disruptive behaviors associated with drug-dependence such as motor-vehicle accidents, unsterile needle use, and unprotected sex. These patients may demonstrate indirect clinical presentations including fractures, human immunodeficiency virus (HIV) infection and hepatitis. Therefore, a high index of clinical suspicion is required to identify such patients. Treatment of medical complications involves treatment of the presenting disease as well as the underlying drug-dependence.

Acute toxic effects may occur on any occasion in which a drug is taken, e.g., fatal depression of respiration resulting from opioid overdosage, while other adverse effects arise only with chronic drug use, e.g., liver cirrhosis induced by heavy alcohol consumption. Table 8.3 gives examples of adverse effects arising from acute or chronic administration of some specific drugs.

A specific category of adverse effects is that produced in the offspring. Information concerning specific drug effects on offspring can be found in Chapter 2 of this textbook.

TABLE 8.3
Main medical complications after acute or chronic exposure to drugs of abuse

Drug	*Acute*	*Chronic*
Alcohol	– gastrointestinal irritation and dehydration (after high concentrated alcohol; > 40%) – respiratory paralysis – delirium	– liver cirrhosis – cardiovascular diseases – cancer of liver, upper digestive tract – gastric disorders – endocrine changes (hypogonadism, feminization) – Wernicke-Korsakoff syndrome
Amphetamine	– increased heart rate and blood pressure, cardiac arrhythmias – restlessness, tremor, insomnia – higher doses – confusion, hyperactivity, paranoia and purposeless, repetitive movements (psychotic state with symptoms like self injury and aggressive and paranoid behavior may appear)	– depletion of dopamine, resulting in posture and movement disorders, depression and amotivated state
Cocaine	– overdose, similar symptoms as amphetamine	– as for amphetamine – intranasal administration may lead to excessive mucus secretion and tissue ulceration in the nose
Inhalants	– depression, hallucinations, confusion, blurred vision, slurred speech – respiratory depression, nausea, vomiting, diarrhea and pain – arrhythmias, leading to cardiac arrest and death	– peripheral and central nervous toxicity (see also Chapter 14) – confusion, slurred speech, movement disorders and intellectual impairment after gasoline sniffing
Marijuana	– anxiety, disorientation, occasionally paranoid reaction	– very high rate use might impair fertility
Nicotine (tobacco use)	– no severe symptoms; very severe in case of nicotine as pesticide	– cardiovascular diseases – cancer of lungs, mouth, throat, larynx, esophagus – chronic lung disease – shortness of breath and excessive coughing
Opiates	– sedation, respiratory depression, lowered heart rate, nausea, vomiting	– sexual impairment – constipation – pinpoint pupils – health problems associated with injections/needle sharing (hepatitis, HIV etc.)

3 Outcome of the treatment of substance use disorders

Several reviews suggested a relatively poor outcome of treatment of substance use disorders. Catalano and colleagues (1988)[9] reported that the relapse rates ranged from 25% to 97% in opioid addicts after treatment. Similarly, 75–80% of patients with tobacco dependence were found to relapse within 1 year after cessation of smoking. Hunt and colleagues (1971)[20] reported that 65–70% of alcoholics, opioid users, and smokers relapsed within the first year of treatment. Relapse was particularly more common within the first 3 months of rehabilitation treatment.

However, these results should be interpreted with caution as most studies suffer from methodological limitations such as inadequate controls, lack of standardized measures of relapse, biases in patient selection, or inadequate follow-up. In addition, resumption of substance use does not always indicate that patients return to their pretreatment levels of substance use and they may still retain some of the improvements in other aspects of psychosocial functioning such as employment status or family functioning[15]. In general, high socioeconomic stability, lack of coexisting psychiatric disorders, negative family history of substance use, having a stable marriage or partner appear to correlate with a more favorable treatment outcome. However, the current consensus also indicates that no single treatment is effective for all patients. Definitive conclusions on the effectiveness of treatment of substance use disorders will have to await the results of well-controlled clinical studies.

Therapeutic interventions for drug intoxication, withdrawal, rehabilitation and relapse prevention, and medical complications are summarized in Table 8.4. Comprehensive review of the pharmacotherapy of substance use disorders is provided elsewhere[31].

TABLE 8.4
Treatment of substance use disorders

Clinical drug problem	*Treatment*
Intoxication	usually self-limited condition, non-pharmacological interventions (e.g., reassurance).
Withdrawal	
– mild to moderate withdrawal	usually self-limited condition, non-pharmacological interventions (e.g., reassurance, nursing care).
– severe withdrawal	non-pharmacological interventions plus pharmacological interventions (e.g., benzodiazepines in severe alcohol withdrawal).
Rehabilitation and relapse prevention	aversive treatments (e.g., disulfiram), questionable clinical efficacy.
	modification of neurotransmitter function (e.g., fluoxetine), promising findings, more research is needed.
	maintenance treatments (e.g., methadone), questionable clinical efficacy.
Medical complications	treatment of the presenting medical problems as well as underlying drug dependence.

4 Summary

A systematic clinical assessment is important for the effective rehabilitation of drug-dependent individuals. Clinical assessment includes the identification of drug use and associated problems, evaluation of response to treatment, and follow-up monitoring to prevent relapse. This is accomplished by a battery of assessment instruments including screening questionnaires (e.g., DAST, CAGE, and MAST), clinical findings, laboratory tests as well as formal diagnostic interviews such as *DSM-IV* and *ICD-10*. Treatment of drug dependence focuses on four clinical problem areas: drug intoxication, withdrawal, rehabilitation and relapse prevention, and medical complications. In each area, treatment may involve pharmacological and/or psychosocial interventions. Future developments in neurosciences should uncover the detailed neurobiological mechanisms of drug-dependence and further improve the treatment of drug- related problems.

Acknowledgment
Dr. Vural Özdemir was the recipient of a fellowship from the Canadian Society for Clinical Pharmacology and the Department of Medicine, University of Toronto.

References

1. Allen, J.P., Litten, R.Z., Anton, R., "Measures of alcohol consumption in perspective." In: Litten, R.Z. and Allen, J.P. (Eds.), *Measuring Alcohol Consumption*. Humana Press, Totowa, NJ, 1992, pp 205–226.
2. AMA, *New Frontiers: Understanding and Treating Alcoholism*. American Medical Association, Chicago, 1992, p 6.
3. APA, *Diagnostic and Statistical Manual of the Mental Disorders*, Fourth edition. American Psychiatric Association, Washington, D.C., 1994.
4. Babor, T.F., Brown, J., Del Boca, F.K., "Validity of self-reports in applied research on addictive behaviors: fact or fiction?" *Addict. Behav.* 1990, **12:** pp 5–32.
5. Babor, T.F., de la Fuente, J.R., Saunders, J.B. et al., *AUDIT: the Alcohol Use Disorders Identification Test. Guidelines For Use in Primary Health Care.* (MNH/76/DAT/89.4) World Health Organization, Geneva, 1989.
6. Beck, A.T., Rush, A.J., Shaw, B.F. et al., *Cognitive Therapy of Depression*. Guilford Press, New York, 1979.
7. Bryant, K., Babor, T., Kranzler, H. et al., "Cross system agreement for substance use disorders: DSM-IIIR, DSM-IV and ICD-10," *Addiction*. 1993, **88:** pp 337–348.
8. Buchan, I.C., Buckley, E.H., Deacon, G.L.S. et al., "Problem drinkers and their problems," *J. R. Coll. Gen. Pract.* 1981, **31:** 151-153.
9. Catalano, R., Howard, M., Hawkins, J. et al., *Relapse in the Addictions: Rates, Determinants, and Promising Prevention Strategies*, 1988 Surgeon General's Report on Health Consequences of Smoking. Office of Smoking and Health, U.S. Government Printing Office, Washington, D.C.
10. Childress, A.R., Ehrman, R., Rohsenow, D.J. et al., "Classically conditioned factors in drug dependence." In: Lowinson, J.H., Ruiz, P., Millman, R.B., Langrod, J.G. (Eds.), *Substance Abuse: A Comprehensive Textbook*. Williams & Wilkins, Baltimore, MD, 1993, pp 56–69.
11. Cone, E.J., "Testing human hair for drugs of abuse. I. individual dose and time profiles of morphine and codeine in plasma, urine and beard compared to drug-induced effects on pupils and behavior," *J. Anal. Toxicol.* 1990, **14:** pp 1–7.
12. Cottler, L.B., Robins, L.N., Grant, B.F. et al., "The CIDI-core substance abuse and dependence questions: cross-cultural and nosological issues," *Br. J. Psychiatry*. 1991, **159:** pp 653–658.
13. Cutler, S.F., Wallace, P.G., Haines, A.P., "Assessing alcohol consumption in general practice patients - a comparison between questionnaire and interview," *Alcohol,* 1988, **23:** pp 441–50.
14. Cyr, M.G. and Wartman, S.A., "The effectiveness of routine screening questions in the detection of alcoholism," *JAMA*. 1988, **259:** pp 51–54.
15. Daley, D.C. and Marlatt, G.A., "Relapse prevention: cognitive and behavioral interventions." In: Lowinson, J.H., Ruiz, P., Millman, R.B., Langrod, J.G. (Eds.), *Substance Abuse: A Comprehensive Textbook*. Williams & Wilkins, Baltimore, MD, 1993, pp 533–542.
16. Davidson, R. and Raistrick, D., "The severity of alcohol dependence questionnaire: its use, reliability and validity," *Br. J. Addict.* 1986, **78:** pp 145–155.
17. First, M.B., Frances, A.J., Pincus, H.A. et al., "Changes in substance-related, schizophrenic, and other primarily adult disorders," *Hosp. Commun. Psychiatry*. 1994, **45(1):** pp 18–20.
18. Fuller, R.K., Branchey, L., Brightwell, D.R., et al., "Disulfiram treatment of alcoholism. A veterans administration cooperative study," *JAMA*. 1986, **256:** pp 1449–1455.

19. Hodgins, D.C. and El-Guebaly, N., "More data on the Addiction Severity Index.: reliability and validity with the mentally ill substance abuser," *J. Nerv. Ment. Dis.* 1992, **180:** pp 197–201.
20. Hunt, W., Barnett, L., Branch, L., "Relapse rates in addiction programs," *J. Clin. Psychol.* 1971, **27:** pp 455–456.
21. Kapur, B.M. and Israel, Y., "A dipstick methodology for rapid determination of alcohol in body fluids," *Clin. Chem.* 1983, **29:** pp 1178.
22. Kintz, P., Tracqui, A., Mangin, P., "Detection of drugs in human hair for clinical and forensic applications," *Int. J. Leg. Med.* 1992, **105:** pp 1–4.
23. Kristensson, H., Lunden, A., Nilsson, B.E., "Fracture incidence and diagnostic roentgen in alcoholics," *Acta Orthop. Scand.* 1980, **51:** pp 205–207.
24. Levine, J., "The relative value of consultation, questionnaires and laboratory investigation in the identification of excessive alcohol consumption," *Alcohol Alcohol.* 1990, **25:** pp 539–553.
25. Maly, R.C., "Early recognition of chemical dependence," *Primary Care.* 1993, **20(1):** pp 33–50.
26. Marlatt, G.A., "Relapse prevention: theoretical rationale and overview of the model." In: Marlatt, G.A. and Gordon, J.R. (Eds.), *Relapse Prevention.* The Guilford Press, New York, 1985, pp 3–70.
27. Mayfield, D., McLeod, G., Hall, P., "The CAGE questionnaire: validation of a new alcoholism screening instrument," *Am. J. Psychiatry.* 1974, **131:** pp 1121–1123.
28. McLellan, A.T., Luborsky, L., Woody, G.E. et al., "An improved diagnostic evaluation instrument for substance abuse patients: the addiction severity index," *J. Nerv .Ment. Dis.* 1980, **168:** pp 26–33.
29. Naranjo, C.A., Ozdemir, V., Bremner, K.E., "Diagnosis and pharmacological treatment of the alcoholic patient," *CNS Drugs.* 1994, **1:** pp 330–340.
30. Naranjo, C.A., Bremner, K.E., Vachon, L. et al., "Long-acting bromocriptine (B) does not reduce relapse in alcoholics," *Clin. Pharmacol. Ther.* 1995, **57:** pp 161.
31. Naranjo, C.A. and Bremner, K.E., "Pharmacotherapy of substance use disorders," *Can. J. Clin. Pharmacol.* 1994, **1:** pp 55–71.
32. Naranjo, C.A. and Kadlec, K.E., "Possible pharmacological probes for predicting and preventing relapse in treated alcoholics," *Alcohol Alcohol. Suppl.* 1991, **1:** pp 523–526.
33. Özdemir, V., Busto, U., Einarson, T.R. et al., "Meta-analysis of the efficacy of alcohol-sensitizing agents in alcohol dependence," *Clin. Pharmacol. Ther.* 1994, **55:** pp 134.
34. Peachey, J.E., Annis, H.M., Bornstein, E.R. et al., "Calcium carbimide in alcoholism treatment. Part 1: a placebo-controlled, double-blind clinical trial of short-term efficacy," *Br. J. Addict.* 1989, 84, pp 877–887.
35. Rawson, R.A. and Ling, W., "Opioid addiction treatment modalities and some guidelines to their optimal use," *J. Psychoact. Drugs.* 1991, **23:** pp 151–163.
36. Robins, L.N., Locke, B.Z., Regier, D.A., "Overview: psychiatric disorders in America," Robins, L.N. and Regier, D.A. (Eds.), *Psychiatric Disorders in America.* Free Press, New York, 1990.
37. Roche, A.M., Guray, C., Saunders, J.B., "General practitioners' experiences of patients with drug and alcohol problems," *Br. J. Addict.* 1991, **86:** pp 263–275.
38. Ross, H.E., Gavin, D.R., Skinner, H.A., "Diagnostic validity of the MAST and the Alcohol Dependence Scale in the assessment of DSM-III alcohol disorders," *J. Stud. Alcohol.* 1990, **51:** pp 506–512.
39. Sanchez-Craig, M., "Random assignment to abstinence or controlled drinking in a cognitive-behavioral program: short-term effects on drinking behavior," *Addict. Behav.* 1980, **5:** pp 35–39.
40. Saunders, J.B. and Aasland, O.G., *World Health Organization Collaborative Project on Identification and Treatment of Persons With Harmful Alcohol Consumption. Report on Phase I. Development of a Screening Instrument.* (MNH/DAT/86.3), World Health Organization, Geneva, 1987.
41. Schuckit, M.A. and Irwin, M., "Diagnosis of alcoholism," *Med. Clin. N. Am.* 1988, **72:** pp 1133–1153.
42. Selzer, M.L., Vinokur, A., van Rooijen, L., "A Self-administered Short Michigan Alcoholism Screening Test (SMAST)," *J. Stud. Alcohol.* 1975, **36:** pp 117–126.
43. Selzer, M.L., "The Michigan Alcoholism Screening Test: the quest for a new diagnostic instrument," *Am. J. Psychiatry.* 1971, **127:** pp 89–94.

44. Skinner, H.A., "Spectrum of drinkers and intervention opportunities." *Can. Med. Assoc. J.* 1990, **143:** pp 1054–1059.
45. Skinner, H.A., Holt, S., Sheu, W.J. et al., "Clinical versus laboratory detection of alcohol abuse: the alcohol clinical index," *BMJ.* 1986, **292:** pp 1703–1708.
46. Skinner, H.A., "The drug abuse screening test," *Addict. Behav.* 1982, **7:** pp 363–371.
47. Skinner, H.A., "Alcohol dependence syndrome: measurement and validation," *J. Abnorm. Psychol.* 1982, **91:** pp 199–209.
48. Skinner, H.A., "Assessment of alcohol problems," *Res. Adv. Alcohol Drug Probl.* 1981, **6:** pp 319–369.
49. Stockwell, T., Murphy, D., Hodgson, R., "The severity of alcohol dependence questionnaire: its use, reliability and validity," *Br. J. Addict.* 1983, **78:** pp 145–155.
50. Volpicelli, J.R., Alterman, A.I., Hayashida, M., O'Brien, C.P., "Naltrexone in the treatment of alcohol dependence," *Arch. Gen. Psychiatry.* **49:** pp 876–880.
51. Whitehead, T.P., Clarke, C.A., Whitfield, A.G.W., "Biochemical and hematological markers of alcohol intake," *Lancet.* 1978, **1:** pp 978–981.
52. Wikler, A., "Recent progress in research on the neurophysiological basis of morphine addiction," *Am. J. Psychiatry.* 1948, **105:** pp 328–338.
53. Woodruff , R.A., Clayton, P.J., Cloninger, R. et al., "A brief method of screening for alcoholism," *Dis. Nerv. Syst.* 1976, **37:** pp 434–435.
54. WHO, *Tenth Revision of the International Classifications of Diseases.* Chapter V. *Mental and Behavioral Disorders.* World Health Organization, Geneva, 1989.
55. WHO, *Composite International Diagnostic Interview.* World Health Organization, Geneva, 1987.
56. Wright, F.D., Beck, A.T., Newman, C.F., et al., "Cognitive therapy of substance abuse: theoretical rationale," *NIDA Res. Monogr.* 1993, **137:** pp 123–146.

Supplementary references

Chick, J., Kreitman, N., Plant, M., "Mean cell volume and gamma-glutamyl transpeptidase as markers of drinking in working men" *Lancet,* 1981, **1:** pp 1249–1251.

Fuller, R.K., Branchey, L., Brightwell, D.R et al., "Disulfiram treatment of alcoholism. A Veterans Administration cooperative study," *JAMA.* 1986, **256:** pp 1449–1455.

Gavin, D.R., Ross, H.E., Skinner, H.A., "Diagnostic validity of the Drug Abuse Screening Test in the assessment of DSM-III drug disorders," *Br. J. Addict.* 1989, **84:** pp 301–307.

Liskow, B.I. and Goodwin, D.W., "Pharmacological treatment of alcohol intoxication, withdrawal and dependence: a critical review," *J. Stud. Alcohol.* 1987, **48:** pp 356–370.

Moore, R.D., Bone, L.R., Geller, G. et al., "Prevalence, detection and treatment of alcoholism in hospitalized patients," *JAMA.* 1989, **261:** pp 403–407.

Naranjo, C.A. and Bremner, K.E., "Behavioral correlates of alcohol intoxication," *Addiction.* 1993, **88:** pp 25–35.

Sellers, E.M., Naranjo, C.A., Harrison, M. et al., "Oral diazepam loading: simplified treatment of alcohol withdrawal," *Clin. Pharmacol. Ther.* 1983, **34:** pp 822–826.

Selzer, M.L., Vinokur, A., van Rooijen, L., "A self-administered Short Michigan Alcoholism Screening Test (SMAST)," *J. Stud. Alcohol.* 1975, **36:** pp 117–126.

Staley, D. and El-Guebaly, N., "Psychometric properties of the Drug Abuse Screening Test in a psychiatric patient population," *Addict. Behav.* 1990, **15:** pp 257–264.

Sullivan, J.T., Sykora, K., Schneiderman, J., et al., "Assessment of alcohol withdrawal: the revised clinical institute withdrawal assessment for alcohol scale (CIWA-Ar)," *Br. J. Addict.* 1989, **84:** pp 1353–1357.

Wallace, P.G. and Haines, A.P., "General practitioner and health promotion: what patients think," *BMJ.* 1984, **289:** pp 534–536.

Whitfield, J.B., Hensley, W.J., Bryden, D. et al., "Some laboratory correlates of drinking habits," *Ann. Clin. Biochem.* 1978, **15:** pp 297–303.

Woody, G., Schuckit, M., Weinrib, R., Yu, E., "A review of the substance use disorders section of the DSM IV," *Psychiatr. Clin. N. Am.* 1993, **16:** pp 21–32.

Contents Chapter 9

New treatment strategies

0-8493-7803-6/99/$0.00+$.50

Chapter 9

New treatment strategies

J.R. Volpicelli, L.A. Volpicelli,
A.E. Kaminski, Y.J. Hong and K.L. Clay

INTRODUCTION

Although many treatment programs for drug and alcohol abuse currently exist, many of these programs are only modestly successful in preventing early relapse and helping a patient to maintain abstinence. Because of the high relapse rates, it is important to find a pharmacological adjunct effective in preventing a return to drug abuse. There are two main pharmacological approaches to helping patients remain off drugs — substitution therapy and biochemical blocking therapy. These approaches increase the effectiveness of treatment because they block the positive and negative reinforcements for returning to drug use. While pharmacological adjuncts are important in treating substance abuse, an effective psychosocial program is also important for helping patients maintain compliance. Therapists and counseling can motivate the patient to remain in treatment. Thus, an effective medication in conjunction with psychosocial treatment is important in preventing relapse and maintaining abstinence from substance abuse.

Treatment for drug dependence can be broken down into two phases:

1. Detoxification: the use of medical treatment to decrease acute withdrawal symptoms and help the patient stop using the drug
2. Rehabilitation: psychosocial and medical treatments to help patients abstain from using the drug

Detoxification involves helping patients cope with physiological and affective withdrawal. Some drugs such as alcohol, sedatives, and tranquilizers produce severe life threatening withdrawal symptoms such as delirium tremens and seizures. It is necessary to support patients as they go through withdrawal from the drug. Typically, medicines are used that block withdrawal symptoms. In order to prevent abuse of the drug used for detoxification, drugs not associated with affective pleasure or medicines with a long halflife are preferred. For example, opiate detoxification often involves the use of the long acting opiate substitute or clonidine, a medicine that blocks the hyperactive sympathetic nervous system. Alcohol detoxification is safely accomplished with the use of benzodiazepines such as oxazepam, a medicine that does not produce affective pleasure and therefore does not lead to dependence. Medicines used during detoxification are given as a person stops drug use and then the medications are gradually tapered over the next several days. Typically in the past, detoxification was accomplished in an inpatient setting. However, contrary to many misconceptions, drug withdrawal does not have to be a terrible experience in a medically supervised setting. Most patients can be safely withdrawn from a drug as an outpatient.

Following detoxification, patients enter the rehabilitation phase of treatment. Rehabilitation can last over a period of a few months. Here, the patients receive various forms of therapy to support them while they are abstaining from drugs. The patients are encouraged to confront their drug problem, and to receive support from family, friends, and other recovering patients. Often therapy consists of group sessions in which they are encouraged to express their feelings and learn to cope with their change of lifestyle. The patients are also taught how to deal with environments in which they may be tempted to use drugs. For example, one behavioral approach to therapy involves exposure to drug reminder cues that can cause craving. The patient is exposed to those internal and external cues associated with drug use. Without the opportunity to use drugs, continual exposure to these reminder cues weakens the association between drug use and the drug reminder cue. In other words, the conditioned response to the stimulus becomes extinct. Over time, drug reminder cues lose the ability to stimulate craving and physiological withdrawal reactions. Thus, the individuals can re-integrate themselves into an environment in which drug reminder cues may exist. Therefore, therapy is used to support and motivate patients during their abstinence.

In addition to psychosocial treatment, pharmacological adjuncts are used to help the patient maintain abstinence from the drug. For example, methadone is used in rehabilitation from opiate dependence. *Methadone* is an opiate substitute, therefore it blocks the negative reinforcement for using opiates such as heroin. Methadone is an improvement from heroin because patients receive the drug in a safe environment such as the treatment center and because it is administered orally as opposed to i.v. injected. Patients no longer feel craving for the heroin, and once adapted to methadone patients do not feel high from the medicine. Another pharmacological approach to treatment of opiate dependence is the use of an opiate blocker such as *naltrexone*. Naltrexone attenuates the high the patients receive from opiate use and thus blocks positive reinforcement for using opiates. Therefore, pharmacological adjuncts for rehabilitation can either be substitution in order to prevent negative reinforcement or biochemical blocking in order to prevent positive reinforcement for using a drug.

1 Treatment for cocaine and amphetamine dependence

While cocaine and amphetamines both produce similar effects such as feelings of euphoria and stimulation, the mechanisms by which they affect the brain are different. Cocaine increases the amount of dopamine by preventing its re-uptake[31]. The feelings of pleasure and stimulation associated with use of cocaine are caused by this increase in dopamine. Amphetamines, however, work by increasing the amount of dopamine released. This increase in dopamine results in feelings of euphoria and stimulation similar to cocaine[52].

1.1 DETOXIFICATION

Cocaine dependent patients can be withdrawn immediately from cocaine without risk. The side effects associated with withdrawal are attributed to lower levels of dopamine. According to the dopamine depletion hypothesis, chronic cocaine abuse results in the reduction of dopamine levels. Thus chronic cocaine users, when not using cocaine, have lower levels of dopamine in comparison to non-cocaine users. The low levels of dopamine caused by acute withdrawal from cocaine result in depression or the "crash"[58]. The dopamine depletion hypothesis inspired the study of dopamine agonists such as *bromocriptine, amantadine,* and *pergolide mesylate* for detoxification from cocaine abuse.

Bromocriptine is a dopamine agonist. A pre-clinical study by Clow et al.[9] that measured brain metabolic activity in cocaine treated rats supports the use of bromocriptine. In this study, rats received either cocaine or saline solution for 14 days and then received 3 days of treatment with either bromocriptine or saline solution. Cerebral metabolic activity was measured in both groups of rats during abstinence from the treatment with cocaine. A decrease in metabolic activity in the rats treated with cocaine was found in comparison to the saline treated rats. However, in the rats that received treatment with bromocriptine, the decrease in metabolic activity was inhibited. Therefore, this pre clinical study suggests that bromocriptine could be effective treatment for cocaine abuse because it blocks the reduction of brain metabolic activity, particularly activity involved in dopamine release, during abstinence.

Clinical studies also support the use of bromocriptine for treatment of cocaine abuse. In a double-blind study by Dackis et al.[11], bromocriptine reduced craving for 13 subjects admitted to the hospital for inpatient treatment of cocaine dependence. In a study done by Extein et al.[15] ten patients addicted to crack cocaine who were given bromocriptine experienced a decrease in withdrawal symptoms. None of the patients experienced aversive side effects except for mild nausea. However, a review of the literature by Tutton[58] revealed that many subjects who receive bromocriptine experience significant side effects such as nausea or headache. Consideration must be given to these side effects before bromocriptine can be used as a pharmacological adjunct for treatment of cocaine abuse.

In contrast to bromocriptine, amantadine does not seem to produce significant adverse side effects. As Tutton[58] explains, amantadine is another dopamine agonist which acts by not only releasing dopamine to the synapses but also by blocking the reuptake of dopamine.

In an open trial study done by Gawin et al.[24] 50 patients given amantadine experienced reduced craving for cocaine soon after administration. However, while many studies support the idea that amantadine decreases craving, the reduction in craving does not seem to last for longer than 3 weeks. For example, Giannini et al.[25] did a study comparing the effectiveness of amantadine and bromocriptine in 30 subjects. Both bromocriptine and amantadine were effective, but the effects of amantadine attenuated after 15 days. In another study comparing the effects of bromocriptine and amantadine[56] all 14 subjects experienced a decrease in craving for cocaine. However, amantadine proved to be a more effective treatment because a few subjects taking bromocriptine dropped out of the study due to adverse side effects[58].

Other studies show less promising results for the effectiveness of amantadine. For example, in a study of 42 subjects[3] amantadine was not found to decrease craving. However, analysis of subject urines showed that abstinence rates for subjects on amantadine were significantly greater than for placebo treated subjects. Thus, results on the effectiveness of amantadine are mixed. Yet, with further study, amantadine might still prove efficacious in treating cocaine abuse.

Another dopamine agonist currently being studied for the treatment of cocaine abuse is pergolide mesylate, a drug currently used for treating Parkinson's disease. Pergolide may be more effective than bromocriptine and amantadine because it acts by binding to D_1 and D_2 dopamine receptors[58]. Also, as explained by Tutton[58], the effects of pergolide seems to be 10–100 times stronger and last longer than the other medications. In one study by Malcolm et al.[42], of 21 patients given pergolide, 16 experienced a decrease in craving for cocaine and an improvement in sleep and concentration. While side effects such as nausea were reported, they occurred less often than with bromocriptine.

Currently there are no medications for detoxification from amphetamines. However, since depression is often noted as a result of amphetamine withdrawal, antidepressants may be an effective treatment.[31]

1.2 REHABILITATION

Reports that many individuals experience depression following cocaine withdrawal inspired research on the use of antidepressants for rehabilitation from cocaine abuse.

In a study by Gawin et al.[21], 72 chronic cocaine abusers were treated with *desipramine, lithium,* or placebo. Patients treated with desipramine had higher rates of abstinence and reported greater decreases in craving compared to subjects treated with either lithium or placebo. The effectiveness of desipramine was also exemplified in a study that compared the effects of amantadine and desipramine.[62] While all 54 patients in this study experienced a decrease in craving and depression, patients on desipramine maintained longer periods of abstinence. While studies show that desipramine is an effective treatment, the effects are not immediate. A study by Kleber et al.[38] also showed that desipramine reduces craving. However, the effects of desipramine did not emerge until 3–4 weeks after treatment began. Patients, however, eventually stopped using cocaine.

In addition to dopamine agonists and antidepressants (that block the negative reinforcement for using cocaine), dopamine antagonists that block the positive reinforcement for using cocaine are also currently being studied. Spealman et al.[54] did a comparative study using squirrel monkeys on the dopamine antagonists *SCH39166* (a D_1 receptor blocker), *YM09151-2* (a D_2 receptor blocker), and *flupenthixol* (a non-selective blocker). When the monkeys received moderate doses of cocaine alone, there was an increase in response rates to stimulus-shocks. When high doses of cocaine were administered, a decrease in response rates was observed. When the monkeys received pre-treatment with SCH39166 and YM09151-2 the rate increase and rate decrease effects of cocaine were blocked 13 fold. Treatment with flupenthixol inhibited the response rates 32 fold. Thus, the dopamine antagonists were successful in blocking the effects of cocaine. In addition to the pre-clinical study, Gawin and Allen[22] studied the use of flupenthixol in a clinical trial on 10 crack cocaine users. The results of this study showed that nine of the subjects experienced a decrease in craving for cocaine. Subjects also remained in treatment for a longer amount of time in comparison to other treatments in which subjects often drop out early.

In a pre-clinical study on 78 rats, Mackey and Kooy[41] found that flupenthixol prevented a preference for amphetamines. In addition, studies have shown that the *Sherring D_1 antagonists* block amphetamine preference[40]. Thus, dopamine antagonists such as flupenthixol and the Sherring D_1 antagonists could prove to be an effective treatment for abuse of amphetamines.

In addition to dopamine antagonists, drugs that block the interaction of *glutamate* with dopamine have been suggested for treatment of cocaine and amphetamine abuse[63]. It is thought that glutamates influence the activity of mesolimbic dopamine. In this study, mice were pre-treated with various glutamate antagonists or placebo and then injected with cocaine or methamphetamine. The results show that the glutamate antagonists reduced the locomotor stimulation caused by cocaine or amphetamines. Further study of glutamate antagonists may show them to be an effective adjunct in the treatment of cocaine and amphetamine abuse.

Recent studies have shown that dopamine agonists such as amantadine, bromocriptine, and pergolide are effective for decreasing craving and withdrawal symptoms for chronic cocaine abusers. However, because of side effects and the short-term efficacy of these medications, antidepressants and dopamine antagonists seem more promising in aiding patients to achieve long term abstinence from cocaine and amphetamines. However, currently no FDA approved pharmacologic adjunct for the treatment of cocaine and amphetamine dependence exists.

2 Treatments for opiate dependence

Opiates act by binding to the mu opioid receptors thus increasing endorphin levels. *Endogenous opioids (endorphins)* are naturally released during periods of stress to reduce physical and emotional pain. This release also causes the feelings of euphoria associated with opiates.

2.1 DETOXIFICATION

During opiate detoxification, two agonist drugs are primarily used to slowly withdraw patients from opioid use. These two agonists are methadone and *clonidine (Catapres®; Catapresan®)*. Both drugs are used for different types of patients and in different settings.

Currently, the only drug approved by the FDA for decreasing doses of opioids is methadone. Methadone is a mu receptor agonist which, in the US, is primarily used in an inpatient setting. Methadone in an outpatient setting seems to be far less effective than inpatient treatment[36]. Methadone is administered orally daily and dosage varies with each patient. Initially, methadone also creates the feelings of euphoria associated with heroin. By recreating the "high" in a controlled environment, opiate abusers have no need to return to their habit. However, over time, Methadone no longer creates the "high" and serves only to prevent craving for heroin.

One of the first studies conducted with methadone used several hundred chronic heroin addicts. This study found a low drop out rate, a major decrease in illicit opioid use, a decrease in criminal behavior, and a gain in legitimate employment[36]. Because of these results, Dole and Nyswander[13] hypothesized that opioid dependence is the cause of psychological problems as opposed to psychological problems being the cause of dependence on opioids.

Another drug used for opioid detoxification is clonidine. Clonidine is a sigma 2 agonist drug currently marketed to treat hypertension. Clonidine is also the only non-opioid drug found to be effective in the detoxification of illicit opioid dependence, but it is not yet approved by the FDA for the treatment of opioid dependence. The use of clonidine as an anti-withdrawal drug began with experiments by Gold et al.[29] They found that an oral dose of clonidine hydrochloride (5 g/kg) suppressed opioid withdrawal symptoms.

Clonidine can be used in inpatient settings, but is utilized more often to facilitate an outpatient detoxification program. Studies have also shown that clonidine is less successful in detoxifying street addicts. The only side effects reported from the use of clonidine are sedation and hypotension.

The apparent suppression of opioid withdrawal symptoms that Gold and his colleagues[30] found were refuted in a study conducted by Charney et al.[8] Charney found that methadone relieves withdrawal symptoms such as anxiety, restlessness, insomnia, and muscular activity. However, methadone did not suppress these symptoms in patients taking clonidine. A study by Jasinski et al.[37] found results that agreed with both Gold and Charney's studies. They compared the effects of clonidine to morphine and to placebo. As Gold and his colleagues found, the placebo was less effective than clonidine in reducing abstinence and sickness scores. In lowering the autonomic signs of withdrawal, clonidine was much more effective than morphine. On the other hand, morphine was much more effective in reducing the subjective discomforts of withdrawal. Therefore, the results of this study support both Gold's and Charney's conclusions.

2.2 REHABILITATION

Another maintenance drug used for opioid addicts is a semi-synthetic lipophilic opioid derived from thebaine called buprenorphine. Buprenorphine is also referred to as Temgesic. This drug is a partial mu receptor agonist that dissociates very slowly from opioid receptors. Buprenorphine produces morphine-like effects, including euphoria as well as side effects similar to morphine-like opioids. Buprenorphine also produces a ceiling effect so that patients can only achieve the same effect as low doses of morphine. Therefore, since the patient cannot induce a greater "high" from taking a greater amount of buprenorphine, the chances for overdosing on the medication are reduced. Another advantage it is that high doses accelerate abstinence symptoms while low doses extinguish withdrawal. Buprenorphine is currently available as an analgesic for use in Europe and Australia, but it is not yet approved to be used as a maintenance drug for opioid dependents.

The only opioid antagonist that is FDA approved for treatment of opioid dependence in the United States is naltrexone which can also be called Trexan. Naltrexone is a relatively pure antagonist that is taken orally. There are two types of dosages given out to patients. The dosages are either 50 mg per day (7 days a week) or 100 mg on Monday and Wednesday and 150 mg on Friday. Therefore, either way, the weekly dosages is 350 mg.

3 Treatments for alcohol dependence

There are no receptors specific for alcohol. However, according to one theory, alcohol increases levels of serotonin in the brain. In addition, it has been hypothesized that alcohol enhances endogenous opioid activity.

3.1 DETOXIFICATION

The *imidazobenzodiazepine* Ro15-4513 has been studied for its ability to hinder intoxication. Ro15-4513 is an inverse agonist at the GABA-benzodiazepine receptor. Thus, it binds to the receptor site and inhibits binding of benzodiazepines. Inverse agonists lower the seizure threshold and produce anxiety[31]. It is believed that they block the immediate effects of alcohol. In a preclinical study[33], Ro15-4513 was shown to prevent the effects of alcohol on coordination in mice. Another study by Suzdak et al.[55] showed that both pre treatment and post treatment with Ro15-4513 antagonized the intoxicating effects of alcohol in rats. Thus, Ro15-4513 could prove to be an effective medication for blocking intoxication by alcohol.

Benzodiazepines are typically used for detoxification because they decrease potentially fatal withdrawal symptoms such as seizures and delirium. One benzodiazepine currently being studied is *oxazepam (Serax®)* that has a low potential for abuse, is not metabolized by the liver, and is an effective anti-seizure medication. *Flunitrazepam*, another benzodiazepine is also being studied for use in alcohol detoxification. In a study by Pycha et al.[45], the sedative effects of flunitrazepam began one to two minutes after i.v. injection. All the patients in the study recovered completely from withdrawal. In addition, except for one patient who had respiratory depression (which was reversed by flumazenil, a benzodiazepine antagonist) no other patients complained of side effects.

Worner et al.[64] studied the effects of *diazepam*, another benzodiazepine used for detoxification from alcohol. None of the subjects taking diazepam experienced withdrawal seizures or delirium tremens. While benzodiazepines are effective in reducing withdrawal symptoms, there are problems associated with their use. For example, there is a potential for a patient to abuse and become addicted to the benzodiazepines. Also, there is evidence that benzodiazepines alter the CNS neurotransmitters. However, benzodiazepines continue to be the safest and most effective treatment for acute alcohol withdrawal.

Anticonvulsants such as carbamazepine (*Tegretol®*) may also aid alcohol detoxification. It is hypothesized that anticonvulsants will prevent seizures and nervous instability that some people experience as a result of withdrawal from alcohol.

3.2 REHABILITATION

Disulfiram, lithium, and serotonin specific reuptake inhibitors (SSRIs) are the most commonly researched pharmacological adjuncts for the treatment of alcoholism.

Disulfiram (antabuse) blocks the metabolism of alcohol which subsequently causes nausea, abdominal pain, and vomiting. The theory behind disulfiram is that patients are conditioned to associate these adverse events with alcohol drinking and are thus motivated to stop drinking. While antabuse is the only currently FDA approved drug for the treatment of alcoholism, its use is limited by low medication compliance. In a study by Fuller et al.[19] compliance was determined by the presence of a riboflavin marker in urine. Using this method, Fuller found that only 20% of subjects regularly took medication as prescribed. Therefore, disulfiram may only be effective for patients who are highly motivated to stop drinking.

Lithium, which is used to treat bipolar and other mood disorders, was also studied as an adjunct to treatment. In theory, lithium would eliminate existing mood disorders which may cause the motivation to drink. However, in a placebo controlled study, there was no significant difference in improvement between depressed and nondepressed patients. Using lithium for the treatment of alcohol dependence may only be effective in a subpopulation of alcoholics with bipolar disorders.

Alvarado et al.[31] showed that alcoholism is associated with low levels of serotonin. This information thus inspired research on serotonin specific reuptake inhibitors (SSRIs) for treatment of alcohol dependence. One limitation of SSRIs, however, is that they decrease appetite for food in addition to that for alcohol. Research on SSRIs as an adjunct to treatment is still in progress.

Another theory on the motivation to drink excessive amounts of alcohol is that alcohol increases the activity of the endogenous opioid receptors in the brain[10,12]. Alcoholics drink in order to get the "high" from increases in beta endorphins. Thus, an antagonist which blocks this "high" would reduce the motivation to drink. The following evidence supports this theory: (1) preclinical animal studies show that alcohol drinking can also stimulate opiate receptor activity, (2) alcohol drinking is influenced by the use of opiates such as morphine, and (3) alcohol drinking is reduced by opiate antagonists, such as naltrexone. Naltrexone shows promising results as an effective and safe drug for treating alcoholism.

Preclinical studies have shown that opiate antagonists reduce alcohol preference in animals. Marfaing-Jallat, Miceli, and Le Magnen[43] found that *naloxone*, a short acting injected opiate antagonist, reduces the preference of alcohol in rats given a choice between water and alcohol. These results have been replicated by numerous researchers in a variety of studies[18,39,51].

In addition, Altshuler and colleagues[4] found that naltrexone decreased lever-pressing for i.v. alcohol in rhesus monkeys. Also, Volpicelli, Davis and Olgin[60] found that rats increase preference for alcohol after sessions of inescapable shock. The research found that naltrexone blocks these stress-induced increases in alcohol drinking.

The preclinical research motivated research on the effectiveness of naltrexone in alcohol dependent humans. For example, Volpicelli et al.[61] administered 50 mg naltrexone or placebo daily to 70 alcohol-dependent male veterans, the majority of whom were African-American, unemployed, and had been drinking heavily for a mean of 20 years. The rates of relapse and the number of days drinking were significantly lower in the naltrexone treated groups compared to those subjects on placebo. In another study, O'Malley et al.[44] administered naltrexone or placebo to a different population than the Volpicelli study. While the majority of the subjects were male, O'Malley also included women. Subjects were predominately Caucasian and were employed full time. The effects of naltrexone in O'Malley's study paralleled Volpicelli's study. For example, the subjects taking naltrexone relapsed less often and drank a lesser amount of alcohol than the subjects taking placebo. These double blind studies suggest that naltrexone treatment may be an effective adjunct to the treatment of alcohol dependence.

4 Treatment for dependence on other drugs of abuse

4.1 NICOTINE

Some of the effects of nicotine are a reduction of anxiety and depression and an increase in the ability to concentrate. As discussed by Vaughan[59], nicotine causes these rewarding effects by stimulating dopamine receptors. Particularly, according to Rosecrans and Karan[49], nicotine affects the nucleus accumbens which contain dopamine neurons. In fact, studies show that nicotine binds with nicotinic-acetylcholinergic receptors that are located on dopamine and serotonin neurons. Thus, the reinforcing properties of nicotine are influenced by the regulation of dopamine.

Current treatment for nicotine addiction involves substitution therapy to reduce withdrawal symptoms such as depression and anxiety. The two FDA approved treatments for nicotine addiction are *nicotine gum* and the *nicotine patch.* Hughes[35] demonstrated that the combination of psychosocial treatment and nicotine gum is more successful than either treatment alone in helping a subject to quit smoking. Abstinence rates are low, however, only10–20 %[31], and are even lower when the gum is not used in conjunction with psychosocial therapy. The low rates of abstinence could be because of[31]: (1) improper use of the gum, (2) incorrect dosage of nicotine, and (3) not using the gum for a long enough time. For example, some patients may need to remain on the gum for 6 months or more.

The nicotine patch appears to be more effective than nicotine gum. The abstinence rates are approximately 25%[31]. In a study by Buchkremer et al.[6] the combination of the patch and therapy resulted in abstinence rates of approximately 35%. The patch may be more effective than the gum because it provides a constant nicotine level in the blood. In a study by Fagerstrom[16], 28 smokers were given either active gum and active patch, placebo gum and active patch, active gum and placebo patch, or placebo gum and patch. The result of the study was that the combination of active gum and active patch was most successful in treating nicotine addiction.

Clonidine is another medication currently being studied for treatment of nicotine addiction because of its ability to reduce withdrawal symptoms such as anxiety. As discussed by Vaughan[59], a study by Glassman et al.[26] found 52% of smokers on clonidine quit smoking compared to 21% on placebo. Long-term abstinence rates for subjects taking clonidine was approximately 27%.

Yet another medication being studied is buspirone, in part because it increases stimulation of serotonin receptors. Buspirone decreases anxiety yet does not have a sedative effect[59]. Gawin et al.[29] studied seven subjects taking buspirone and found that buspirone treatment reduced anxiety symptoms for all subjects, who also reported a reduction in number of cigarettes smoked.

Because depression is one of the acute withdrawal symptoms associated with nicotine, anti-depressants such as *doxepin*, *fluoxetine*, and *trazodone* are being studied as possible adjuncts for the treatment of nicotine addiction. In a study of 19 smokers by Edwards et al.[14], all subjects on doxepin quit smoking after 1 week (as discussed by Vaughan[59]). Seven of the nine subjects on doxepin reported to remain abstinent after nine weeks in comparison to one of ten subjects on placebo. Doxepin is not without problems, however. One major side effect is significant weight gain. Fluoxetine and trazodone may be more promising treatment for nicotine addiction because they have been found to cause weight loss, rather than weight gain.

4.2 BENZODIAZEPINES

Benzodiazepines cause muscle relaxation and sedation. These effects are caused by binding to specific receptors and enhancing the transmission of *GABA*. Because GABA is an inhibitory neurotransmitter, synaptic transmission is subsequently decreased. Withdrawal symptoms are caused by the reduction of GABA neurotransmission that subsequently lowers the seizure threshold.

The abuse of benzodiazepines is typically combined with polydrug abuse. Treatment for benzodiazepine abuse, therefore, must also take into consideration other drugs being abused. For example, patients who abuse both sedatives and opiates should receive a low dose of morphine until there are no withdrawal symptoms from benzodiazepines.

Withdrawal from benzodiazepines is potentially fatal. Patients should be monitored closely for cardiac arrhythmias or seizures. Smith[53] describes the most effective procedure for detoxification from benzodiazepines as a slow tapering of the dosage of benzodiazepine with the substitution of a long-acting benzodiazepine such as *chlordiazepoxide*, followed by the substitution of phenobarbital, a long-acting sedative with a low potential for abuse.

Detoxification can be also accomplished within a few weeks with the use of diazepam. However, since diazepam can potentially be abused, phenobarbital is another option. In addition, a study of the effectiveness of carbamazepine[48] suggested that this may be a possible treatment because it is non-addictive and fast acting.

Following acute withdrawal, substitution treatment in order to decrease negative reinforcement for continuation of abuse is a possibility for treatment. Treatment with benzodiazepine antagonists is also a possible treatment for rehabilitation from benzodiazepine abuse. Flumazenil is one such benzodiazepine antagonist. Currently, however, flumazenil is only available as short-acting, i.v. injected medication. With chemical modification into a longer-acting oral medication, flumazenil may be effective for treating benzodiazepine abuse[40].

4.3 BARBITURATES

Withdrawal from barbiturates can produce seizures or hallucinations and can be fatal. During detoxification of a barbiturate-dependent patient, it is important to slowly taper the usual dose, similar to benzodiazepine detoxification. In addition, substituting the barbiturate with the sedative *phenobarbital* is effective because it does not produce the "high", but relieves withdrawal symptoms. Today, barbiturates are not often prescribed for medical problems and there are low rates of abuse. Therefore little focus has been given recently to the treatment of barbiturate dependence.[32]

4.4 MARIJUANA

It is hypothesized that marijuana affects brain reward systems by stimulating the release of dopamine[20]. In addition, Howlett et al.[34] discussed the existence of specific receptor sites in the brain that bind THC, the active chemical in marijuana. These sites are found particularly in regions of the brain where dopamine receptors originate and where they protrude.[20] Therefore, these sites that bind THC may have a significant influence on the rewarding effects of marijuana via dopamine release.

Although there are no current treatments for marijuana dependence, a preclinical study by Gardner et al.[20] found that naloxone antagonized the effects of THC on electrical brain stimulation. In particular, they found that naloxone impedes the effects of THC on dopamine in the nucleus accumbens. Thus, these studies show that naloxone can potentially block the rewarding effects of THC. In addition, benzodiazepines can be used to treat the severe anxiety sometimes caused by marijuana intoxication.

4.5 HALLUCINOGENS

Many studies show that hallucinogens such as *LSD, mescaline,* and *psilocybin* bind to the *5-HT2 (serotonin)* receptors in the cerebral cortex[27,50,57]. It is hypothesized that the cognitive distortions associated with hallucinogens can be attributed to the activity of these receptors. In addition, hallucinogens affect the *locus coeruleus* which receives cues from external and internal stimuli[1,7,17]. Thus, the changes in perception of the environment associated with hallucinations could be the result of the affect of hallucinogens on the locus coeruleus.[2] Hallucinogens have been reported to cause severe paranoia, panic, depression, and anxiety. These symptoms, otherwise known as a "bad trip" can be alleviated using diazepam (Valium)[32]. In addition, 5-HT_2 antagonists may block the effects of hallucinogens. A review of the literature by Sadzot et al.[50] showed that in a pre clinical study on rats by Glennon et al.[28], the 5-HT_2 antagonists *ketanserin* and *pirenperone* blocked the effects of the hallucinogens LSD, DOI and DOB. Also, Rasmussen and Aghajanian[46] showed that rat locus coeruleus firing as a result of LSD and DOM (another hallucinogen) was antagonized by ritanserin, another 5-HT_2 antagonist. They also found that antipsychotic drugs which bind to the 5-HT_2 receptors block the effects of hallucinogens[47]. Therefore, these studies show that 5-HT_2 receptor antagonists can potentially block the effects of hallucinogens in humans.

5 Conclusions

Currently, pharmacological adjuncts to treatment for drug dependence are limited. More research needs to be conducted to determine the efficacy of these medications. In addition, medications and various forms of treatment may be particularly helpful for certain subpopulations of patients. Research has found, however, that psychosocial treatment in conjunction with pharmacological adjuncts is the most effective method for treating drug dependence.

References

1. Aghajanian, G.K., "Mescaline and LSD facilitate the activation of locus coeruleus neurons by peripheral stimuli," *Brain Res.* 1980, **186:** pp 492–498.
2. Aghajanian, G.K., "Serotonin and the action of LSD in the brain," *Psychiatr. Ann.* 1994, **24:** pp 137–141.
3. Alterman, A.I., Droba, M., Antelo, R.E., Cornish, J.W., "Amantadine may facilitate detoxification of cocaine addicts," *Drug Alcohol Depend.* 1992, 31, pp 19–29.
4. Altshuler, H.L., Phillips, P.E., Feinhandler, D.A., "Alterations of ethanol self-administration by naltrexone," *Life Sci.* 1980, **26:** pp 679–688.
5. Alvarado, R., Contreras, S., Segovia-Riquelme, N., Mardones, J., "Effects of serotonin uptake blockers and of 5-hydroxytryptophan on the voluntary consumption of ethanol, water and solid food by UCHa and UChB rats," *Alcohol.* 1990, **7:** pp 315–319.
6. Buchkremer, G., Minneker, E., Block, M., "Smoking-cessation treatment combining transdermal nicotine substitution with behavioral therapy," *Pharmacopsychiatry.* 1991, **24:** pp 96–102.
7. Cedarbaum, J.M. and Aghajanian, G.K., "Activation of locus coeruleus neurons by peripheral stimuli: modulation by a collateral inhibitory mechanism," *Life Sci.*, 1978, **23:** pp 1383–1391.
8. Charney, D.S., Sternberg, D.E., Kleber, H.D., Heninger, G.R., Redmond Jr., D.E., "The clinical use of clonidine in abrupt withdrawal from methadone," *Arch. Gen. Psychiatry.* 1981, **38:** pp 1273–1277.
9. Clow, D.W. and Hammer, R.P., "Cocaine abstinence following chronic treatment alters cerebral metabolism in dopaminergic reward regions: bromocriptine enhances recovery," *Neuropsychopharmacology.* 1991, **4:** pp 71–75.
10. Cohen, G. and Collins, M.A., "Alkaloids from catecholamines in adrenal tissue: possible role in alcoholism," *Science.* 1970, **167:** pp 1749–1751.
11. Dackis, C.A., Gold, M.S., Sweeney, D.R., Byron, J.P., "Single-dose bromocriptine reverses cocaine craving," *Psychiatry Res.* 1987, **20:** pp 261–264.
12. Davis, V.E. and Walsh, M.D., "Alcohol, amines, and alkaloids: a possible basis for alcohol addiction. *Science.* 1970, **167:** pp 1005–1007.
13. Dole, V.P. and Nyswander, M.E., "Heroin addiction — a metabolic disease," *Arch. Int. Med.* 1967, **120:** pp 19–24.
14. Edwards, N.B., Murphy, J.K., Downs, A.P., Ackerman, B.J., "Doxepin as an adjunct to smoking cessation: a double-blind pilot study," *Am. J. Psychiatry.* 1989, **146:** pp 373–376.
15. Extein, I., Gross, D.A., Gold, M.S., "Bromocriptine treatment for cocaine withdrawal symptoms [letter]," *Am. J. Psychiatry.* 1989, **146:** p 403.
16. Fagerstrom, K.O., Schneider, N.G., Luneel, E., "Effectiveness of nicotine patch and nicotine gum as individual versus combined treatments for tobacco withdrawal symptoms," *Psychopharmacology.* 1993, **111:** pp 271–277.
17. Foote, S.L., Aston-Jones, G., Bloom, F.E., "Impulse activity of locus coeruleus neurons in awake rats and monkeys is a function of sensory stimulation and arousal," *Proc. Nat. Acad. Sci.* 1980, **77:** pp 3033–3037.
18. Froehlich, J.C., Harts, J., Lumeng, L., Li, T-K., "Naloxone attenuates voluntary ethanol intake in rats selectively bred for high ethanol preference," *Pharmacol. Biochem. Behav.* 1990, **35:** pp 385–390.
19. Fuller, R.K., Branchey, L., Brightwell, D.R., Derman, R.M., Emrick, C.D., Iber, F.L., James, K.E., Lacoursiere, R.B., Lee, K.K., Lowenstam, I., Maany, I., Neiderhiser, D., Nocks, J.J., Shaw, S., "Disulfiram treatment of alcoholism: a Veteran's Administration cooperative study," *JAMA.* 1986, **256:** pp 1449–1455.
20. Gardner, E.L. and Lowinson, J.H., "Marijuana's interacton with brain reward systems: update 1991," *Pharmacol. Biochem. Behav.* 1991, **40:** pp 571–580.
21. Gawin, F.H., Kleber, H.D., Byck, R., Rounsaville, B.J., "Desipramine facilitation of initial cocaine abstinence," *Arch. Gen. Psychiatry.* 1989, **46:** pp 117–121.
22. Gawin, F.H., Allen, D., Humblestone, B., "Outpatient treatment of 'crack' cocaine smoking with flupenthixol decanoate," *Arch. Gen. Psychiatry.* 1989, **46:** pp 322–325.

23. Gawin, F.H., Compton, M., Buck, R., "Buspirone reduces smoking," *Arch. Gen. Psychiatry.* 1989, **46:** pp 288–289.
24. Gawin, F.H., "Chronic neuropharmacology of cocaine: progress in pharmacotherapy," *J. Clin. Psychiatry.* 1988, **49:** pp 11–16.
25. Giannini, A.J., Folts, D.J., Feather, J.N., Sullivan, B.S., "Bromocriptine and amantadine in cocaine detoxificaiton," *Psychiatry Res.* 1989, **29:** pp 11–16.
26. Glassman, A.H., "Heavy smokers, smoking cessation, and clonidine: results of a double-blind, randomized trial," *JAMA.* 1988, **259:** pp 2863–2866.
27. Glennon, R.A., Titeler, M., McKennay, J.D., "Evidence for 5-HT2 involvement in the mechanism of action of hallucinogenic agents," *Life Sci.* 1984, **35:** pp 2505–2511.
28. Glennon, R.A., Young, R., Rosecrans, J.A., "Antagonism of the effects of the hallucionogen DOM and the purported 5-HT agonist quipazine by 5-HT2 antagonists," *Eur. J. Pharmacol.* 1983, **91:** pp 189–196.
29. Gold, M.S., Redmond, D.E., Kleber, H.D., "Clonidine blocks acute opiate withdrawal symptoms," *Lancet.* 1978, **2:** pp 599–602.
30. Gold, M.S., Redmond, D.E., Kleber, H.D., "Clonidine in opiate withdrawal," *Lancet.* 1978, **1:** pp 929–930.
31. Gorelick, D.A., "Overview of pharmacologic treatment approaches for alcohol and other drug addiction: intoxication, withdrawal, and relapse prevention," *Psychiatr. Clin. N. Am.* 1993, **16:** pp 141–157.
32. Grinspoon, L. and Bakalar, J.B., "Drug abuse and addiction." In: *The Harvard Medical School: Mental Health Review*, President and Fellows of Harvard College, 1993, pp 1–37.
33. Hoffman, P.L., Tabakoff, B., Szabo, G., Suzdak, P.D., Paul, S.M., "Effect of an imidazo-benzodiazepine, Ro15-4513, on the incoordination and hypothermina produced by ethanol and pentobarbital," *Life Sci.* 1987, **41:** pp 611–619.
34. Howlett, A.C., Bidaut-Russell, M., Devane, W.A., Melvin, L.S., "The cannabinoid receptor: biochemical, anatomical, and behavioral characterization," *Trends Neurosci.* 1990, **13:** pp 420–423.
35. Hughes, J.R., Gust, S.W., Keenan, R., Fenwick, J.W., Skoog, K., Higgins, S.T., "Long-term use of nicotine vs. placebo gum," *Arch. Int. Med.* 1991, **151:** pp 1993–1998.
36. Jaffe, J.H., "Opioids." In: Francis, A.J. and Hales, R.E. (Eds.), *Psychiatry Update: American Psychiatric Association Annual Review*, Vol. 5. American Psychiatric Press, Washington D.C., 1986, pp 137–159.
37. Jasinski, D.R., Johnson, R.E., Kocher, T.R., "Clonidine in morphine withdrawal," *Arch. Gen. Psychiatry.* 1985, **42:** pp 1063–1066.
38. Kleber, H.D. and Gawin, F.H., "Cocaine abuse: a review of current and experimental treatments," *NIDA Res. Monogr. Series.* 1984, **50:** pp 111–129.
39. Kornet, M., Goosen, C., Van Ree, J.M., "Effect of naltrexone on alcohol consumption during chronic alcohol drinking and after a period of imposed abstinence in free choice drinking rhesus monkeys," *Psychopharmacology.* 1991, **104:** pp 367–376.
40. Kosten, T.A. and Kosten, T.R., "Pharmacological blocking agents for treating substance abuse," *J. Nerv. Ment. Dis.* 1991, **179:** pp 583–592.
41. Mackey, W.B. and Van der Kooy, D.A., "Neuroleptics block the positive reinforcing effects of amphetamine but not of morphine as measured by place conditioning," *Pharmacol. Biochem. Behav.* 1985, **22:** pp 101–105.
42. Malcolm, R., "Pergolide mesylate treatment of cocaine withdrawal," *J. Clin. Psychiatry.* 1991, **52:** pp 39–40.
43. Marfaing-Jallat, P., Micheli, D., Le Magnen, J., "Decrease in ethanol consumption by naloxone in naive and dependent rats," *Pharmacol. Biochem. Behav.* 1983, **18:** pp 537–539.
44. O'Malley, S.S., Jaffe, A.J., Chang, G., Schottenfeld, R.S., "Naltrexone and coping skills therapy for alcohol dependence: a controlled study," *Arch. Gen. Psychiatry.* 1992, **49:** pp 881–887.
45. Pycha, R., Miller, C., Barnas, C., Hummer, M., "Intravenous flunitrazepam in the treatment of alcohol withdrawal delirium," *Alcohol Clin. Exp. Res.* 1993, **17:** pp 753–757.
46. Rasmussen, K. and Aghajanian, G.K., "Effect of hallucinogens on spontaneous and sensory-evoked locus coeruleus unit activity in the rat: reversal by selective 5-HT2 antagonists," *Brain Res.* 1986, **385:** pp 395–400.

47. Rasmussen, K. and Aghajanian, G.K., "Potency of antipsychotics in reversing the effects of a hallucinogenic drug on locus coeruleus neurons correlates with 5-HT2 binding affinity," *Neuropsychopharmacology.* 1988, **1:** pp 101–107.
48. Ries, R., Cullison, S., Horn, R., Ward, N., "Benzodiazepine withdrawal: clinicians' ratings of carbamazepine treatment versus traditional taper methods," *J. Psychoact. Drugs.* 1991, **23:** pp 73–76.
49. Rosecrans, J.A. and Karan, L.D., "Neurobehavioral mechanisms of nicotine action: role in the initiation and maintenance of tobacco dependence," *J. Subst. Abuse Treat.* 1993, **10:** pp 161–170.
50. Sadzot, B., Baraban, J.M., Glennon, R.A., Lyon, R.A., "Hallucinogenic drug interactions at human brain 5-HT2 receptors: applications for treating LSD-induced hallucinogenesis," *Psychopharmacology.* 1998, **98:** pp 495–499.
51. Samson, H.H. and Doyle, T.F., "Oral ethanol self-administration in the rat: effect of naloxone," *Pharmacol. Biochem. Behav.* 1985, **22:** pp 91–99.
52. Sanchez-Ramos, J.R., "Psychostimulants," *Neurol. Clin.* 1993, **11:** pp 535–553.
53. Smith, D.E. and Landry, M.J., "Benzodiazepine dependency discontinuation: focus on the chemical dependency detoxification setting and benzodiazepine-polydrug abuse," *J. Psychiatr. Res.* 1990, **24:** pp 145–156.
54. Spealman, R.D., "Antagonism of behavioral effects of cocaine by selective dopamine receptor blockers," *Psychoparmacology.* 1990, **101:** pp 142–145.
55. Suzdak, P.D., Glowa, J.R., Crawley, J.N., Schwartz, R.D., "A selective imidazobenzodiazepine antagonist of ethanol in the rat," *Science.* 1988, **234:** pp 1253–1247.
56. Tennant, F.S. and Sagherian, A.A., "Double-blind comparison of amantadine and bromocriptine for ambulatory withdrawal from cocaine dependence," *Arch. Int. Med.* 1987, **147:** pp 109–112.
57. Titeler, M., Lyon, R.A., Glennon, R.A., "Radioligand binding evidence implicates the brain 5-HT2 receptor as a site of action for LSD and phenylisopropylamine hallucinogens," *Psychopharmacology.* 1988, **94:** pp 213–216.
58. Tutton, C.S. and Crayton, J.W., "Current pharmacotherapies for cocaine abuse: a review," *J. Addict. Dis.* 1993, **12:** pp 109–127.
59. Vaughan, D.A., "Frontiers in pharmacologic treatment of alcohol, cocaine, and nicotine dependence," *Psychiat. Ann.* 1990, **20:** pp 695–710.
60. Volpicelli, J.R., Davis, M.A., Olgin, J.E., "Naltrexone blocks the post-shock increase of ethanol consumption," *Life Sci.* 1986, **38:** pp 841–847.
61. Volpicelli, J.R., Alterman, A.I., Hayashida, M., O'Brien, C.P., "Naltrexone in the treatment of alcohol dependence," *Arch. Gen. Psychiatry.* 1992, **49:** pp 876–880.
62. Weddington, W.W., Brown, B.S., Haertzen, C.A., Hess, J.M., "Comparison of amantadine and desipramine combined with psychotherapy for treatment of cocaine dependence.,"*Am. J. Drug Alcohol Abuse.* 1991, **17:** pp 137–152.
63. Witikin, J.M., "Blockade of the locomotor stimulant effects of cocaine and methamphetamine by glutamate antagonists," *Life Sci.* 1993, **53:** pp 405–410.
64. Worner, T.M., "Propranolol versus diazepam in the management of the alcohol withdrawal syndrome: double-blind controlled trial," *Am. J. Drug Alcohol Abuse.* 1994, **20:** pp 115–124.

Contents Chapter 10

Interactions between food and addiction

0-8493-7803-6/99/$0.00+$.50

Chapter 10

Interactions between food and addiction

Marilyn E. Carroll

INTRODUCTION

The purpose of this review is to discuss interactions between food and drugs of abuse at a number of levels. Both animal and human data are reviewed. Clinical and experimental findings are compared. For instance, the comorbidity between eating disorders and drug addiction in women is discussed as well as effects of restricted feeding on drug self-administration in animals. Behavioral, economic and neurobiological explanations are applied to the commonalities between feeding and drug self-administration. The effects of dietary manipulations on drug-seeking behavior is emphasized. In contrast, the effects of drugs of abuse on feeding behavior is beyond the scope of this review. The effects of dietary manipulations are considered at all levels of the addiction process from predicting vulnerability to drug addiction to acquisition of drug self-administration and to maintenance or steady state levels, withdrawal, and relapse after a period of abstinence.

A few assumptions that are made in this review are worth noting. First, it is assumed that food and drugs function as positive reinforcers in animals and humans, and it is the rewarding effect of these substances that increases the probability of drug-seeking behavior. This has been clearly demonstrated in operant conditioning paradigms, and many similarities have emerged between food and drugs as reinforcing stimuli. Behavior maintained by these two classes of substances is controlled by the same variables such as schedule, magnitude of reinforcement, and behavioral history. Behavioral (e.g., food restriction or availability of alternative reinforcers) and pharmacological interventions affect behavior maintained by food and drug reinforcers in similar ways.

A second assumption is that animal laboratory models are a good representation of human behavior. With few exceptions drugs that are abused by humans are self-administered by animals[35], and drug intake in the human and animal laboratory is controlled by a similar set of variables (e.g., dose, schedule of access, route of administration, and drug history). Treatment attempts that are successful in reducing drug self-administration in the laboratory usually generalize to the clinical setting.

A third assumption is that the presence or absence of food greatly alters the reinforcing effects of drugs, such as during self-administration, withdrawal, or relapse, not strictly through its physical characteristics and nourishing capacity, but through its ability to function as an alternative reinforcer to drugs. Food without caloric value (e.g., saccharin) and other nondrug reinforcers (e.g., money) produce interactions with drugs that are similar to the food-drug interactions, although noncaloric alternative reinforcers have not been as extensively studied. These reinforcers will be considered along with food. Overall, the present review suggests that our understanding of drug dependence and the resulting self-inflicted toxic drug effects of drugs may be advanced by systematically studying and implementing the use of food and other nondrug reinforcers in treatment programs.

1 Eating disorders in women and the development of drug addiction

One of the classic examples of a possible interaction between food and drug addiction is found in the comorbidity between eating disorders and drug addiction. Eating disorders including bulimia nervosa and anorexia are highly prevalent among young women. There is considerable human literature that suggests bulimia nervosa and drug abuse are correlated[46]. Rates of substance abuse in bulimic patients (25–60%) and rates of bulimia in women with drug abuse problems (20–40%) are much higher than the occurrence of these disorders in the general population (1–5%)[8]. A high correlation between eating disorders and drug abuse does not imply that one disorder has caused the order, but generally, in women diagnosed with both disorders, eating disorders occur before drug abuse problems[38].

In an attempt to investigate the sequence of bulimia and drug abuse, Bulik and colleagues (1992)[8] analyzed temporal patterns of binging, purging, alcohol, cigarette, and laxative use. They studied 42 women with bulimia nervosa and 29 women with anorexia nervosa who were hospitalized for the duration of the study. Their results showed that these behaviors increased throughout the day and peaked during the evening hours. Thus, they observed a sequential relationship: as binging and purging increased, drug use increased. They suggested that the repeated dieting and fasting associated with eating disorders may subsequently enhance the reinforcing effects of drugs. There is a related case report regarding three young women diagnosed with bulimia nervosa who abused toxic levels of caffeine as an appetite suppressant[29]. In this case drug use appeared to be secondary to the eating disorder. These studies suggest that fasting increases the reinforcing effects of drugs, and this may account for the high rate of drug abuse in bulimic patients. However, carefully designed prospective studies with long-term follow-up are needed to identify causal relations. Fasting may not be the only important variable, as anorexia nervosa patients also fast, but prevalence of drug abuse in that population is low.

2 Use of food and alternative nondrug reinforcers in the treatment of drug abuse

While some studies suggest that altered feeding patterns may be responsible for increased drug abuse, others show that food and other nondrug alternative reinforcers can function therapeutically for drug abusers. For instance, alcoholics and opiate addicts report cravings for sweets, and they consume large quantities of sweet foods and beverages when they are abstinent from alcohol or opiates. A study of imprisoned heroin addicts revealed that when craving for drug occurred, the substance that they would prefer next would be a liquid or solid sweet, such as candy, soft drinks, cake or pastry[71]. Some addicts were able to reduce drug craving by imagining ingestion of the preferred sweet. Another study of recovering, abstinent alcoholics revealed that those who stayed sober longest chose diets with more carbohydrates and twice as much sugar added to beverages than those who relapsed sooner[73]. In fact, there is a quote from the Alcoholics Anonymous literature (1987)[2] that emphasizes the importance of dietary sweets to maintaining abstinence: "Thousands of us — even many who said they never liked sweets — have found that eating or drinking something sweet allays the urge to drink." Alcoholics Anonymous also uses the acronym HALT to remind members to avoid becoming hungry, angry, lonely or tired, as these conditions are related to relapse. It is also interesting that patients with eating disorders generally prefer sweet foods, and fat- or sweet-containing foods are often the food of choice during bulimic binges.

The apparent substitution of preferred foods for drugs of abuse in these studies may be an example of a more general phenomenon, reinforcer interaction. Any nondrug alternative reinforcer may substitute for drug. The are several examples from the clinical treatment literature. Cocaine abusers have been treated recently in a study of the use of competing nondrug reinforcers[41]. Subjects were asked to supply urine samples at random intervals. They were rewarded for drug-free urines with tokens that were exchangeable for nondrug rewards. The tokens could be exchanged for retail items or tickets to cultural or sporting events, and subjects assist in constructing a list of alternative nondrug reinforcers. The number of tokens awarded increases cumulatively with repeated drug-free urines. When the group undergoing this community reinforcement approach was compared to a control group that experienced a standard 12-step program, the community reinforcement approach was able to retain more patients drug-free for a longer period of time. Other variations in the use of alternative nondrug reinforcers have been reported in the treatment of alcohol, cocaine, benzodiazepines, opioids, amphetamine, and polydrug abuse.

There have been several recent human laboratory studies indicating that alternative reinforcers suppress drug self-administration. Zacny and coworkers[75] found that money reduced beer consumption in human volunteers, and the effect was enhanced as the amount of money increased. Hatsukami and coworkers[39] offered human subjects a mutually exclusive choice between tokens worth $2, $3, $5, or $7, or the opportunity to smoke 5 mg, 0.2 mg/kg or 0.4 mg/kg cocaine base. The low dose (5 mg) served as an inactive control dose, and was generally not chosen over tokens (except at the $2 token value). At the two higher cocaine unit doses, money interfered more with drug self-administration as token value increased. The importance of food or other nondrug rewards in these studies is that it seems to, at least temporarily, replace the rewarding aspects of drug abuse. It has been suggested that drugs of abuse and food activate the same brain reward mechanisms.

3 Dietary preferences as a predictor of vulnerability to drug addiction

Given the correlation between eating disorders and drug abuse and the influence of dietary manipulations in abstinent drug abusers, the question arises whether there are individual differences in sensitivity to foods and drugs. Recently, several investigators have reported results of animal studies suggesting that dietary preferences may predict the amount of drugs consumed. For instance, rats that drink large amounts of a palatable saccharin solution have higher ethanol consumption than rats that do not have a high preference for saccharin. As an example, in one experiment Gosnell and Krahn[33] divided 52 male rats into high, medium, or low saccharin drinkers based on their saccharin intake, measured over 3 successive days. Their ethanol intake (2–8% wt/vol) was then evaluated. Under conditions of limited food access, all three groups consumed similar amounts, but under free feeding conditions, the medium and high saccharin drinkers consumed more ethanol than the low saccharin drinkers. Since water intakes did not vary as a function of saccharin preference the results could not be explained by differences in general liquid intake. These results have been reproduced in genetically heterogeneous, as well as inbred rat strains. Also, Bell and coworkers[4] extended the results to operant-based behavior in which rats were required to work on a lever for an opportunity to drink ethanol. They examined ethanol-reinforced behavior over a range of concentrations (1–8% wt/vol), and found that at 25 of 32 parameters tested, the high, medium, and low saccharin-preferring rats were ranked in the same order on ethanol-reinforced behavior. Recently, Gosnell et al.[34] have demonstrated that rats selected for high saccharin preference self-administered more morphine i.v. than rats selected for low-preference.

Others have demonstrated that rats selected on the basis of high and low ethanol consumption showed corresponding high and low levels of saccharin intake, respectively. A recent report by Hyyatiä and Sinclair[45] compared a genetic strain of alcohol-preferring rats (AA; see also Chapter 4), and Wistar rats on oral etonitazene, cocaine, and quinine consumption. The AA group drank significantly more of each solution than the other group. These investigators have also reported that AA rats consume more saline and citric acid solution than other strains suggesting that increased drug intake may be explained by a greater acceptability of tastants. In a related study, Nichols and Hsiao[60] bred rats for high and low morphine consumption and found that they had corresponding high and low levels of ethanol intake. This data suggests that selectivity for high and low consumption of drugs and dietary substances may be based on a sensitivity to bitter taste, as most drugs have a bitter taste. However, other data show that increased intake of substances that are not bitter (e.g., fat) predict morphine[56] and alcohol intake.

A more general interpretation that would include this data is that preferred dietary components, such as sweets, salt, and fats, may produce rewarding effects similar to those of alcohol, morphine, and other drugs of abuse. Those animals predisposed to seek one form of reinforcement might more readily accept other substances that produce similar effects. There is additional data that supports this notion. For instance, it is known that intracranial self-stimulation (ICSS) is reinforcing to rats. Rats that were bred for high rates of ICSS had higher saccharin intakes than those bred for low rates of ICSS. The significance of these findings is that an increased consumption of preferred dietary substances may be used to predict individual vulnerability to drug abuse, and to implement prevention strategies. Alternatively, if preferred dietary substances compete with the rewarding effects of drugs, it may be possible to use preferred dietary substances systematically in the treatment of drug abuse.

4 Food deprivation increases acquisition of drug self-administration

There are parallels in animal studies for the increased drug abuse that is associated with the eating disorders in humans. The aspect of eating disorders that may be most closely associated with increased drug use is fasting. Food deprivation has been used since drug self-administration studies began. Initially it was used in a nonsystematic way to train the lever-pressing response in rats. In some cases, food was withheld and rats were trained to press a lever for food, and then later for drug. In other studies, the lever was baited with food, and when it was contacted, a drug-infusion resulted. Drug-reinforced lever pressing usually persisted without baiting the lever. Later, when the large effects of altered feeding on drug self-administration were recognized[13], feeding conditions were controlled in a more systematic manner.

Studies have been conducted with rats[17] and monkeys[15], indicating that partial food deprivation enhances the speed of acquisition, as well as the maintenance levels of drug self-administration. In a rat experiment, intravenously delivered cocaine was available contingent upon a fixed-ratio (FR) 1 schedule. One lever press yielded one infusion. Most rats began responding at a moderate rate during the first few 24-hour sessions. However, every third day, when the rats' daily food regimen was reduced from an unlimited amount to 8 g, cocaine infusions increased dramatically. Rates of lever pressing returned to lower levels on intervening satiated days.

Later experiments with rhesus monkeys and orally delivered phencyclidine also resulted in a more rapid acquisition of drug self-administration during food deprivation. Food deprivation not only increased the total amount of phencyclidine consumed, but it changed the animal's patterns of responding during the 3-hour drug session. During food deprivation responding began immediately at session onset and continued at high rates in one long drinking bout for about an hour. During food satiation responding did not always begin at session onset, and there was considerably pausing between smaller drinking bouts. Drinking occurred intermittently throughout the 3-hour session. The differing patterns of responding, rapid during food deprivation and sporadic during food satiation, may suggest that feeding conditions produce different motivational states that determine the strength of drug-seeking behavior.

5 Preferred dietary substances prevent and reduce acquisition of drug-reinforced behavior

Since feeding conditions and the availability of preferred dietary substances have a marked effect on drug self-administration in laboratory animals, it is possible to manipulate this variable to produce an animal model of drug abuse prevention. Substantial effort has been directed to the development of prevention strategies to reduce the prevalence of drug addiction. Prevention is currently one of the major funding goals of the United States government. Animal models are useful in this regard because first-time use of drugs can be experimentally controlled in animals, but it is not ethically possible to manipulate the initiation of drug abuse in humans.

There are several reports indicating that preferred dietary substances alter the acquisition of drug self-administration. The effects of preferred dietary substances on acquisition of i.v. cocaine self-administration have been examined in a series of experiments. The first was previously described: when rats were food deprived every third day, cocaine infusions were two- to three-fold more frequent than on the 2 intervening food-satiation days.

One interpretation of this data was that the novel effects of food deprivation were paired with the reinforcing effects of the drug, thereby increasing their salience as a discriminative stimulus. To test this novelty hypothesis, another group of rats were preexposed to food deprivation so it would not be novel when cocaine was introduced. They did not show the dramatic increases in responding during the deprivation sessions. Based on these results, it was hypothesized that the specific pairing of the novel interceptive effects of food deprivation with the novel and reinforcing effects of cocaine allowed food deprivation to function as a discriminative stimulus for drug reward, and to set the occasion for greater responding and drug intake.

A subsequent investigation was conducted to determine if a preferable dietary substance would slow down acquisition of cocaine self-administration. The preferred substance was a palatable solution of glucose and saccharin (G+S)[16]. This solution had been reported to be so palatable that rats would drink the equivalent of their body weight in 24 hours. An autoshaping procedure was used to provide an objective, quantitative index of acquisition. This procedure had been used by others to establish food-rewarded behavior[58]. According to this method, for 6 hours each day 60 cocaine infusions were automatically delivered at random intervals. Before each infusion a retractable lever extended into the cage for 15 seconds. Subsequently, there was another 6-hour self-administration component when the lever remained extended and each response was reinforced by a cocaine infusion under a FR1 schedule. When the rats had self-administered an average of 100 infusions for 5 consecutive days during the second 6-hour component, they had met the criterion for cocaine acquisition.

TABLE 10.1
Cocaine acquisition experimental design

Operant chamber during autoshaping	*Availability of nondrug alternative reinforcer, G+S[+] home cage before autoshaping*	
	G+S	H_2O
G+S	Group 1*	Group 2*
H_2O	Group 3*	Group 4* Group 5**

[+] 3% glucose + 0.125% Sacc (wt/vol).
* Groups 1-4 had access to 20 g food and consumed a mean of 16.7 g.
** Group 5 had acces to unlimited food and consumed a mean of 20.5 g.

Five groups of 12 to 14 rats were compared, as shown in Table 10.1. The first four groups were divided according to whether they had (1) G+S or water from automatic drinking spouts during cocaine acquisition, or G+S or water 3 weeks before and during cocaine acquisition. The first four groups received 20 g of food daily. A fifth group had no G+S and unlimited access to food. Table 10.2 shows that the two groups that received G+S were the slowest to acquire i.v. cocaine self-administration. The group with longer G+S access (home cage and during acquisition) and higher G+S intake had the slowest acquisition. In this latter group, only 50% of the rats attained the criterion within the 30-day limit. In the G+S group that had exposure only during cocaine acquisition, 83% met the criterion. Both the group having no G+S, as well as the group having only a 3-week G+S history before cocaine acquisition began, showed rapid acquisition with 100% of the group acquiring. Thus, history of G+S alone did not interfere with cocaine acquisition. The group that received *ad lib* access to food showed delayed acquisition, with only 71.4% attaining criterion. These results suggested that food and palatable dietary substances delayed and/or prevented acquisition of cocaine-reinforced behavior.

TABLE 10.2
Results from a cocaine acquisition experiment

	*Mean days to meet acquisition criterion (±S.E.)***	*Animals meeting criterion*
Group 1	22.8 (±2.8)*	50%
Group 2	20.3 (±2.2)*	83%
Group 3	9.08 (±1.6)	100%
Group 4	9.5 (±1.3)	100%
Group 5	16.1 (±2.6)*	71.4%

* Significantly different than Group 4 ($p < 0.05$), survival analysis.
** Acquisition criterion = 5 consecutive days with a mean > 100 infusions during the 6-h self-administration period. The experiment was terminated if acquisition criterion was not met in 30 days.

A follow-up experiment was conducted to determine if it was the caloric value of G+S or the palatability that interfered with acquisition of cocaine self-administration[51]. Others have demonstrated that G+S is consumed for its palatability[67]. Similar autoshaping methods were used, and six groups of rats were compared. Three of the groups had saccharin (0.2% wt/wt) added to their daily allotment of ground rat chow and the other three groups consumed standard ground rat chow. The groups were also divided according to total amount of food available per day: 10 g, 20 g, or *ad libitum* food. Figure 10.1 shows that acquisition rates were delayed by the addition of saccharin to food at all conditions, and acquisition was further delayed as the amount of food increased. Saccharin was more effective at slowing acquisition as food access increased. The data from this series of experiments clearly shows that increased food and increased palatability of food are factors that may, separately or together, reduce acquisition of cocaine self-administration.

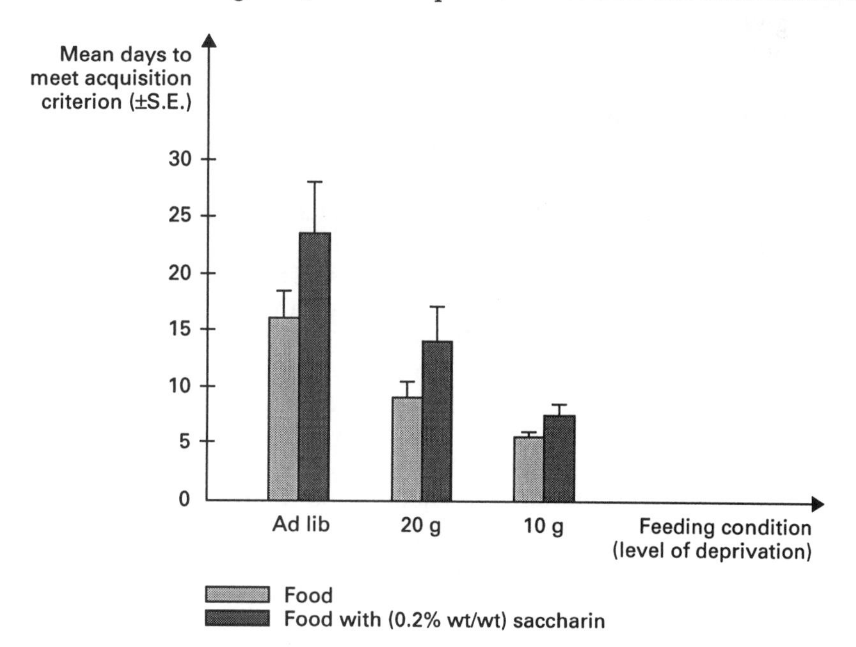

FIGURE 10.1

Mean (± SEM) number of days to complete acquisition criterion (mean of 100 infusions over 5 consecutive days) for six groups of rats receiving *ad libitum*, 20 or 10 g food (ground laboratory chow) with or without the addition of (0.2% wt/wt) saccharin

Light bars indicate ground laboratory chow was available, and dark bars indicate that saccharin was mixed with ground chow. The number of subjects per group ranged between 6 and 17. The experiment was terminated if the acquisition criterion was not met within 30 days. There were 3 out of 13 rats in the regular *ad lib* food group and 4 out of 6 rats in the saccharin *ad lib* food group that did not meet acquisition criteria.

6 Food deprivation increases drug self-administration in the maintenance phase

The previous section suggested that transition states of behavior such as acquisition are particularly vulnerable to the effects of feeding conditions. Feeding conditions and alternative nondrug reinforcers may also dramatically change drug intake later after drug self-administration has stabilized. Animal studies have shown that drug self-administration markedly increases when daily food rations are even slightly reduced[13,14]. In contrast, drug intake is reduced when animals are offered free access to food. In these studies food access was generally controlled by varying the amount of daily food provided to the animals, but the results have recently been extended to a condition in which food availability was contingent upon an operant response.

This inverse relationship between feeding and drug self-administration is not related to the effects of specific drugs on appetite. It occurs with virtually all drugs that are abused by humans, including those that stimulate feeding (e.g., barbiturates, opioids), those that suppress feeding (e.g., amphetamine, caffeine, cocaine, nicotine), and those that may have mixed effects depending on dose (e.g., alcohol, cannabis, and phencyclidine). The generality of the food-drug self-administration interaction has been extended to several species, schedules of reinforcement, and routes of self-administration.

One of the most dramatic demonstrations of the effects of food on drug self-administration was that of Aigner and Balster[1] who reported that monkeys given a mutually exclusive choice between food and cocaine chose cocaine most of the time. The monkeys lost weight and their health was threatened by their nearly exclusive choice of cocaine. Recent data indicate that a more balanced selection of food and cocaine could be obtained by increasing the number of food pellets available at the end of each trial[59]. As the number of pellets was increased from 1 to 16, the frequency of cocaine choice and total cocaine intake declined. Other conditions of food availability also influenced cocaine choice. For instance, when food access was restricted to only what could be earned during the session (closed economy) (see Table 10.3), cocaine choice was lower than when food earned during the session was supplemented by post-session feedings (open economy). These findings indicate that the amount of food and the contingencies attached to its availability are important variables in determining the extent to which food will interfere with drug self-administration.

One notable exception to the food deprivation effect is that in human subjects food deprivation has not produced large increase in drug intake when tested in laboratory studies (e.g., see ref. 10, 75). This discrepancy may be due to methodological limitations in the human laboratory. In some studies low-calorie sweetened liquids were given to deprived subjects to control for stomach contents; however animal studies indicate that noncaloric sweetened liquids suppress drug self-administration. Also, this discrepancy may be due to the fact that human subjects are not taking the drugs in their accustomed environment.

In the natural setting there have been a few anecdotal reports of increased drug self-administration due to reduced food intake in humans. An early report of coca leaf chewing among Peruvian Indians indicated that feeding high-calorie well-balanced meals reduced coca-leaf chewing, while a return to the normal sparse diet increased chewing[38]. Starvation studies conducted during World War II, with conscientious objectors as subjects, revealed excess consumption of caffeine and nicotine[33]. Cigarette smoking is one form of drug abuse that appears to be affected by food deprivation. A recent investigation of weight loss and smoking reported a significant increased in saliva cotinine levels (an indicator of smoking) in dieting subjects as well as number of cigarettes smoked[18]. Hall and colleagues[37] also reported in a follow-up of a smoking cessation program that female quitters who maintained their pre-smoking weight were more likely to relapse than those whose weight increased after smoking cessation. Body weight generally increases after smoking cessation due to lowered metabolic rates and increased snacking[61]. The significance of this research is that dieting during the early phases of smoking cessation may be contraindicated.

TABLE 10.3
Behavioral Economic Terms

Unit price – Number of responses required to obtain a reinforcer/unit amount of reinforcer delivered (e.g., responses/mg of drug or responses/g of food).

Consumption – The amount of reinforcer earned (e.g., money) or consumed (e.g., g of food, mg of drug) in a specified time period.

Economy

Open – Earned reinforcer is supplemented by experimenter.
Closed – Subject earns all reinforcers.

Demand – The effects of unit price on consumption usually plotted as a log /log scale. With drugs as reinforcers, the demand curve is usually a positively decelerating function.

Elasticity of demand – The degree to which consumption changes (proportionally) as unit price changes (in log space, elasticity = slope) (Hursh and Bauman, 1987).

Elastic demand – When consumption decreases at a proportionally greater rate than increases in unit price.

Inelastic demand – When consumption decreases at a proportionally lower rate than increases in unit price.

Intensity of demand – A parallel shift downward or upward in the demand curve.

Essential commodity – When consumption varies little with increases in unit price (e.g., water, fuel).

Luxury item – When consumption decreases greatly with increases in unit price (e.g., entertainment).

P-max – The unit price at which maximum responding occurs.

Income – The amount of funds, goods, services or resources available to an organism during a specific time interval.

Reinforcer interactions

Substitution – When consumption of one reinforcer increases as price of another reinforcer increases (and consumption decreases).

Complementarity – When consumption of one reinforcer decreases with increases in price (and decreases in consumption) of another reinforcer.

Independence – When there is no functional relationship between consumption of one reinforcer (commodity) and the price and consumption of another.

6.1 LIMITATIONS OF THE FOOD DEPRIVATION EFFECT

There are several limitations to the food-deprivation effect that may reveal aspects of the mechanism of action. Food deprivation does not induce self-administration of drugs that have equivocal reinforcing effects in animals, such as delta-9-THC and diazepam, even though these drugs are abused by humans. The phenomenon seems to be limited to drugs that can be established as reinforcers in animals and are abused by humans[13]. Reduced food intake may facilitate self-administration of drugs that are highly addictive in humans but not easily established as reinforcers in animals, such as nicotine or caffeine, and very low doses of drugs that are easily established as reinforcers at high doses, such as cocaine and amphetamine. That drug reinforcement is a necessary condition for the interaction suggests that food deprivation specifically interacts with reinforcing effects of drugs.

Another general observation about the food-drug interaction is that food deprivation has a much greater enhancing effect on low drug doses than on high doses. Since higher doses are expected to have greater reinforcing effects than lower doses, it might be assumed that there is a ceiling on increasing reinforcing effects that cannot be elevated by reduced feeding. It is consistent with these findings that cocaine is one of the most reinforcing drugs of abuse, yet food deprivation does not always have as large an effect on self-administration of cocaine and other psychomotor stimulants as it does on self-administration of some other drugs of abuse. This has been especially apparent in recent studies of the effect of food deprivation on cocaine-base (crack) smoking. Humans report that crack smoking produces more intense effects than other routes of self-administration. When food deprivation and satiation conditions were compared, there was slightly less responding for drug at all FRs indicating that additional food weakened the reinforcing effects of cocaine-base smoking. However, overall the magnitude of the food deprivation effect was not nearly as great as that reported for i.v. cocaine self-administration in rats.

6.2 INTERPRETATIONS OF THE FOOD DEPRIVATION EFFECT

Many alternative explanations of the food-drug self-administration interaction have been ruled out by the increased generality of the finding. Control procedures have been implemented to show that increased drug self-administration during reduced feeding is not due to an increase in general activity, pharmacokinetic changes, palatability of the drugs, differences in tolerance development, or caloric replacement (as might be the case with alcohol). Evidence against hypotheses based on metabolic factors comes from studies in which food deprivation increased drug-rewarded behavior at times when animals have no drug in their system. For example, food deprivation increased responding in monkeys when all drug was presented at the end of the session, and the amount of drug available was fixed and not contingent upon the rate of responding[23]. In experiments with rats self-administering intravenously delivered cocaine, increased rates of responding due to food deprivation also occurred during a drug-free extinction period.

Increased drug-reinforced behavior during food deprivation can be interpreted in several ways. It may be that the reinforcing effects are reduced during deprivation, and more drug is needed to achieve effects comparable to those during food satiation. In contrast, food deprivation may increase the sensitivity to a drug's reinforcing effects, as would be comparable to a higher unit dose. In either case, increased responding could result. Another problem with interpretation of the increased responding is that the direct rate-increasing or rate-decreasing effects of drugs may influence behavior independently of reinforcing effects.

In a few studies direct effects of drugs have been minimized by separating the response component from the time of drug intake. One means of accomplishing this separation was to use a second-order schedule in which a substantial amount of responding preceded drug delivery. In that situation, responding leading to drug access was greatly enhanced as a result of food deprivation even when no drug was available during responding and the amount of drug available later was the same during food deprivation and satiation. A second method that was used to separate responding and drug-intake was the extinction method. Rats were trained to self-administer intravenously delivered cocaine, and then saline replaced cocaine. After 12 days of extinction, the rats were food deprived every third day, and responding reinforced by saline increased up to seven-fold when the rats were tested on these deprivation days.

Another method of minimizing the direct effects of drugs on drug-reinforced responding is the progressive-ratio schedule in which each drug delivery is contingent upon an increasingly higher response requirement. At higher FRs the interdrug interval becomes quite long, thus minimizing cumulative effects. The FR at which responding ceases (break point) is used as a measure of reinforcing efficacy. The break point for food reinforcement becomes higher with higher levels of food deprivation. This method has been recently applied to the effects of food deprivation on drug self-administration. Increases in break point occurred during food deprivation[65]. A method that is similar to the PR schedule is to compare peak responding as the FR values are gradually changed and behavior is allowed to stabilize under each FR over several days. Figure 10.2 illustrates a comparison of the point at which peak responding (P-max) for phencyclidine or ethanol occurred during food satiation and deprivation. For both drugs P-max values were higher during food deprivation conditions.

Food deprivation enhances other drug effects in addition to the reinforcing effects. Drug-induced psychomotor effects[26], opioid-mediated analgesic effects, and stereotyped behavior are increased during food deprivation. However, discriminative stimulus effects of drugs are not altered by food deprivation[69].

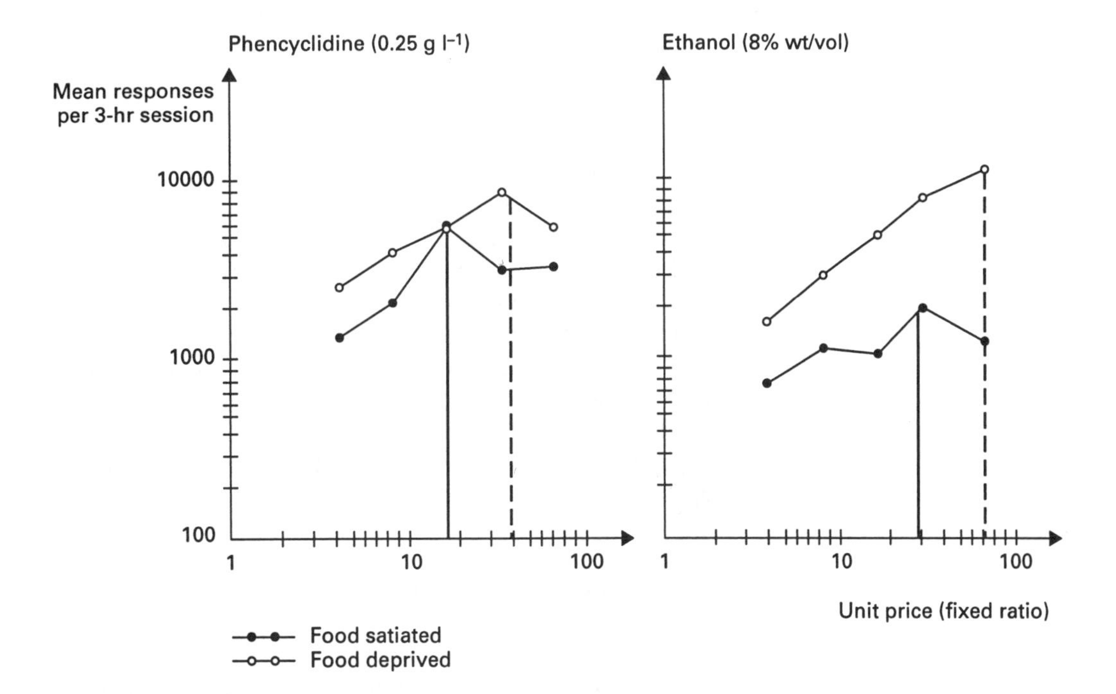

FIGURE 10.2

The effect of food satiation and deprivation on self-administration of orally delivered phencyclidine (left) and ethanol (right) is illustrated

Open circles refer to conditions under which monkeys were maintained at 85% of their free-feeding body weights, and filled circles represent the free feeding condition. For both drugs the FR at which maximum responding occurred (P-max) was higher during food deprivation than during food satiation. The theoretical peaks of each curve are indicated by where the dashed lines intersect the x-axis. Each point represents a mean for 5 monkeys over the last 5 days of stable behavior at each FR.

Food deprivation also increases behavior maintained by other reinforcers, such as intracranial self-stimulation, wheel running, light onset, and saccharin presentation. These diverse findings suggest that increased intake due to food deprivation is a subset of a larger phenomenon. Food deprivation may produce changes in functions of the central nervous system that modulate reward or reinforcement from a variety of sources.

There is some evidence that even past experience with food deprivation during development in rats may lead to unusual feeding and drug-taking patterns later in life. Hagan and Moss[36] briefly restricted feeding in female rats when they were 30, 90, and 140 days old. When they were later injected with butorphanol tartrate, an opioid drug that stimulates feeding, both groups' food intake increased. However, the group with the food deprivation history consumed significantly more food than a control group that did not have that history. The increased feeding was not due to more rapid eating, but the rats with food deprivation history continued to eat longer (for 2 hours after the control rats had decreased their food intake). Coscina and Dixon[21] had previously reported similar results for consumption of palatable foods in rats that had a history of a 4-day fast. They found that subsequent exposure to highly palatable foods elicited overeating and exaggerated obesity. It is not known if fasting early in life in humans leads to subsequent eating disorders or obesity, however, this procedure may serve as an animal model for binge eating or bulimia, and may be useful for further studies of interactions between eating disorders and drug abuse.

The Hagan and Moss[356] study was subsequently replicated to establish binge-eating rats, and to examine whether the food deprivation history altered the rate of i.v. cocaine self-administration. The group with deprivation history acquired cocaine self-administration more rapidly than controls. The percent of rats meeting the acquisition criterion was 86% for the group with deprivation history and 69% for the controls. These findings may be interpreted in terms of increased stress in the group with food deprivation history. Alternatively, the group with free access to food had a more enriched environment. In another study rats raised in enriched environments had lower amphetamine and barbital preference and lower total drug intake than rats raised in socially deprived environments. In a similar study of environmental deprivation vs. enrichment, Fowler and coworkers[31] found that rats reared and housed in a socially enriched environment were more sensitive to cocaine in cocaine-saline discrimination training and to the toxic effects of cocaine than those reared and housed in deprived environments. These findings suggest that food deprivation may decrease the rats' sensitivity to drugs, such that more drug is needed for reinforcement to occur.

7 Reduction of drug self-administration by palatable dietary substances

Skinner[66] discussed the advantages of reducing undesirable behavior by reinforcing alternative behavior rather than by using extinction or punishment. Enrichment of the environment by preferred dietary substances and other reinforcers has the same suppressant effect on drug self-administration as food satiation[11]. For instance, self-administration of orally delivered phencyclidine in monkeys was reduced by making a palatable saccharin solution concurrently available. The effects of saccharin, like food satiation, were given at low drug concentrations. Similar results have been reported with G+S and i.v. cocaine self-administration, sucrose and oral intake of amphetamine, morphine, ethanol in rats, and oral pentobarbital self-administration in monkeys. Recent studies revealed that the harder monkeys work for drug (increased FR) the more drug intake is suppressed by the concurrent availability of saccharin. These results are summarized in Figure 10.3. Overall, the results of these studies indicate that the effectiveness of alternative reinforcers in suppressing drug self-administration depends upon amount of effort expended to obtain the drug reinforcers.

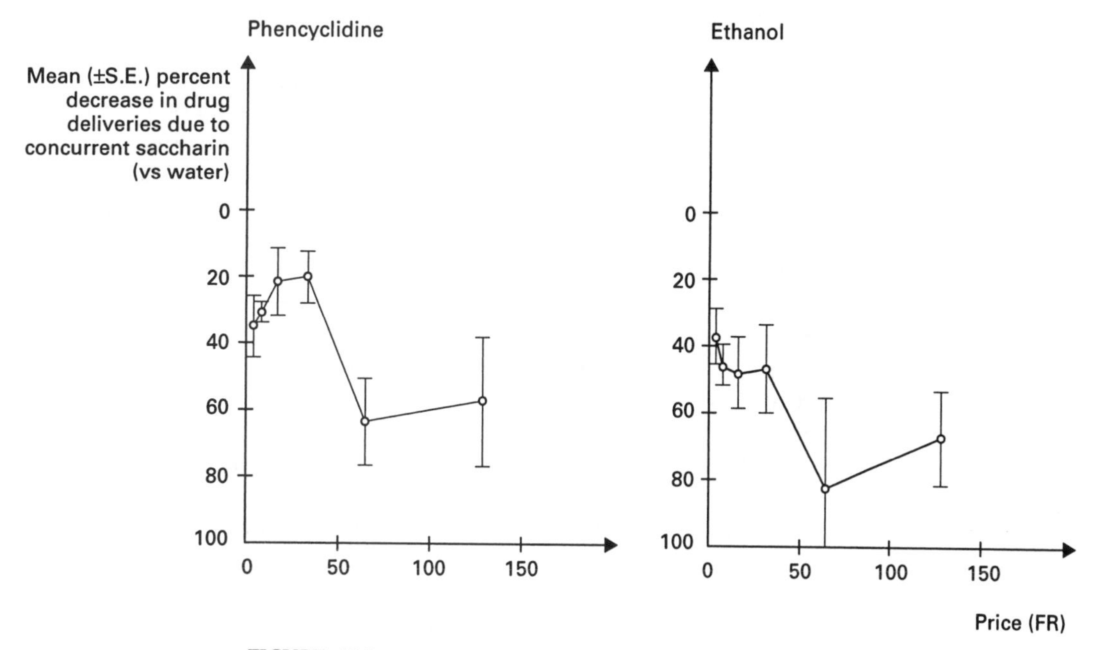

FIGURE 10.3
The effect of saccharin on drug self-administration is presented for ethanol (left frame) and phencyclidine (right frame)
Mean drug deliveries due to concurrent saccharin is presented as a percentage of the drug deliveries obtained when water was concurrently available. Concurrent saccharin produced a 20 to 40% decrease in drug intake when the price (FR) of drug was low and a greater decrease (60 to 80%) occurred when drug price was high.

Other studies have shown that the addition of a nondrug reinforcer to the environment suppressed drug intake, or removal of the nondrug reinforcer from the environment increased drug intake[70]. These have been previously summarized[11], but the range of self-administered drugs includes amphetamine, caffeine, cigarettes, cocaine, ethanol, etonitazene, heroin, morphine, nicotine, pentobarbital, phencyclidine, and THC. The range of nondrug reinforcers that suppressed drug intake includes food, fat, carbohydrates, glucose, sucrose, saccharin, video-game playing, and money.

8 Economic analysis of the food-drug interaction: does access to preferred foods change the demand for drug?

The effect of dietary substances on the reinforcing effect of drugs can be quantitatively and objectively evaluated by using some basic principles of behavioral economics[43,44] (see Table 10.3). For instance, demand for a drug can be quantified by presenting the drug at different prices (FR values) and plotting consumption (mg/day) as a function of unit price (responses/mg). When plotted on log coordinates the result is usually a positively decelerating function with demand decreasing faster as price increases. If the absolute value of the slope of that function is less than 1, then demand is defined as inelastic (i.e., there is little change with increased price). If the absolute value of the slope is greater than 1, the demand is defined as elastic (e.g., there is a more rapid decay in consumption as price increases). Demand is typically inelastic (or resistant to price changes) for essential commodities such as food, water, and gasoline; however, it is elastic for luxury items such as lobster, wine, and vacation travel.

It is of interest to analyze the demand for drug to determine the baseline elasticity upon which factors that reduce or enhance demand can be evaluated. Initial studies indicate that the demand for drug is relatively inelastic[5]; that is, it is highly resistant to price changes like an essential commodity (e.g., food or water). Drugs are not like luxury items that can easily be ignored when prices increase. Since the economic model predicts that consumption of an inelastic commodity will persist even at high prices, it is essential to find interventions that can increase the elasticity of demand for drug. Nondrug alternative reinforcers, such as food, are a class of variables that interfere significantly with drug self-administration; thus, one line of research has examined the effect of dietary substances on demand for drug using economic analyses. An ideal treatment would increase the elasticity of demand for drug.

In a recent attempt to alter the demand for drug, six rhesus monkeys self-administered phencyclidine (0.25 mg/ml) and saccharin (0.3% wt/vol) or water under concurrent FR schedules that varied as follows: 4, 8, 16, 32, 64, and 128. Saccharin and water deliveries were contingent upon an FR 32 schedule. The absolute value of the slope of the phencyclidine demand curve was greater (4.4) when saccharin was concurrently available than when water was present (3.6), indicating increased elasticity of demand. The intensity of demand was also markedly reduced by concurrent saccharin. Intensity is reflected by a parallel shift downward in the demand curve. Another measure of the strength of responding is the unit price at which the maximum response output occurred. This is referred to as P-max[44]. Concurrent saccharin substantially reduced the P-max value for phencyclidine from 415 (with concurrent water) to 160 responses per mg[16]. P-max is analogous to break point on a progressive-ratio schedule; it is the FR at which responding peaks and decreases rapidly thereafter. This data concurs with the recent finding that alternative nondrug reinforcers decrease the break point when a drug is functioning as a reinforcer under a progressive ratio schedule[65].

The effect of adding concurrently available saccharin (vs. water) was also examined with monkeys trained to self-administer orally delivered ethanol (8% wt/vol), and similar results were obtained. Saccharin produced a small increase in elasticity and a large decrease in the intensity of demand for ethanol. This type of demand analysis was also extended to a model of cocaine base smoking in rhesus monkeys[20]. The demand curve was obtained over a range of FRs (64-1024) with and without concurrent saccharin available during daily 3-hour sessions. Saccharin had little effect on the demand for cocaine. These findings are consistent with the limited effects of food deprivation/satiation on cocaine-base smoking[20].

In these studies, the relative decrease in drug-maintained responding was much greater at higher unit prices (responses per mg) (see Figure 10.3). For example, across the phencyclidine and ethanol studies the reduction in drug deliveries due to saccharin was 20–45% at FR 4, while the reduction was 60–90% at FR 128. Since the constituents of unit price (responses/mg) are response requirement and dose, unit price can be altered by changing FR, as in these studies, or drug concentration. The results of the studies investigating demand for phencyclidine are consistent with earlier studies in which concurrent saccharin had a greater suppressant effect at lower phencyclidine concentrations (higher unit prices-responses/mg).

9 Does food substitute for drug reinforcement?

One possible explanation for the effect of food on drug-reinforced behavior is that the rewarding effects of food substitute for those of the drug and vice versa. Substitution can be quantified in economic terms by demonstrating an inverse relationship between the demand for drug and food. If there is substitution, as the demand for drug decreases, as price increases, the demand for food (with a fixed price) should increase, resulting in a positive slope or elasticity coefficient[6,54]. In previous studies in which a drug with a changing price (FR 4-128) and saccharin were concurrently available with fixed price (FR 32), substitution was evaluated. There was increased consumption of saccharin as the phencyclidine price increased, and the cross-price elasticity coefficient was 0.22, suggesting a substitution.

A substitution effect was reported by Nader and Woolverton[59]. In monkeys that were given a discrete-trials choice between food and i.v. injections of cocaine, increasing the FR for one reinforcer increased the frequency of choice of the other reinforcer. In this case the substitution effect was symmetrical, but it was not in the previous study. Increasing the price of saccharin had no effect on intake of phencyclidine. Comer and colleagues[20] found an asymmetrical substitution effect when saccharin and smoked cocaine base were concurrently available to monkeys. Saccharin had no effect on cocaine self-administration, but when the price (FR) of cocaine was increased and consumption decreased, saccharin intake more than doubled (175-440 ml). The results were not due to increased thirst because water intake did not significantly change as cocaine smoking decreased.

9.1 EFFECTS OF AVAILABLE RESOURCES (INCOME) ON DRUG-FOOD PREFERENCE

In behavioral economic terms income is defined as the amount of resources available to purchase a commodity such as a drug or nondrug reinforcer. It is another environmental variable that modifies the interaction between food or other nondrug reinforcers and drug self-administration. For human drug users, income is usually money. In the human laboratory, money, tokens, or time are used to study income effects. In the animal laboratory, income is manipulated by altering the amount of time allowed to earn drug or nondrug reinforcers or the number of reinforcers available. Earlier studies of human consumer behavior have shown that different income levels change consumer preferences for various commodities, and laboratory studies with animals similarly indicate there are changes in dietary preference as a function of income.

There are only a few studies of income effects involving drugs or drug-food interactions, but these initial reports suggest that income is a major variable that determines choice of competing drug vs. food reinforcers. In a drug vs. drug comparison, DeGrandpre and colleagues[25] compared consumption of subjects' more expensive "own-brand" cigarettes with less expensive "other-brand" cigarettes under varied income conditions. Income was regulated by the amount of money allocated to subjects at the beginning of the session. At the low income condition, there was a preference for the less-expensive other brand; however, the preference was reversed at the higher income. Interestingly, although subjects had the opportunity to increase nicotine consumption by purchasing more of the "other brand" at high income, nicotine intake did not change much with changes in income.

The few studies of income effects that have been reported to date have mixed results, but they illustrate marked reversals in preference due to income changes. In one experiment, baboons were offered a choice between food or heroin, and income was defined as the number of trials[28]. When income was high (many trials) heroin was preferred to food, but when income was low (few trials), food was preferred to heroin. The reversal of preference for drug over food at low income could reveal which is the more essential commodity, in this case food. The income preference study may be a useful means of comparing reinforcing efficacy between two commodities. In two recent investigations of income rhesus monkeys were offered concurrent access to drug (phencyclidine or ethanol) and saccharin. When the income levels were high (180 min), saccharin was preferred to phencyclidine; however, at low (20 min) income levels drug was preferred to saccharin. In this case saccharin functioned as the expendable, luxury item and drug functioned as the essential commodity. In summary, the income studied to date reveal that allocation of resources is an important factor to consider in terms of manipulating drug self-administration by enriching the environment with alternative nondrug reinforcers.

9.2 EFFECTS OF FOOD REINFORCERS ON DRUG DEPENDENCE AND WITHDRAWAL

Behavior during withdrawal is another transition state that may be particularly sensitive to altered feeding conditions. Subtle behavioral changes due to withdrawal can be measured with operant techniques and the effect of feeding conditions on withdrawal behavior has been studied. In these studies, either the animal is trained to self-administer, orally or intravenously delivered drug, or drugs are regularly administered by the experimenter. Food is typically available contingent upon a lever-pressing response under a high-ratio schedule (e.g., FR 100). When a water or saline vehicle is substituted for drug over several days, and the animals are in withdrawal, there is a decrease in food-reinforced responding. This is a very sensitive procedure because it detects withdrawal in animals that are showing no physical signs of illness, and in animals treated with doses that are too low to result in a physical withdrawal syndrome. This procedure detects withdrawal to drugs such as caffeine, cocaine, and nicotine, that produce no observable withdrawal effects. This procedure has also produced withdrawal disruptions when drug access occurs only briefly each day or every other day.

9.3 EFFORT TO OBTAIN FOOD AFFECTS THE SEVERITY OF WITHDRAWAL

Using these operant procedures, the interaction between feeding and magnitude of a drug withdrawal effect was examined. Oral phencyclidine self-administration was established in six rhesus monkeys, and then drug was replaced by water for eight-day periods. Drug access was then restored and behavior was allowed to return to baseline conditions after each withdrawal period. The FR for food was increased across a range of values (64, 128, 256, 512 and 1024), and phencyclidine withdrawal was assessed at each one. The demand for food was compared during phencyclidine self-administration and withdrawal, and it was found that drug withdrawal lowered the intensity of demand for food. As the FR increased from 64 to 256, elasticity of demand for food increased indicating that food was a less valued commodity during drug withdrawal.

9.4 LEVEL OF FOOD DEPRIVATION AFFECTS THE SEVERITY OF WITHDRAWAL

In the second portion of this study, phencyclidine withdrawal was compared under conditions of food deprivation and satiation. The food FR was held constant at 1024 which resulted in relatively low food intake. This is referred to as a closed economy in which the animal earns all of its daily food during session (Table 10.3). Under these conditions, phencyclidine withdrawal did not alter food-maintained responding. However, if the monkeys were given 100 g of food in addition to the food earned under the FR 1024 schedule (open economy), the withdrawal-induced suppression of food-maintained responding returned. These results suggest that the type of economic arrangement and motivational variables associated with the procurement of food may allow a drug withdrawal effect to be revealed under some conditions and not under others. It appears that drug withdrawal is not simply a physical illness characterized by decreased food intake, but it has complex effects on motivational systems.

9.5 EFFECTS OF FOOD DEPRIVATION ON DRUG RELAPSE

There are anecdotal and retrospective reports that food deprivation triggers relapse in abstinent alcoholics (e.g., see ref. 2) and in those abstinent from other drugs[37]. However, there have been few controlled laboratory studies of relapse in either animal or human subjects. There would be ethical considerations in manipulating relapse in abstinent human subjects, but fortunately there is a useful animal model of reinstatement of behavior that was previously rewarded by drug (relapse)[68]. In this model, rats were trained to self-administer intravenously delivered drugs. Saline was then substituted for the drug, and when behavior extinguished, one priming dose of the drug was injected by the experimenter through the i.v. cannula. Behavior (lever pressing) that was previously reinforced by drug typically increased in a dose–dependent manner after the priming injection[68]. Exteroceptive stimuli associated with drug-self-administration, such as lights and sounds, have also been presented during extinction, and they also reinstate responding.

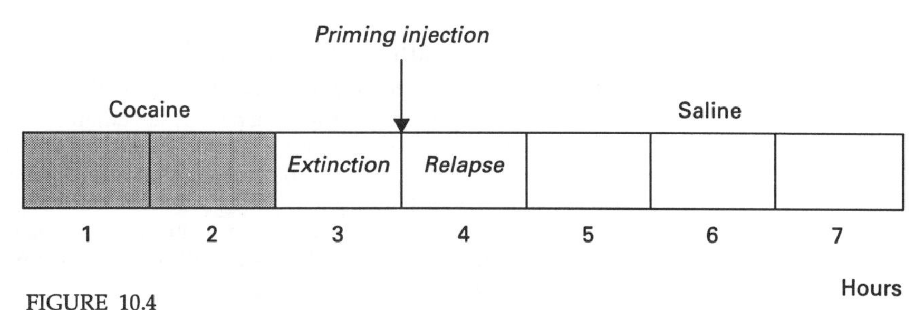

FIGURE 10.4
Diagram of the priming procedure

The reinstatement model described by Stewart and Wise[68] has been used to investigate the effect of food deprivation on relapse. Comer and coworkers[19] trained rats to self-administer i.v. cocaine during daily 2-hour sessions followed by 5-hour saline extinction sessions. A priming dose of cocaine was administered 1 hour after drug access was terminated, and responding was monitored for the subsequent 4 hours. The experimental procedure is illustrated in Figure 10.4. Three feeding conditions were studied: *ad lib*, 20 g or 8–12 g in separate groups. Figure 10.5 shows that responding during the 1-hour extinction period and for the hour after the priming injection (relapse) increased as the level of food deprivation increased.

There were also systematic increases in reinstated responding as the priming dose increased from saline (0) to 0.32, 1 and 3.2 mg/kg cocaine (see Figure 10.5) The differences due to food deprivation during extinction and after priming were not due to different baseline levels of cocaine intake (hours 1 and 2). Table 10.4 shows the number of cocaine infusions during hour 2 did not change as a function of deprivation level of priming dose. Enhancement of behavior previously reinforced by cocaine (extinction) and reinstatement of that behavior after a priming injection by food deprivation suggests that dieting would be contraindicated in patients seeking to remain abstinent from drugs. This suggestion has been made in a recent review of smoking cessation and weight gain[61].

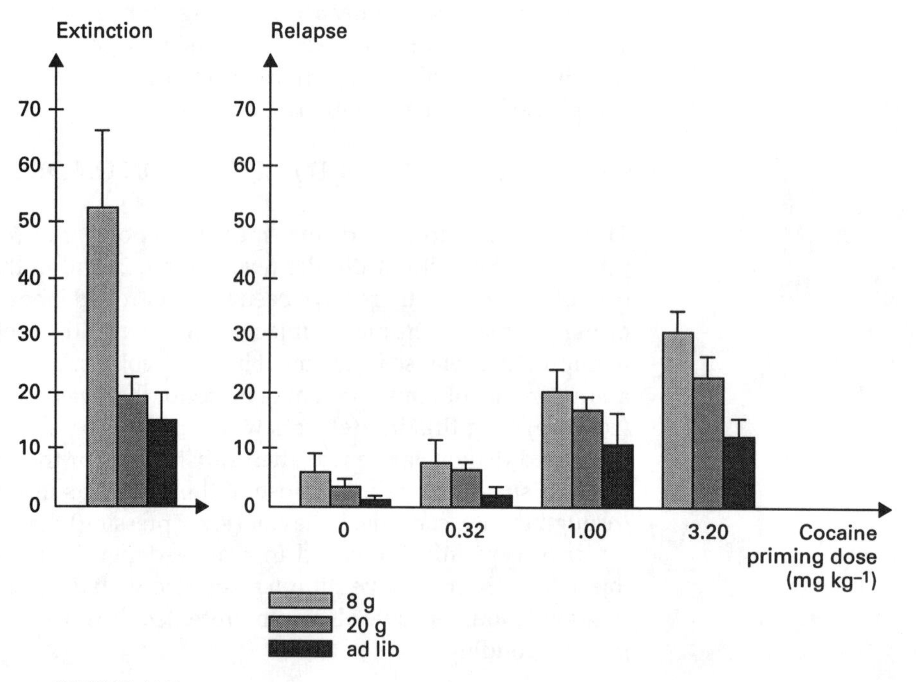

FIGURE 10.5

The number of saline infusions (±SEM) self-administered is presented for the third (extinction) and fourth (relapse) hour of the relapse session

During the first 2 hours cocaine (0.4 mg per kg) self-administration occurred, and then cocaine was replaced with saline at the beginning of hour 3. Light bars indicate that 8 g of food was available each day. Medium bars refer to a 20 g per day allotment. Dark bars represent the unlimited (*ad lib*) food condition. Saline reinforced responding during hour 3 or extinction responding was high and related to level of food deprivation. During hour 4 (right panel) when various doses of cocaine priming injections were given there was a dose dependent increase in saline infusions as food access decreased.

The results of this study concurred with an earlier report of the increased extinction responding during food deprivation. That study showed that the increases in drug-seeking behavior due to food deprivation persisted for several weeks. Rats were given 24-hour access to i.v. cocaine for 11 days and food deprived (8 g) every 3 days. Saline was then substituted, and after 12 days of extinction, they were again food deprived every third day. Responding on the lever that previously yielded cocaine infusions (now reinforced by saline) increased up to seven-fold on deprivation days compared to satiation days. This reinstatement of responding produced by food deprivation persisted until the 41st day when the experiment was terminated. Overall, the results of these experiments indicate that feeding conditions have as powerful an influence

TABLE 10.4

Reinstatement of responding previously reinforced by i.v. cocaine (0.4 mg per kg)

(Data for hours 3 and 4 are also shown in Figure 10.5.)

Deprivation level	*Infusion*	*Hour*	*Priming doses (mg/kg)*			
			0	0.32	1	3.2
			Number of infusions			
ad lib food	cocaine	2*	28.5	27.3	27.3	26.6
	saline	3**	18.5	15.8	17.3	10.0
	saline	4***	1.3	1.8	10.5	11.6
20 g food	cocaine	2*	32.3	22.9	29.9	22.3
	saline	3**	22.0	10.6	17.0	27.8
	saline	4***	3.6	6.3	16.3	21.7
8 g food	cocaine	2*	24.7	22.3	23.5	21.5
	saline	3**	45.9	67.2	52.7	39.7
	saline	4***	6.0	7.3	19.3	29.3

* Cocaine self-administration (0.4 mg/kg)

** Saline self-administration

*** Saline self-administration after cocaine priming injection; the priming injection occurred at the beginning of hour 4.

over reinstatement of drug-seeking behavior as they do during acquisition, maintenance, and withdrawal phases of the addictive process. Furthermore, food deprivation reinstates drug-seeking behavior during an extinction phase before a priming injection of drug was given.

10 Neurobiological mechanisms of food-drug reinforcer interactions

Interactions between food and addiction may be explained simply on a behavioral level, such as by choice theory[43,63]. When access to one substance is restricted, choice shifts to available alternative reinforcers. This hypothesis was suggested by Premack and Premack[64] in their study of increased wheel-running during food deprivation. Evidence that supports this notion is the symmetrical substitution that has been found with reinforcers. For example, Premack and Premack[64] and others[53] also found that deprivation of wheel-running increased food intake. Similar reciprocal interactions have been found with food and drug reinforcers such as food and morphine, and G+S and cocaine. However, recent efforts to examine mechanisms of the interaction between dietary variables and drugs of abuse have focused upon neurobiological mechanisms (e.g., see ref. 3, 42, 47). Although there are excellent reviews of the neurobiological substrates of drug addiction (e.g., see ref. 49, 50) there is not yet a substantial body of literature devoted to the neurobiological basis for commonalities between food and drug reward.

There are several lines of evidence suggesting that common reward mechanisms mediate feeding and drug abuse. One theory is that there are common biological substrates for food and drugs that are based on unconditioned approach behaviors that precede positive reinforcement[72]. Electrical stimulation of the medial forebrain bundle and associated anatomical structures results in forward locomotion. The same type of increased locomotion is associated with positive reinforcement from food or access to a conspecific of the opposite sex.

Similarly, lesion studies have illustrated that the same anatomical structures (nucleus accumbens, frontal cortex, and other forebrain structures) are critical for the positive reinforcing effects of drugs. Most drugs that are abused by humans interact with these structures either directly or indirectly through dopaminergic pathways. Psychomotor stimulants such as amphetamine and cocaine release or block reuptake of dopamine in these regions. Opioids such as heroin or morphine activate the system at the dopaminergic synapse and at sites afferent to the dopaminergic neurons. Other stimulants such as caffeine and nicotine, general depressants such as ethanol, and dissociative anesthetics such as phencyclidine alter dopamine turnover that affects these brain regions. All of these drugs are positive reinforcers for animals and produce locomotor activation at low doses. While food deprivation does not always increase the general activity level of rats[13], it does enhance the motor activating effects of amphetamine and morphine. Other drugs such as barbiturates do not have as clear a role in the dopaminergic link. They do produce locomotor activating effects and they function as reinforcers under some conditions in animals and humans.

Other lines of research suggest that neural dopaminergic activity is related to the interaction between food deprivation and drug effects. In microdialysis studies, the tastes of saccharin released dopamine, and when the saccharin taste was made aversive by taste-aversion conditioning, less dopamine was released[42]. Microdialysis studies with drugs of abuse similarly resulted in increased extracellular dopamine. There are many other parallel findings regarding neurochemical changes that are beyond the scope of this review. Overall, the similarities in neurochemical outcomes using several rewarding events, such as food, drugs, and electrical brain stimulation as stimuli support the notion of common reward mechanisms. However, further research is needed in which these diverse stimuli are manipulated within the same experiments and within subjects. In addition to dopamine-based theories, there are also several lines of evidence suggesting that common reward mechanisms for food and drugs may be based upon the opioid system and serotonergic neurotransmitter systems[47,52]. Neuroendocrine interactions have also been implicated[26]. An advantage of this approach is that a better understanding of neurochemical changes may lead to drug treatment for both eating disorders and drug abuse.

The suppressant effects of palatable dietary substances on self-administration of drugs of abuse may be due to release of endogenous opioids that substitute for the reinforcing effects of drugs. One line of evidence supporting this hypothesis is the demonstration of an enhancement of the analgesic effects of morphine after acute exposure to sweetened liquids. This may be due to additive effects of endogenous and exogenous opioid levels. Chronic exposure to sweetened liquids has been demonstrated to produce tolerance to morphine-induced analgesia. These findings also suggest increased levels of endogenous opioids resulting from exposure to sweets. There is additional evidence for opioid mechanisms from a study by Dum and colleagues[27] showing that sucrose increases release and breakdown of hypothalmic β-endorphin; Marks-Kaufman and others[57] have reported that sucrose increases opiate receptor binding in rats.

In addition to the effects of sweetened foods on opioid systems, there is also evidence that presentation and deprivation of standard laboratory chow alters endogenous opioids. For instance, food deprivation alters pituitary and brain levels of opioid peptides and opioid receptor binding in rats. Hodgson and Bond[41] reported that it is not only the feeding condition that controls opioid effects, but the animal's controllability over food. They found that food delivery that is contingent upon lever-press responding results in a nonopioid analgesia, and food that is delivered noncontingently upon responding produces opioid-mediated analgesia.

Further evidence for opioid involvement in food-drug interactions comes from studies indicating that acute administration of opiate receptor antagonists (e.g., naltrexone) reduce the intake of palatable substances. Specifically, antagonists at μ and κ opioid receptors reduce intake of high-fat, palatable diets, but delta receptor antagonists are not implicated. In drug studies, acute administration of mu antagonists (e.g., naltrexone) reduces cocaine[24] and alcohol self-administration in rats and humans, respectively.

Other avenues of research have implicated serotonin or 5-hydroxytryptophan (5HT) in the interaction between food- and drug-rewarded behavior (cf. see refs. 3, 42). Microdialysis studies reveal higher extracellular levels of 5HT during feeding and prior to feeding when animals smelled food[42]. Treatment with serotonergic drugs similarly reduced food intake[7], drug abuse in humans, and drug self-administration in animals.

Another mechanism that may modulate the enhanced effects of drugs due to food deprivation is corticosterone. Coveney and coworkers[22] demonstrated that food deprivation increased the bioavailability of phencyclidine and produced an enhanced corticosterone response. The enhanced bioavailability of phencyclidine may explain the increased responding maintained by that drug during food deprivation. A recent study by Deroche and colleagues[26] implicated corticosterone secretion in the modulation of food deprivation-induced increases in sensitivity to drugs. Previous work indicated that corticosterone potentiated the reinforcing effects of stimulant drugs[62], and that food deprivation increased corticosterone levels. Thus, the goal of the Deroche et al.[26] study was to determine if food deprivation-induced increases in sensitivity to morphine and amphetamine could be blocked by preventing corticosterone levels from rising in response to food deprivation.

Increases in corticosterone were blocked by adrenalectomizing one group of rats and restoring basal levels of corticosterone by pellet implantation. The pelleted group had normal levels of corticosterone, but they did not have a mechanism to increase those levels in response to food deprivation. Another group of rats was sham operated, and their corticosterone response was intact. The results of the experiment were enhanced sensitivity to the locomotor effects of morphine and amphetamine only in the sham-operated controls and not in the adrenalectomized group. These results suggested that corticosterone secretion is one of the mechanisms by which food restriction amplifies the behavioral response to drugs. Since increased locomotor activity is related to the reinforcing effects of drugs, corticosterone may modulate reinforcing effects of drugs as well. There are recent data that support this hypothesis. Opioid self-administration is increased in rats that have elevated corticosterone levels due to restraint stress. Also, blockade of corticosterone release by pretreatment with metyrapone suppressed cocaine self-administration (Goeders and Piazza, personal communication, 1994).

In summary, it is evident from the range of mechanisms proposed, and the recency of much of the supporting data that there is not yet a clear explanation of the neurobiological basis for food-drug interactions. However, several neurotransmitter, neuropeptide, and neuroendocrine systems seem to be involved, and the interactions among these systems are undoubtedly complex.

References

1. Aigner, T.G. and Balster, R.L., "Behavioral effects of chronic oral administration of levo-alpha-acetylmethadol in the rat," *Pharmacol. Biochem. Behav.* 1978, **8:** pp 593–596.
2. Alcoholics Anonymous, *Living Sober*. Alcoholics Anonymous World Services, Inc., New York, 1987, p 23.
3. Asghar, K., "Role of dietary and environmental factors in drug abuse," *Alcohol Drug Res.* 1986, **7:** pp 61–83.
4. Bell, S.M., Gosnell, B.A., Krah, D.D., Meisch, R.A., "Ethanol reinforcement and its relationship to saccharin preference in Wistar rats," *Alcohol.* 1994, **11:** pp 141–145.
5. Bickel, W.K., DeGrandpre, R.J., Higgins, S.T., Hughes, J.R., "Behavioral economics of drug self-administration. I. Functional equivalence of response requirement and drug dose," *Life Sci.* 1990, **47:** pp 1501–1510.
6. Bickel, W., Hughes, J.R., DeGrandpre, R.J., Higgins, S.T., Rizzuto, P., "Behavioral economics of drug self-administration IV. The effects of response requirement on the consumption of and interaction between concurrently available coffee and cigarettes," *Psychopharmacology.* 1992, **107:** pp 211–216.
7. Blundell, J.E., "Serotonin and appetite," *Neuropharmacology.* 1984, **23:** pp 1537–1551.
8. Bulik, C.M., Sullivan, P.F., Epstein, L.H., McKee, M., Kaye, M., Dahl, R.E., Weltzin, T.E., "Drug use in women with anorexia and bulimia nervosa," *Int. J. Eating Disord.* 1992, **3:** pp 213–225.
9. Bulik, C.M., "Abuse of drugs associated with eating disorders," *J. Subst. Abuse.* 1992, 4, pp 69–90.
10. Bulik ,C.M. and Brinded, E.C., "The effect of food deprivation on alcohol consumption in bulimic and control women," *Addiction.* 1993, **88:** pp 1545–1551.
11. Carroll, M.E., "Reducing drug abuse by enriching the environment with alternative reinforcers." In: Green, L. and Kagel, J.H. (Eds.), *Advances in Behavioral Economics*, Vol. 3. Abley Publishing, Norwood, NJ, 1996.
12. Carroll, M.E. "The economic context of drug and nondrug reinforcers affects acquisition and maintenance of drug-reinforced behavior and withdrawal effects," *Drug Alcohol Depend.* 1993, **33:** pp 201–210.
13. Carroll, M.E. and Meisch, R.A., "Increased drug-reinforced behavior due to food deprivation," *Adv. Behav. Pharmacol.* 1984, **4:** pp 47–88.
14. Carroll, M.E., Stitzer, M.L., Strain, E., Meisch, R.A., "The behavioral pharmacology of alcohol and other drugs." Galanter, M. (Ed.), *Recent Developments in Alcoholism*, Vol. 8. Plenum Press, New York, 1990, pp 5–46.
15. Carroll, M.E., "Rapid acquisition of oral phencyclidine self-administration in food-deprived and food-satiated rhesus monkeys: concurrent phencyclidine and water choice," *Pharmacol. Biochem. Behav.* 1982, **17:** pp 341–346.
16. Carroll, M.E. and Lac, S.T., "Autoshaping i.v. cocaine self-administration in rats: effects of nondrug alternative reinforcers on acquisition," *Psychopharmacology.* 1993, **110:** pp 5-12.
17. Carroll, M.E., "The role of food deprivation in the maintenance and reinstatement of cocaine-seeking behavior in rats," *Drug Alcohol Depend.* 1985, **16:** pp 95–109.
18. Cheskin, L.J., Wiersema, L., Goldsborough, D., Tayback,, Henningfield, J., Gorelick, D., "Caloric restriction increases nicotine consumption in cigarette smokers." In: Harris, L.S. (Ed.), *Problems of Drug Dependence, 1994: Proceedings of the 56th Annual Scientific Meeting*, The College on Problems of Drug Dependence, U.S. Dept. of Health and Human Services, Rockville, MD, 1994, p 262.
19. Comer, S.D., Lac, S.T., Curtis, L.K., Carroll, M.E., "Food deprivation affects extinction and reinstatement of responding in rats," *Psychopharmacology.* 1995, **121:** pp 150–157.
20. Comer, S.D., Hunt, V.R., Carroll, M.E., "Effects of concurrent saccharin availability and buprenorphine pretreatment on demand for smoked cocaine base in rhesus monkeys," *Psychopharmacology.* 1994, **115:** pp 15–23.
21. Coscina, D.V. and Dixon, L.M., "Body weight regulation in anorexia nervosa: insights from animal model," Padraig, L. (Ed.), *Anorexia Nervosa: Recent Developments in Research.* New York, NY, 1983, pp 207–219.

22. Coveney, J.R., Neal, B.S., Sparber, S.B., "Food deprivation alters behavioral and plasma corticosterone responses to phencyclidine in rats," *Pharmacol. Biochem. Behav.* 1990, **36:** pp 451–456.
23. de la Garza, R., Bergman, J., Hartel, C.R., "Food deprivation and cocaine self-administration," *Pharmacol. Biochem. Behav.* 1981, **15:** pp 141–144.
24. De Vry, J., Donselaar, I., Van Ree, J.M., "Food deprivation and acquisition of intravenous cocaine self-administration in rats: effect of naltrexone and haloperidol," *J. Pharmacol. Exp. Ther.* 1989, **251:** pp 735–740.
25. DeGrandpre, R.J., Bickel, W.K., Rizvi, S.A.T., Hughes, J.R., "The behavioral economics of drug self-administration. VII. Effects of income on drug choice in humans," *J. Exp. Anal. Behav.* 1993, 59, pp 483–500.
26. Deroche, V., Piazza, P.V., Casolini, P., LeMoal, M., Simon, H., "Sensitization to the psychomotor effects of amphetamine and morphine induced by food restriction depends on corticosterone secretion," *Brain Res.* 1993, **611:** pp 352–356.
27. Dum, J., Gramsch, C.H., Herz, A., "Activation of hypothalamic beta-endorphine pools by reward induced by highly palatable foods," *Pharmacol. Biochem. Behav.* 1983, **18:** pp 443–448.
28. Elsmore, T.F., Fletcher, G.V., Conrad, D.G., Sodetz, F.J., "Reduction of heroin intake in baboons by an economic constraint," *Pharmacol. Biochem. Behav.* 1980, **13:** pp 729–731.
29. Fahy, T.A. and Treasure, J., "Caffeine abuse in bulimia nervosa," *Int. J. Eating Disord.* 1991, **10:** pp 373–377.
30. Filstead, W., Parrella, D., Ebitt, J., "High risk situations for engaging in substance abuse and binge-eating behaviors," *J. Stud. Alcohol.* 1988, **49:** pp 136–141.
31. Fowler, S.C., Johnson, J.S., Kallman, M.J., Lou, J.R., Wilson, M.C., Hikal, A.H., "In a drug discrimination procedure isolation-reared rats generalize to lower doses of cocaine and amphetamine than rats reared in an enriched environment," *Psychopharmacology.* 1993, **110:** pp 115–118.
32. Franklin, J.C., Shiele, B.C., Brozek, J., Keys, A., "Observations on human behavior in experimental semistarvation and rehabilitation," *J. Clin. Psychol.* 1948, **4:** pp 28–45.
33. Gosnell, B.A. and Krahn, D.D., "The relationship between saccharin and alcohol intake in rats," *Alcohol.* 1992, **9:** pp 203–206.
34. Gosnell, B.A., Lane, K.E., Bell, S.M., Krahn, D.D., "Intravenous self-administration by rats with low vs. high saccharine preferences," *Psychopharmacology.* 1995, **117:** pp 248 –252.
35. Griffiths, R.R., Bigelow, G.E., Henningfield, J.E., "Similarities in animal and human drug taking behavior." In: Mello, N.K. (Ed.), *Advances in Substance Abuse*, Vol. 1. JAI Press, Greenwich, CT, 1980, pp 1–90.
36. Hagan, M.M. and Moss, D.E., "An animal model of bulimia nervosa: opioid sensitivity to fasting episodes," *Pharmacol. Biochem. Behav.* 1991, **39(2):** pp 421–422.
37. Hall, S.M., Ginsberg, D., Jones, R.T., "Smoking cessation and weight gain," *J. Counsel. Clin. Psychol..* 1986, **54:** pp 342–346.
38. Hanna, J.M. and Hornick, C.A., "Use of coca leaf in southern Peru: adaptation or addiction," *Bull. Narcot.* 1977, **29:** pp 63–74.
39. Hatsukami, D.K., Thompson, T., Pentel, P.R., Flygaare, B., Carroll, M.E., "Self-administration of smoked cocaine," *Exp. Clin. Psychopharmacol.* 1994, **2:** pp 115–125.
40. Higgins S.T., Delaney, D.D., Budney, A.J., Bickel, W.K., Hughes, J.R., Foerg, F., Fenwick, J.W., "A behavioral approach to achieving initial cocaine abstinence," *Am. J. Psychiatry.* 1991, **148:** pp 1218–1224.
41. Hodgson, D.M. and Bond, N.W., "Control of food delivery in food-deprived rats mediates analgesia," *Behav. Brain Res.* 1991, **44:** pp 205–209.
42. Hoebel, B.G., Hernandez, L., Schwartz, D.H., Mark, G.P., Hunter, G.A., "Microdialysis studies of brain norepinephrine, serotonin, and dopamine release during ingestive behavior," *Ann. N.Y. Acad. Sci.* 1989, **575:** pp 171–191.
43. Hursh, S.R., Bauman, R.A., Green L, Kagel, J.H. (Eds.), *Advances in Behavioral Economics*, Vol. 1. Abley Publishing Corp., Norwood, NJ, 1987, pp 117–165.
44. Hursh, S.R., "Behavioral economics of drug self-administration and drug abuse policy," *J. Exp. Anal. Behav.* 1991, **56:** pp 377–393.

45. Hyyatiä, P. and Sinclair, J.D., "Oral etonitazene and cocaine consumption by AA, ANA and Wistar rats," *Psychopharmacology.* 1993, **111:** pp 409–414.
46. Jonas, J.M., "Do substance-abuse, including alcoholism, and bulimia covary?" In: Reid, L.D. (Ed.), *Opioids, Bulimia, and Alcoholism.* Springer Verlag, New York, 1990, pp 247–258.
47. Kanarek, R.B. and Marks-Kaufman, R., "Animal models of appetitive behavior: interaction of nutritional factors and drug seeking behavior." In: Winick, M. (Ed.), *Control of Appetite.* John Wiley & Sons, New York, 1988, pp 1–26.
48. Kanarek, R.B. and Marks-Kaufman, R., "Dietary modulation of oral amphetamine intake in rats," *Physiol. Behav.* 1988, **44:** pp 501-505.
49. Koob, G.F. and Bloom, F.E., "Cellular and molecular mechanisms of drug dependence," *Science.* 1988, **242:** pp 715–723.
50. Koob, G.F. and Goeders, N.E., (1989) "Neuroanatomical substrates of drug self-administration." In: Liebman, J.M. and Cooper, S.J. (Eds.), *The Neuropharmacological Basis of Reward.* Oxford University Press, New York, 1989, pp 214–263.
51. Lac, S.T. and Carroll, M.E., "Cocaine acquisition in rats: effect of feeding conditions and palatability," *Pharmcol. Biochem. Behav.* 1994, **48:** pp 836.
52. Leibowitz, S.F. and Shor-Posner, G., "Hypothalamic monoamine systems for control of food intake: analysis of meal patterns and macronutrient selection." In: Carruba, M.O. and Blundell, J.E. (Eds.), *Pharmacology of Eating Disorders: Theoretical and Clinical Development.* Raven, New York, 1986, pp 29–49.
53. Looy, H. and Eikelboom, R., "Wheel running, food intake, and body weight in male rats. *Physiol. Behav.* 1989, **45:** pp 403–405.
54. Mansfield, E., *Microeconomics: theory/applications,* Sixth edition. W.W. Norton, New York, 1988.
55. Markou, A., Weiss, F., Gold, L.H., Caine, S.B., Schulteis, G., Koob, G.F., "Animal models of drug craving," *Psychopharmacology.* 1993, **112:** pp 163–182.
56. Marks-Kaufman, R. and Lipeles, B.J., "Patterns of nutrient selection in rats orally self-administering morphine," *Nutr. Behav.* 1982, **1:** pp 33–46.
57. Marks-Kaufman, R., Hamm, M.W., Barbato, G.F., "The effects of dietary sucrose on opiate receptor binding in genetically obese (ob/ob) and lean mice," *J. Am. Coll. Nutr.* 1989, **8(1):** pp 9–14.
58. Messing, R.B., Kleven, M.S., Sparber, S.B., "Delaying reinforcement in an autoshaping task generates adjunctive and superstitious behaviors," *Behav. Proc.* 1986, **13:** pp 327–339.
59. Nader, M.A. and Woolverton, W.L., "Effects of increasing response requirement on choice between cocaine and food in rhesus monkeys," *Psychopharmacology.* 1992, **108:** pp 295–300.
60. Nichols, J.R. and Hsiao, S., "Addiction liability of albino rats: breeding for quantitative differences in morphine drinking," *Science,* 1967, **157:** pp 561–563.
61. Perkins, K.A., (1994) "Issues in the prevention of weight gain after smoking cessation," *Ann. Behav. Med.* 1994, **16:** pp 46 –52.
62. Piazza, P.V., Derninieve, J.M., Le Moal, M., Mormede, P., Simon, H., "Factors that predict individual vulnerability to amphetamine self-administration," *Science.* 1991, **245:** pp 1511–1513.
63. Pierce, W.D. and Epling, W.F., "Choice, matching and human behavior: a review of the literature," *Behav. Anal.* 1983, **6:** pp 57–76.
64. Premack, D. and Premack, A.J., "Increased eating in rats deprived of running," *J. Exp. Anal. Behav.* 1963, **6:** pp 209–212.
65. Rodefer, J.S. and Carroll, M.E., "Progressive ratio and behavioral economic evaluation of the reinforcing efficacy of orally-delivered phencyclidine and ethanol in monkeys: effects of feeding conditions," *Psychopharmacology* (in press).
66. Skinner, B.F., *Science and Human Behavior.* MacMillan, New York, 1953.
67. Smith, J.C. and Foster, D.F., "Some determinants of intake of glucose and saccharin solutions," *Physiol. Behav.* 1980, **25:** pp 127–133.
68. Stewart, J. and Wise, R.A., "Reinstatement of heroin self-administration habits: morphine prompts and naltrexone discourages renewed responding after extinction," *Psychopharmacology.* 1992, **108:** pp 79–84.
69. Ukai, M. and Holtzman, S.G., "Restricted feeding does not modify discriminative stimulus effects of morphine in the rat," *Pharmacol. Biochem. Behav.* 1988, **29:** pp 201–203.

70. Vuchinich, R.E. and Tucker, J.A., "Behavioral theories of choice as a framework for studying drinking behavior," *J. Abnorm. Psychol.* 1983, **92:** pp 408–416.
71. Weiss, G., "Food fantasies of incarcerated drug users," *Int. J. Addict.* 1982, **17(5):** pp 905–912.
72. Wise, R.A., "A psychomotor stimulant theory of addiction," *Psychol. Rev.* 1987, **94:** pp 469–492.
73. Yung, L., Gordis, E., Holt, J., "Dietary choices and likelihood of abstinence among alcoholic patients in an outpatient clinic," *Drug Alcohol Depend.* 1983, **12:** pp 355–362.
74. Zacny, J.P., Divane, W.T., de Wit, H., "Assessment of magnitude and availability of a non-drug reinforcer on preference for a drug reinforcer," *Hum. Psychopharmacol.* 1992, **7:** pp 281–286.
75. Zacny, J.P. and de Wit, H., "Effects of food deprivation on subjective responses to *d*-amphetamine in humans," *Pharmacol. Biochem. Behav.* 1989, **34:** pp 791–795.

Glossary

abrupto placenta: premature detachment of the placenta, often attended by maternal systemic reactions in the form of shock, oliguria, and coagulation abnormalities.

abuse liability: a drug's ability to maintain self-administration regardless of whether or not this is associated with evidence of physical dependence.

acetylcholine (ACh): one of the major neurotransmitters of the nervous system, it is involved in the arousal of the organism.

acetylcholinesterase (AChE): enzyme that catalyzes the cleavage of acetylcholine to choline and acetate; it is found in the CNS, particularly in gray matter of nerve tissue, in red blood cells and in motor endplates of skeletal muscle.

achievement test: a measure of knowledge or proficiency. The term is usually applied to an examination on outcomes of school instruction.

acquisition: usually defined in operational terms as a change in response measure and typically is used to refer to that portion of any learning process during which there is a consistent increase in responsiveness and the change has a certain permanence.

Addiction Research Center Inventory (ARCI): questionnaire used in addiction research consisting of 550 questions of the yes/no type to distinguish different types of subjective drug effects. If a drug of unknown effects is given, ARCI results can be used to categorize the drug on he basis of its subjective effects.

alcohol abuse, see drug abuse

Alcohol Accepting rats: strain of rats, specially bred for their quality to consume alcohol when administered to their drinking water.

alcohol dehydrogenase: an enzyme that catalyzes the reversible oxidation of primary or secondary alcohols to aldehydes.

alcohol dependence, see drug dependence

Alcohol non-Accepting rats: strain of rats, specially bred for their quality not to consume drinking water when alcohol has been administered to it.

alcoholism: the personality and behavioral syndrome characteristic of a person who abuses alcohol.

aldehyde dehydrogenase: an enzyme that catalyzes the oxidation of various aldehydes, using NAD^+ as an electron acceptor.

alkaloids: organic substances existing in combination with organic acids in great variety in many plants, and to which many drugs owe their medicinal properties.

amethystic agent: agent that prevents drunkenness. The Greeks believed that the stone and plant of that name to be preventives of intoxication.

amino acids: organic compounds containing an amino group ($-NH_2$) and a carboxyl group (-COOH). Amino acids are the fundamental constituents of all proteins.

amnesic syndrome: an organic mental disorder characterized by the impairment of memory, both retrograde and anterograde amnesia, occurring in a normal state of consciousness. This syndrome may be caused by thiamine deficiency, but also result from any pathological process causing bilateral damage to certain structures in the medial temporal lobe and diencephalon (e.g., the hippocampal formation). Causes include head trauma, brain tumors, infarction, cerebral hypoxia, and carbon monoxide poisoning.

amniocentesis: withdrawal of a sample of the amniotic fluid surrounding the embryo in the uterus by piercing the amniotic sac through the abdominal wall.

amperozide (calcium homotaurinate): medicine used for the pharmacotherapy of alcoholism. It is believed that this drug decreases craving for alcohol in alcoholics.

amphetamines: a class of drugs, including Benzedrine®, Dexedrine®, and Methedrine® that act as central-nervous-system stimulants. Amphetamines suppress appetite, increase heart rate and blood pressure, and, in larger doses, produce a feeling of euphoria and power. Therapeutically, they are used to alleviate depression, control appetite, relieve narcolepsy and paradoxically, to control childhood hyperkinesis. Amphetamine abuse is common and chronic use leads to a model paranoid psychosis.

amygdala (nucleus amygdalae): one of the basal ganglia: a rough almond shaped mass of gray matter deep inside each cerebral hemisphere. It has extensive connections with the olfactory system and sends nerve fibers to the hypothalamus, its functions are apparently concerned with mood, feeling, instinct, and possibly memory for recent events.

analgesic: a drug that relieves pain.

analysis of covariance: extension of the analysis of variance, used when there is improper control of one or more of the variables. The procedure allows for statistical control of the uncontrolled variations so that normal analysis techniques may be carried out without distorting the results.

anandamide: endogenous, marijuana-like compound.

anhedonia: condition marked by a general lack of interest in living, in the pleasures of life. A loss of the ability to enjoy things.

anorectic: drug that induces loss of appetite.

anorexia nervosa, psychological illness, most common in female adolescents, in which the patients starve themselves or use other techniques, such as vomiting or taking laxatives, to induce weight loss.

anorexia: loss of appetite

antacids: drugs that neutralize the hydrochloric acid secreted in the digestive juices of the stomach.

antisocial personality disorder (ASPD): personality disorder characterized by a history of chronic antisocial behavior (often observed in childhood as a conduct disorder) the essential feature of which is the violation of the rights of others. Strictly spoken the term is only used for cases in which the onset is before age 15 and continues in later life. Typical patterns of behavior include: truancy from school,

0-8493-7803-6/99/$0.00+$.50

inability to hold a job, lying, stealing, aggressive sexual behavior, drug and alcohol abuse, vagrancy and a high rate of criminality.

anxiety disorders: a group of disorders characterized by unrealistic fear, panic, or avoidance behavior. These disorders include (among others) panic attacks, phobias, obsessive-compulsive disorder, and generalized anxiety disorder.

area postrema: a small tongue-shaped area on the lateral wall of the fourth ventricle.

asphyxia: pathological changes caused by lack of oxygen in respired air, resulting in hypoxia and hypercapnia.

asphyxiation: the causing of, or the state of asphyxia.

assessment batteries of cognition: test battery to measure cognitive functions (see cognition).

assessment: any procedure that measures or describes. Most often refers to integration of several pieces of information, in order to reach a judgment about an individual's health or fitness for an assignment. Sometimes a study of a group or a situation.

astrocyte: type of cell with numerous sheet-like processes extending from its cell-body, found throughout the central nervous system. It is one of the several different types of cells that make up the neuroglia. The cells have been ascribed the function of providing nutrients for neurons and possibly of taking part in information storage processing.

astroglia, see astrocyte; neuroglia

ataxia: failure of muscular coordination; irregularity of muscular action.

athetoid: relating to athelosis

athetosis: form of involuntary movement which is repetitive, slow, and writhing or undulating in character.

atropine: an alkaloid derived from belladonna (from the deadly night-shade plant). It is a respiratory and circulatory stimulant and counteracts parasympathetic stimulation.

attention deficit disorder (ADD): general psychiatric syndrome characterized by the child displaying developmentally inappropriate failures of attention and a pervasive impulsivity. Often there is excessive motor activity.

attention-deficit hyperactivity disorder (ADHD), see Attention Deficit Disorder

auto shaping: process in which organisms come to direct operant behavior (e.g., key picking; lever pressing) toward stimuli regularly associated with the presentation of reinforcement.

aversion: a repugnance or dislike for something, an internal negative reaction.

axon: thread-like nerve cell extension functioning to conduct information. At the specialized endings, the synapses, information is transmitted to other cells.

axonopathy: axonal degeneration with the axon itself being the primary focus of injury.

basal ganglia: several large masses of gray matter embedded deep within the white matter of the cerebrum. They include the caudate and lenticular nuclei, together known as the corpus striatum, and the amygdaloid nucleus. The lenticular nucleus consists of the putamen and globus pallidus. The basal ganglia have complex neural connections with both the cerebral cortex and thalamus: they are involved with the regulation of voluntary movements and at a subconscious level.

behavioral rating scale: any device used to assist a rater in making ratings of behavior. Rating behavior is the use of rating techniques in a fairly restricted manner such that only overt, objectively observable behaviors enter into the assessment.

benzodiazepines: a group of pharmacologically active compounds used as minor tranquillizers and hypnotics.

Betz's cells: large pyramidal ganglion cells found in the internal pyramidal layer of the cerebral cortex; also called giant pyramids and giant pyramidal cells.

bias: an inclination toward a position or conclusion: a prejudice. A distortion of the facts caused by errors in selecting or classifying the subjects of a study.

biochemical marker: a biochemical measure that designates the presence or absence of distinctive features.

blood-brain-barrier: a selectively permeable physiological 'barrier' between the circulating blood and the brain which functions by preventing some substances from reaching the brain.

bolus dose: a concentrated mass of a pharmaceutical preparation given (intravenously) within a minimal amount of time; often used for diagnostic purposes, e.g., in the form of a radioactive isotope.

Bonferroni correction: in statistical analysis, multiple tests or comparisons greatly increase the Type I error rate compared to the single-test situation. The most direct way to obtain multiple-test error rates is to use a Bonferroni correction, in which the single-test alpha is multiplied by the number of independent (non redundant) comparisons (k) to yield a multiple-test alpha that protects against even one fortuitous association.

2-bottle choice test: test in which an animal has free access to two drinking bottles. Often one of these bottles contains tap water (drinking water) and the other bottle a solution of a drug, e.g., alcohol; sometimes both bottles contain a solution of the drug, but in different concentrations. By measuring the amount of the fluid consumed by the animal during a certain period, its preference for the drug over tap water is measured.

bradyphrenia: slowness of thought or fatigability of initiative, resulting from depression or CNS disease.

brain stimulation reward: the pleasant or rewarding sensations that a subject perceives when special centers in the brain are activated by an electric stimulus.

bromocriptine: a drug derived from ergot, that has actions similar to those of dopamine. It is used in the treatment of parkinsonism and to prevent lactation by inhibiting the secretion of the hormone prolactin by the pituitary gland.

bulimia nervosa: mental disorder occurring predominantly in females, with onset usually in adolescence or early adulthood, characterized by episodes of binge eating that continue until terminated by abdominal pain, sleep, or self-induced vomiting, by awareness that the binges are abnormal

buprenorphine (Temgesic™): analgesic that demonstrates narcotic agonist -antagonist properties.

buspirone: a tranquilizer used to relieve the symptoms of anxiety.

caffeine: 1,3,7-trimethylxanthine, a purine derivative present e.g., in tea leaves and coffee beans.

calcarine fissure (sulcus calcarinus): a sulcus on the medial surface of the occipital lobe, separating the cuneus from the lingual gyrus.

cannabinoids: group of C_{21} compounds typical of and present in *Cannabis sativa*. More than 60 different cannabinoids have been identified in cannabis.

cannabis: the generic name given to plants of the species *Cannabis sativa*.

canulla: tube implanted into an organism, usually in a vein, through which drugs may be injected.

carbamazepine (Tegretol®): analgesic and anticonvulsant substance.

catecholamines: a group of physiologically important substances, including adrenaline, noradrenaline and dopamine, having various different roles (mainly as neurotransmitters) in the functioning of the sympathetic and central nervous system. Chemically, all contain a benzene ring with adjacent hydroxyl groups (catechol) and an amine group on a side chain.

challenge: administration of a chemical substance to a patient for observation of whether the normal physiological response occurs.

chi-square: a statistical test that allows tests of differences between independent samples using frequency data or between a sample and some set of expected scores.

chlordiazepoxide: a minor tranquillizer used in the treatment of anxiety.

choice response: refers to tests that offer fixed response options (e.g., multiple-choice, true-false, like-dislike ...).

choline acetyl transferase (ChAT): mitochondrial enzyme that synthesizes the neurotransmitter acetylcholine from its precursors acetyl CoA and choline.

chorea: the ceaseless occurrence of a wide variety of rapid, highly complex, jerky, dyskinetic movements that appear to be well coordinated but are performed involuntary.

choreiform: resembling chorea.

choreoathethosis: a condition marked by choreic and athetoid movements

chorioid plexus: a rich network of blood vessels, derived from those of the pia mater, in each of the brain's ventricles. It is responsible for the production of cerebrospinal fluid.

citalopram: a drug that increases the amount of serotonin at the synapse and has been tested in the medication of alcoholism.

classical conditioning: basic form of learning in which stimuli initially incapable of evoking certain responses acquire the ability to do so through repeated pairing with other stimuli that are able to elicit such responses. The learning in classical conditioning consists of acquiring responsiveness (conditioned response) to a stimulus that originally was ineffective (conditioned stimulus) by pairing it with a stimulus (unconditioned stimulus) that elicits an overt response (unconditioned response).

classification: broadly any decision in which alternative descriptive labels or courses of action are available, and one of them is chosen for the individual.

clearance (renal clearance): a quantitative measure of the rate at which waste products are removed from the blood by the kidney; expressed in terms of the volume of blood that could be completely cleared of a particular substance in one minute.

clinical investigations: investigations founded on actual observation of patients.

clinical trial: a large scale plan for testing and evaluating the effectiveness of some drug or therapeutic procedure in human subjects.

clonic: pertaining to or of the nature of clonus.

clonus: rhythmical contraction of a muscle in response to a suddenly applied and then sustained stretch stimulus. It is caused by an exaggeration of the stretch reflexes and is usually a sign of disease in the brain or spinal cord. The term is most commonly used to describe the rhythmical limb movements in convulsive epilepsy.

cognition: the mental process by which knowledge is acquired. These include perception, reasoning, acts of creativity, problem-solving, and possibly intuition.

cognitive (spatial) maps: a term describing the theoretical interpretation of the behavior of an animal learning a (spatial) maze. The original author, E.C. Tolman, argued that the animal was developing a set of spatial relationships – a cognitive 'map' – rather than merely learning a chain of overt responses.

cognitive tests: tests to measure cognitive functions (see cognition).

Cohen's kappa test: statistical method to measure the intraclass correlation between two methods; a kappa > 0.75 means that the similarity between both tests is excellent, 0.4 < kappa < 0.75 means that the similarity is moderate and a kappa < 0.40 means a bad similarity.

cohort study: epidemiological method used to search for the cause(s) of disease. Groups of people whose members differ on one or more characteristic(s) suspected of causing disease are followed over time to see if the people with the characteristic(s) get more disease.

collateral data: accompanying data.

compliance: yielding to others; overt behavior of one person that conforms to the wishes or behaviors of others. That is, they use it so that there is no notion that the compliant person necessarily believes in what he or she is doing.

computer axial tomography (CAT): brain imaging technique in which a series of X-rays of the brain is computerized and built up into a three-dimensional picture.

computer-based assessment: using computers as a medium for the presentation of psychological tests.

computerized tomography: diagnostic radiologic technique for the examination of the soft tissues of the body. It involves the recording of slices of the body with an X-ray scanner (CT-scanner); these records are then integrated by computer to give a cross-sectional image. This investigation is without risk to the patient (cf. PET). Within the skull it can be used to diagnose pathological conditions of the brain, such as tumors, abscesses and hematomas.

concordance rate: a measure of the extent to which one member of a twin pair will express a trait if the other member of the twin pair expresses that trait.

conditioned avoidance responding: the flight response evoked by a conditioned stimulus.

conditioned flavor aversion: a negative reaction towards a flavor, resulting from pairing the flavor with some painful or unpleasant stimulus.

conditioned stimulus: a stimulus which acquires the capacity to evoke particular responses through repeated pairing with another stimulus capable of eliciting such reactions.

conditioned suppression: a reduction or suppression in responding in the presence of a previously neutral stimulus. The suppression is produced by pairing the neutral stimulus with a noxious stimulus. For example, if a 1-minute light signal is repeatedly followed by a shock, the organism will come to suppress responding during the full minute that the light is on.

conduct disorder: general psychiatric classification encompassing a variety of behavior patterns in which the individual repetitively and persistently violates the rights, privileges and privacy of others.

confidence interval: the margin of the error calculated for a risk estimate or an effect. 95% confidence intervals are the most common, meaning that there is a 95% probability that the risk, or the effect, is no higher or lower than the range of values included in this interval.

confounder: a cause of disease that is not under investigation, but that distorts the cause-effect relationship under study.

control group: a group in an experiment which is not exposed to the independent variable under investigation. The behavior of subjects in this condition is used as a baseline against which to evaluate the effects of experimental treatments.

coprine (1-cyclopropanol-1-N-glutamine): mushroom (*Coprinus atramentarius)* toxin from which the metabolite cyclopropanone hydrate elicits a disulfiram like response to alcohol consumption for up to three days after eating the mushroom.

corpus striatum: the striate body, one of the components of the basal nuclei.

correlation coefficient: a statistic which indicates, on a scale from –1.00 to +1.00, the degree of relationship between two or more variables. The larger the correlation, the stronger the observed relationship.

correlational method: a research method in which variables of interest are observed in a careful and systemic manner in order to determine whether changes in one are associated with changes in the other.

cotinine: the major urinary metabolite of nicotine

covariance analysis: an extension of the ANOVA used when there is improper control of one or more of the variables.

crack: the most common street name for freebase cocaine, probably called this because a crackling sound is made both when cocaine hydrochloride powder ('snow') is mixed with water and sodium bicarbonate to make the freebase cocaine, and when crack is smoked.

craving: the desire to experience the effect(s) of a previously used psycho-active substance; not all drug craving is based on withdrawal, since craving can often occur in the absence of withdrawal. It is not an easily defined or quantified term because it is a state of mind.

crest: a projection or a projecting structure or ridge.

criterion: standard against which the success of prediction is judged. For a diagnosis, the criterion is usually an independent diagnosis considered to be rather trustworthy but impractical for routine use.

crossover design: variation on the double blind (see double blind) design, in which as an added control, the conditions are crossed in the middle of the experiment. For example a group of subjects receives one week placebo and the second week a drug, the control group receives the drug in the first week and the placebo during the second week.

cross-sectional study: epidemiological method used to determine whether a problem exists that warrants further study. A population of interest is identified and the individuals asked about current illnesses and current exposures.

cross-tolerance: a condition in which tolerance to a certain drug results in tolerance to another drug, mostly the latter are from the same class.

delirium tremens: psychosis caused by alcoholism, usually seen as a withdrawal syndrome in chronic alcoholics. Typically it is precipitated by a head injury or an acute infection causing abstinence from alcohol. Features include anxiety, tremor, sweating, and vivid and terrifying visual and sensory hallucinations, often of animals and insects. Severe cases may end fatally.

dementia pugilistica: boxer's dementia.

dementia: progressive decline in mental function, in memory, and in acquired intellectual skills.

demyelination: disease process selectively damaging the myelin sheets surrounding the nerve fibers in the central and peripheral nervous system. This in turn affects the functioning of the nerve fibers, which the myelin normally supports. Demyelination may be the primary disorder, as in multiple sclerosis, or it may occur after injury of the nervous system.

dendrite: thread-like nerve cell extension functioning to receive information.

depressive disorders: a group of disorders including various forms of depression and manic-depression.

detection threshold: the statistically determined point along a stimulus continuum at which the energy level is just sufficient for one to detect the presence of the stimulus.

developmental milestones: significant behaviors which are used to mark the progress of development. Examples are: saying phrases, turning pages, carrying out requests, pointing to body parts, holding a pencil, imitating a drawn circle, catching a ball.

developmental studies: tests for evaluating the developmental stage of infants and preschoolers.

diagnosis: narrowly, choosing one of a set of labels that best fits an individual's disorder or disability. Broadly, developing an understanding of the individual's difficulties, and insofar possible, their origins.

Diagnostic and Statistical Manual of Mental Disorders (DSM): the official system for classification of psychological and psychiatric disorders prepared and published by the American Psychiatric Association. The current version is DSM-IV.

differential rate reinforcement (DRR): in operant conditioning a term for a class of schedules of reinforcement in which the delivery of reinforcement depends on the immediately preceding rate of responding. *DRL (differential reinforcement of low rates)*: a class of schedules based on a specified rate of responses which must not be exceeded for reinforcement to occur. Thus, in DRL 5 (seconds), 5 seconds must pass between responses or no reinforcement is delivered. *DRH (differential reinforcement of high rates)*: in contrast to DRL, here the rate must exceed some set value for reinforcement to occur. DRH 5 (seconds) means that the interresponse time must be less than 5 seconds.

diprenorphine: a synthetic opioid agonist, derived from morphine.

dipsomania: 19th century term for morbid and insatiable craving for alcohol, occurring in paroxysms. Only a small proportion of alcoholics show this symptom.

discrimination learning, see discrimination training procedure.

discrimination training procedures: class of experimental procedures in which a subject learns to judge between two (or more) stimuli. In *operant-conditioning* experiments, responses in the presence of one stimulus (S1) are reinforced but responses in the presence of another (Sx) are not. Such training leads to the emitting of responses in the presence of Sx, but not in the presence of S1. In *classical conditioning*, in the presence of one stimulus the conditioned stimulus (CS) and the unconditioned stimulus (UCS) are paired, but in the presence of another stimulus they are not. This leads to elicitation of the conditioned response (CR) in the presence of the conditioned stimulus, but not in the presence of another stimulus.

discrimination: the ability to perceive differences between two or more stimuli.

dissociative anesthetic: a class of drug that was developed as anesthetics that do not cause significant respiratory depression. Although these drugs can cause hallucinations, the profile of their effects is different from drugs such as LSD. Two of these drugs are phencyclidine (PCP) and ketamine. Ketamine is used in clinical medicine and both, ketamine and phencyclidine are used as veterinary anesthetics.

disulfiram: an antioxidant which inhibits the oxidation of the acetaldehyde metabolized from alcohol, resulting in high concentrations of acetaldehyde in the body. Extremely uncomfortable symptoms occur when alcohol is ingested; therefore this drug is used as an aversion therapy for alcoholism.

divergent thinking: intellectual fluency; finding a variety of possible solutions to a problem.

DOB (2,5-dimethoxy-4-bromoamphetamine): a phenylalkylamine with hallucinogenic and psychotogenic activity

domoic acid: algal toxin responsible for the amnestic shellfish poisoning syndrome.

dopaminergic: characterizing or pertaining to pathways, fibers or neurons in which dopamine is a neurotransmitter.

dorsal root ganglion (ganglion spinale, sensory ganglion): spinal ganglion: the ganglion found on the posterior root of each spinal nerve, composed of the unipolar nerve cell bodies of the sensory neurons of the nerve.

double-blind procedure: experimental procedure in which neither the subject nor the person administering the experimental procedure knows what are considered to be the crucial aspects of the experiment. In the case of drug studies neither the subject nor the administering person knows whether the drug concerned or a placebo is administered.

DRL (differential reinforcement of low rates), see differential rate reinforcement.

drug abuse: improper use of drugs. The usual connotation is that of excessive, irresponsible and self-damaging use of psychoactive and/or addictive drugs.

drug dependence: a term used in scientific writings that is favored over the terms addiction and habituation. In its simplest way it can be said that a person has developed dependence on a drug when there is a strong, compelling desire to continue taking it. Dependence on a drug may in origin be largely psychological or physiological. For more scientific definitions, see the DSM-IV classification, or the ICD-10.

drug discrimination test: drug detection procedure whereby animals are trained to recognize or discriminate the stimulus effects of a given dose of a particular training drug from those of a different dose of the same training drug, a different training drug, or more commonly, saline/vehicle. The most commonly employed apparatus for conducting drug discrimination studies is a two-lever operant chamber. The drug discrimination paradigm cannot be used to completely characterize a novel agent; however it can be used to investigate a wide variety of pharmacological aspects relating to the stimulus properties of a drug.

drug self-administration: model in which subjects can self-administer drugs and in which they have control over the infusion apparatus. Frequently animals are trained to lever press in a food-reward set-up before being tested for drug self-administration. Different protocols may be engaged, for instance, intravenous, intracerebral or oral and animals may be drug naive or drug dependent.

DSP-4 (N-chloroethyl-N-ethyl-2-bromobenzylamine hydrochloride): a potent noradrenaline neurotoxin that only affects noradrenergic neurons. Neurobiologists employ DSP-4 to destroy specific groups of noradrenergic neurons, using stereotactic injections in specific anatomic sites.

dynorphin: any of a family of opioid peptides found throughout the central and peripheral nervous system. Most of them are agonists at opioid receptor sites.

echoencephalography: radiographic method demonstrating the intracranial fluid containing spaces using the echo obtained from beams of ultrasonic waves directed through the cranium.

electoconvulsive shock (ECS): brief electrical shock applied to the head that produces full-body seizure, convulsions, and usually loss of consciousness; used in animals to study the neurobiology of memory and in patients as a therapeutic procedure for psychiatric disorders (*electroconvulsive treatment, ECT*)

electroencephalogram (EEG): record of the changes in electrical potential of the brain. The pattern of the EEG reflects the state of the patient's brain and his level of consciousness in a characteristic manner.

electroshock, see electoconvulsive shock (ECS)

embryo: organism in its early stages of development. In humans the first 2 months after conception.

embryogenesis: the development of the embryo.

encephalopathy: any degenerative disease of the brain.

β-endorphin: one of the endogenous endorphins; it is self-administered in animals and possesses opioid activity.

endorphins: any of three neuropeptides, amino acid residues of β-lipotropin. They bind to opioid receptors in the brain and have potent analgesic activity.

enkephalin: peptide naturally occurring in the brain and having effects resembling those of opiates like morphine

entorhinal cortex: part of the cortex near the entorhinal fissure that links the neocortex with the limbic system. The entorhinal cortex receives its input from areas of the association cortex and sends its information to the hippocampus by way of the perforant path.

eosinophilia: an increase in the number of eosinophils in the blood. Eosinophilia occurs in response to certain drugs and in a variety of diseases, including allergies, parasitic infestations and certain forms of leukemia.

ependyma: the extremely thin membrane, composed of cells of the neuroglia (ependymal cells), that lines the ventricles of the brain and of the chorioid plexus. It helps to form the cerebrospinal fluid.

ependymal cell: neuroglial cells that line the ventricles (cavities) of the brain and the central canal of the spinal cord (see ependyma).

epidemiology: the study of how and why diseases and other conditions are distributed within the population.

epiphysis, see pineal gland.

equine: pertaining to, characteristic of, or derived from the horse.

ergot alkaloids (ergolines): alkaloids obtained primarily from *Claviceps species*, they are also produced by other fungi and by certain plants. Natural ergot alkaloids stimulate smooth muscle, especially that of peripheral vessels causing vasoconstriction and of the pregnant uterus at term. Certain hydrogenated derivatives have the reverse effect and cause vasodilatation. There are two classes of ergot alkaloids: the clavine alkaloids and the lysergic acid derivatives.

etonitazene, a narcotic analgesic

event related potential (ERP): change in potential recorded from many neurons in response to a sensory stimulus. Typically lasts for several hundred milliseconds and consists of a number of positive and negative waves. The potentials are generally recorded on the scalp, and it is very difficult to localize the internal sources of these waves. The term is used when the author is referring to a specific evoked potential that occurs in response to a specific known stimulus, the event.

evoked potentials: regular pattern of electrical activity recorded from neural tissue evoked by a controlled stimulus. The stimulus may be auditory (brainstem auditory evoked potential, BAEP), visual (visual evoked potential, VEP) or somatosensory (somatosensory evoked potential, SSEP).The term specifically applies to potentials from the brain (see event related potential).

excitotoxic amino acid, see excitotoxin

excitotoxicity: toxicity induced by excitotoxin (see excitotoxin).

excitotoxin: neurotoxic substances analogous to glutamic acid, that mimic glutamic acids excitatory effect on neurons of the CNS as well as producing lesions on the perikarya.

extinction: the process through which conditioned responses are weakened and eventually eliminated.

fabulation (confabulation): unconscious filling in of gaps in memory with fabricated facts and experiences, commonly seen in organic amnestic syndromes. It differs from lying in that the patient has no intention to deceive and believes the fabricated memories to be real.

fantasticas or phantasticants: a substance that produces illusions (a hallunicogenic substance).

fenfluramine: an adrenergic used as an anorexic in the short-term treatment of exogenous obesity. Also used as an antidepressant.

Fentanyl: a narcotic analgesic derivative of piperidine.

Fetal Alcohol Syndrome (FAS): cluster of abnormal developmental features of a fetus resulting from severe alcoholism in the mother. The features may include: microcephaly, growth deficiencies, mental retardation, hyperactivity, heart murmurs and skeletal malformations.

fixed interval performance: response following the passage of a fixed period of time in a fixed interval schedule of reinforcement.

fixed interval schedule of reinforcement: a schedule in which the first response following the passage of a fixed interval of time yields reinforcement.

flumazenil: a benzodiazepine antagonist

flunitrazepam: benzodiazepin hypnotic and induction agent in anesthesia

fluoxetine (Prozac®): an antidepressant, antiobsessional, and antibulimic drug. The actions of fluoxetine are presumed to be linked to its ability to inhibit neuronal re-uptake of serotonin.

food-reinforced alternation performance: response in a test in which the reversal learning is reinforced with food.

fornix (fornix cerebri): a triangular anatomical structure of white matter in the brain, situated between the hippocampus and hypothalamus.

free-choice paradigm: experimental situation in which a subject can choose out of more than one possibilities.

frontal lobe tests: tests to assess impairment of functions that are anatomically located in the frontal lobe of the brain. These tests mainly assess cognitive functioning, mood and personality.

frontal lobe: area of the neocortex, involved in executive functioning. Executive functioning can be conceptualized as having four components: goal formulation, planning, carrying out goal directed plans and effective performance.

generalized anxiety disorder: a disorder characterized by at least 6 months of persistent and excessive anxiety and worry not resulting from exposure to a drug or medication.

glial cells, see neuroglia.

guanabenz: an α_2-adrenergic agonist that stimulates the α_2-adrenergic receptors of the central nervous system, resulting in a reduction of sympathetic outflow to the heart and peripheral vascular system.

habituation: decrease in the behavioral response to a repeatedly presented stimulus.

half-life (half-time): the time taken for the concentration of drug in the blood or plasma to decline to half its original value.

haloperidol: a butyrophenone antipsychotic drug used to relieve anxiety and tension in the treatment of schizophrenia and other psychiatric disorders.

harmine: an alkaloid with hallucinogenic activity. It was first isolated from seeds of *Perganum harmala*, a plant from the Asian steppes which has spread from the Mediterranean to Southeast Asia. Seed preparations have been used traditionally for a variety of purposes. Harmine and related alkaloids are also the active principles of psychoactive preparations from *Banisteriopsis caapi* and related species used by Amazonian Indians in social and religious festivals. There is cross tolerance with lysergide and psylocybin, suggesting a common site of action. Harmine is a MAO-inhibitor. At one time it was used in the treatment of Parkinson's disease.

hepatomegaly: an increase in liver size

hepatotoxic: damaging or destroying liver cells.

High Alcohol Drinking rats: strain of rats, specially bred for their quality to consume alcohol when administered to their drinking water. The rats were selectively bred based upon their daily intake of an 8% alcohol solution.

higher order conditioning: a process in which previously established conditioned stimuli serve as the basis for further conditioning.

hippocampal formation: a curved band of cortex lying within each cerebral hemisphere: in evolutionary terms one of the brain's most primitive parts. It forms a portion of the limbic system and is involved in the complex physical aspects of behavior governed by emotion and instinct.

hippocampus: a swelling of the floor of the lateral ventricle of the brain. It contains complex foldings of cortical tissue and is involved, with other connections of the hippocampal formation, in the workings of the limbic system.

HOME-scale: scale assessing quality of the home environment of subjects undergoing a psychological test. The home environment seems to be an important confounder in neuropsychological studies using developmental psychological tests.

homovanillic acid (HVA): a product of the catecholamine metabolism.

human leukocyte antigen system (HLA system): This is a series of four gene families that code for polymorphic proteins expresses on the surface of most nucleated cells. Individuals inherit from each parent one gene, or set of genes, for each subdivision of the HLA system. If two individuals have identical HLA types, they are said to be histocompatible. Successful tissue transplantation requires a minimum of HLA differences between donor and recipient tissues.

hyperkinetic syndrome: mental disorder, usually of children, characterized by a grossly excessive level of activity and a marked impairment of the ability to attend. Learning is impaired as a result, and behavior is disruptive and may be defiant or aggressive. The syndrome is more common in the intellectually subnormal, the epileptic and the brain damaged. Treatment usually involves drugs (stimulants) and behavior therapy. The terms attention deficit disorder and hyperkinetic syndrome are often erroneously used indifferently.

hypothalamus: the region in the forebrain in the floor of the third ventricle, linked with the thalamus above and the pituitary gland below. It contains important centers controlling body temperature, thirst, hunger, and eating, water balance and sexual functions. It is, via the limbic system, closely connected with emotional activity and sleep and functions as the integration center for hormonal and autonomic nervous activity through its control of the pituitary secretions.

hypothesis: a proposition that seeks to place certain facts (or variables) within a construct that will explain or predict relationships between these facts. A prediction regarding the relationship between two variables is tested by conducting research: if the findings offer support for the hypothesis, confidence in accuracy may increase, while if findings fail to offer such support, confidence in its accuracy may be reduced.

ICI-174,864: a selective δ-opioid receptor antagonist.

ICS 205-930: a $5HT_3/5HT_4$ receptor antagonist.

idiopathic: denoting a disease or condition the cause of which is not known or that arises spontaneously.

illusinogens: a drug that produces a false or misinterpreted sensory impression (a hallunicogenic substance).

imaging techniques: the technique of producing images of organs or tissues using radiological procedures, particularly by using scanning techniques. Examples of brain imaging techniques are: CAT (see computer axial tomography), MRI (see magnetic resonance imaging), PET (see positron emission tomography).

immediate recall: the process of bringing a memory of an immediately preceding event or activity into consciousness.

immunolabeling procedures: procedures in which endogenous molecules are labeled with immunoactive substances that have been labeled with a detectable site, e.g., a stain or a fluorescent molecule. The procedure makes it possible to localize certain systems within the brain, e.g., the dopaminergic system, or certain peptidergic (endorphins) systems.

impotentia coeundi: lack of copulative power in the male due to initiate an erection or to maintain an erection until ejaculation caused by psychogenic or organic dysfunction.

inbreeding: the production of offspring by parents who are closely related; for example who are first cousins or siblings.

incidence rate: a fraction expressing the rate of new cases of disease in a population over a period of time: the numerator of the rate is the number of new cases during a specified time period and the denominator is the population at risk during the period.

incidence: the rate at which a certain event occurs, e.g., the number of new cases of a specific disease occurring during a certain period.

inebriety: 19th century term for habitual drunkenness

informed consent: permission to carry out a research or medical procedure where the subject is given information including the nature of the procedures, potential risks and benefits, any other alternative procedures that are available and acknowledgment that such consent is voluntary.

instrumental conditioning: a basic form of learning in which responses that yield positive (i.e., desirable) consequences or lead to escape from avoidance of negative (i.e., undesirable) outcomes are strengthened.

intelligence quotient (IQ): age related measure of intelligence level.

intermittent reinforcement: a term for all those schedules of reinforcement in which some of the responses made go unreinforced; that is all those schedules of reinforcement other than continuous reinforcement and extinction. Also called partial reinforcement.

International Classification of Diseases (ICD): system of classification of diseases developed under the supervision of the World Health Organization. Over the years, many revisions have been made, the most recent system is the 10th (ICD-10). The ICD has an extensive section on psychiatric and psychological disorders, which is in wide use in many countries.

intracerebroventricular: directly into the cerebral ventricles

intragastric: situated or occurring within the stomach. Intragastric administration means a form of (drug)administration in which the substance is directly administered into the stomach, using a cannula or catheter.

intramuscular: within the substance of a muscle.

intraperitoneal: within the peritoneal cavity.

inventory: a questionnaire, typically one that represents many questions about each aspect of personality that is under investigation. Directions may ask for a self-description of an acquaintance who is being assessed.

inverse agonist: a drug that produces a response that is opposite to the response of the agonist, e.g., if the agonist increases blood pressure, the inverse agonist will decrease it.

inverted U-shaped curve: dose effect curve in which medium doses have large effects, small and high doses have relatively little effect; also called bell-shaped curve.

IPPO: isopropylbicyclophosphate, a picrotoxin ligand.

ipsapirone: a non-benzodiazepine anxiolytic, a 5-HT_1 receptor agonist.

IQ tests: any test that purports to measure an intelligence quotient. Generally such tests consist of a graded series of tasks each of which have been standardized with a large, representative population of individuals.

kainate receptor: one of the four types of NMDA receptors. It binds the glutamate agonist kainate and regulates a channel permeable to Na^+ and K^+.

Kava: fermented drink made by Polynesian people of the South Pacific from the stem and root of the shrub *Piper methysticum;* it produces mild stimulation followed by drowsiness.

ketanserine: serotonin antagonist

kinesin: large soluble cytoplasmic protein that binds tightly to microtubules and transports vesicles and particles along them using energy from ATP hydrolysis.

kinetics: mathematical study of the changes of the concentration of a substance or its metabolites in the human or animal body after exposure or administration.

Korsakoff syndrome: syndrome of anterograde and retrograde amnesia; currently used synonymously with 'amnestic syndrome' or, more narrowly, to refer to the amnestic component of the Wernicke-Korsakoff syndrome.

laudanum: hydroalcoholic solution containing 1% morphine, prepared from macerated raw opium; formerly widely used as a narcotic analgesic, administered orally.

***l*-cathinone**: dopamine agonist, an addictive stimulant (by chewing the leaves)

***l*-dopa (levodopa)**: a naturally occurring amino acid administered orally to treat parkinsonism.

learned helplessness: in learning experiments, a subject's passive response to stress after being placed in situations in which there is no way to avoid an electric shock.

limbic system: complex system of nerve pathways and networks in the brain, involving several different nuclei; involved in the expression of instinct and mood in activities of the endocrine and motor systems of the body. Among brain regions involved are the amygdala, hippocampal formation, and hypothalamus.

β-lipotropin: a precursor molecule for different endogenous peptides, such as the endorphins and ACTH.

lithium carbonate: a white granular powder used in the treatment of acute manic states and in the prophylaxis of recurrent affective disorders manifested by depression or mania only.

lithium, see lithium carbonate

liver transaminases: a sub-subclass of enzymes of the transferase class within the liver that catalyze the transfer of an amino group from a donor (usually an amino acid) to an acceptor (generally a 2-keto acid). Also called aminotransferases.

locomotor activity tests: observational tests involving an environment in which the activity of an animal is recorded as it moves from place to place.

locus coeruleus: small pigmented region in the floor of the fourth ventricle of the brain.

long-term exposure: continuous or repeated exposure to a substance over a long period of time, usually several years in man, and of the greater part of the total life-span in animals.

***l*-tryptophan**: amino acid existing in proteins from which it is set free by triptic digestion. It is essential for optimal growth in infants and for nitrogen equilibrium in human adults. It is also the precursor for serotonin.

magnetic resonance imaging (MRI): in this technique, the brain is bombarded with radio waves; molecules in the neurons respond by producing radio waves of their own, and these are recorded, computerized, and assembled into a three dimensional picture.

major depressive episode: a period of at least 2 weeks during which there is either depressed mood or the loss of interest or pleasure in nearly all activities.

major histocompatibility complex: a set of linked genetic loci which dominates the control of modification of the immune response. The genes and their products are designated by prefixes (e.g., HLA in man, ChLA in chimpanzee) MHC gene products occur on cell surfaces and serve as markers which help to distinguish 'self' from 'non-self' tissue.

mania: a period of abnormally and persistently elevated, expansive, or irritable mood.

manic-depressive: characterized by alternating manic and depressive episodes.

Marchiafava-Bignami syndrome (disease): A rare condition, originally described in Italian drinkers of crude red wine, occurs occasionally in other alcoholic patients. It is characterized clinically by disorders of emotional control and cognitive function followed by variable delirium, fits, tremors, rigidity, and paralysis; most patients eventually become comatose and die within a few months. Symmetrical demyelination with subsequent cavitation and axonal destruction is found in the corpus callosum and often, in varying degree, in the central white matter of the cerebral hemispheres, the optic chiasm, and in the middle cerebellar peduncles.

marijuana: dried leaves and flowers of the cannabis plant, usually smoked in a cigarette or pipe, sometimes baked in cookies and cakes; it is one of the many forms in which substances of cannabis are used.

Marlatt: the relapse prevention (RP) model of Marlatt focuses on the maintenance phase of the habit change process, rather than overemphasizing initial habit change. Relapse is not viewed merely as an indicator of treatment failure. Instead, potential and actual episodes are key targets for both proactive and reactive intervention strategies. RP treatment procedures include specific intervention techniques designed to teach the individual to effectively anticipate and cope with potential relapse situations. Also included are more global lifestyle interventions aimed at improving overall coping skills and promoting health and well being.

MDI, mental development index (see Bayley Scales of Infant Development).

mean: the arithmetic average of a set of data. The mean is computed by adding all of the data together and then dividing by the number of scores in the set.

medial forebrain bundle: a pathway in the limbic system leading from the precommissural fornix and the olfactory bulb through the preoptic mesencephalon. The pathway has been found to produce highly reinforcing in self-stimulation experiments.

median: the midpoint of a set of scores. Fifty percent of the scores fall above the median, 50 percent below.

mental age: a measure of intellectual ability obtained on early experimenters' tests of intelligence. An individual's mental age was assumed to reflect his or her level of intellectual maturity.

mental arithmetic: assessment test for measuring skills necessary for academic achievement. In the WAIS-R mediocre value as measures of general ability, assesses knowledge and ability to apply arithmetic operations only.

mescaline: alkaloid obtained from *Lophophora williamsii*. It is a CNS depressant, and causes visual hallucinations. Mescaline is the hallucinatory principle of peyote.

mesocortical dopamine pathway: one of the four major dopaminergic tracts in the brain; the mesocortical pathway runs from the ventral tegmental area to the neocortex, especially prefrontal areas. The other three are the *nigostriatal*, from substantia nigra to caudate and putamen, the *tuberoinfundubular*, from the arcuate nucleus of the hypothalamus and the *mesolimbic pathway*, from the ventral tegmental area to many components of the limbic system.

mesolimbic system, see mesocortical dopamine pathway.

meta-analysis: systematic method that uses statistical analysis to integrate the data from a number of independent studies.

methadone: synthetic opiate drug which can be taken orally and will prevent withdrawal symptoms in a heroin addict for 24 hours. It is used as a maintenance and detoxification drug for heroin addicts.

***N*-methyl-D-aspartate (NMDA)**: a neurotransmitter similar to glutamate, found in the central nervous system; a synthetic preparation is used experimentally to study the excitatory mechanisms of glutamate transmitters.

methylenedioxyamphetamine (MDA): a psychoactive substance, chemically related to mescaline and amphetamine, first synthesized in the 1930s. MDA is obtained from safrol, a psychoactive oil found in nutmeg that is chemically related to amphetamine.

methylphenidate: a central stimulant used in the treatment of hyperkinetic children, various types of depression and narcolepsy.

Meynert's nucleus (nucleus basalis of Meynert): a group of neurons situated in the basal part of the forebrain that has wide projections to the neocortex and is rich in acetylcholine and choline acetyltransferase. It undergoes degeneration in Parkinson's disease and in Alzheimer's disease.

microcephalic: small headed, this term is reserved for those cases in which the abnormality is so great that retardation results.

microcyte: one of the cells of the neuroglia, responsible for phagocytizing the waste products of nerve tissue.

microdialysis: technique used to locally measure the rate of synthesis and/or secretion of neurotransmitters in the brain.

microglia, see microcyte.

microsomal ethanol-oxidizing system: one of the 3 enzymatic systems that metabolize ethanol in the body. The other two are catalase and alcohol dehydrogenase.

minimal brain dysfunction (MBD): cover term for a variety of behavioral, cognitive and affective abnormalities observed in young children. The term is typically reserved for such cases in which the patterns of thought and action are such that one would expect to find some organic abnormality but none is apparent. The term is often used as though there were an identifiable MBD syndrome, a collection of fairly specific disorders that could be taken as hallmarks of some underlying neurological causal mechanism. However, the evidence to support a single MBD syndrome is largely unconvincing

miosis: constriction of the pupil of the eye to contract.

mnemonic: relating to memory.

mood disorders: depressive and manic disorders.

mood questionnaire: questionnaire to assess the state of mood.

mood: a relatively short-lived, low intensity emotional state.

morphine: principal alkaloid of opium. Morphine and its salts are very valuable analgesic drugs, but are highly addictive. In addition to suppression of pain, morphine causes constipation, decreases pupillary size and depresses respiration. Only the (+)-stereoisomer is biologically active.

morphinism: morphine addiction

motivation: a hypothetical internal process that provides the energy for behavioral and directs it towards a specific goal.

motive: an acquired motivational system.

MPTP: 1-methyl-4-phenyl-1,2,3,4-tetrahydropyridine, a byproduct in the manufacture of MPPP due to inadequate technique. It produces severe and irreversible Parkinson's disease.

multiple correlation: the relationship between one dependent variable and two or more independent variables. A multiple correlation coefficient yields an estimate of the combined influence of the independent variables on the dependent.

multiple FI-FR schedules: a schedule of reinforcement (see schedule of reinforcement), in which a fixed-interval (see fixed interval (FI) schedule of reinforcement) and a fixed ratio schedule (see fixed ratio (FR) schedule of reinforcement) are combined into a compound form.

multiple regression: statistical technique that is an extension of simple regression and allows one to make predictions about performance on one variable (called the criterion variable) based on performance on two or more variables (called the predictor variables). If the regression equation is in standard score form then the relative weights or contributions of each of the predictor variables may be assessed (see multiple correlation).

multivariate analysis: any statistical test that is designed to analyze data from more than one variable.

myalgia: pain in the muscles.

myelin: the substance of the cell membrane of Schwann's cells that coils to form the myelin sheath; it has a high proportion of lipid to protein and serves as an electrical insulator.

nalmefene: structural analog of naltrexone with opiate antagonist activity.

naloxone: short acting opiate antagonist

naltrexone: long acting (= long $t_{1/2}$)opiate antagonist

naltrindole: selective, non-peptide, δ-opioid receptor antagonist; also active when peripherally administered.

negative reinforcement: the process by which a person learns to avoid behavior that causes unpleasant sensations.

negative reinforcer: stimulus that increases the frequency of behavior that prevents its presentation.

nerve growth factor (NGF): one of the neurotrophic factors that control neuronal growth during nervous system development and during regeneration (see neurotrophic factor).

neuralgia: severe burning or stabbing pain often following the course of a nerve

neurofibrillary tangles: bundles of abnormal filaments within neurons. The filaments of these cytoskeletal abnormalities, as seen in Alzheimer's disease, consist of two thin filaments arranged in a helix (paired helical filament).These structures do not resemble normal cytoskeletal proteins of neurons.

neuroglia: the supporting structure of the nervous system. It consists of a fine web of tissue made up of modified ectodermal elements, in which are enclosed peculiar branched cells known as neuroglial cells or glial cells. The neuroglial cells are of three types: astrocytes and oligodendrocytes (astroglia and oligodendroglia), which appear to play a role in myelin formation, transport of material to neurons, and maintenance of the ionic environment of neurons, and microcytes (microglia), which phagocytize waste products of nerve tissue.

neuroglial cells, see neuroglia.

neuropathy: a functional disturbance or pathological change in the peripheral nervous system.

neurotrophic factor: proteins that ensure the survival and maintenance of defined populations of neurons. In addition, neurotrophic factors have the competence to promote the outgrowth of neurites. To date, three neurotrophic proteins have been purified: nerve growth factor (NGF), brain-derived neurotrophic factor (BDNF) and ciliary neurotrophic factor (CNTF).

nociception: perception of painful/damaging stimuli.

non contingent responding: responding not under the control of the environmental contingencies of the task.

no-observed-effect-level (NOEL): experimentally determined dose at which no statistically or biologically significant indication of the toxic effect of concern is observed.

norepinephrine: synaptic neurotransmitter, also called noradrenaline. It is one of the monoamines; lowered levels are associated with depression. Norepinephrine is also released from the adrenal medulla as part of the peripheral arousal response.

nucleus accumbens (NA): collection of pleomorphic cells in the caudal part of the anterior horn of the lateral ventricle of the olfactory tubercle, lying between the head of the caudate nucleus and the anterior perforated substance.

numbness: anesthesia.

nystagmus: an involuntary, rapid, rhythmic movement of the eyeball, which may be horizontal, vertical, rotatory, or mixed.

objective test: a personality test involving structured items and a limited set of responses (such as TRUE - FALSE)

obsessive compulsive behavior: characteristic behavior as performed in an obsessive-compulsive disorder.

obsessive compulsive disorder: a subclass of anxiety disorders with two essential characteristics: recurrent and persistent thoughts, ideas and feelings and repetitive, ritualized behaviors. Attempts to resist a compulsion produce mounting tension and anxiety, which are relieved immediately by giving into it. The term is not properly used for behaviors like excessive drinking, gambling, eating etc. on the grounds that the compulsive gambler actually derives considerable pleasure from gambling, it is the losing that hurts.

olfaction: the sense of smell.

olfactory perception threshold: the minimal concentration of a substance that evokes smell.

oligodendrocyte: one of the cells of the neuroglia, responsible for producing myelin sheaths of the neurons of the central nervous system and therefore equivalent to the Schwann cells of the peripheral nerves.

oligodendroglia, see oligodendrocyte; neuroglia

open field: observational test method to measure the behavior of animals, especially rats. An animal is placed in the center of a field and number of squares crossed in a limited time is measured. Other posture/activities that can be measured are: rearing (in the middle and against the wall), defecations, urination etc. The test has been automated and devices are available that record locomotion, rearing, and even grooming, as separate acts. Many modifications of this test are introduced, such as introducing (unknown) objects in the field.

operant conditioning: instrumental conditioning. Learning by association of an organism's own behavior with a subsequent reinforcing or punishing environmental event.

operant paradigm: paradigm in which the actions of the subject are voluntary and are not elicited by a discrete, identifiable stimulus.

operant response: response, for example a behavior, which is voluntary and is not elicited by a discrete, identifiable stimulus.

opioid agonist: (1) any synthetic narcotic agonist of the opioid receptor that has opiate-like activity but is not derived from opium; (2) any of the group of naturally occurring peptides that bind at or otherwise influence opiate receptors of cell membranes, they have opiate-like effects.

opioid antagonist: (1) any synthetic narcotic antagonist of the opioid receptor but is not derived from opium; (2) any of the group of naturally occurring peptides that bind at or otherwise influence opiate receptors of cell membranes, they have opiate antagonist effects.

opioidergic: pertaining to opioids or opioid-like.

organum subfornicale, see subfornical organ.

outbreeding: the mating of totally unrelated individuals, which frequently results in the production of offspring that show more vigor, as measured in terms of growth, survival, and fertility, than the parents.

oxazepam (Serax®): benzodiazepine drug used to relieve anxiety and tension and for the treatment of alcoholism; orally administered and commonly causes drowsiness.

palpebral closure: closure of the eyelids.

pandemic: epidemic, so widely spread that vast numbers of people in different countries are affected.

panic disorder: a disorder characterized by sudden, unexpected, and persistent episodes of intense fear, accompanied by a sense of imminent danger and an urge to escape.

Paregoric: preparation of powdered opium, anise oil, benzoic acid, camphor, diluted alcohol and glycerin, each 100 ml. of which yields 35-45 mg. of anhydrous morphine; used as antiperistaltic in the treatment of diarrhea.

peer rating: a classmate, fellow soldier, fellow student, or other acquaintance marks a rating scale or inventory to describe the target person. Usually, the average of several reports is taken as the target's peer rating.

performance test: among ability tests the term is usually applied to those where the respondent is to execute an appropriate physical action – tracing a maze path for example; although it can also be verbal as in the Wechsler Performance section. Among personality measures, the term is usually applied to observations of response in a standardized situation, usually a situation that arouses strong motives.

periamygdalian cortex: neocortex situated near nucleus amygdalae.

perikaryon: the cell body of a neuron, as distinguished from the nucleus and the processes.

perinatal: period before birth, birth and just after birth. In man pertaining to the period extending from the 28th week of gestation to the 28th day after birth.

personality indices: the items in questionnaires and inventories that are used to assess a subject's personality.

personality questionnaire: questionnaire to assess a subject's personality.

personality: the characteristic way in which a person thinks, feels, and behaves; the relatively stable and predictable part of a person's thought and behavior; it includes conscious attitudes, values, and styles as well as unconscious conflicts and defense mechanisms.

pharmacokinetic phase: process in which a drug is absorbed, distributed, biotransformed in, and excreted out of the body.

pharmacotreatment: any form of medical therapy in which medicines are used.

phenelzine sulfate: a monoamine oxidase inhibitor used as an antidepressant, administered orally.

phenobarbital: sedative barbiturate with low reinforcing properties

phentermine: an adrenergic isomeric with amphetamine, used as an anorexic.

Phi-coefficient: an index of the relationship between any two sets of scores, provided that both sets of scores can be represented on ordered, binary dimensions, e.g., male-female; married-single.

physical craving: the behavior that accompanies craving; the physical characteristics that can be observed during the process of craving.

physical dependence: physiological state of adaptation to a drug normally following the development of tolerance, and resulting in a characteristic withdrawal syndrome peculiar to the drug following absence.

piloerection: erection of the hair.

pineal gland or pineal body (epiphysis): a pea-sized mass of nerve tissue attached by a stalk to the posterior wall of the third ventricle of the brain, deep between the cerebral hemispheres at the back of the skull. It functions as a gland, secreting the hormone melatonin. The gland becomes calcified as age progresses.

pirenperone: serotonin antagonist

placebo effect: changes in behavior stemming from conditions or procedures which accompany, but are not directly related to, independent variables in an experiment. For example, changes in behavior following injections of a specific drug may result from the act of being injected, rather than from the drug itself.

polydrug abuse: pattern of multiple drug use. It is very common for e.g., heroin users to take drugs such as benzodiazepines, barbiturates and alcohol. Sometimes these drugs are combined specifically ('speedball' is a combination of a heroin/cocaine mixture and antihistamine-opiate combinations).

poor metabolizer: large variations occur in drug metabolism in humans. This often is due to a genetic predisposition. Persons in which the rate of metabolism is slow are called poor metabolizers.

positive reinforcement: the process by which a person learns to repeat rewarding behavior.

positive reinforcer: stimulus that increases the frequency of behavior that leads to its presentation.

positron emission tomography (PET): imaging technique involving the injection of radioactive glucose in the bloodstream. Glucose is used as energy source by brain cells, the most active cells taking up more glucose. The radioactive particles emitted by the glucose are picked up by an array of detectors around the head, and after computer-analysis give an overall picture of brain cell activity.

predictive value: the validity is the extent to which a model measures what it purports to measure. In the case of high predictive value a model demonstrates concordance between predictions based on the effects of a chemical agent on the animal model and effects of the agent on humans, in essence an ex post facto validation for the model.

prefrontal cortex: superstructure above all other parts of the cortex, situated in the anterior part of the frontal lobe of the brain.

prenatal: period before birth. In man pertaining to the period between the last menstrual cycle and birth of the child.

prevalence rate: the number of people in a population who have a disease at a given time: the numerator of the rate is the number of existing cases of disease at a point in time and the denominator is the total population.

prevalence: the number of cases of a disease that are present in a population at one point in time.

projective test: a personality test involving ambiguous stimuli. A subject's responses are supported to reveal aspects of the unconscious.

pro-opiomelanocortin: the prohormone that is the precursor of ACTH, the lipotropins, melanocyte stimulating hormones and the endorphins.

proprioceptive sensation, see proprioceptor.

proprioceptor: specialized sensory nerve ending that monitors internal changes in the body brought about by movement and muscular activity. Proprioceptors located in muscles and tendions transmit information that is used to coordinate muscular activity.

prospective study, see cohort study.

psychedelics (syn. hallucinogen): pertaining to or characterized by visual hallucinations, intensified perception, and, sometimes, behavior similar to that seen in psychosis; also a drug that produces such effects.

psychiatric rating scale: device used to assist a psychiatrist in making a diagnosis of a patient.

psychodysleptics: substance inducing a dreamlike or delusional state of mind (see psychedelic).

psychological (behavioral) dependence: the strong need to continue to experience a drug's effect; condition characterized by an emotional or mental drive to continue to take a drug whose effects the user feels are necessary to maintain a sense of optimal well-being. Psychological dependence is manifested by marked craving for the drug so that its continued use is mandated either to reproduce pleasure or to avoid discomfort. Compulsive drug-seeking behavior results.

psychomotor: refers to abilities that require coordinated adaptation of muscular actions. (Example: aiming at a target that moves irregularly.)

psychotaraxics: mind disrupting hallucinogenic drug.

psychotogen: a drug that produces a state of psychosis.

psychotomimetics: pertaining to, characterized by, or producing manifestations resembling those of a psychosis; a drug that produces such effects.

psylocibin: hallucinogenic from the mushrooms of the *Psylocybe* spp.

ptosis: drooping of the upper eyelid.

punisher: stimulus that decreases the frequency of behavior that leads to its presentation.

putamen: part of the lenticular nucleus, one of the basal ganglia.

quantitative trait loci (QTL): the collection of genes that are positively correlated with a certain deviation or illness.

Quinpirole (= SKF-38393): selective dopamine D-2 receptor agonist.

racemic: made up of two enantiomorphic isomers and therefore optically inactive.

raclopride: an atypical neuroleptic; selective dopamine D-2 receptor antagonist.

radioimmunoassay: any assay procedure that employs an immune reaction, in which either the antigen or the antibody is labeled with a radionuclide to permit accurate quantification.

radioreceptor assay: a radioimmunoassay to localize specific receptors.

range: the difference between the highest and lowest scores in a distribution, or set of data.

raphe nuclei: a system of nuclei in the brain that controls sleep.

rate-independent threshold procedure: reward threshold levels are determined in intracranial self-stimulation designs (ICSS), using a discrete trial task in which the presentation of stimulus 1 (S1) signals the availability of an identical electrical pulse of the same intensity. An operant response within 7.5 sec. after S1 results in the immediate delivery of the second stimulus (S2), which is the reinforcement. The rate independent procedure includes that responses made during an interval institute a 30 sec. delay or time-out period before the onset of the next trial to punish unsolicited responding. Therefore, it is to the subjects advantage to make responses (discrete wheel turns) only in response to stimuli that are rewarding.

reaction time (RT) paradigms: tests to measure the minimum time between the presentation of a stimulus and the subject's response to it. One of the oldest dependent variables in experimental psychology, specialized types have been studied. The introduction of computerized assessment has facilitated the recording and calculating of reaction times. (1) *choice reaction time* is the reaction time between the onset of one of a set of possible stimuli and the completion of the set of responses associated with the stimulus. (2) *simple reaction time* is the time between the onset of a stimulus and the completion of a response to it. Only one stimulus and one response are possible.

receptor antagonist: a substance that binds to a specific receptor site and blocks the effect of the agonist, e.g., a neurotransmitter.

receptor binding studies: studies to measure the binding of certain substances to a specific receptor.

reference memory: processes involved in storing information which does not change and which is retained for long periods of time.

reinforcer: *(primary)* a stimulus whose reinforcing effect appears to be inherent (unconditioned, i.e., does not have to be acquired through conditioning or learning); *(secondary)* a stimulus whose reinforcing effect has been acquired through its previous association with a primary reinforcer (also called conditioned reinforcer).

relapse model: model in drug abuse research to study the underlying mechanisms of relapse or to study new treatment strategies to prevent relapse.

relapse prevention therapy: therapy used to prevent relapse in drug abuse.

relapse: the falling back into a previous state or into an earlier behavior pattern.

relapse-like drinking: experimental (animal) model for reimplementation of alcohol consumption at first renewed contact.

repeated measure design: statistical design of a study in which a certain variable is measured more than one time in the same subject, e.g., at different time intervals.

respiration depression (RD50): measure to quantify the irritant potency of solvents.

retention gradient: in a delayed matching test, the correct matching of a sample stimulus with the comparison stimuli decreases with the length of the delay, the function relating accuracy to delay is the retention gradient (see delayed matching).

retention: the process of holding onto or retaining a thing. Most commonly used with respect to issues surrounding the retention of information, where the basic presumption is that some 'mental content' persists from the time of initial exposure to the material or initial learning of a response until some later request for recall or re-performance.

reticular activating system (RAS): the system of nerve pathways in the brain concerned with the level of consciousness, from the state of sleep, drowsiness and relaxation to full alertness and attention. The system integrates information from all of the senses and from the cerebrum and cerebellum and determines the overall activity of the brain and the autonomic nervous system and patterns of behavior during waking and sleeping.

reverse dialysis: brain dialysis involves the implantation of a dialysis tube into the brain tissue. Substances in the extracellular fluid will diffuse into the perfusate, enabling monitoring of levels of neurotransmitters and their metabolites in the extracellular fluid *in vivo*. In addition, substances included in the perfusate can diffuse into the tissue ('reverse').

reward pathways: a system of mainly dopaminergic fibers in the central nervous system that is involved in the reinforcing effects of e.g., electric self-stimulation.

rhinorrhoea: persistent watery mucus discharge from the nose, as in the common cold.

risperidone: a putative atypical neuroleptic; a 5-HT_2/D-2 receptor antagonist.

Ritanserin®: a trademark for preparations of methylphenidate hydrochloride

RO 15-4513: GABA antagonist and benzodiapine partial inverse agonist; also ethanol antagonist; anxiogenic.

SCH-23390: a selective D-1 receptor antagonist.

schedule induced polydipsia: an organism can be induced to drink enormous quantities of water simply by delivering small quantities of food on a regular basis. The food may be a reinforcement for responding on a FI schedule, or it may not be contingent on any behavior.

schedule of reinforcement: program determining relationship between responses and the occurrence of reinforcing stimuli.

schedule-controlled behavior: behavior that has been placed on a schedule of reinforcement. Following an animal's responses with reinforcing stimuli according to an explicit schedule produces orderly patterns of behavior.

second messenger: organic molecule acting within a cell to initiate the response to a signal carried by a chemical messenger (e.g., a neurotransmitter or a hormone) that does not itself enter the cell. Examples of second messengers are inositol triphosphate and cyclic AMP.

second-order schedule: a schedule of a schedule. The defining feature of a second order schedule is that the component schedule (the first order schedule) is treated as if it were a single response by the second order schedule. For example, a second order FR4 (FI 1 min.) schedule would require the completion of four 1-min. Fixed Intervals prior to delivery of the reinforcer. The FI schedule is the first-order schedule and the FR schedule is the second-order schedule.

self-administration model: model in which subjects can self-administer drugs and in which they have control over the infusion apparatus. Frequently animals are trained to lever press in a food-reward set-up before being tested for drug self-administration. Different protocols may be engaged, for instance, intravenous, intracerebral or oral and animals may be drug naive or drug dependent.

self-administration, see drug self-administration.

self-stimulation reward: the pleasant feeling that a subject undergoes when special centers in the brain are activated by an electric stimulus.

senile plaques: pathological hallmark of Alzheimer's disease formed by an extracellularly accumulation of amyloid and irregular, loosely arranged aggregate of neuronal and glial processes. They are found most characteristically in the gray matter of the neocortex and hippocampus, but also occur in the basal ganglia, thalamus and cerebellum.

sensitivity: the degree of responsiveness to weak stimuli; having a low threshold.

sensitization: increased response after a challenge induced by previous exposure to the substance.

sensorimotor tests: tests to assess basic central and peripheral nervous system integrity of sensory and motor nerves.

sensory ganglion, see dorsal root ganglion.

sensory latency: the time between the onset of a stimulus and the actual perception of the stimulus.

septum pellucidum (= septum lucidum): triangular double membrane separating the anterior horns of the lateral ventricles of the brain. It is part of the limbic system and controls aggressive behavior.

septum, see septum pellucidum.

Serax™, see oxazepam

set: the psychological expectation or anticipation by a user as to the nature of a drug's effect.

setting: the physical environment in which a drug is used.

sexual dimorphic area (SDN): nucleus in the hypothalamus for which a difference in cell number has been found between men and women.

shock-elicited fighting: agonistic behavior by a rat that is stressed by an electric shock given through his paws.

short term memory: a memory system subject to decay and in which information must be maintained by rehearsal.

shuttle-box: apparatus employed in the investigation of escape or avoidance conditioning. Usually, organisms are placed in one side of the apparatus and must jump to the other in order to escape or avoid painful electric shocks.

side-effects: actions of a drug other than those desired for a beneficial pharmacological effect.

Sidman avoidance (temporal avoidance conditioning): an operant-conditioning procedure in which an aversive stimulus is presented at regular intervals. When the organism makes the proper response the shock is delayed some fixed amount of time. The procedure thus has two independent variables, the shock-shock and the shock-interval. Typically, good temporal avoidance conditioning develops and the animal learns to avoid the noxious events without the presence of any external stimulus.

single-blind procedure: an experimental technique in which the subjects are ignorant of the experimental conditions, but the researcher running the experiment knows them.

SKF-38393 [Quinpirole]: a dopamine D-1 receptor agonist.

SLC-90: 90-item symptom checklist, provides a measure of mood and psychopathological condition.

smack: one of the street names for heroin.

specificity: preciseness, uniqueness. The term is used to the connotation that the thing so characterized is relevant to but a single phenomenon or event.

spider naevi: telangiectatic arterioles in the skin with radiating capillary branches simulating the legs of a spider

spinal ganglion, see dorsal root ganglion.

spiperone: a tranquilizer used in the treatment of schizophrenia.

stimulus discrimination: the ability to tell the difference between two or more stimuli.

stimulus property of a drug: the ability of a drug to act as a stimulant to induce or change a behavior.

stimulus: any action or agent that causes or changes an activity in an organism, organ, or part.

Student t-test: any of a number of statistical tests of significance based upon the t statistic.

subarachnoid: situated or occurring between the arachnoid and the pia mater.

subchronic study: animal experiment serving to study the effects produced by the test material when administered in repeated doses (or continually in food, drinking water, air) over a period of up to about 90 days.

subependymal: situated beneath the ependyma.

subfornical organ (organum subfornicale): a group of specialized ependymal cells, similar to those of the subcommissural organ, projecting toward the cavity of the third ventricle from its anterior wall between the columns of the fornix.

substantia innominata: nerve tissue immediately caudate to the anterior perforated substance and ventral to the globus pallidus and ansa lenticularis.

substantia nigra: a midbrain nucleus, the origin of the nigro-striatal dopamine pathway whose degeneration causes Parkinson's disease.

succinylcholine: synthetic muscle relaxant with a curare-like activity, it blocks acetylcholine receptors.

sucrose substitution technique: using an operant design, rats are gradually habituated to alcohol. Initially sucrose solution is provided as reinforcement and subsequently, after established conditioning, alcohol is added to the solution, in gradually ascending concentrations of alcohol.

Sudden Infant Death Syndrome (SIDS): the sudden and unexpected death of an apparently healthy infant, typically occurring between the ages of 3 weeks and 5 months and not explained by careful postmortem studies; also called crib or cot death.

supersensitivity: abnormally increased sensitivity; increased activity of a neural pathway following chronic exposure to an antagonist caused by changes in postsynaptic receptors.

surveys: a method of research in which the opinions of a large number of individuals are measured. Provided the sampling is genuinely representative, the results of surveys can often be employed to predict product sales, public reactions to various events, or the results of political elections.

survival analysis: statistical technique for the analysis of longitudinal data on the occurrence and timing of events.

sustained attention: the ability to stay alert during a longer lasting test or situation.

tachyphylaxis: rapidly decreasing response to a drug or physiologically active agent after administration of a few doses.

tachypnea: excessive rapidity of respiration.

tail-flick test: rodent test to measure changes in nociception (pain sensitivity) by placing the distal part of an animal's (mouse or rat) tail in warm water and noting the emergence of the tail from the water.

tardive dyskinesia: iatrogen disorder due to long-term treatment with dopamine antagonists (phenothiazines, butyro-phenones), manifested by abnormal involuntary movements especially of the face and tongue, usually temporary, but it can be perma-nent. The cause is an alteration in dopaminergic receptors causing hypersensitivity to dopamine and dopamine agonists.

Tegretol®, see carbamazepine

Tetrahydroisoquinolines: class of alkaloids, formed via conden-sation of a catecholamine (e.g., dopamine) with acetaldehyde found in the cerebrospinal fluid of alcoholics. It is postulated that these acetaldehyde-amine-derived alkaloids are involved in alcoholism.

THBC: a serotonin agonist used as analgetic.

TIQ-challenge: a challenge treatment with tetrahydroisoquinoline

T-maze, see maze.

tolerance: a decreasing response to repeated constant doses of a drug or the need for increasing doses to maintain a constant response.

tonic movement: a movement marked by continuous tension (contraction) of muscles; spasm.

tranylcypromine: monoamine oxidase inhibiter structurally related to amphetamine.

tremor: an involuntary trembling or quivering.

tremorgenic mycotoxin: fungal toxin that induces tremor.

tremulousness: behavior characterized by tremor.

trophic: of or pertaining to nutrition.

tropic: turning toward.

tropical ataxic neuropathy: toxic, distal axonopathy probably due to chronic consumption of improperly prepared cassava. Clinical features comprise central and peripheral manifestations of a central-peripheral distal axonopathy, characterized by painful paresthesiae of the feet with numbness of the hand, with sight and hearing loss when the disease progresses.

tropical spastic paraparesis: chronic progressive myelopathy.

t-tests: any of a number of statistical tests of significance based upon the t statistic.

twin studies: studies carried out on monozygotic and dizygotic twins raised together or apart from each other. The focus of these studies is to sort out the relative contributions of heredity and environment to human behavior.

unavoidable shock procedure: a procedure in which a subject (mostly a rat) receives an electrical shock that it cannot avoid. The procedure is often used to study stress or stress-related phenomena.

U-shaped curve: dose effect curve in which small and high doses have relatively large effects and medium doses have relatively low effects or no effect at all, the curve has the form of an 'u'.

vagus nerve: the 10th cranial nerve which supplies motor nerve fibers to the muscles of swallowing and parasympathetic fibers to the heart and organs of the crest cavity and abdomen. Sensory branches of the vagus carry impulses from the viscera and the sensation of taste from the mouth.

vas deferens (ductus deferens): the excretory duct of the testis, which unites with the excretory duct of the seminal vesicle to form the ejaculatory duct.

ventral pallidum: ventral part of the pallidum, one of the large subcortical nuclei that make up the basal ganglia.

ventral striatum: the ventral part of the corpus striatum.

ventral tegmental area (VTA): area in the central nervous system from which two main dopaminergic pathways start, the mesocortical dopamine pathway and the mesolimbic pathway.

vigilance: alertness, watchfulness.

Wernicke-Korsakoff syndrome: behavioral disorder caused by thiamin deficiency, most commonly due to chronic alcohol abuse and associated with other nutritional polyneuro-pathies. Wernicke's encephalopathy (confusion, ataxia of gait, nystagmus, and ophtalmoplegia) occurs as an acute attack and is reversible, except for some residual ataxia or nystagmus, by administration of thiamine. Korsakoff's syndrome (severe anterograde and retrograde amnesia) may occur in conjunction with Wernicke's encephalopathy or may become apparent later: only 20% of patients recover completely from the amnesia.

withdrawal symptoms: any of the effects of the cessation of administration of a substance upon which one has built up a dependence.

withdrawal: a pattern of behavior characterized by the person removing him or herself from normal day-to-day functioning.

Zimeldine®: a trademark for zimelidine hydrochloride, a serotonergic antidepressant.

Z-scores: a statistical score which has been standardized by being expressed in terms of relative position in the full distribution of scores. A z-score is always expressed relative to the mean of the distribution and in standard deviation units.

Table of (neuro)psychological tests

Addiction Research Center Inventory (ARCI): questionnaire used in addiction research consisting of 550 questions of the yes/no type to distinguish different types of subjective drug effects. If a drug of unknown effects is given, ARCI results can be used to categorize the drug on he basis of its subjective effects.

behavioral rating scale: any device used to assist a rater in making ratings of behavior. Rating behavior is the use of rating techniques in a fairly restricted manner such that only overt, objectively observable behaviors enter into the assessment.

Brief Psychiatric Rating Scale: scale providing a measure for a psychiatric status of a patient.

Brazelton's Neonatal Behavioral Assessment Scale (BNBAS): method for assessing infant (ages 3 days to 4 weeks) behavior by its response to environmental stimuli.

British Ability Scales (BASC or BAS): test to assess mental ability of children from 2.5-17.5 years of age; process areas included are: speed of problem solving, reasoning, spatial imagery, perceptual matching, short term memory and retrieval and application of knowledge.

child behavior checklist: a rating scale for internalizing and externalizing behavior disorders in children between 7 and 16 years of age.

choice response: refers to tests that offer fixed response options (e.g., multiple-choice, true-false, like-dislike…).

clinical trial: a large scale plan for testing and evaluating the effectiveness of some drug or therapeutic procedure in human subjects.

cognitive tests: tests to measure cognitive functions (→ cognition).

Denver Developmental Screening Test: assessment test for developmental delays in children aged 0-6 years. The test concerns four areas: motor, social and/or language delays.

developmental mile stones: significant behaviors which are used to mark the progress of development. Examples are: saying phrases, turning pages, carrying out requests, pointing to body parts, holding a pencil, imitating a drawn circle, catching a ball.

diagnosis: narrowly choosing one of a set of labels that best fits an individual's disorder or disability. Broadly, developing an understanding of the individual's difficulties, and insofar possible, their origins.

Diagnostic and Statistical Manual (Diagnostic and Statistical Manual of Mental Disorders; DSM): the official system for classification of psychological and psychiatric disorders prepared and published by the American Psychiatric Association. The current version is DSM-IV.

Drug Use Screening Inventory: a self-report test (paper/pencil or computerized) that scores 10 domains of health, psychiatric and psychosocial severity in patients from 10 years on and gives a profile of problem severity.

Finnegan Scoring System: clinical assessment test for neonatal abstinence symptoms.

International Classification of Diseases (ICD): system of classification of diseases developed under the supervision of the World Health Organization. Over the years, many revisions have been made, the most recent system is the 10th (ICD-10). The ICD has an extensive section on psychiatric and psychological disorders, which is in wide use in many countries.

inventory: a questionnaire, typically one that represents many questions about each aspect of personality that is under investigation. Directions may ask for a self-description of an acquaintance who is being assessed.

IQ tests: any test that purports to measure an intelligence quotient. Generally such tests consist of a graded series of tasks each of which have been standardized with a large, representative population of individuals.

McCarthy Scales of Children's Abilities: a set of developmental scales for assessing the abilities of preschoolers between ages of 2 and 8. There are 18 separate tests combined into six distinct scales: verbal, perceptual, quantitative, memory, motor and general cognitive. The general cognitive scale is based on all but three of the tests and is generally treated as a measure of intellectual development.

MDI, mental development index (See: Bayley Scales of Infant Development).

0-8493-7803-6/99/$0.00+$.50

mental age: a measure of intellectual ability obtained on early experimenters' tests of intelligence. An individual's mental age was assumed to reflect his or her level of intellectual maturity.

MMPI (Minnesota Multiphasic Personality Inventory): an objective test in questionnaire form which compares a particular subject's responses to those of individuals in various diagnostic categories. The 400-item true/false questionnaire provides a profile of personal and social adjustment.

mood questionnaire: questionnaire to assess the state of mood.

objective test: a personality test involving structured items and a limited set of responses (such as TRUE - FALSE)

peer rating: a classmate, fellow soldier, fellow student, or other acquaintance marks a rating scale or inventory to describe the target person. Usually, the average of several reports is taken as the target's peer rating.

performance test: among ability tests the term is usually applied to those where the respondent is to execute an appropriate physical action-tracing a maze path for example; although it can also be verbal as in the Wechsler Performance section. Among personality measures, the term is usually applied to observations of response in a standardized situation, usually a situation that arouses strong motives.

Personality Development Index (PDI), see Bayley Scales of Infant Development.

personality indices: the items in questionnaires and inventories that are used to assess a subject's personality.

personality questionnaire: questionnaire to assess a subject's personality.

PIAT: Peabody Individual Achievement Test.

Profile of Mood States (POMS): a questionnaire containing adjectives describing different mood states, e.g., tired, happy, etceteras. The subject indicates on a four point scale the degree which the appropriate description of his or her own feelings.

projective test: a personality test involving ambiguous stimuli. A subject's responses are supported to reveal aspects of the unconscious.

psychiatric rating scale: device used to assist a psychiatrist in making a diagnosis of a patient.

sensorimotor tests: tests to assess basic central and peripheral nervous system integrity of sensory and motor nerves.

stimulus discrimination: the ability to tell the difference between two or more stimuli.

surveys: a method of research in which the opinions of a large number of individuals are measured. Provided the sampling is genuinely representative, the results of surveys can often be employed to predict product sales, public reactions to various events, or the results of political elections.

test battery: a collection of tests the results of which can be combined to produce a single score. Such batteries are used on the assumption that the errors inherent in each separate test will cancel each other out and the single score obtained will be maximally valid. In neurobehavioral toxicology it is often a collection of behavioral tests which makes it possible to test for more than one functional behavioral disturbance.

Visual Analog Scales (VAS): test to assess drug-induced momentary changes in, for example, affect. In a visual analog scale, a 100-mm line is presented with at the ends showing their opposing adjectives (e.g., happy-sad) or labels "not at all" and "an awful lot". Subjects are instructed to rate how they feel along that dimension by making a mark anywhere along the line. Various VASs are used, but they should include: "stimulated", "anxious:", "liking", "sedated", "high", "down", and "hungry".

WAIS-R (Wechsler Adult Intelligence Scale-Revised): a group of tests for assessment of intellectual functioning in adults. The output of this test is the performance on 10 subtests (verbal and performance tests) measuring various dimensions of intellectual functioning.

Wechsler Intelligence Scale for Children – Revised (WISC-R): a group of tests for assessment of intellectual functioning, thought processes and ego functioning in children ages 5 to 15. The output of this test is the performance on 10 subtests measuring various dimensions of intellectual functioning.

Wechsler Memory Scale: a psychiatric examination of memory ability to assess verbal and nonverbal memory in adults. It is a battery of tests designed to asses different aspects of verbal and visual short-term memory functions in brain-damaged populations consisting of 7 subtests: information/orientation, mental control, logical memory, memory span, visual reproduction and paired associates.

Wechsler Preschool and Primary Scale of Intelligence (WPPSI): assessment scale for a variety of cognitive abilities in children between 4 and 6 years of age.

Werry-Weiss-Peters Activity Scale: activity rating scale to assess (hyper)activity in children; often used to test hyperactivity and aggression in child psychiatry.

Wide Range Achievement Test (WRAT): assessment test for reading, spelling and arithmetic skills in children from 5 years on and in adults.

working memory task: equivalent to short-term memory.

List of commonly used abbreviations

AA, alcohol accepting strain
AA, Alcoholics Anonymous
ACh, acetylcholine
AChE, acetylcholinesterase
ACTH, adrenocorticotrophic hormone
AD, Alzheimer's disease
ADAMHA, Alcohol, Drug Abuse and Mental Health Administration
ADD, attention deficit disorder
ADH, alcohol dehydrogenase
ADHD, Attention-Deficit Hyperactivity Disorder
ADS, Alcohol Dependence Scale
ALAT, alanine transaminase
ALDH, aldehyde dehydrogenase
AN, arcuate nucleus
ANA, alcohol non-accepting strain
ANOVA, analysis of variance
AP, amphetamine
APO, apomorphine
APSD, antisocial personality disorder
ARBD, Alcohol Related Birth Defects,
ARCI, Addiction Research Center Inventory
A-scale, amphetamine scale of ARCI
ASI, Addiction Severity Index
ASPD, antisocial personality disorder
AST, aspartate amine transferase, a liver enzyme.
AUDIT, Alcohol Use Disorders Identification Test
AWS, alcohol withdrawal syndrome
BAL, blood alcohol level
BDNF, brain-derived neurotrophic factor
BG scale, benzedrine group scale of ARCI
BNAS, Brazelton Neonatal Assessment Scale,
BPRS, Brief Psychiatric Rating Scale
CA, Cocaine Anonymous
CAGE, screening test for alcoholism
CCK, cholecystokinin
cGMP, cyclisch guanosine monophosphate
ChAT, choline acetyl transferase
CIDI, Composite International Diagnostic Interview
CNS, central nervous system
CNTF, ciliary neurotrophic factor
CPP, conditioned place preference
CRF, corticotropin-releasing factor
CSF, cerebrospinal fluid
CT, computerized tomography
CTA, conditioned taste aversion
DA, dopamine
DAST, Drug Abuse Screening Test
2-DG, 2-[^{14}C]deoxyglucose
DOB, 2,5-dimethoxy-4-bromoamphetamine
DOI, 1-(2,5-dimethoxy-4-iodophenyl)-2-aminopropane
DSM, Diagnostic and Statistical Manual
DZ, dizygotic twins
EAA, excitatory amino acid
ECS, electoconvulsive shock
ECT, electroconvulsive treatment
ER, endoplasmic reticulum
ERP, event-related potentials
ESM, Experience Sampling Methodology
ETZ, etonitizine
FAE, Fetal Alcohol Effects,
FAS, fetal alcohol syndrome
FDA, Food and Drug Administration (US)
FHP, Family History Positive
FI, fixed interval
FR, fixed ratio schedule
FSS, Finnegan Scaling System
G × E, genotype by environmental interaction
GABA, γ-amino butyric acid
GAD, glutamate decarboxylase
gamma-GT, gamma glutamyl transferase
GD, gestational day
GFAP, glial fibrillary acidic protein.
GGT, γ-glutamyltransferase
GTP, guanosine triphosphate
HAD, high alcohol drinking strain
HALT, hungry, angry, lonely, or tired
5-HIAA, 5-hydroxyindoleacetic acid
HPA-axis, hypothalamic-pituitary-adrenal-axis
5-HT, 5-hydroxytryptamine (= serotonine)
HVA, homovanillic acid
i.c.v., intracerebroventricular
i.g., intragastric
i.m., intramuscular
i.p., intraperitoneal
ICD, International Classification of Diseases
ICSS, intracranial self–stimulation
IUGR, Intra Uterine Growth Retardation,
LAD, low alcohol drinking strain
LB, Lewis bodies
LC, locus coeruleus
LHRH, luteinizing hormone-releasing hormone
LS, Long Sleep strain
LSD, Lysergic acid diethylamine
LTP, long term potentiation
MAG, myelin associated glycoprotein
MAM, methyl azoxymethanol
MAO, monoamine oxidase
MAOI, monoamino oxidase inhibitor
MAST, Michigan Alcoholism Screening Test

0-8493-7803-6/99/$0.00+$.50

MBD, minimal brain dysfunction
MBG, morphine-benzedrine group scale of ARCI
MBP, myelin basic protein
MCV, mean corpuscular volume
MDA, methylenedioxyamphetamine
MDMA, 3,4-Methylene-dioxy-methyl-amphetamine (ecstasy, XTC)
METH, Methamphetamine
MHC, major histocompatibility complex.
MMC, methylmercury
MOA, monoamine oxidase
MPP^+, 1-methyl-4-phenylpyridinium
MPPP, 1-methyl-4-phenylpropionoxypiperidine
MPTP, 1,2,3,6-tetrahydro-1-methyl-4-phenyl-pyridine1.4.3
MT, microtubules
MZ, monozygotic twins
NA, Narcotic Anonymous
NA, nucleus accumbens
NBAS, Neonatal Behavioral Assessment Scale
NCAM, neural cell adhesion molecule
NCTB, Neurobehavioral Core Test Battery
NDMA, N-methyl-D-aspartate
NF, neurofilaments
NFT, neurofibrillary tangles
NGF, nerve growth factor
NIT, Naltrindole
NMDA, N-methyl-D-aspartate
NP, alcohol non preferring strain
P, alcohol preferring strain
PCAG, pentobarbital-chlorpromazine alcohol group of ARCI
PCP, phencyclidine
PD, Parkinson's disease
PD, postnatal day
PET, positon emission tomography
PKU, phenylketonuria
PLP, proteolipid protein
PNS, peripheral nervous system
POMC, pro-opiomelanocortine
POMS, Profile of Mood States
PR, progressive ratio (schedule)
QTL, quantitative trait loci
RAS, reticular activating system
REM, rapid eye movement
RI, recombinant inbred strain
s.c., subcutaneous
SADQ, Severity of Alcohol Dependence Questionnaire
SDH, succinate dehydrogenase
SER, smooth endoplasmic reticulum
SGA, Small for Gestational Age,
-SH group, thiol group
SIDS, Sudden Infant Death Syndrome
SN, substantia nigra
SOD, superoxide dismutase
SP, senile plaque
SPES, Swedish Performance Evaluation System
THC, Δ^9-tetrahydrocannabinol
THC, tetrahydrocannabinol
TIQ, tetrahydroisoquinolines
TRH, thyrotropin-releasing hormone
VAS, Visual Analog Scales
VTA, ventral tegmental area
WHO, World Health Organization
WSP, withdrawal seizure prone strain
WSR, withdrawal seizure resistant strain

Index